Statistics for Biology and Health

Series Editors:
M. Gail
K. Krickeberg
J. Samet
A. Tsiatis
W. Wong

Statistics for Biology and Health

Bacchieri/Cioppa: Fundamentals of Clinical Research

Borchers/Buckland/Zucchini: Estimating Animal Abundance: Closed Populations

Burzykowski/Molenberghs/Buyse: The Evaluation of Surrogate Endpoints

Duchateau/Janssen: The Frailty Model

Everitt/Rabe-Hesketh: Analyzing Medical Data Using S-PLUS

Ewens/Grant: Statistical Methods in Bioinformatics: An Introduction, 2nd ed.

Gentleman/Carey/Huber/Irizarry/Dudoit: Bioinformatics and Computational Biology Solutions Using R and Bioconductor

Hougaard: Analysis of Multivariate Survival Data

Keyfitz/Caswell: Applied Mathematical Demography, 3rd ed.

Klein/Moeschberger: Survival Analysis: Techniques for Censored and Truncated Data, 2nd ed.

Kleinbaum/Klein: Survival Analysis: A Self-Learning Text, 2nd ed.

Kleinbaum/Klein: Logistic Regression: A Self-Learning Text, 2nd ed.

Lange: Mathematical and Statistical Methods for Genetic Analysis, 2nd ed.

Manton/Singer/Suzman: Forecasting the Health of Elderly Populations

Martinussen/Scheike: Dynamic Regression Models for Survival Data

Moyé: Multiple Analyses in Clinical Trials: Fundamentals for Investigators

Nielsen: Statistical Methods in Molecular Evolution

O'Quigley: Proportional Hazards Regression

Parmigiani/Garrett/Irizarry/Zeger: The Analysis of Gene Expression Data: Methods and Software

Proschan/LanWittes: Statistical Monitoring of Clinical Trials: A Unified Approach

Siegmund/Yakir: The Statistics of Gene Mapping

Simon/Korn/McShane/Radmacher/Wright/Zhao: Design and Analysis of DNA Microarray Investigations

Sorensen/Gianola: Likelihood, Bayesian, and MCMC Methods in Quantitative Genetics

Stallard/Manton/Cohen: Forecasting Product Liability Claims: Epidemiology and Modeling in the Manville Asbestos Case

Sun: The Statistical Analysis of Interval-censored Failure Time Data

Therneau/Grambsch: Modeling Survival Data: Extending the Cox Model

Ting: Dose Finding in Drug Development

Vittinghoff/Glidden/Shiboski/McCulloch: Regression Methods in Biostatistics: Linear, Logistic, Survival, and Repeated Measures Models

Wu/Ma/Casella: Statistical Genetics of Quantitative Traits: Linkage, Maps, and QTL

Zhang/Singer: Recursive Partitioning in the Health Sciences

Zuur/Ieno/Smith: Analysing Ecological Data

John O'Quigley

Proportional Hazards Regression

 Springer

John O'Quigley
Institut Curie
26 rue d'Ulm
75005 Paris, France

Series Editors

M. Gail
National Cancer Institute
Rockville, MD 20892
USA

K. Krickeberg
Le Chatelet
F-63270 Manglieu
France

J. Sarnet
Department of Epidemiology
School of Public Health
Johns Hopkins University
615 Wolfe Street
Baltimore, MD 21205-2103
USA

A. Tsiatis
Department of Statistics
North Carolina State University
Raleigh, NC 27695
USA

W. Wong
Department of Statistics
Stanford University
Stanford, CA 94305-4065
USA

ISBN 978-0-387-25148-6   e-ISBN 978-0-387-68639-4
DOI: 10.1007/978-0-387-68639-4

Library of Congress Control Number: 2008920048

9 8 7 6 5 4 3 2 1

springer.com

To Máire Proinsias

Preface

Proportional hazards models and their extensions (models with time dependent covariates, models with time dependent regression coefficients, models with random coefficients and any mixture of these) can be used to characterize just about any applied problem to which the techniques of survival analysis are appropriate. This simple observation enables us to find an elegant statistical expression for all plausible practical situations arising in the analysis of survival data. We have a single unifying framework. In consequence, a solid understanding of the framework itself offers the statistician the ability to tackle the thorniest of questions which may arise when dealing with survival data.

The main goal of this text is not to present or review the very substantial amount of research that has been carried out on proportional hazards and related models. Rather, the goal is to consider the many questions which are of interest in a regression analysis of survival data (prediction, goodness of fit, model construction, inference and interpretation in the presence of misspecified models) from the standpoint of the proportional hazards and the non-proportional hazards models.

This standpoint is essentially mathematical in that the aim is to put all of the inferential questions on a firm conceptual footing. However, unlike the current widely accepted approach based almost entirely on counting processes, stochastic integrals and the martingale central limit theorem for multivariate counting processes, we mostly work with much more classic and better known central limit theorems. In particular we appeal to theorems dealing with sums of independent but not necessarily identically distributed univariate random variables and, of particular interest, the functional central limit theorem establishing the Brownian motion limit for standardized univariate sums. Delicate measure theoretic arguments borrowed from mathematical analysis can then be wholly avoided. Admittedly, some tricky situations can still be more readily resolved by making an appeal to the martingale central limit theorem and it may also be true that the use of martingale techniques for multivariate counting processes affords greater generality. Nonetheless, in the author's view, a very high percentage of practical problems can be tackled by making an appeal to the arguably less general but more standard and well known central limit theorems for univariate sums.

Mathematicians always strive for the greatest generality and, in the author's view - at least as far as survival analysis is concerned - this undertaking has not been without some unfortunate drawbacks. The measure theoretic underpinning of the counting processes and martingale approach is quite opaque. The subject is very difficult and while outstanding efforts have been made across the globe in leading statistics and biostatistics departments to explain the essential ideas behind the material, few would claim that students, other than the small minority already well steeled in mathematical analysis, ever really fully grasp just what is going on. This is a situation that we need be concerned about. The author, having taught such courses in a number of institutions, speculates that even for the most successful students entering research careers and publishing articles in our leading journals it is not easy for them to do other than reiterate well rehearsed near inscrutable arguments. Their work, reviewed by their peers - alumni survivors of similar courses - may clear the publishing hurdle and achieve technical excellence but somehow, along the way, creativity is stifled.

In brief, there is a real danger of the subject of survival analysis sustaining itself from within and reluctant to absorb input from without. The pressure to focus so much attention on the resolution of mathematical subtleties by non-mathematicians has led us away from those areas where we have traditionally done well ... abstract modeling of medical, biological, physical and social phenomena. The technical demands of those in the area of survival analysis are such that it is becoming difficult for them to construct or challenge models via considerations other than those pertaining to the correct application of an abstruse theory. Those not in the area, a large and diverse pool of potential contributors, will typically throw up their hands and say that they are not sufficiently comfortable with survival analysis to make criticism of substance. A somewhat ambitious goal, or hope, for this text is to help change this state of affairs.

Work on this book began during the author's thesis. I would like to acknowledge the input of Dr Salah Rashid, a visiting surgeon to the University of Leeds Medical School, for asking a lot of awkward questions to a, then, very inexperienced statistician. Among these questions were 'how much of the variation is explained by the predictors', 'why would you assume that the strength of effect remains the same through time' and 'what is the relative importance of biological measurements to clinical measurements.' I believe that I can now attempt an answer to some of these questions although I fear, taking rather longer than expected to answer the clinician's concern - in this case some twenty odd years - that my good friend Dr Rashid may have moved on to other questions. Much of my thesis was based on collaborative work with Dr Rashid and his comments and questions then, and for decades to follow, provided an invaluable source of food for thought. I share the debt we all owe to Professor Sir David Cox for his great vision and scientific imagination, making all of this work possible, but also a personal debt to Sir David for having so very kindly agreed to be the external examinor on my own Ph.d thesis and for having patiently explained issues which, alone, I was unable to resolve.

My career at the Institut National de la Santé et de la Recherche Médicale in France was made possible thanks to the unfailing support of Professor Daniel Schwartz, one of the founders of the modern theory of clinical trials and I thank him warmly for that as well as for numerous discussions on parametric survival models, especially as they relate to problems in human fertility. A number of Professor Schwartz's colleagues, Joseph Lellouch, Denis Hémon, Alfred Spira in particular, were of great assistance to me in gaining understanding of the role played by survival analysis in quantitative epidemiology. However, my good fortune did not end there and I would like to offer my warm appreciation for the support, help and advice offered by Ross Prentice in inviting me to work at the Fred Hutchinson Cancer Research Center, Seattle during the late eighties, an opportunity which brought me into contact with a remarkable number of major contributors to the area of survival analysis. Among these I would like to express my gratitude to

Ross himself alongside Norman Breslow, John Crowley, Tom Fleming, Suresh Moolgavkar, Margaret Pepe and Steve Self, all of whom showed great generosity and forbearance in discussing general concepts along with their own ideas on different aspects of survival analysis.

Competing with Seattle as the world's leader in survival analysis is the Department of Biostatistics in the Harvard School of Public Health. I was given the opportunity of spending several months there in 1999 and would like to thank Nan Laird, the then department chair, for that. Many visits and collaborations followed and, although these were not in the area of survival analysis, I took advantage of the proximity to talk to those who have left quite a mark in the field. In particular I would like to offer my thanks to Victor DeGruttola, Dave Harrington, Michael Hughes, Steve Lagakos, David Schoenfeld, L.J. Wei, Marvin Zelen, all of whom, from very demanding schedules, gave time to the exchange of ideas.

I have been very fortunate in coming into contact with quite a number of the most creative researchers in this area, so many of whom have shown such scholarly patience in explaining their own views to a keen, always enthusiastic but often slow listener. I hesitate to list them since I will surely miss out some names and then, of course, if I were to credit all of the writings which have greatly helped me, this preface would take a good third of the whole book. But let me include a special mention for Janez Stare and Ronghui Xu who will recognize within these pages much which stems from our extensive collaborations. And, as an extension to this special mention, let me thank those colleagues whose kindness, as well as exceptional talent, helped provide part of a hard-to-define support structure without which this enduring task would most likely have been abandoned many years ago. I have in mind Jacques Bénichou, Claude Chastang (whose colorful view of statistics as well as life in general is so sorely missed), Michel Chavance, Philippe Flandre, Catherine Hill, Joe Ibrahim, Richard Kay, John Kent, Susanne May, Thierry Moreau, Loki Natarajan, Fabienne Pessione, Maja Pohar, Catherine Quantin, Peter Sassieni, Michael Schemper, Martin Schumacher, Lesley Struthers and Joe Whittaker. The many students who followed my course on Survival Analysis in the Department of Mathematics at the University of California at San Diego, the course upon which the skeleton of this book ended up being based, are sincerely thanked for their enthusiasm and obstinate questioning.

Finally, although the very word statistics, let alone proportional hazards regression, would leave them quite at a loss, this work owes its greatest debt to those closest to me - my nearest and dearest - for a contribution which involved untold patience and tender indulgence. My warmest gratitude goes to them.

Paris, February 2007.

Contents

Chapter 1

Introduction

1.1 Summary

In keeping with the general pattern of this book each chapter is summarized by a brief description of its contents. This section (Section 1.1) is the first such summary in this book. Following these summaries is usually a section headed "Motivation" which describes the key reasons for our interest in the topic. In this Introduction we consider a number of examples in which the methods of survival analysis, proportional and non-proportional hazards model in particular, have been used to advantage. We then describe the outline of the book as a whole and point out some particular areas in which, at the time of writing, there is no established consensus about the best way of tackling them.

1.2 Motivation

The use of the methods of survival analysis and the proportional hazards model appear to be ever widening; some recent areas of interest including unemployment data, lengths of time on and off welfare, for example, and some unusual applications such as interviewer bias in sociological surveys. The more classic motivating examples behind the bulk of theoretical advances made in the area have come from reliability problems in engineering and, especially, clinical studies in chronic diseases such as cancer and AIDS. Many of the examples given in this text come from clinical research. Parallels for these examples from other disciplines are usually readily transposable. Proportional hazards regression is now widely appreciated as a very powerful technique for the statistical analysis of broad classes of problems. Novel uses as well as innovatory applications of the basic approach continue to grow. The following examples represent a small selection of problems for which the proportional hazards model has been shown to be useful as an analytic tool.

Randomized clinical trials

Patients, identified as having some chronic illness under study, are recruited to a clinical trial. A new, proposed, treatment, if effective, should prolong survival. The idea is that instead of aiming for a "cure" we aim to improve survival. In cancer research it has been common practice to equate cure with five years survival without relapse. Such a quantity may be difficult to estimate if some patients leave the study or die from unrelated causes during the five year period. Survival methods address such difficulties. Proportional and non-proportional hazards models enable us to investigate the dependence of survival on controlled and uncontrolled sources of variation.

Endpoint in randomized clinical trials

In the example described above it could occur that therapy A and therapy B have the same five year survival rates but different survival curves. As an illustration consider curative treatments based on surgery having very high initial risks but, conditional upon having survived some length of time, a reduced long-term risk. Such differences in overall survival are important and would be considered relevant when weighing the relative advantages and disadvantages of any proposed new treatment. In actual studies patients may be lost to follow-up at any time. Proportional hazards and non-proportional hazards modeling can allow us to correctly use the partial information provided by such patients.

Studies in epidemiology

A large number of individuals are studied across a relatively wide distribution of ages. Incidence of some target disease is measured. This incidence will typically vary with age and, in the majority of cases, the single most important source of variability in the observed incidence rates comes from age itself. Cancer is a particularly striking example.

Here the variable age plays the role of the time variable in a survival model. Prolonging survival amounts to living to a longer age. How does incidence relate to risk variables, obvious examples being smoking, asbestos exposure or other industrial exposures. In practice the amount of time needed for such prospective studies may be too great and we may prefer to use a case-control study where, rather than compare the age distributions for different exposure categories, we compare the exposure distributions at different ages.

In doing this our purpose is to address some key feature of the conditional distribution of age given covariate information via a study of the conditional distribution of the covariates given age. The question arises as to the validity of this and to the efficiency of such a study design. Additional complications arise when the exposure categories change for some individuals. At any given age a particular individual can have a unique and complex exposure history. It would not be possible to study the relationship between any unique exposure history and disease outcome without assuming some structure, i.e., the potential ways in which the different components comprising the exposure history relate to one another and, more importantly, to outcome. The way to achieve this is via a statistical model, specifically, in the context of this work, the proportional hazards regression model.

Clinical studies involving time

Not only can there be problems with patients going off study or being removed from the study due to causes believed to be unrelated to the main outcome under study, it is also not unusual for subjects to enter the study at different times. In chronic diseases such as AIDS patients may leave the study because they are too sick to continue participation. Such loss to follow-up may be of a different nature to that described above and may, of itself, provide information on the risk of being incident for the outcome. In technical terms we would described the situation as being one where the censoring mechanism is not independent of the failure mechanism. In order to avoid potentially misleading biases it would be necessary to appeal to workable models to describe this dependence.

The complexity of potential problems can easily grow beyond the reach of available methods and continues to promote and drive the interest for further methodological developments. One example, again from AIDS, where the methods presented here can provide insight, concerns the situation where the time frame is lengthy. Patients may have very involved time-dependent exposure histories to the different treatments, the class of available treatments themselves having evolved through time as well as within any given patient.

In prognostic studies, many variables are measured on individuals and we would like to know how individually and collectively these variables influence survival (prognosis). It may be possible to use such information to stratify future clinical trials, the goal being to increase precision, or, possibly, to use the information in a context of clinical patient management. Further complexity arises with correlated data. For recurrent conditions such as asthma the same patient may be studied for more than one principal event. Cooperative clinical trial groups may use many centers to recruit patients, the center effects themselves not being of any intrinsic interest, but nonetheless contributing a correlation factor for within-center observations. Similar considerations arise in familial data in which the presence of genetic characteristics may influence the survival probabilities collectively. Proportional hazards models including random component coefficients, equivalently non-proportional hazards models, can be used to efficiently analyze such data structures.

Industrial setting

In an industrial setting we may be interested in the reliability of components. How long do they last on average. A useful experiment may be to put a sample of components on test and end testing after a certain number of failures. In economics there are many possibilities. How long does it take for a price index to increase to a certain value, to regain a previous value or to change by some given amount? In this and other examples the technical term "failure" may in a more common everyday sense correspond to success. Depending on the context we may wish to prolong or reduce the average time it takes to observe some kind of event. Failure models have been used to study, under differing circumstances, the average time taken for a welfare recipient to come off welfare, an event in social terms that we would regard as a success.

An area of application where the law of large numbers comes directly into play is that of insurance. The calculus of probability together with very large object populations enable effective prediction. We may wish to know how demographic and other variables influence insurance risk. Further examples arise in car insurance; how long on average does it take to be involved in a traffic accident and, more importantly, which variables influence this time. Such information will enable the insurer to correctly price policies for different driver profiles.

Reliability

One of the early applications of survival methods concerns the reliability of components as well as mechanical, electrical, or electronic systems. Careful modeling can enable us to use various sources of information to predict the probability of failure before some given time. Many consumers will have first-hand experience of such modeling whereby it is possible to keep small the number of manufactured products failing before the expiration of the guarantee while, at the same time, ensuring that the probability of survival well beyond the guarantee is not so large as to ultimately impact consumer demand. It can be of interest to model more than one endpoint, tracing out the lifetime of a product. The first event may be something minor and subsequent events of increasing degrees of seriousness, or involving different components, until the product is finally deemed of no further use. The many different paths through these states which any product may follow in the course of a lifetime can be very complex. Models can usefully shed light on this.

Financial analysis

Mathematical techniques of stochastic integration have seen wide application in recent years in the study of financial products such as derivatives, futures, and other pricing schemes for certain kinds of options. An alternative and potentially more flexible approach to many of these questions is via regression modeling. The increasing volatility of many of the markets has also given rise to the use of survival modeling as a tool to identify, among the large-scale borrowers, large industries, countries and even financial institutions themselves the ones which are most likely to fail on their repayment schedule. Modeling techniques can make such analyses more precise, identifying just which factors are most strongly predictive.

1.3   Objectives

Main goal

The purpose of this book is to provide the structure necessary to the building of a coherent framework in which to view proportional and non-proportional hazards regression. The essential idea is that of prediction, a feature common to all regression models, but one that sometimes slips from sight amid the wealth of methodological innovations that has characterized research in this area. Our motivation should mostly derive from the need to obtain insights into the complex data sets that can arise in the survival setting, keeping in mind the key notion that the outcome time, however measured, and its dependence on other factors, is at the center of our concerns.

The predictive power of any model, in this case the proportional hazards model, is an area that we pay a lot of attention to. It is necessary to investigate how we can obtain predictions in the absence of information (often referred to as explanatory variables or covariate information) which relate to survival and, subsequently, obtaining predictions in the presence of such information. How best such information can be summarized brings us into the whole area of model adequacy and model performance. Measures of explained variation and explained randomness can indicate to what extent our prediction accuracy improves when we include additional covariate information into any model.

In order to give the reader, new to the area, a feeling as to how the Cox model fits into the general statistical literature we provide some discussion of the original paper of Professor Cox in 1972, some of the background leading up to that seminal paper, and some of the scientific discussion that ensued. The early successes of the model in characterizing and generalizing several classes of statistics are described.

As is true of much of science - statistics is no exception - important ideas can be understood on many levels. Some researchers, with a limited training in even the most basic statistical methods, can still appreciate the guiding principles behind proportional hazards regression. Others are mainly interested in some of the deeper inferential mathematical questions raised by the estimation techniques employed. Hopefully both kinds of reader will find something in this book to their taste. The aim of the book is to achieve a good balance, a necessary compromise, between the theoretical and the applied. This will necessarily be too theoretical for some potential readers, not theoretical enough for others. Hopefully the average gap is not too great.

The preliminary chapters aim to fill in many of the basic ideas from probability and statistics which are needed throughout the text. These chapters are short and do not therefore cover extensively either probability or statistics. But any of the key ideas needed throughout the book can be found in these chapters. The text could therefore serve as a graduate text for students with a relatively limited background in either probability or statistics. Advanced measure theory is not really necessary either in terms of understanding the proportional hazards model or for gaining insight into applied problems. It is not emphasized in this book and this is something of a departure from a number of other available texts which deal with these and related topics. Proofs of results are clearly of importance, partly to be reassured as to the validity of the techniques we apply, but also in their own right and of interest to those focusing mostly on the methods. In order not to interrupt the text with proofs, we give theorems, corollaries, and lemmas, but leave the proofs to be gathered together in a chapter of their own at the end of the text. The interested reader will find them there in a concise form. Those not interested in the more formal presentation in terms of theorems and proofs will nonetheless, it is hoped, find the style helpful in that, by omitting the proofs at the time of development, the necessary results are organized and brought out in a sharper focus.

1.4 Controversies

This goal of this book is not to present an undifferentiated, dispassionate review of the literature. The bulk of the discussed methodology converges around one or two key central notions and these come together to represent a particular viewpoint. Certain significant areas of the literature are not covered. As an example, local smoothing of the baseline hazard, or of the residuals, is not given any space. The same goes for kernel estimates of the density function or the use of polynomial splines. The view taken here is that, for the great majority of applications, cumulative quantities do "enough" smoothing. However, such work is valuable and certainly of interest. An absence of coverage in this text should not be taken as implying any diminished importance for such techniques. The choice of how much weight to afford different topics, including the assignment of a zero weight, is often a subjective one - a matter of taste. On other topics, a stronger position is sometimes adopted and one that can represent something of a departure from the literature. Relatively little weight is given, for instance, to the study of frailty models, outside that of goodness of fit. Random effects models do have use and this is described in the text but for frailty models, if we are to understand the term "frailty" as meaning a random effect for an individual (without reapeated measurements), as opposed to a group (random effects models), then this is just another way of expressing lack of fit. In other words a proportional hazards model with a frailty term is equivalent to a non proportional hazards model. Tests, described in the literature, for heterogeneity of individuals in this context can be wholly misleading. This has been described in O'Quigley and Stare (2002).

The term "partial likelihood" is an important one in this area. We use it mostly as a name, or label, for a particular form of a statistic first introduced by Cox (1972). The concept of partial likelihood as a general technique of inference in its own right, and for problems other than inference for the proportional hazards model, has never really been thoroughly developed. One difficulty with the concept, as outlined in Cox (1975), is that, for given problems, the partial likelihood would not be unique. For general situations then it may not be clear as to the best way to proceed. For these reasons we do not study partial likelihood as a tool for inference and the concept is not given particular weight. This is a departure from several other available texts on survival analysis.

One important area where there may exist disagreement between the position of this text and the position taken by a number of workers in the area is that of explained variation. Considerable confusion on this topic is prevalent among statisticians and yet the main ideas are very simple. The origin of the confusion for many was in a paper of Draper (1984), echoed subsequently by Healy (1984), which was strongly critical of R^2. Several authors (Kvalseth 1985; Scott and Wild 1991; Willett and Singer 1988) followed in the wake of these two papers, adding further nails, it was argued, to the coffin of explained variation. However, Draper's paper was in error. He was certainly the first to notice this and he wrote a retraction (Draper 1985) - changing entirely the conclusion of his earlier paper from "R^2 is misleading" to "R^2 *is* a useful indicator" (his italics). By one of those curious twists governing the survival mechanisms of scientific ideas, few workers in the area, and apparently none of those building on Draper's 1984 paper, seem to be aware of the retraction. The original contribution is still frequently quoted and its erroneous conclusions are widely believed to be valid.

Indeed, there is little mystery to the idea of explained variation. Note that, by virtue of the Chebyshev-Bienaymé inequality, explained variation directly quantifies predictive ability. In this text we develop a basic theory of explained variation in order to have a solid foundation upon which to appeal when we wish to consider the particular case of explained variation for survival models. In the light of this theory and the main results concerning inference it is relatively straightforward to develop suitable measures of explained variation (O'Quigley and Flandre 1994, O'Quigley and Xu 2001). In this book the aim is to appeal to elementary concepts, constructing our inferential techniques, if necessary, from scratch. A theory for explained variation in terms of random variables alone, outside of the context of any model, is required and we build upon that.

The related topic, explained randomness, in view of its direct connection to likelihood, is also given consideration. In the same way, although leaning on the concepts of entropy rather than the Chebyshev-Bienaymé inequality, explained randomness also translates predictive ability. For normal models the population quantities coincide. For non-normal models, when it might be argued that the variance measure is less compelling, a case can be made for preferring explained randomness over explained variation. We return to these topics later in the text.

Among the users of statistical models it can be argued that there are two main schools of thought; the first sees a model as an approximation to some, infinitely more complex, reality: that is the model is taken as a more-or-less refined means to achieving some specific end, usually that of prediction of some quantity of interest. Any model is then simply judged by its predictive performance. The second school sees the statistician's job as tracking down the "true" model that can be considered to have generated the data. The position taken in this work is closer to the first than the second. This means that certain well-studied concepts, such as efficiency, a concept which assumes our models are correct, are given little attention. The regression parameter β in our model is typically taken as some sort of average. The model is then viewed as a working model and not some absolute truth. The proportional hazards model stipulates that effects, as quantified by β, do not change through time. In reality the effects must surely change, hopefully not too much, but absolute constancy of effects is too strong an assumption to hold up. The working model enables us to estimate useful quantities, one of them being average regression effect, the average taken through time. Interestingly, the usual partial likelihood estimator in the situation of changing regression effects does not estimate an average effect, as is often believed. Even so, we can estimate an average effect but we do require an estimator different to that commonly used. We return to this in those chapters dealing with inference.

1.5   Data sets

The importance of working through the calculations on actual data cannot be overemphasized. The reader is encouraged to use their own or any of the many available data sets though, for example, *statlib* as well as simulated data sets corresponding to specific conditions. In the text we often refer to the Freireich data, described by several authors including Kalbfleisch and Prentice (1980) and Professor Cox in his 1972 founding paper. These data arose in the context of a balanced comparison of two treatment groups. We also refer to breast cancer data which were gathered at the Institut Curie in Paris, France over a near thirty year period. Finally, another important data set used here to illustrate many concepts was obtained in the context of a survival study in gastric cancer carried out at the St James Hospital, Leeds, U.K. (see for example Rashid et al; 1982). An interesting case of non-proportional hazards arose in a clinical trial studied by Stablein et al (1980). The regression effect appeared to change direction during the study. We refer to this in the text as the Stablein data.

1.6   Use as a graduate text

The text can be used as support for an introductory graduate course in survival analysis with particular emphasis on proportional and non-proportional hazards models. The approach to inference is more classical than often given in such courses, steering mostly away from the measure-theoretic difficulties associated with multivariate counting processes and stochastic integrals and focusing instead on the more classical, and well known, results of empirical processes. Brownian motion and functions of Brownian motion play a central role. Exercises are provided in order to reinforce the course work. Their aim is not so much to help develop facility with analytic calculation but

more to build insight into the important features of models in this setting. Some emphasis then is given to practical work carried out using
a computer. No particular knowledge of any specific software package
is assumed.

1.7 Exercises and class projects

1. For the several examples described in Section 1.2 write down those
features of the data which appear common to all examples. Which
features are distinctive?

2. In Section 1.4 it is claimed that cumulative quantities may do enough
smoothing in practice. How do you understand this idea of "enough
smoothing?" What does a statistician aim to achieve by smoothing.

3. Suppose that the methods of survival analysis were not available
to us. Suggest how we might analyze a randomized clinical trial using
(i) multiple linear regression, (ii) multiple logistic regression.

4. For Exercise 3, describe the shortcomings that we expect to be
associated with either type of analysis. How are these shortcomings
amplified by the presence of increasing censoring, and in what way
do we anticipate the techniques of survival analysis to address these
shortcomings.

5. Consider a long-term survival study in which the outcome of interest
is survival itself and we wish to study any possible dependence of this
upon two other variables, one binary and one continuous. The marginal
distribution of survival is unknown but suspected to be very skew. For
the data at hand there is no censoring, i.e., all the actual survival times
have been observed. Describe at least one possible approach, aside from
those of survival analysis, to analyzing such data. Do you think more
could been obtained from the data using survival techniques? Explain.

6. How would you describe to a nonscientist the purpose of a statistical
model, explaining the issues involved with misspecified models. How
would you describe to a physicist the difference between statistical
models which may be employed in epidemiology and statistical models that he or she may be using to elaborate theories in quantum
mechanics.

Chapter 2

Background: Probability

2.1 Summary

We review the fundamental tools used to establish the inferential basis for our models. Results are stated as theorems, lemmas and corollaries. Most of the key proofs are provided in Chapter 16 although, sometimes, when useful to the general development, proofs are given within the text itself. The main ideas of stochastic processes, in particular Brownian motion and functions of Brownian motion, are explained in non-measure-theoretic terms. The background to this, i.e., distribution theory and large sample results, is recalled. Rank invariance is an important concept, i.e., the ability to transform some variable, usually time, via monotonic increasing transformations without having an impact on inference. These ideas hinge on the theory of order statistics and the basic notions of this theory are recalled. An outline of the theory of counting processes and martingales is presented without leaning upon measure-theoretic constructions. The important concepts of explained variation and explained randomness are outlined in elementary terms, i.e., only with reference to random variables and, at least initially, making no explicit appeal to any particular model. This is important since the concepts are hardly any less fundamental than a concept such as variance itself. They ought therefore stand alone, and not require derivation as a particular feature of some model. In practice, of course, we may need estimate conditional distributions and making an appeal to a model at this point is quite natural.

2.2   Motivation

The last few decades have seen the topic of survival analysis become increasingly specialized, having a supporting structure based on large numbers of theorems and results which appear to have little application outside of the field. Many recently trained specialists, lacking a good enough grasp of how the field relates to many others, are left with little option but to push this specialization yet further. The result is a field which is becoming largely inaccessible to statisticians from other areas. A key motivation of this work, and this chapter in particular, is to put some brakes on this trend by leaning on classical results. Most of these are well known, others less so, and in this chapter we cover the main techniques from probability and statistics which we will need. Results are not simply presented and the aim is to motivate them from elementary principles known to those with a rudimentary background in calculus.

2.3   Integration and measure

The reader is assumed to have some elementary knowledge of set theory and calculus. We do not recall here any of the basic notions concerning limits, continuity, differentiability, convergence of infinite series, Taylor series and so on and the rusty reader may want to refer to any of the many standard calculus texts when necessary. One central result which is frequently called upon is the mean value theorem. This can be deduced as an immediate consequence to the following result known as Rolle's theorem.

Theorem 2.1 *If $f(x)$ is continuously differentiable at all interior points of the interval $[a, b]$ and $f(a) = f(b)$, then there exists a real number $\xi \in (a, b)$ such that $f'(\xi) = 0$.*

A simple sketch would back up our intuition that the theorem would be correct. Simple though the result appears to be, it has many powerful implications including;

Theorem 2.2 *If $f(x)$ is continuously differentiable on the interval $[a, b]$, then there exists a real number $\xi \in (a, b)$ such that*

$$f(b) = f(a) + (b - a)f'(\xi).$$

When $f(x)$ is monotone then ξ is unique. This elementary theorem can form the basis for approximation theory and series expansions such as the Edgeworth and Cornish-Fisher (see Section 2.9). For example, a further immediate corollary to the above theorem obtains by expanding in turn $f'(\xi)$ about $f'(a)$ whereby:

Corollary 2.1 *If $f(x)$ is at least twice differentiable on the interval $[a, b]$ then there exists a real number $\xi \in (a, b)$ such that*

$$f(b) = f(a) + (b - a)f'(a) + \frac{(b - a)^2}{2}f''(\xi).$$

The ξ of the theorems and corollary would not typically be the same and we can clearly continue the process, resulting in an expansion of $m + 1$ terms, the last term being the mth derivative of $f(x)$, evaluated at some point $\xi \in (a, b)$ and multiplied by $(b - a)^m/m!$. An understanding of Riemann integrals as limits of sums, definite and indefinite integrals, is mostly all that is required to follow the text. It is enough to know that we can often interchange the limiting processes of integration and differentiation. The precise conditions for this to be valid are not emphasized. Indeed, we almost entirely avoid the tools of real analysis. The Lebesgue theory of measure and integration is on occasion referred to, but a lack of knowledge of this will not hinder the reader. Likewise we will not dig deeply into the measure-theoretic aspects of the Riemann-Stieltjes integral apart from the following extremely useful construction:

Definition 2.1 *The Riemann integral of the function $f(x)$ with respect to x, on the interval $[a, b]$, is the limit of a sum $\sum \Delta_i f(x_{i-1})$, where $\Delta_i = x_i - x_{i-1} > 0$, for an increasing partition of $[a, b]$ in which $\max \Delta_i$ goes to zero.*

The limit is written $\int_a^b f(x)dx$ and can be seen to be the area under the curve $f(x)$ between a and b. If $b = \infty$ then we understand the integral to exist if the limit exists for any $b > 0$, the result itself converging to a limit as $b \to \infty$. Similarly for $a = -\infty$. Now, instead of only considering small increments in x, i.e., integrating with respect to x, we can make use of a more general definition. We have:

Definition 2.2 *The Riemann-Stieltjes integral of the function $f(x)$ with respect to $g(x)$ is the limit of a sum $\sum \{g(x_i) - g(x_{i-1})\}f(x_{i-1})$, for an increasing partition of $[a, b]$ in which, once again, $\max \Delta_i$ goes to zero.*

The limit is written $\int_a^b f(x)dg(x)$ and, in the special case where $g(x) = x$, reduces to the usual Riemann integral. For functions, necessarily continuous, whereby $g(x)$ is an antiderivative of, say, $h(x)$ and can be written $g(x) = \int_{-\infty}^x h(u)du$ then the Stieltjes integral coincides with the Riemann integral $\int f(x)h(x)dx$. On the other hand whenever $g(x)$ is a step function with a finite or a countable number of discontinuities then $\int f(x)dg(x)$ reduces to a sum, the only contributions arising at the discontinuities themselves. This is of great importance in statistical applications where step functions naturally arise as estimators of key functions. A clear example of a step function of central importance is the empirical distribution function, $F_n(x)$ (this is discussed in detail in Chapter 3). We can then write the sample mean $\bar{x} = \int udF_n(u)$ and the population mean $\mu = \int udF(u)$, highlighting an important concept, that fluctuations in the sample mean can be considered a consequence of fluctuations in $F_n(x)$ as an estimate of $F(x)$. Consider the following theorem, somewhat out of sequence in the text but worth seeing here for its motivational value. The reader may wish to take a glance ahead at Sections 2.4 and 3.5.

Theorem 2.3 *For every bounded continuous function $h(x)$, if $F_n(x)$ converges in distribution to $F(x)$, then $\int h(x)dF_n(x)$ converges in distribution to $\int h(x)dF(x)$.*

This is the Helly-Bray theorem. The theorem will also hold (see the Exercises) when $h(x)$ is unbounded provided that some broad conditions are met. A deep study of $F_n(x)$ as an estimator of $F(x)$ is then all that is needed to obtain insight into the sample behavior of the empirical mean, the empirical variance and many other quantities. Of particular importance for the applications of interest to us here, and developed, albeit very briefly, in Section 2.12, is the fact that, letting $M(x) = F_n(x) - F(x)$, then

$$E\left\{\int h(x)dM(x)\right\} = \int h(x)dF(x) - \int h(x)dF(x) = 0, \qquad (2.1)$$

a seemingly somewhat innocuous result until we interchange the order of integration (expectation, denoted by E being an integral operator) and, under some very mild conditions on $h(x)$ described in Section 2.12, we obtain a formulation of great generality and into which can be fit many statistical problems arising in the context of stochastic processes (see Section 2.12).

2.4 Random variables and probability measure

The possible outcomes of any experiment are called events where any event represents some subset of the sample space. The sample space is the collection of all events, in particular the set of elementary events. A random variable X is a function from the set of outcomes to the real line. A probability measure is a function on some subset of the real line to the interval [0,1]. Kolmogorov (1933) provided axioms which enable us to identify any measure as being a probability measure. These axioms appear very reasonable and almost self-evident, apart from the last, which concerns assigning probability measure to infinite collections of events. There is, in a well defined sense, many more members in the set of all subsets of any infinite set than in the original set itself, an example being the set of all subsets of the positive integers which has as many members as the real line. This fact would have hampered the development of probability without the inclusion of Kolmogorov's third axiom which, broadly says that the random variable is measurable, or, in other words, that the sample space upon which the probability function is defined is restricted in such a way that the probability we associate with the sum of an infinite collection of mutually exclusive events is the same as the sum of the probabilities associated with each composing event.

A great deal of modern probability theory is based on measure-theoretic questions, questions that essentially arise from the applicability or otherwise of Kolmogorov's third axiom in any given context. This is an area that is highly technical and relatively inaccessible to non-mathematicians, or even to mathematicians lacking a firm grounding in real analysis. The influence of measure theory has been strongly felt in the area of survival analysis over the last 20 or so years and much modern work is now of a very technical nature. Even so, none of the main statistical ideas, or any of the needed demonstrations in this text, require such knowledge. We can therefore largely avoid measure-theoretic arguments, although some of the key ideas that underpin important concepts in stochastic processes are touched upon whenever necessary. The reader is expected to understand the meaning of the term *random variable* on some level.

Observations or outcomes as random variables and, via models, the probabilities we will associate with them are all part of a theoretical, and therefore artificial, construction. The hope is that these probabilities will throw light on real applied problems and it is useful to keep in mind that, in given contexts, there may be more than one way to set things up. Conditional expectation is a recurring central topic but can arise in ways that we did not originally anticipate. We may naturally think of the conditional expected survival time given that a subject begins the study under, say, some treatment. It may be less natural to think of the conditional expectation of the random variable we use as a treatment indicator given some value of time after the beginning of treatment. Yet, this latter conditional expectation, as we shall see, turns out to be the more relevant for many situations.

Convergence for random variables

Simple geometrical constructions (intervals, balls) are all that are necessary to formalize the concept of convergence of a sequence in real and complex analysis. For random variables there are a number of different kinds of convergence, depending upon which aspect of the random variable we are looking at. Consider any real value Z and the sequence $U_n = Z/n$. We can easily show that $U_n \to 0$ as $n \to \infty$. Now let U_n be defined as before except for values of n that are prime. Whenever n is a prime number then $U_n = 1$. Even though, as n becomes large, U_n is almost always arbitrarily close to zero, a simple definition of convergence would not be adequate and we need consider more carefully the sizes of the relevant sets in order to accurately describe this. Now, suppose that Z is a uniform random variable on the interval (0,1). We can readily calculate the probability that the distance between U_n and 0 is greater than any arbitrarily small positive number ϵ and this number goes to zero with n. We have convergence in probability. Nonetheless there is something slightly erratic about such convergence, large deviations occurring each time that n is prime. When possible, we usually prefer a stronger type of convergence. If, for all integer values m greater than n and as n becomes large, we can assert that the probability of the distance between U_m and 0 being greater than some arbitrarily small positive number goes to zero, then such a mode of convergence is called strong convergence. This stronger convergence is also called convergence with probability one or almost sure convergence. Consider also $(n + 3)U_n$. This random variable will converge almost surely to the random variable Z. But, also, we can say that the distribution of $\log_e(n+3)U_n$, at all point of continuity z, becomes arbitrarily close to that of a standard exponential distribution. This is called convergence in distribution. The three modes of convergence are related by:

Theorem 2.4 *Convergence with probability one implies convergence in probability. Convergence in probability implies convergence in distribution.*

Also, for a sequence that converges in probability, there exists a subsequence that converges with probability one. This latter result requires the tools of measure theory and is not of wide practical applicability since we may not have any obvious way of identifying such a subsequence. In theoretical work it can sometimes be easier to obtain results for weak rather than strong convergence. However, in practical applications, we usually need strong (almost sure, "with probability one") convergence since this corresponds in a more abstract language to the important idea that, as our information increases, our inferences becomes more precise.

Convergence of functions of random variables

In constructing models and establishing inference for them we will frequently appeal to two other sets of results relating to convergence. The first of these is that, for a continuous function $g(z)$, if Z_n converges in probability to c, then $g(Z_n)$ converges in probability to $g(c)$ and, if Z_n converges in distribution to Z, then $g(Z_n)$ converges in distribution to $g(Z)$. The second set, Slutsky's theorem (a proof is given in Randles and Wolf 1979), enables us to combine modes of convergence. In particular, for modeling purposes, if a convergence in distribution result holds when the parameters are known, then it will continue to hold when those same parameters are replaced by consistent estimators. This has great practical value.

2.5 Distributions and densities

We anticipate that most readers will have some familiarity with the basic ideas of a distribution function $F(t) = \Pr(T < t)$, a density function $f(t) = dF(t)/dt$, expectation and conditional expectation, the moments of a random variable and other basic tools. Nonetheless we will go over these elementary notions in the context of survival in the next chapter. We write

$$E\,\psi(T) = \int \psi(t)f(t)dt = \int \psi(t)dF(t)$$

for the expected value of the function $\psi(T)$. Such an expression leaves much unsaid, that $\psi(t)$ is a function of t and therefore $\psi(T)$ itself random, that the integrals exist, the domain of definition of the function being left implicit, and that the density $f(t)$ is an anti-derivative of the cumulative distribution $F(t)$ (in fact, a slightly weaker mathematical construct, absolute continuity, is enough but we do not feel the stronger assumption has any significant cost attached to it). There is a wealth of solid references for the rusty reader on these topics, among which Billingsley (1968), Rao (1973), and Serfling (1980) are particularly outstanding. It is very common to wish to consider some transformation of a random variable, the simplest situation being that of a change in origin or scale. The distribution of sums of random variables arises by extension to the bivariate and multivariate cases.

Theorem 2.5 *Suppose that the distribution of X is $F(x)$ and that $F'(x) = f(x)$. Suppose that $y = \phi(x)$ is a monotonic function of x and that $\phi^{-1}(y) = x$. Then, if the distribution of Y is $G(y)$ and $G'(y) = g(y)$,*

$$G(y) = F\{\phi^{-1}(y)\}; \quad g(y) = f\{\phi^{-1}(y)\}\left|\frac{d\phi(x)}{dx}\right|^{-1}_{x=\phi^{-1}(y)}. \quad (2.2)$$

Theorem 2.6 *Let X and Y have joint density $f(x, y)$. Then the density $g(w)$ of $W = X + Y$ is given by*

$$g(w) = \int_{-\infty}^{\infty} f(x, w - x)dx = \int_{-\infty}^{\infty} f(w - y, y)dy. \quad (2.3)$$

A result for $W = X - Y$ follows immediately and, in the case of X and Y being independent, the corresponding expression can also be written down readily as a product of the two respective densities. Similar results hold for the product or ratio of random variables (see Rohatgi 1984, Section 8.4) but, since we have no call for them in this work, we do not write them down here. An immediate corollary that can give an angle on small sample behavior of statistics that are written as sums is;

Corollary 2.2 *Let $X_1, \ldots, X_n$ be independent, not always identically distributed, continuous random variables with densities $f_1(x)$ to $f_n(s)$ respectively. Let $S_n = \sum_{j=1}^{n} X_j$. Then the density, $g_n(s)$, of S_n is given by*

$$g_n(s) = \int_{-\infty}^{\infty} g_{n-1}(s - x)f_n(x)dx.$$

This result can be used iteratively building up successive solutions by carrying out the integration. The integration itself will mostly be not particularly tractable and can be evaluated using numerical routines. Note the difference between making a large sample statistical approximation to the sum and that of a numerical approximation to the integral. The integral expression itself is an exact result.

Normal distribution

A random variable X is taken to be a a normal variate with parameters μ and σ when we write $X \sim \mathcal{N}(\mu, \sigma^2)$. The parameters μ and σ^2 are the mean and variance respectively, so that $\sigma^{-1}(X - \mu) \sim \mathcal{N}(0, 1)$. The distribution $\mathcal{N}(0, 1)$ is called the standard normal. The density of the standard normal variate, that is, having mean zero and variance one, is typically denoted $\phi(x)$ and the cumulative distribution $\Phi(x)$. The density $f(x)$, for $x \in (-\infty, \infty)$ is given by

$$f(x) = \phi(x) = \frac{1}{\sqrt{2\pi}\sigma} \exp\left[-\frac{1}{2}\left(\frac{x - \mu}{\sigma}\right)^2\right].$$

For stochastic processes described below, Brownian motion relates to a Gaussian process, that is, it has been standardized, in an analogous way that the standard normal relates to any other normal distribution. For the normal distribution, all cumulants greater than 2 are equal to zero. Simple calculations (Johnson and Kotz, 1970) show that, for $X \sim \mathcal{N}(0, 1)$, then $E(X^r) = (r - 1)(r - 3)\ldots 3.1$. Thus, all odd moments are equal to zero and all even moments are expressible in terms of the variance. The normal distribution is of very great interest in view of it frequently being the large sample limiting distribution for sums of random variables. These arise naturally via simple estimating equations. These topics are looked at in greater detail below.

The multivariate normal can be characterized in various ways. If and only if all marginal distributions and all conditional distributions are normal then we have multivariate normality. If and only if all linear combinations are univariate normal then we have multivariate normality. It is only necessary to be able to evaluate the standard normal integral, $\Phi(x) = 1 - \int_x^\infty \phi(x)dx$, since any other normal distribution, $f(x)$, can be put in this form via the linear transformation $(X - \mu)/\sigma$. Tables, calculator, and computer routines can approximate the numerical integral. Otherwise, it is worth bearing in mind the following;

Lemma 2.1 *Upper and lower bounds for the normal integral can be obtained from*

$$\frac{x}{1+x^2}\, e^{-x^2/2} < \int_x^\infty e^{-u^2/2}du < \frac{1}{x}\, e^{-x^2/2}.$$

The lemma tells us that we expect $1 - \Phi(x)$ to behave like $\phi(x)/x$ as x increases. The ratio $\phi(x)/x$ is known as Mill's ratio. Approximate calculations are then possible without the need to resort to sophisticated algorithms, although, in modern statistical analysis, it is now so commonplace to routinely use computers that the value of the lemma is rather limited. The normal distribution plays an important role in view of the central limit theorem described below but also note the interesting theorem of Cramer (1937) whereby, if a finite sum of independent random variables is normal, then each variable itself is normal. Cramer's theorem might be contrasted with central limit theorems whereby sums of random variables, under broad conditions, approach the normal as the sum becomes infinitely large. These limit results are looked at later. The normal distribution is important since it provides the basis to Brownian motion and this is the key tool that we will use for inference throughout this text.

Uniform distribution and the probability integral transform

For the standard uniform distribution in which $u \in [0, 1]$, $f(u) = 1$ and $F(u) = u$. Uniform distributions on the interval $[a, b]$ correspond to the density $f(u) = 1/(b - a)$ but much more important is the fact that for any continuous distribution, $G(t)$, we can say:

Theorem 2.7 *For the random variable T, having distribution $G(t)$, letting $U_1 = G(T)$ and $U_2 = 1 - G(T)$, then both U_1 and U_2 have a standard uniform distribution.*

This central result, underpinning a substantial body of work on simulation and re-sampling, is known as the probability integral transform. Whenever we can invert the function G, denoted G^{-1}, then, from a single uniform variate U we obtain the two variates $G^{-1}(U)$ and $G^{-1}(1 - U)$ which have the distribution G. The two variates are of course not independent but, in view of the strong linearity property of expectation (the expectation of a linear function of random variables is the same linear function of the expectations), we can often use this to our advantage to improve precision when simulating. Another interesting consequence of the probability integral transform is that there exists a transformation of a variate T, with any given distribution, into a variate having any other chosen distribution. Specifically, we have:

Corollary 2.3 *For any given continuously invertible distribution function H, and continuous distribution $G(t)$, the variate $H^{-1}\{G(T)\}$ has distribution H.*

In particular, it is interesting to consider the transformation $\Phi^{-1}\{G_n(T)\}$ where G_n is the empirical estimate (discussed below) of G. This transformation, which preserves the ordering, makes the observed distribution of observations as close to normal as possible. Note that since the ordering is preserved, use of the transformation makes subsequent procedures nonparametric in as much as the original distribution of T has no impact. For the problems of interest to us in survival analysis we can use this in one of two ways: firstly, to transform the response variable time in order to eliminate the impact of its distribution and, secondly, in the context of regression problems, to transform the distribution of regressors as a way to obtain greater robustness by reducing the impact of outliers.

Exponential distribution and cumulative hazard transformation

The standard exponential distribution is defined on the positive real line $(0, \infty)$. We have, for $u \in (0, \infty)$, $f(u) = \exp(-u)$ and $F(u) = 1 - \exp(-u)$. An exponential distribution with mean $1/\alpha$ and variance $1/\alpha^2$ has density $f(u) = \alpha \exp(-\alpha u)$ and cumulative distribution $F(u) = 1 - \exp(-\alpha u)$. The density of a sum of m independent exponential variates having mean $1/\alpha$, is an Erlang density whereby $f(u) = \alpha(\alpha u)^{m-1} \exp(-\alpha u)/\Gamma(m)$ and where $\Gamma(m) = \int_0^\infty \exp(-u)u^{m-1}du$. The gamma distribution has the same form as the Erlang although, for the gamma, the parameter m can be any real positive number and is not restricted to being an integer. An exponential variate U can be characterized as a power transformation on a Weibull variate in which $F(t) = 1 - \exp[(-\alpha t)^k]$. Finally, we have the important result:

Theorem 2.8 *For any continuous positive random variable T, with distribution function $F(t)$, the variate $U = \int_0^T f(u)/[1 - F(u)]du$ has a standard exponential distribution.*

This result is important in survival modeling and we return to it later. The function $f(t)/[1 - F(t)]$ is known as the hazard function and

$\int_0^t f(u)/[1 - F(u)]du$ as the cumulative hazard function. The transformation is called the cumulative hazard transformation.

2.6 Expectation

It is worth saying a word or two more about expectation as a fundamental aspect of studies in probability. Indeed it is possible for the whole theory to be constructed with expectation as a starting point rather than the now classical axiomatic structure to probability. For a function of a random variable T, $\psi(T)$ say, as stated at the beginning of the previous section, we write, $E(\psi(T)$ of this function via

$$E \psi(T) = \int \psi(t)f(t)dt = \int \psi(t)dF(t),$$

where the integrals, viewed as limiting processes, are all assumed to converge. The normal distribution function for a random variable X is completely specified by $E(X)$ and $E(X^2)$. In more general situations we can assume a unique correspondence between the moments of X, $E(X^r), r = 1, 2, \ldots$, and the distribution functions as long as these moments all exist. While it is true that the distribution function determines the moments the converse is not always true. However, it is almost always true (Stuart and Ord 1994, page 111) and, for all the distributions of interest to us here, the assumption can be made without risk. It can then be helpful to view each moment, beginning with $E(X)$, as providing information about $F(x)$. This information typically diminishes quickly with increasing r. We can use this idea to improve inference for small samples when large sample approximations may not be sufficiently accurate. Moments can be obtained from the moment generating function, $M(t) = E\{\exp(tX)\}$ since we have:

Lemma 2.2 *If $\int \exp(tx)f(x)dx < \infty$ then*

$$E(X^r) = \left\{ \frac{\partial^r M(t)}{\partial t^r} \right\}_{t=0}, \quad \text{for all } r.$$

In Section 2.8 we consider the variance function which is also an expectation and is of particular interest to one of our central goals here, that of constructing useful measures of the predictive strength of any model. At the root of the construction lie two important inequalities, the Chebyshev-Bienaymé inequality (described in Section 2.8 and Jensen's inequality described below. For this we first need:

Definition 2.3 *The real-valued function $w(x)$ is called "convex" on some interval I (an infinite set and not just a point) whenever, for $x_1, x_2 \in I$ and for $0 \leq \lambda \leq 1$, we have*

$$w[\lambda x_1 + (1 - \lambda)x_2] \leq \lambda w(x_1) + (1 - \lambda)w(x_2).$$

It is usually sufficient to take convexity to mean that $w'(x)$ and $w''(x)$ are greater than or equal to zero at all interior points of I since this is a consequence of the definition. We have (Jensen's inequality):

Lemma 2.3 *If w is convex on I then, assuming expectations exist on this interval, $w[E(X)] \leq E[w(X)]$. If w is linear in X throughout I, that is, $w''(x) = 0$ when twice differentiable, then equality holds.*

For the variance function we see that $w(x) = x^2$ is a convex function and so the variance is always positive. The further away from the mean, on average, the observations are to be found, then the greater the variance. We return to this in Section 2.8. Although very useful, the moment-generating function, $M(t) = E\{\exp(tX)\}$ has a theoretical weakness in that the integrals may not always converge. It is for this, mainly theoretical, reason that it is common to study instead the characteristic function, which has an almost identical definition, the only difference being the introduction of complex numbers into the setting. The characteristic function, denoted by $\phi(t)$, always exists and is defined as:

$$\phi(t) = M(it) = \int_{-\infty}^{\infty} \exp(itx)dF(x), \quad i^2 = -1.$$

Note that the contour integral in the complex plane is restricted to the whole real axis. Analogous to the above lemma concerning the moment-generating function we have

$$E(X^r) = (-i)^r \left\{ \frac{\partial^r \phi(t)}{\partial t^r} \right\}_{t=0}, \quad \text{for all } r.$$

This is important in that it allows us to anticipate the cumulative generating function which turns out to be of particular importance in obtaining improved approximations to those provided by assuming normality. We return to this below in Section 2.9. If we expand the exponential function then we can write;

$$\phi(t) = \int_{-\infty}^{\infty} \exp(itx)dF(x) = \exp\left\{ \sum_{r=1}^{\infty} \kappa_r (it)^r / r! \right\}$$

and, identifying κ_r as the coefficient of $(it)^r/r!$ in the expansion of $\log \phi(t)$. The function $\psi(t) = \log \phi(t)$ is called the cumulative generating function. When this function can be found then the density $f(x)$ can be defined in terms of it. We have the important relation

$$f(x) = \frac{1}{2\pi} \int_{-\infty}^{\infty} e^{-itx} \phi(t) dt, \quad \phi(t) = \int_{-\infty}^{\infty} e^{itx} f(x) dx.$$

It is possible to approximate the density $f(x)$ by working with i.i.d. observations $X_1, \cdots, X_n$ and the empirical characteristic function $\phi(t) = n^{-1} \sum_{i=1}^{n} \exp(itx_i)$ which can then be inverted. It is also possible to approximate the integral using a method of numerical analysis, the so-called method of steepest descent, to obtain a saddlepoint approximation (Daniels, 1954). We return to this approximation below in Section 2.9.

2.7   Order statistics and their expectations

The normal distribution and other parametric distributions described in the next chapter play a major role in survival modeling. However, robustness of any inferential technique to particular parametric assumptions is always a concern. Hopefully, inference is relatively insensitive to departures from parametric assumptions or is applicable to whole families of parametric assumptions. The most common way to ensure this latter property is via the theory of order statistics which we recall here. Consider the n independent identically distributed (i.i.d.) random variables: $X_1, X_2, \ldots, X_n$ and a single realization of these that we can order from the smallest to the largest: $X_{(1)} \leq X_{(2)} \leq \cdots \leq X_{(n)}$. Since the X_i are random, so also are the $X_{(i)}$, and the interesting question concerns what we can say about the probability structure of the $X_{(i)}$ on the basis of knowledge of the parent distribution of X_i. In fact, we can readily obtain many useful results which, although often cumbersome to write down, are in fact straightforward. Firstly we have:

Theorem 2.9 *Taking $P(x) = \Pr(X \leq x)$ and $F_r(x) = \Pr(X_{(r)} \leq x)$ then:*

$$F_r(x) = \sum_{i=r}^{n} \binom{n}{i} P^i(x)[1 - P(x)]^{n-i}. \tag{2.4}$$

This important result has two immediate and well known corollaries dealing with the maximum and minimum of a sample of size n.

Corollary 2.4

$$F_n(x) = P^n(x) , \quad F_1(x) = 1 - [1 - P(x)]^n \qquad (2.5)$$

In practice, in order to evaluate $F_r(x)$ for other than very small n, we exploit the equivalence between partial binomial sums and the incomplete beta function. Thus, if, for $a > 0$, $b > 0$, $B(a,b) = \int_0^1 t^{a-1}(1-t)^{b-1}dt$ and $I_\pi(a,b) = \int_0^\pi t^{a-1}(1-t)^{b-1}dt/B(a,b)$, then putting $P(x) = \pi$, we have that $F_r(x) = I_\pi(r, n-r+1)$. These functions are widely tabulated and also available via numerical algorithms to a high level of approximation. An alternative, although less satisfying, approximation would be to use the DeMoivre-Laplace normal approximation to the binomial sums. Differentiation of (2.4) provides the density which can be written as

$$f_r(x) = \frac{1}{B(r, n-r+1)} P^{r-1}(x)[1 - P(x)]^{n-r}p(x). \qquad (2.6)$$

Since we have a relatively straightforward expression for the distribution function itself, then this expression for the density is not often needed. It can come in handy in cases where we need to condition and apply the law of total probability. Expressions for $f_1(x)$ and $f_n(x)$ are particularly simple and we have

Corollary 2.5

$$f_1(x) = n[1 - P(x)]^{n-1}p(x) , \quad f_n(x) = nP^{n-1}(x)p(x). \qquad (2.7)$$

More generally it is also straightforward to obtain

Theorem 2.10 *For any subset of the n order statistics: X_{n_1}, X_{n_2}, ..., X_{n_k}, $1 \le n_1 \le \ldots \le n_k$, the joint distribution $f(x_1, \ldots, x_2)$ is expressed as*

$$f(x_1, \ldots, x_k) = n! \left[\prod_{j=1}^k p(x_j) \right] \prod_{j=0}^k \left\{ \frac{[P(x_{j+1}) - P(x_j)]^{n_{j+1}-n_j-1}}{(n_{j+1} - n_j - 1)!} \right\} \qquad (2.8)$$

in which $p(x) = P'(x)$. This rather involved expression leads to many useful results including the following corollaries:

Corollary 2.6 *The joint distribution of $X_{(r)}$ and $X_{(s)}$ is*

$$F_{rs}(x, y) = \sum_{j=s}^{n} \sum_{i=r}^{j} \frac{n!}{i!(j-i)!(n-j)!} P^i(x)$$

$$[P(y) - P(x)]^{j-i}[1 - P(y)]^{n-j}.$$

The joint distribution of $X_{(r)}$ and $X_{(s)}$ is useful in establishing a number of practical results such as the distribution of the range, the distribution of the interquartile range and an estimate for the median among others. Using the result (Section 2.5) for the distribution of a difference, a simple integration then leads to the following:

Corollary 2.7 *Letting $W_{rs} = X_{(s)} - X_{(r)}$ then: in the special case of a parent uniform distribution we have*

$$f(w_{rs}) = \frac{1}{B(s-r, n-s+r+1)} w_{rs}^{s-r-1}(1 - w_{rs})^{n-s+r}. \qquad (2.9)$$

Taking $s = n$ and $r = 1$, recalling that $B(\alpha, \beta) = \Gamma(\alpha)\Gamma(\beta)/\Gamma(\alpha + \beta)$ and that $\Gamma(n) = n!$, then we have the distribution of the range for the uniform.

Corollary 2.8 *Letting $w = U_{(n)} - U_{(1)}$ be the range for a random sample of size n from the standard uniform distribution, then the cumulative distribution is given by*

$$F_U(w) = nw^{n-1} - (n-1)w^n. \qquad (2.10)$$

Straightforward differentiation gives $f_U(w) = n(n-1)w^{n-2}(1-w)$, a simple and useful result. For an arbitrary distribution, $F(\cdot)$ we can either carry out the same kind of calculations from scratch or, making use once more of the probability integral transform (see Section 2.4), use the above result for the uniform and transform into arbitrary F. Even this is not that straightforward since, for some fixed interval (w_1, w_2), corresponding to $w = w_2 - w_1$ from the uniform, the corresponding $F^{-1}(w_2) - F^{-1}(w_1)$ depends not only on $w_2 - w_1$ but on w_1 itself. Again we can appeal to the law of total probability, integrating over all values of w_1 from 0 to $1 - w$. In practice, it may be good enough to divide the interval $(0, 1-w)$ into a number of equally spaced points, ten would suffice, and simply take the average. Interval estimates for any given quantile, defined by $P(\xi_\alpha) = \alpha$, follow from the basic result and we have:

Corollary 2.9 *In the continuous case, for $r < s$, the pair $(X_{(r)}, X_{(s)})$ covers ξ_α with probability given by $I_\pi(r, n - r + 1) - I_\pi(r, n - s + 1)$.*

Theorem 2.11 *For the special case in which $n_1 = 1$, $n_2 = 2$, ... $n_n = n$, then*

$$f(x_1, \ldots, x_n) = n! \prod_{j=1}^{n} p(x_j). \tag{2.11}$$

A characterization of order statistics: Markov property

The particularly simple results for the exponential distribution lead to a very useful and powerful characterization of order statistics. If $Z_1, \ldots, Z_n$ are i.i.d. exponential variates with parameter λ, then an application of Corollary 2.4 shows that the minimum of Z_1 to Z_n has itself an exponential distribution with parameter $n\lambda$. We can define the random variable Y_1 to be the gap time between 0 and the first observation, $Z_{(1)}$. The distribution of Y_1 (equivalently $Z_{(1)}$) is exponential with parameter $n\lambda$. Next, we can define Y_2 to be the gap $Z_{(2)} - Z_{(1)}$. In view of the lack of memory property of the exponential distribution, once $Z_{(1)}$ is observed, the conditional distribution of each of the remaining $(n-1)$ variables, given that they are all greater than the observed time $Z_{(1)}$, remains exponential with parameter λ. The variable Y_2 is then the minimum of $(n-1)$ i.i.d. exponential variates with parameter λ. The distribution of Y_2 is therefore, once again, exponential, this time with parameter $(n-1)\lambda$. More generally we have the following lemma:

Lemma 2.4 *If $Z_{(1)}, \ldots, Z_{(n)}$ are the order statistics from a sample of size n of standard exponential variates, then, defining $Z_{(0)} = 0$,*

$$Y_i = Z_{(i)} - Z_{(i-1)}, \quad i = 1, \ldots, n$$

are n independent exponential variates in which $E(Y_i) = 1/(n-i+1)$.

This elementary result is very important in that it relates the order statistics directly to sums of simple independent random variables which are not themselves order statistics. Specifically we can write

$$Z_{(r)} = \sum_{i=1}^{r} \{Z_{(i)} - Z_{(i-1)}\} = \sum_{i=1}^{r} Y_i ,$$

leading to the immediate further lemma:

Lemma 2.5 *For a sample of size n from the standard exponential distribution and letting $\alpha_i = 1/(n - i + 1)$, we have:*

$$E[Z_{(r)}] = \sum_{i=1}^{r} E(Y_i) = \sum_{i=1}^{r} \alpha_i, \quad \text{Var}\,[Z_{(r)}] = \sum_{i=1}^{r} \text{Var}\,(Y_i) = \sum_{i=1}^{r} \alpha_i^2.$$

The general flavor of the above result applies more generally than just to the exponential and, applying the probability integral transform (Section 2.5), we have:

Lemma 2.6 *For an i.i.d. sample of size n from an arbitrary distribution, $G(x)$, the rth largest order statistic, $X_{(r)}$ can be written*

$$X_{(r)} = G^{-1}\{1 - \exp(-Y_1 - Y_2 - \cdots - Y_r)\},$$

where the Y_i are independent exponential variates in which $E(Y_i) = 1/(n - i + 1)$.

One immediate conclusion that we can make from the above expression is that the order statistics from an arbitrary distribution form a Markov chain. The conditional distribution of $X_{(r+1)}$ given $X_{(1)}$, $X_{(2)}, \ldots, X_{(r)}$ depends only on the observed value of $X_{(r)}$ and the distribution of Y_{r+1}. This conditional distribution is clearly the same as that for $X_{(r+1)}$ given $X_{(r)}$ alone, hence the Markov property. If needed we can obtain the joint density, f_{rs}, of $X_{(r)}$ and $X_{(s)}$, $(1 \leq r < s \leq n)$ by a simple application of Theorem 2.10. We then write:

$$f_{rs}(x, y) = \frac{n!\, P^{r-1}(x)p(x)p(y)[P(y) - P(x)]^{s-r-1}[1 - P(y)]^{n-s}}{(r - 1)!(s - r - 1)!(n - s)!}.$$

From this we can immediately deduce the conditional distribution of $X_{(s)}$ given that $X_{(r)} = x$ as:

$$f_{s|r}(y|x) = \frac{(n - r)!}{(s - r - 1)!(n - s)!} \frac{p(y)[P(y) - P(x)]^{s-r-1}[1 - P(y)]^{n-s}}{[1 - P(x)]^{n-r}}.$$

A simple visual inspection of this formula confirms again the Markov property. Given that $X_{(r)} = x$ we can view the distribution of the remaining $(n - r)$ order statistics as an ordered sample of size $(n - r)$ from the conditional distribution $P(u|u > x)$.

Expected values of order statistics

Given the distribution of any given order statistic we can, at least in principle, calculate any moments, in particular the mean, by applying the basic definition. In practice, this may be involved and there may be no explicit analytic solution. Integrals can be evaluated numerically but, in the majority of applications, it can be good enough to work with accurate approximations. The results of the above subsection, together with some elementary approximation techniques are all that we need. Denoting the distribution of X as $P(x)$, then the probability integral transform (Section 2.5) provides that $U = P(X)$ has a uniform distribution. The moments of the order statistics from a uniform distribution are particularly simple so that $E\{U_{(r)}\} = p_r = r/(n+1))$. Denoting the inverse transformation by $Q = P^{-1}$, then

$$X_{(r)} = P^{-1}\{U_{(r)}\} = Q\{U_{(r)}\}.$$

Next, we can use a Taylor series development of the function $X_{(r)}$ about the p_r so that

$$X_{(r)} = Q(p_r) + \{U_{(r)} - p_r\}Q'(p_r) + \{U_{(r)} - p_r\}^2 Q''(p_r)/2 + \cdots$$

and, taking expectations, term by term, we have

$$E\{X_{(r)}\} \approx Q(p_r) + \frac{p_r q_r}{2(n+2)} Q''(p_r)$$
$$+ \frac{p_r q_r}{(n+2)^2} \left\{ \frac{1}{3}(q_r - p_r)Q'''(p_r) + \frac{1}{8}p_r q_r Q''''(p_r) \right\}$$

and

$$\mathrm{Var}\,\{X_{(r)}\} = \frac{p_r q_r}{2(n+2)}[Q'(p_r)]^2 + \frac{p_r q_r}{(n+2)^2}\left\{2(q_r - p_r)Q'(p_r)Q''(p_r)\right.$$
$$\left. + p_r q_r \left(Q'(p_r)Q'''(p_r) + [Q''(p_r)]^2\right)\right\}.$$

It is straightforward to establish some relationships between the moments of the order statistics and the moments from the parent distribution. Firstly note that

$$E\left\{\sum_{r=1}^{n} X_{(r)}^{k}\right\}^{m} = E\left\{\sum_{r=1}^{n} X_{r}^{k}\right\}^{m},$$

so that, if μ and σ^2 are the mean and variance in the parent population, then $\sum_{r=1}^{n} \mu_r = n\mu$ and $\sum_{r=1}^{n} E\{X_{(r)}^2\} = nE(X^2) = n(\mu^2 + \sigma^2)$.

Normal parent distribution

For the case of a normal parent the expected values can be evaluated precisely for small samples and the approximations themselves are relatively tractable for larger sample sizes. One approach to data analysis in which it may be desirable to have a marginal normal distribution in at least one of the variables under study is to replace the observations by the expectations of the order statistics. These are sometimes called normal scores, typically denoted by $\xi_{rn} = E(X_{(r)})$ for a random sample of size n from a standard normal parent with distribution function $\Phi(x)$ and density $\phi(x)$. For a random sample of size n from a normal distribution with mean μ and variance σ^2 we can reduce everything to the standard case since $E(X_{(r)}) = \mu + \xi_{rn}\sigma$. Note that, if n is odd, then, by symmetry, it is immediately clear that $E(X_{(r)}) = 0$ for all r that are odd. We can see that $E(X_{(r)}) = -E(X_{(n-r+1)})$. For n as small as, say, 5 we can use integration by parts to evaluate ξ_{r5} for different values of r. For example, $\xi_{55} = 5 \int 4\Phi^3(x)\phi^2(x)dx$ which then simplifies to: $\xi_{55} = 5\pi^{-1/2}/4 + 15\pi^{-3/2}\sin^{-1}(1/3)/2 = 1.16296$. Also, $\xi_{45} = 5\pi^{-1/2}/2 - 15\pi^{-3/2}\sin^{-1}(1/3) = 0.49502$ and $\xi_{35} = 0$. Finally, $\xi_{15} = -1.16296$ and $\xi_{25} = -0.49502$. For larger sample sizes in which the integration becomes too fastidious we can appeal to the above approximations using the fact that

$$Q'(p_r) = \frac{1}{\phi(Q)}, \quad Q''(p_r) = \frac{Q}{\phi^2(Q)}, \quad Q'''(p_r) = \frac{1+2Q^2}{\phi^3(Q)},$$

$$Q''''(p_r) = \frac{Q(7+6Q^2)}{\phi^4(Q)}.$$

The above results arise from straightforward differentiation. Analogous calculations can be used to obtain exact or approximate expressions for $\text{Cov}\{X_{(r)}, X_{(s)}\}$.

2.8   Entropy and variance

In view of the mathematical equivalence of the density, distribution function and the hazard, we can be satisfied knowing any one of these functions for a variable T of interest. In the majority of areas of application of statistics, theoretical physics, and, possibly, biophysics being potential exceptions, we cannot really know much about these functions. Our usual strategy will be to collect data that enables the estimation of one or more of the functions, with any additional plausible assumptions about the nature of these functions making this task that much easier. Paucity of data, or a need to only know the most important features of a distribution, will often lead us to restricting our attention to some simple summary measures. The most common summary measures are those of location and variance. For a measure of location we usually take the mean μ or the median $\xi_{0.5}$. They tell us something about where the most likely values of T occur. An idea of just how "likely" these "likely" values are, in other words how concentrated is the distribution around the location measure, is most often provided by the variance or the square root of this, the standard deviation. The variance σ^2 is defined by

$$\sigma^2 = E\{T - E(T)\}^2 = \int (t - \mu)^2 f(t)dt = \int (t - \mu)^2 dF(t). \quad (2.12)$$

An important insight into just why σ^2 provides a good measure of precision, in other terms predictability, is given by:

Theorem 2.12 *For every positive constant a*

$$\Pr\{|T - \mu| \geq a\sigma\} \leq 1/a^2. \quad (2.13)$$

This famous inequality, known as the Bienaymé-Chebyshev inequality, underlines the fact that the smaller σ^2 the better we can predict. A lesser used, although equally useful, measure of concentration is the so-called entropy of the distribution. Apart from a negative sign, this is also called the information of the distribution which is defined by $V(f, f)$ where

$$V(g, h) = E \log g(T) = \int \log g(t)h(t)dt. \quad (2.14)$$

The entropy is just $-V(g, h)$. Note that the integral operator E in $E \log g(T)$ is with respect to the density $h(t)$, this added generality being needed in the regression context. For univariate study the information is simply $V(f, f)$ and would be written V since the arguments are implicit. Our intuition is good for σ^2, since it is clear that the further away, on average, are the values of T, then the larger will be σ^2. The same is true, although less obvious, for V. As T becomes concentrated around its mode (value of t, taken to be unique, at which $f(t)$ assumes its greatest value), then, since $\int f(t)dt$, the area under the curve, is fixed at one, $f(t)$ itself becomes larger at and around the mode. In the limit, as all the information becomes concentrated at a single point t_0, then $f(t_0)$, as well as $E \log f(T)$, tends to positive infinity. The more spread out are the values of T then the closer to zero will tend to be $E \log f(T)$. Intermediary values of $E \log f(T)$ then can be taken to correspond to different degrees of dispersion. Consider also the following which is true for any number of random variables and which, for the purposes of illustration, we limit to X_1, X_2 and X_3. We have

$$E \log f(X_1, X_2, X_3) = E \log f(X_3 | X_2, X_1) + E \log f(X_2 | X_1)$$
$$+ E \log f(X_1),$$

so that the total information can be decomposed into sequential orthogonal contributions, each adding to the total amount of information so far. Note also, since we can interchange the X_i, the order in which the total information is put together has no impact on the final result. This is of course a desirable property. The information measure, as an indicator of precision, is well known in communication theory (Shannon and Weaver 1949) and statistical ecology, but is not so well known in biostatistics. It is also worth considering the fact that the most commonly used estimating technique, maximum likelihood, is best viewed as an empirical version of information. This follows since the usual log-likelihood divided by the sample size (which can be taken as a fixed constant) provides a consistent estimate of the information. Both the variance and the information are of particular interest when we condition on some other variable Z, possibly a vector. This is the regression setting where we focus on the impact of explanatory variables on some response variable of interest. The information gain would consider the distance between the distribution $f(t)$ and $f(t|z)$. In the above construction the function $g(t)$ is first equated with $f(t)$ and subsequently to $f(t|z)$, whereas $h(t)$ remains fixed at $f(t,z)$. Note also, that in this case, the integral is over the space of T and Z. This enables the construction of a simple and powerful measure of predictability. The amount by which the variance, or information, changes following such conditioning provides a direct quantification of the predictive strength of Z. We look at this more closely in the following subsection.

Explained randomness and explained variation

Any models we work with are simply tools to enable us to efficiently construct conditional distributions. Validity of our models is an important issue, upon which we dwell later, but, for now, let us suppose our models are good enough to accurately reproduce the conditional

distributions of T given Z where Z may be a vector The improvement
in our predictive ability, given Z, can be quantified in view of the above
Bienaymé-Chebyshev inequality and the variance decomposition

$$\text{Var}(T) = \text{Var}\,E(T|Z) + E\,\text{Var}(T|Z). \qquad (2.15)$$

The total variance, $\text{Var}(T)$, breaks down into two parts, one of which
we can interpret as the signal, $\text{Var}\,E(T|Z)$, and one as the pure noise
$E\,\text{Var}(T|Z)$. The percentage of $\text{Var}(T)$ that is taken up by $\text{Var}\,E(T|Z)$
is the amount of the total variance that can be explained by Z. This
translates directly the predictive power of Z so that the percentage of
explained variance, is then quite central to efforts at quantifying how
well our models do. We define it as

$$\Omega^2 = \frac{\text{Var}\,E(T|Z)}{\text{Var}(T)} = \frac{\text{Var}(T) - E\,\text{Var}(T|Z)}{\text{Var}(T)}. \qquad (2.16)$$

The quantity Ω^2 in its own right is not well developed in the literature and we devote Section 3.9 to studying its importance. Following
Draper (1984), there have been a number of challenges to Ω^2 as a useful concept (Healy 1984: Kvalseth 1985: Scott and Wild 1991: Willett
and Singer 1988). However Draper's paper of 1984 was flawed and its
conclusions did not hold up (Draper 1985). As a result, this subsequent
work, having taken Draper's 1984 paper as its starting point, inherits
the same logical errors.

Explained randomness, as opposed to explained variation, arises
from a less transparent construction. We can use a monotonic transform of the expected information (expectation taken with respect to
the distribution of Z) and, taking $D(T) = \exp -2E\,V\{f(t), f(t|Z)\}$:
$D(T|Z) = \exp -2E\,V\{f(t|Z), f(t|Z)\}$, we define the explained randomness ρ^2 to be

$$\rho^2 = \frac{D(T) - D(T|Z)}{D(T)}. \qquad (2.17)$$

We interpret ρ^2 as the proportion of explained randomness in T attributable to Z. We also have the following important lemma that
could, in its own right, be taken as a reason for studying explained
randomness, but which, in any event, underlines a useful relationship
between explained variation and explained randomness:

Lemma 2.7 *If the pair (T, Z) are bivariate normal then $\Omega^2 = \rho^2$.*

The lemma provides further motivation for being interested in ρ^2, in
that, for the more familiar classic regression case of a bivariate normal,

we obtain the same results by considering explained randomness that
we obtain by considering explained variation. For other distributions,
where variance itself may not be the best measure of dispersion, the
concept explained randomness, based on entropy, might be viewed as
having more generality. Our own experience in practical data analysis
suggests that, as far as hierarchical model building or the quantification
of partial or multiple effects is concerned, there does not appear to be
anything to really choose between the two measures. For operational
purposes we can take the two measures to be essentially equivalent,
the use of one rather than the other being more a question of taste
rather than one based on any real advantages or disadvantages.

2.9   Approximations

Approximations to means and variances for functions of T (δ-method)

Consider some differentiable monotonic function of X, say $\psi(X)$. Our
particular concern often relates to parameter estimates in which case
the random variable X would be some function of the n i.i.d. data
values, say θ_n as an estimator of the parameter θ. In the cases of
interest, θ_n converges with probability one to θ and so also does $\psi(\theta_n)$
to $\psi(\theta)$. Although θ_n may not be unbiased for θ, for large samples, the
sequence $E(\theta_n)$ converges to $E(\theta) = \theta$. Similarly $E[\psi(\theta_n)]$ converges
to $\psi(\theta)$. The mean value theorem (Section 2.2) enables us to write

$$\phi(\theta_n) = \psi(\theta) + (\theta_n - \theta)\phi'(\theta) + \frac{(\theta_n - \theta)^2}{2}\psi''(\xi) \qquad (2.18)$$

for $\xi \in (\theta \pm \theta_n)$ Rearranging this expression, ignoring the third term
on the right hand side, and taking expectations we obtain

$$\text{Var}\{\psi(\theta_n)\} \approx E\{\psi(\theta_n) - \psi(\theta)\}^2 \approx \{\psi'(\theta)\}^2\text{Var}(\theta_n) \approx \{\psi'(\theta_n)\}^2\text{Var}(\theta_n)$$

as an approximation to the variance. The approximation, once obtained in any given setting, is best studied on a case-by-case basis. It
is an exact result for linear functions. For these, the second derivative
is equal to zero and, more generally, the smaller the absolute value of
this second derivative, the better we might anticipate the approximation to be. For θ_n close to θ the squared term will be small in absolute
value when compared with the linear term, an additional motivation

to neglecting the third term. For the mean, the second term of Equation (2.18) is zero when θ_n is unbiased, otherwise close to zero and, this time, ignoring this second term, we obtain

$$E\{\psi(\theta_n)\} \approx \psi(\theta_n) + \frac{1}{2}\mathrm{Var}\,(\theta_n)\psi''(\theta_n) \qquad (2.19)$$

as an improvement over the rougher approximation based on the first term alone of the above expression. Extensions of these expressions to the case of a consistent estimator $\psi(\theta_n) = \psi(\theta_{1n}, \ldots, \theta_{pn})$ of $\psi(\theta)$ proceeds in the very same way, only this time based on a multivariate version of Taylor's theorem. These are:

$$\mathrm{Var}\,\{\psi(\theta_n)\} \approx \sum_{j=1}^{p}\sum_{m\geq j}^{p} \frac{\partial\psi(\theta)}{\partial\theta_j}\frac{\partial\psi(\theta)}{\partial\theta_m}\,\mathrm{Cov}\,(\theta_{jn}, \theta_{mn}),$$

$$E\{\psi(\theta_n)\} \approx \psi(\theta_{1n}, \ldots, \theta_{pn}) + \frac{1}{2}\sum_j\sum_m \frac{\partial^2\psi(\theta_n)}{\partial\theta_j\partial\theta_m}\,\mathrm{Cov}\,(\theta_{jn}, \theta_{mn}).$$

When $p = 1$ then the previous expressions are recovered as special cases. Again, the result is an exact one in the case where $\psi(\cdot)$ is a linear combination of the components θ_j and this helps guide us in situations where the purpose is that of confidence interval construction. If, for example, our interest is on ψ and some strictly monotonic transformation of this, say ψ^*, is either linear or close to linear in the θ_j, then it may well pay, in terms of accuracy of interval coverage, to use the delta-method on ψ^*, obtaining the end points of the confidence interval for ψ^* and subsequently inverting these, knowing the relationship between ψ and ψ^*, in order to obtain the interval of interest for ψ. Since ψ and ψ^* are related by one-to-one transformations then the coverage properties of an interval for ψ^* will be identical to those of its image for ψ. Examples in this book include confidence intervals for the conditional survivorship function, given covariate information, based on a proportional hazards model as well as confidence intervals for indices of predictability and multiple coefficients of explained variation.

Cornish-Fisher approximations

In the construction of confidence intervals, the δ-method makes a normality approximation to the unknown distribution and then replaces the first two moments by local linearization. A different approach, while still working with a normal density $\phi(x) = (2\pi)^{-1/2}\exp(-x^2/2)$, in a way somewhat analogous to the construction of a Taylor series, is to express the density of interest, $f(x)$, in terms of a linear combination of $\phi(x)$ and derivatives of $\phi(x)$. Normal distributions with nonzero means and variances not equal to one are obtained by the usual simple linear transformation and, in practical work, the simplest approach is to standardize the random variable X so that the mean and variance corresponding to the density $f(x)$ are zero and one, respectively.

The derivatives of $\phi(x)$ are well known, arising in many fields of mathematical physics and numerical approximations. Since $\phi(x)$ is simply a constant multiplying an exponential term it follows immediately that all derivatives of $\phi(x)$ are of the form of a polynomial that multiplies $\phi(x)$ itself. These polynomials (apart from an alternating sign coefficient $(-1)^i$) are the Hermite polynomials, $H_i(x)$, $i = 0, 1, \ldots$, and we have

$$H_0 = 1, \quad H_1 = x, \quad H_2 = x^2 - 1, \quad H_3 = x^3 - 3x, \quad H_4 = x^4 - 6x^2 + 3,$$

with H_5 and higher terms being calculated by simple differentiation. The polynomials are of importance in their own right, belonging to the class of orthogonal polynomials and useful in numerical integration. Indeed, we have that

$$\int_{-\infty}^{\infty} H_i^2(x)\phi(x)dx = i!, \; i = 0, \ldots : \int_{-\infty}^{\infty} H_i(x)H_j(x)\phi(x)dx = 0, \; i \neq j.$$

This orthogonality property is exploited in order for us to obtain explicit expressions for the coefficients in our expansion. Returning to our original problem we wish to determine the coefficients c_i in the expansion

$$f(x) = \sum_{i=0}^{\infty} c_i H_i(x)\phi(x) \tag{2.20}$$

and, in order to achieve this we multiply both sides of equation (2.20) by $H_j(x)$, subsequently integrating to obtain the coefficients

$$c_j = \frac{1}{j!} \int_{-\infty}^{\infty} f(x)H_j(x)dx. \tag{2.21}$$

Note that the polynomial $H_j(x)$ is of order j so that the right-hand side of equation (2.21) is a linear combination of the moments, (up to the jth), of the random variable X having associated density $f(x)$.

These can be calculated step-by-step For many standard densities several of the lower-order moments have been worked out and are available. Thus, it is relatively straightforward to approximate some given density $f(x)$ in terms of a linear combination of $\phi(x)$.

The expansion of Equation (2.20) can be used in theoretical investigations as a means to study the impact of ignoring higher-order terms when we make a normal approximation to the density of X. We will use the expansion in an attempt to obtain more accurate inference for proportional hazards models fitted using small samples. Here the large sample normal assumption may not be sufficiently accurate and the approximating equation is used to motivate potential improvements obtained by taking into account moments of higher order than just the first and second. When dealing with actual data, the performance of any such adjustments needs to be evaluated on a case-by-case basis. This is because theoretical moments will have to be replaced by observed moments and the statistical error involved in that can be of the same order, or greater, than the error involved in the initial normal approximation. If we know or are able to calculate the moments of the distribution, then the c_i are immediately obtained. When the mean is zero we can write down the first four terms as

$$c_0 = 1\,,\ c_1 = 0\,,\ c_2 = (\mu_2 - 1)/2\,,\ c_3 = \mu_3/6\,,\ c_4 = (\mu_4 - 6\mu_2 + 3)/24\,,$$

from which we can write down an expansion in terms of $\phi(x)$ as

$$f(x) = \phi(x)\{1 + (\mu_2 - 1)H_2(x)/2 + \mu_3 H_3(x)/6$$
$$+ (\mu_4 - 6\mu_2 + 3)H_4(x)/24 + \cdots \}.$$

This series is known as the Gram-Charlier series, and stopping the development at the fourth term corresponds to making corrections for skewness and kurtosis. In our later development of the properties of estimators in the proportional hazards model we will see that making corrections for skewness can help make inference more accurate, whereas, at least in that particular application, corrections for kurtosis appear to have little impact (Chapter 11).

Saddlepoint approximations

A different, although quite closely related, approach to the above uses saddlepoint approximations. Theoretical and practical work on these approximations indicate them to be surprisingly accurate for the tails of a distribution. We work with the inversion formula for the cumulant generating function, a function that is defined in the complex plane, and in this two-dimensional plane, around the point of interest (which is typically a mean or a parameter estimate) the function looks like a minimum in one direction and a maximum in an orthogonal direction: hence the name "saddlepoint." Referring back to Section 2.6 recall that we identified κ_r as the coefficient of $(it)^r/r!$ in the expansion of the cumulant generating function $K(t) = \log \phi(t)$ where $\phi(t)$ is the characteristic function. We can exploit the relationship between $\phi(t)$ and $f(x)$; that is,

$$f(x) = \frac{1}{2\pi} \int_{-\infty}^{\infty} e^{-itx} \phi(t) dt \,, \quad \phi(t) = \int_{-\infty}^{\infty} e^{itx} f(x) dx \,.$$

to approximate $f(x)$ by approximating the integral. The numerical technique that enables this approximation to be carried out is called the method of steepest descent and is described in Daniels (1954). The approximation to $f(x)$ is simply denoted as $f_s(x)$ and, carrying through the calculations, we find that

$$f_s(x) = \left\{ \frac{n}{2\pi K''(\lambda_x)} \right\}^{1/2} \exp[n\{K(\lambda_x) - x\lambda_x\}] \tag{2.22}$$

in which the solution to the differential equation in λ, $K'(\lambda) = x$ is given by λ_x. Our notation here of x as a realization of some random variable X is not specifically referring to our usual use of X as the minimum of survival time T and the censoring time C. It is simply the variable of interest and that variable, in our context, will be the score statistic (Chapter 11). For now, we assume the score to be composed of n contributions so that we view x as a mean based on n observations. Since, mostly, we are interested in the tails of the distribution, it can often help to approximate the cumulative distribution directly rather than make a subsequent appeal to numerical integration. Denoting the saddlepoint approximation to the cumulative distribution by $F_s(x)$, we write

$$F_s(x) = \Phi(u_x) + \phi(u_x)(u_x^{-1} + v_x^{-1}) \tag{2.23}$$

where $\phi(x)$ indicates the standard normal density, $\Phi(x) = \int_{-\infty}^{x} \phi(u) du$, the cumulative normal, $u_x = [2n\{x\lambda_x - K(\lambda_x)\}]^{1/2}\text{sgn}(\lambda_x)$, and $v_x = \lambda_x\{nK''(\lambda_x)\}^{1/2}$. Since we are only concerned with tail probabilities we need not pay attention to what occurs around the mean. If we do wish to consider $F_s(x)$, evaluated at the mean, the approximation is slightly modified and the reader is referred to Daniels (1987).

2.10 Stochastic processes

We define a stochastic process to be a collection of random variables indexed by $t \in T$. We write these as $X(t)$ and take t to be fixed. If the set T has only a finite or a countably infinite number of elements then $X(t)$ is referred to as a discrete-time process. We will be most interested in continuous-time processes. In applications we can standardize by the greatest value of t in the set T that can be observed, and so we usually take $\sup\{t : t \in T\} = 1$. We also take $\inf\{t : t \in T\} = 0$. We will be especially interested in observations on any given process between 0 and t. We call this the sample path.

Independent increments and stationarity

Consider some partition of (0,1) in which $0 = t_0 < t_1 < t_2 < \cdots < t_n = 1$. If the set of random variables $X(t_i) - X(t_{i-1})\, i = 1, \ldots, n$ are independent then the stochastic process $X(t)$ is said to have independent increments. Another important property is that of stationarity. We say that a stochastic process $X(t)$ has stationary increments if $X(s+t) - X(s)$ has the same distribution for all values of s. Stationarity indicates, in as much as probabilistic properties are concerned, that when we look forward, from the point s, a distance t, the only relevant quantity is how far forward t we look. Our starting point itself is irrelevant. As we progress through time, everything that we have learned is summarized by the current position. It can also be of value to consider a process with a slighter weaker property, the so-called second-order stationarity. Rather than insist on a requirement for the whole distribution we limit our attention to the first two moments and the covariance between $X(s+t)$ and $X(s)$ which depends only upon $|t|$. Our main focus is on Gaussian processes which, when they have the property of second-order stationarity, will in consequence be stationary processes. Also, simple transformations can produce stationary processes from nonstationary ones, an example being the transformation of the Brownian bridge into an Ornstein-Uhlenbeck process.

Gaussian processes

If for every partition of (0,1), $0 = t_0 < t_1 < t_2 < \cdots < t_n = 1$, the set of random variables $X(t_1), \ldots, X(t_n)$ has a multivariate normal distribution, then the process $X(t)$ is called a Gaussian process. Brownian motion, described below, can be thought of as simply a standardized Gaussian process. A Gaussian process being uniquely determined by the multivariate means and covariances it follows that such a process will have the property of stationarity if for any pair $(s, t : t > s)$, $\mathrm{Cov}\{X(s), X(t)\}$ depends only on $(t - s)$. In practical studies we will often deal with sums indexed by t and the usual central limit theorem will often underlie the construction of Gaussian processes.

2.11 Brownian motion

Consider a stochastic process $X(t)$ on $(0, 1)$ with the following three properties:

1. $X(0) = 0$, i.e., at time $t = 0$ the starting value of X is fixed at 0.

2. $X(t), t \in (0, 1)$ has independent stationary increments.

3. At each $t \in (0, 1)$ the distribution of $X(t)$ is $\mathcal{N}(0, t)$.

This simple set of conditions completely describes a uniquely determined stochastic process called Brownian motion. It is also called the Wiener process or Wiener measure. It has many important properties and is of fundamental interest as a limiting process for a large class of sums of random variables on the interval (0,1). An important property is described in Theorem 2.13 below. Firstly we make an attempt to describe just what a single realization of such a process might look like. Later we will recognize the same process as being the limit of a sum of independent random contributions. The process is continuous and so, approximating it by any drawing, there cannot be any gaps. At the same time, in a sense that can be made more mathematically precise, the process is infinitely jumpy. Nowhere does a derivative exist. Figure 2.1 illustrates this via a simulated approximation. The right-hand figure is obtained from the left-hand one by homing in on the small interval (0.20, 0.21), subtracting off the value observed at $t = 0.20$, and rescaling to the interval (0,1). The point we are trying to make is that the resulting process itself looks like (and indeed is) a realization of Brownian motion. Theoretically, this could be repeated

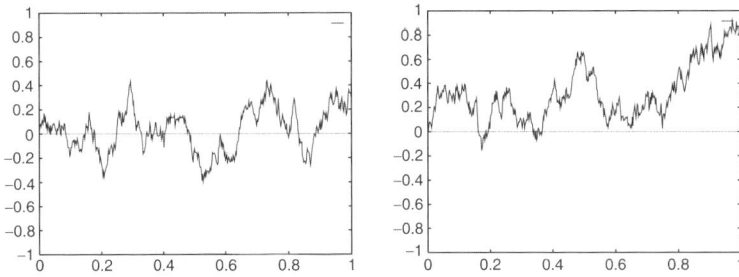

Figure 2.1: Two simulated independent realizations of a Brownian motion process.

without limit which allows us to understand in some way how infinitely jumpy is the process. In practical examples we can only ever approximate the process by linearly connecting up adjacent simulated points.

Theorem 2.13 *Conditioning on a given path we have*

$$\Pr\{X(t+s) > x | X(s) = x_s, X(u), \ 0 \le u < s\}$$
$$= \Pr\{X(t+s) > x | X(s) = x_s.\}$$

So, when looking ahead from time point s to time point $t+s$, the previous history indicating how we arrived at s is not relevant. The only thing that matters is the point at which we find ourselves at time point s. This is referred to as the Markov property. The joint density of $X(t_1), \ldots, X(t_n)$ can be written as

$$f(x_1, x_2, \ldots, x_n) = f_{t_1}(x_1) f_{t_1-t_2}(x_2 - x_1) \cdots f_{t_n-t_{n-1}}(x_n - x_{n-1})$$

This follows from the independent stationary increment condition. A consequence of the above result is that we can readily evaluate the conditional distribution of $X(s)$ given some future value $X(t)$ ($t > s$). Applying the definition for conditional probability we have the following.

Corollary 2.10 *The conditional distribution of $X(s)$ given $X(t)$ ($t > s$) is normal with a mean and a variance given by,*

$$E\{X(s)|X(t) = w\} = ws/t, \quad \mathrm{Var}\{X(s)|X(t) = w\} = s(t-s)/t.$$

This result helps provide insight into another useful process, the Brownian bridge described below. Other important processes arise as simple transformations of Brownian motion. The most obvious to consider is where we have a Gaussian process satisfying conditions (1) and (2) for Brownian motion but where, instead of the variance increasing linearly, i.e., $\mathrm{Var}\, X(t) = t$, the variance increases either too quickly or too slowly so that $\mathrm{Var}\, X(t) = \phi(t)$ where $\phi(\cdot)$ is some monotonic increasing function of t. Then we can transform the time axis using $\phi(\cdot)$ to produce a process satisfying all three conditions for Brownian motion. Consider also the transformation

$$V(t) = \exp(-\alpha t/2)X\{\exp(\alpha t)\}$$

where $X(t)$ is Brownian motion. This is the Ornstein-Uhlenbeck process. It is readily seen that:

Corollary 2.11 *The process $V(t)$ is a Gaussian process in which* $E\{V(t)\} = 0$ *and* $\mathrm{Cov}\,\{V(t), V(s)\} = \exp\{-\alpha(t - s)/2\}$.

Time-transformed Brownian motion

Consider a process, $X^{\psi}(t)$, defined via the following three conditions, for some continuous ψ such that, $\psi(t') > \psi(t)$ $(t' > t)$; (1) $X^{\psi}(0) = 0$ (2) $X^{\psi}(t), t \in (0, 1)$ has independent stationary increments; (3) at each $t \in (0, 1)$ the distribution of $X^{\psi}(t)$ is $\mathcal{N}\{0, \psi(t)\}$. The usual Brownian motion described above is exactly this process when $\psi(t) = t$. However, in view of the continuity and monotonicity of ψ, there exists an inverse function ψ^{-1} such that $\psi^{-1}\{\psi(t)\} = t$. Clearly, we can transform the process $X^{\psi}(t)$ by multiplying, at each t, by $\sqrt{t/\psi(t)}$, and, defining $\sqrt{0/\psi(0)} = 0$. The resulting process we can call $X(t)$ and it is readily seen that this process is standard Brownian motion. Thus, the only crucial assumption in Brownian motion is that of independent increments. Once we can assert this to be the case, it is only a question of scale and location to obtain standard Brownian motion.

Brownian bridge

Let $W(t)$ be Brownian motion. We know that $W(0) = 0$. We also know that with probability one the process $W(t)$ will return at some point to the origin. Let's choose a point, and in particular the point $t = 1$ and consider the conditional process $W^{0}(t)$, defined to be Brownian motion

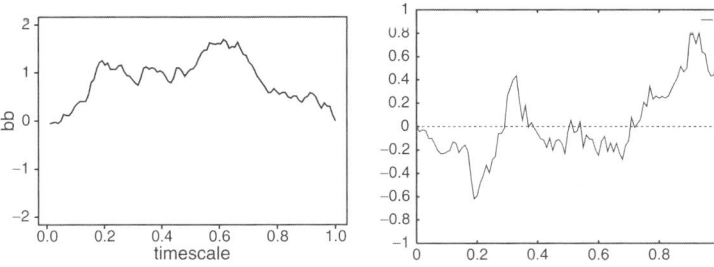

Figure 2.2: Two transformations of simulated Brownian motion by conditioning on $W(1)$. The first has $W(1) = 0$ (Brownian bridge); the second has $W(1) = 0.5$.

conditioned by the fact that $W(1) = 0$. For small t this process will look very much like the Brownian motion from which it is derived. As t goes to one the process is pulled back to the origin since at $t = 1$ we have that $W^0(1) = 0$ and $W(t)$ is continuous. Also $W^0(0) = W(0) = 0$. Such a process is called tied down Brownian motion or the Brownian bridge. Figure 2.2 illustrates a realization of a Brownian bridge and a realizatin of a Brownian motion constrained to assume a value other than zero at $t = 1$. We will see below that realizations of a Brownian bridge can be viewed as linearly transformed realizations of Brownian motion itself, and vice versa. From the results of above the section we can investigate the properties of $W^0(t)$. The process is a Gaussian process so we only need consider the mean and covariance function for the process to be completely determined. We have

$$E\{W(s)|W(1) = 0\} = 0 \quad \text{for } s < t.$$

This comes immediately from the above result. Next we have:

Theorem 2.14

$$\text{Cov}(W(s), W(t)|W(1) = 0) = s(1 - t). \tag{2.24}$$

This provides a simple definition of the Brownian bridge as being a Gaussian process having mean zero and covariance function $s(1 - t)$, $s < t$. An alternative way of constructing the Brownian bridge is to consider the process defined as

$$W^0(t) = W(t) - tW(1), \quad 0 \le t \le 1.$$

Clearly $W^0(t)$ is a Gaussian process. We see that

$$E\{W(0)\} = W(0) = E\{W(1)\} = W(1) = E\{W(t)\} = 0$$

so that the only remaining question is the covariance function for the process to be completely and uniquely determined. The following corollary is all we need.

Corollary 2.12 *The covariance function for the process defined as* $W^0(t)$ *is,*

$$\text{Cov}\left\{W^0(s), W^0(t)\right\} = s(1-t) \quad s < t.$$

This is the covariance function for the Brownian bridge developed above and, by uniqueness, the process is therefore itself the Brownian bridge. Such a covariance function is characteristic of many observed phenomena. The covariance decreases linearly with distance from s. As for Brownian motion, should the covariance function decrease monotonically rather than linearly, then a suitable transformation of the time scale enables us to write the covariance in this form. At $t = s$ we recover the usual binomial expression $s(1-s)$.

Notice that not only can we go from Brownian motion to a Brownian bridge via the simple transformation

$$W^0(t) = W(t) - tW(1), \quad 0 \le t \le 1,$$

but the converse is also true, i.e., we can recover Brownian motion, $X(t)$, from the Brownian bridge, $Z(t)$, via the transformation

$$X(t) = (t+1)Z\left(\frac{t}{t+1}\right). \tag{2.25}$$

To see this, first note that, assuming $Z(t)$ to be a Brownian bridge, then $X(t)$ is a Gaussian process. It will be completely determined by its covariance process $\text{Cov}\{X(s), X(t)\}$. All we then require is the following lemma:

Lemma 2.8 *For the process defined in (2.25),* $\text{Cov}\{X(s), X(t)\} = s$.

The three processes: Brownian motion, the Brownian bridge, and the Ornstein-Uhlenbeck are then closely related and are those used in the majority of applications. Two further related processes are also of use in our particular applications: integrated Brownian motion and reflected Brownian motion.

Integrated Brownian motion

The process $Z(t)$ defined by $Z(t) = \int_0^t W(u)du$, where $W(t)$ is Brownian motion is called integrated Brownian motion. Note that $dZ(t)/dt = W(t)$ so that, for example, in the context of a model of interest, should we be able to construct a process converging in distribution to a process equivalent to Brownian motion, then the integrated process will converge in distribution to a process equivalent to integrated Brownian motion. We can see (by interchanging limits) that $Z(t)$ can be viewed as the limit of a sum of Gaussian processes and is therefore Gaussian. Its nature is completely determined by its mean and covariance function. We have that

$$E\{Z(t)\} = E\left\{\int_0^t W(u)du\right\} = \int_0^t E\{W(u)\}du = 0. \qquad (2.26)$$

For $s < t$ we have:

Lemma 2.9 *The covariance function for $Z(s)$ and $Z(t)$ is*

$$\mathrm{Cov}\left\{Z(s), Z(t)\right\} = s^2\left(t/2 - s/6\right). \qquad (2.27)$$

Lemma 2.10 *The covariance function for $Z(t)$ and $W(t)$ is*

$$\mathrm{Cov}\left\{Z(t), W(t)\right\} = t^2/2. \qquad (2.28)$$

For a model in which inference derives from cumulative sums, this would provide a way of examining how reasonable are the underlying assumptions if repetitions are available. Repetitions can be obtained by bootstrap resampling if only a single observed process is available. Having standardized, a plot of the log-covariance function between the process and the integrated process against log-time ought be linear with slope of two and intercept of minus log 2 assuming that model assumptions hold.

Reflected Brownian motion

Suppose we choose some positive value r and then define the process $W_r(t)$ as a function of Brownian motion, $W(t)$, in the following way: If $W(t) < r$ then $W_r(t) = W(t)$. If $W(t) \geq r$ then $W_r(t) = 2r - W(t)$. We have:

Lemma 2.11 $W_r(t)$ *is a Gaussian process,* $EW_r(t) = 0$, $\mathrm{Cov}\{W_r(s), W_r(t)\} = s$ *when* $s < t$.

Thus, $W_r(t)$ is also Brownian motion. Choosing r to be negative and defining $W_r(t)$ so that, when $W(t) > r$ then $W_r(t) = W(t)$. If $W(t) \leq r$ then $W_r(t) = 2r - W(t)$. accordingly we have the same result. The process $W_r(t)$ coincides exactly with $W(t)$ until such a time as a barrier is reached. We can imagine this barrier as a mirror, and beyond the barrier the process $W_r(t)$ is a simple reflection of $W(t)$. The interesting thing is that the resulting process is itself Brownian motion. One way of conceptualizing the idea is to imagine a large number of realizations of a completed Brownian motion process sampled independently. Imagine then these same realizations with a reflection applied. Then, whatever the point of reflection, if we consider the two collected sets of realizations, our overall impression of the behavior of the two processes will be the same. The value of this construction is to be seen in situations where, at some point in time, corresponding to some expected point of reflection under a hypothesis of drift, the drift changes direction. Under the hypothesis of Brownian motion, both Brownian motion, and Brownian motion reflected at some point, will look alike and will obey the same probability laws. Under an alternative hypothesis of drift however (see below), the behaviors will look quite different. This observation enables a simple construction with which to address the problem of crossing hazards.

Maximum of a Brownian motion

A useful further result can be immediately obtained from the preceding one dealing with reflected Brownian motion. Suppose that $W(t)$ is a Brownian motion. We might wish to consider the process $M(t) = \sup_{u \in (0,t)} W(u)$, which is the greatest value obtained by the process $W(u)$ in the interval $(0, t)$. The greatest absolute distance is also of interest but, by symmetry arguments, this can be obtained immediately from the distribution of $M(t)$. Another related question, useful in interim analyzes, is the distribution of $W(t)$ given the maximum $M(t)$ obtained up until that time point. We have the following:

Lemma 2.12 *If $W(t)$ is standard Brownian motion and $M(t)$ the maximum value attained on the interval $(0, t)$, i.e., $M(t) = \sup_{u \in (0,t)} W(u)$, then*

$$\Pr\{M(t) > a\} = 2\Pr\{W(t) > a\}.$$

This is a simple and elegant result and enables us to make simultaneous inference very readily. Sometimes, when using a Brownian motion approximation for a process, we may want to, for example, describe an approximate confidence interval for the whole process rather than just a confidence interval at a single point t. In such a case the above result comes into play immediately. The joint distribution is equally simple and we make use of the following.

Lemma 2.13 *If $W(t)$ is standard Brownian motion and $M(t)$ the maximum value attained on the interval $(0, t)$, i.e., $M(t) = \sup_{u \in (0,t)} W(u)$, then*

$$\Pr\{W(t) < a - b, M(t) > a\} = \Pr\{W(t) > a + b\}.$$

The conditional distribution $\Pr\{W(t) < a - b \,|\, M(t) > a\}$ can then be derived immediately by using the results of the two lemmas.

Brownian motion with drift

We will see that simple Brownian motion provides a good model for describing score statistics, or estimating equations, once standardized. This is because we can visualize these sums as approximating a limiting process arising from summing increments, for which the expected value is equal to zero. The setting in which we study such sums is typically that of evaluating some null hypothesis, often one of some given effect, $H_0 : \beta = \beta_0$, but sometimes a less obvious one, in the goodness-of-fit context, for example, whereby we can have, $H_0 : \beta(t) = \hat{\beta}$. Almost invariably, when we consider a null hypothesis, we have an alternative in mind, frequently a local or first alternative to the null. For a null hypothesis of Brownian motion, a natural and immediate alternative is that of Brownian motion with drift. Consider then the stochastic process $X(t)$ defined by

$$X(t) = W(t) + \mu t$$

where $W(t)$ is Brownian motion. We can immediately see that $E\{X(t)\} = \mu t$ and $\text{Var}\{X(t)\} = t$ As for Brownian motion $\text{Cov}\{X(s), X(t)\} = s$, $s < t$. Alternatively we can define the process in a way analogous to our definition for Brownian motion as a process having the following three properties:

1. $X(0) = 0$.

2. $X(t), t \in (0, 1)$ has independent stationary increments.

3. At each $t \in (0, 1)$, $X(t)$ is $\mathcal{N}(\mu t, t)$.

Clearly, if $X(t)$ is Brownian motion with drift parameter μ, then the process $X(t) - \mu t$ is standard Brownian motion. Also, for the more common situation in which the mean may change non-linearly with time, provided the increments are independent, we can always construct a standard Brownian motion by first subtracting the mean at time t, then transforming the timescale in order to achieve a linearly increasing variance.

Probability results for Brownian motion

There are a number of well-established and useful results for Brownian motion and related processes. The arcsine law can be helpful in comparing processes. Defining $X^+(t)$ to be the time elapsed from the origin that the Brownian process remains positive, i.e., $\sup\{t : X(s) > 0 : 0 < s < t\}$ then $\Pr(X^+ < x) = (2/\pi)\sin^{-1}\sqrt{x}$ This law can be helpful in comparing processes and also in examining underlying hypotheses. For the Brownian bridge the largest distance from the origin in absolute value has a known distribution given in a theorem of Kolmogorov:

$$\Pr\left\{\sup_t |W_0(t)| \leq \alpha\right\} \to 1 - 2\sum_{k=1}^{\infty}(-1)^{k+1}\exp(-2k^2\alpha^2), \quad \alpha \geq 0. \quad (2.29)$$

The sum can be seen to be convergent since this is an alternating sign series in which the kth term goes to zero. Furthermore, the error in ignoring all terms higher than the nth is less, in absolute value, than the size of the $(n+1)$th term. Given that the variance of $W_0(t)$ depends on t it is also of interest to study the standardized distribution $B_0(t) = W_0(t)/\sqrt{t(1-t)}$. This is, in fact, the Ornstein-Uhlenbeck process. Simple results for the supremum of this are not possible since the process becomes unbounded at $t = 0$ and $t = 1$. Nonetheless, if we are prepared to reduce the interval from $(0, 1)$ to $(\varepsilon_1, \varepsilon_2)$ where $\varepsilon_1 > 0$ and $\varepsilon_2 < 1$ then we have an approximation due to Miller and Siegmund (1982):

$$\Pr\left\{\sup_t |B_0(t)| \geq \alpha\right\} \approx \frac{4\phi(\alpha)}{\alpha} + \phi(\alpha)\left(\alpha - \frac{1}{\alpha}\right)\log\left\{\frac{\varepsilon_2(1-\varepsilon_1)}{\varepsilon_1(1-\varepsilon_2)}\right\}, \quad (2.30)$$

where $\phi(x)$ denotes the standard normal density. This enables us to construct confidence intervals for a bridged process with limits themselves going to zero at the endpoints. To obtain these we use the fact that $\Pr\{W_0(t) > \alpha\} = \Pr\{\sqrt{t(1-t)}B_0(t) > \alpha\}$. For most practical purposes though it is good enough to work with Equation 2.29 and approximate the infinite sum by curtailing summation for values of k greater than 2.

2.12 Counting processes and martingales

Although not automatically our first choice for inference, the use of counting processes and martingales for inference in survival problems currently dominates this subject area. We will look at inference based on counting processes and martingales in Section 3.6, for some general results, and in Chapter 10 for the specific application to the proportional hazards model. In this chapter we aim to provide some understanding to the probability structure upon which the theory is based.

Martingales and stochastic integrals

Recalling the discussion of Section 2.3 and that, for a bounded function $H(x)$ and the empirical distribution function $F_n(x)$, we have, by virtue of the Helly-Bray theorem, that $\int H(x)dF_n(x)$ converges in distribution to $\int H(x)dF(x)$. If we define $M(x) = F_n(x) - F(x)$ and change the order of integration, i.e., move the expectation operator, E, outside the integral, then

$$E\left\{\int H(x)dM(x)\right\} = 0.$$

This expression is worth dwelling upon. We think of E as being an integral operator or as defining some property of a random variable, specifically a measure of location. The random variable of relevance is not immediately apparent but can be seen to be $F_n(x)$, an $n-$dimensional function from the observations to the interval $[0, 1]$. We can suppose, at least initially, the functions $F(x)$ and $H(x)$ to be fixed and known. Our conceptual model allows the possibility of being able to obtain repetitions of the experiment, each time taking n independent observations. Thus, for some fixed given x, the value of $F_n(x)$ will generally vary from one experiment to the next. We view x as an argument to a function, and $F_n(x)$ as being random having a distribution studied below in Section 3.3. Recalling Section 2.3 on integration, note that we can rewrite the above equation as:

$$E \lim_{\max \Delta_i \to 0} \sum \{M(x_i) - M(x_{i-1})\}H(x_{i-1}) = 0, \qquad (2.31)$$

where $\Delta_i = x_i - x_{i-1} > 0$ and where, as described in Section 2.3 the summation is understood to be over an increasing partition in which $\Delta_i > 0$ and $\max \Delta_i$ goes to zero. Now, changing the order of taking limits, the above expression becomes

$$\lim_{\max \Delta_i \to 0} \sum E\{[M(x_i) - M(x_{i-1})]H(x_{i-1})\} = 0, \qquad (2.32)$$

a result which looks simple enough but that has a lot of force when each of the infinite number of expectations can be readily evaluated. Let's view Equation 2.32 in a different light, one that highlights the sequential and ordered nature of the partition. Rather than focus on the collection of $M(x_i)$ and $H(x_i)$, we can focus our attention on the increments $M(x_i) - M(x_{i-1})$ themselves, the increments being multiplied by $H(x_{i-1})$, and, rather than work with the overall expectation implied by the operator E, we will set up a sequence of conditional expectations. Also, for greater clarity, we will omit the term $\lim_{\max \Delta_i \to 0}$ altogether. We will put it back when it suits us. This lightens the notation and helps to make certain ideas more transparent. Later, we will equate the effect of adding back in the term $\lim_{\max \Delta_i \to 0}$ to that of replacing finite differences by infinitesimal differences. Consider then

$$U = \sum \{M(x_i) - M(x_{i-1})\}H(x_{i-1}) \qquad (2.33)$$

and, unlike the preceding two equations, we are able to greatly relax the requirement that $H(x)$ be a known function or that $M(x)$ be restricted to being the difference between the empirical distribution function and the distribution function. By sequential conditioning upon $\mathcal{F}(x_i)$ where $\mathcal{F}(x_i)$ are increasing sequence of sets denoting observations on $M(x)$ and $H(x)$, for all values of x less than or equal to x_i, we can derive results of wide applicability. In particular, we can now take $M(x)$ and $H(x)$ to be stochastic processes. Some restrictions are still needed for $M(x)$, in particular that the incremental means and variances exist. We will suppose that

$$E\{M(x_i) - M(x_{i-1})|\mathcal{F}(x_{i-1})\} = 0, \qquad (2.34)$$

in words, when given $\mathcal{F}(x_{i-1})$, the quantity $M(x_{i-1})$ is fixed and known and the expected size of the increment is zero. This is not a strong requirement and only supposes the existence of the mean since, should the expected size of the increment be other than zero, then we can subtract this difference to recover the desired property. Furthermore, given $\mathcal{F}(x)$, the quantity $H(x)$ is fixed. The trick is then to exploit the device of double expectation whereby for events, $\mathcal{A}$ and $\mathcal{B}$, it is always true that $E(\mathcal{A}) = EE(\mathcal{A}|\mathcal{B})$. In the context of this expression, $\mathcal{B} = \mathcal{F}(x_{i-1})$, leading to

$$E(U) = \sum H(x_{i-1})E\{M(x_i) - M(x_{i-1})|\mathcal{F}(x_{i-1})\} = 0 \quad (2.35)$$

and, under the assumption that the increments are uncorrelated we have the variance is the sum of the variance of each component to the sum. Thus

$$\text{Var}(U) = \sum E\{H^2(x_{i-1})[M(x_i) - M(x_{i-1})]^2|\mathcal{F}(x_{i-1})\}. \quad (2.36)$$

In order to keep the presentation uncluttered we use a single operator E in the above expressions, but there are some subtleties that ought not go unremarked. For instance, in Equation 2.36, the inner expectation is taken with respect to repetitions over all possible outcomes in which the set $\mathcal{F}(x_{i-1})$ remains unchanged, whereas the outer expectation is taken with respect to all possible repetitions. In Equation 2.35 the outer expectation, taken with respect to the distribution of all potential realizations of all the sets $\mathcal{F}(x_{i-1})$, is not written and is necessarily zero since all of the inner expectations are zero. The analogous device to double expectation for the variance is not so simple since $\text{Var}(Y) = E\,\text{Var}(Y|Z) + \text{Var}\,E(Y|Z)$. Applying this we have

$$\text{Var}\,\{M(x_i) - M(x_{i-1})\} = E\,\text{Var}\{M(x_i) - M(x_{i-1})|\mathcal{F}(x_{i-1})\} \quad (2.37)$$

since $\text{Var}\,E\{M(x_i) - M(x_{i-1})|\mathcal{F}(x_{i-1})\}$ is equal to zero, this being the case because each term is itself equal to the constant zero. The first term also requires a little thought, the outer expectation indicated by E being taken with respect to the distribution of $\mathcal{F}(x_{i-1})$, i.e., all the conditional distributions $M(x)$ and $H(x)$ where $x \leq x_{i-1}$. The next key point arises through the sequential nesting. These outer expectations, taken with respect to the distribution of $\mathcal{F}(x_{i-1})$ are the same as those taken with respect to the distribution of any $\mathcal{F}(x)$ for which $x \geq x_{i-1}$. This is an immediate consequence of the fact that the lower-dimensional distribution results from integrating out all the additional terms in the higher-dimensional distribution. Thus, if $x_{\max}$ is the greatest value of x for which observations are made then we can consider that all of these outer expectations are taken with respect to $\mathcal{F}(x_{\max})$. Each time that we condition upon $\mathcal{F}(x_{i-1})$ we will treat $H(x_{i-1})$ as a fixed constant and so it can be simply squared and moved

outside the inner expectation. It is still governed by the outer expectation which, for all elements of the sum, we will take to be with respect
to the distribution of $\mathcal{F}(x_{\max})$. Equation 2.36 then follows.

Making a normal approximation for U, and from the theory of
estimating equations, given any set of observations, that U depends
monotonically on some parameter β, then it is very straightforward
to set up hypothesis tests for $\beta = \beta_0$. Many situations, including that
of proportional hazards regression, lead to estimating equations of the
form of U. The above set-up, which is further developed below in a
continuous form, i.e., after having "added in" the term $\lim_{\max \Delta_i \to 0}$,
applies very broadly. We need the concept of a process, usually indexed
by time t, the conditional means and variances of the increments, given
the accumulated information up until time t.

We have restricted our attention here to the Riemann-Stieltjes definition of the integral. The broader Lebesgue definition allows the
inclusion of subsets of t tolerating serious violations of our conditions
such as conditional means and variances not existing. The conditioning
sets can be also very much more involved. Only in a very small number
of applications has this extra generality been exploited. Given that it
considerably obscures the main ideas to all but those well steeled in
measure theory, it seems preferable to avoid it altogether. Also avoided
here is the martingale central limit theorem. This theorem is much
quoted in the survival analysis context and, again, since there are so
few applications in which the needed large sample normality cannot be
obtained via more standard central limit theorems, a lack of knowledge
of this theorem will not handicap the reader.

Counting processes

The above discussion started off with some consideration of the empirical cumulative distribution function $F_n(t)$ which is discussed in much
more detail in Section 3.5. Let's consider the function $N(t) = \{nF_n(t) :
0 \leq t \leq 1\}$. We can view this as a stochastic process, indexed by time
t so that, given any t we can consider $N(t)$ to be a random variable
taking values from 0 to n. We include here a restriction that we generally make which is that time has some upper limit, without loss of
generality, we call this 1. This restriction can easily be avoided but
it implies no practical constraint and is often convenient in practical
applications. We can broaden the definition of $N(t)$ beyond that of
$nF_n(t)$ and we have:

Definition 2.4 *A counting process $N = \{N(t) : 0 \leq t \leq 1\}$ is a stochastic process that can be thought of as counting the occurrences (as time t proceeds) of certain type of events. We suppose these events occur singly.*

Very often $N(t)$ can be expressed as the sum of n individual counting processes, $N_i(t)$, each one counting no more than a single event. In this case $N_i(t)$ is a simple step function, taking the value zero at $t = 0$ and jumping to the value one at the time of an event. The realizations of $N(t)$ are integer-valued step functions with jumps of size $+1$ only. These functions are right-continuous and $N(t)$ is the (random) number of events in the time interval $[0, t]$. We associate with the stochastic process $N(t)$ an intensity function $\alpha(t)$. The intensity function serves the purpose of standardizing the increments to have zero mean. In order to better grasp what is happening here, the reader might look back to Equation 2.34 and the two sentences following that equation. The mean is not determined in advance but depends upon $\mathcal{F}_{t-}$ where, in a continuous framework, $\mathcal{F}_{t-}$ is to $\mathcal{F}_t$ what $\mathcal{F}(x_{i-1})$ is to $\mathcal{F}(x_i)$. In technical terms:

Definition 2.5 *A filtration, $\mathcal{F}_t$, is an increasing right continuous family of sub-sigma algebras.*

This definition may not be very transparent to those unfamiliar with the requirement of sigma additivity for probability spaces and there is no real need to expand on it here. The requirement is a theoretical one which imposes a mathematical restriction on the size, in an infinite sense, of the set of subsets of $\mathcal{F}_t$. The restriction guarantees that the probability we can associate with any infinite sum of disjoint sets is simply the sum of the probabilities associated with those sets composing the sum. For our purposes, the only key idea of importance is that $\mathcal{F}_{t-}$ is a set containing all the accumulated information (hence "increasing") on all processes contained in the past up until but not including the time point t (hence "right continuous"). We write, $\alpha = \{\alpha(t) : 0 \leq t \leq 1\}$ where

$$\alpha(t)dt \quad = \quad \Pr\{N(t) \text{ jumps in } [t, t+dt]|\mathcal{F}_{t-}\} = E\{dN(t)|\mathcal{F}_{t-}\},$$

the equality being understood in an infinitesimal sense, i.e., the functional part of the left-hand side, $\alpha(t)$, is the limit of the right-hand side divided by $dt > 0$ as dt goes to zero. In the chapter on survival analysis we will see that the hazard function, $\lambda(t)$, expressible as the ratio of the density, $f(t)$, to the survivorship function, $S(t)$, i.e., $f(t)/S(t)$, can be expressed in fundamental terms by first letting $Y(t) = I(T \geq t)$. Understanding, once again, the equality sign as described in the previous sentences but one, we have

$$\lambda(t)dt = \Pr\{N(t) \text{ jumps in } [t, t+dt)|Y(t)=1\} = E\{dN(t)|Y(t)=1\}.$$

It is instructive to compare the above definitions of $\alpha(t)$ and $\lambda(t)$. The first definition is the more general since, choosing the sets $\mathcal{F}_t$ to be defined from the at-risk function $Y(t)$ when it takes the value one, enables the first definition to reduce to a definition equivalent to the second. The difference is an important one in that if we do not provide a value for $I(T \geq t)$ then this is a $(0,1)$ random variable and, in consequence, $\alpha(t)$ is a $(0, \lambda(t))$ random variable. For this particular case we can express this idea succinctly via the formula

$$\alpha(t)dt = Y(t)\lambda(t)dt. \tag{2.38}$$

Replacing $Y(t)$ by a more general "at risk" indicator variable will allow for great flexibility, including the ability to obtain a simple expression for the intensity in the presence of censoring as well as the ability to take on-board multistate problems where the transitions are not simply from alive to dead but from, say, state j to state k summarized via $\alpha_{jk}(t)dt = Y_{jk}(t)\lambda_{jk}(t)dt$ in which $Y_{jk}(t)$ is left continuous and therefore equal to the limit $Y_{jk}(t-\epsilon)$ as $\epsilon > 0$ goes to zero through positive values, an indicator variable taking the value one if the subject is in state j and available to make a transition to state k at time $t - \epsilon$ as $\epsilon \to 0$. The hazards $\lambda_{jk}(t)$ are known in advance, i.e., at $t = 0$ for all t, whereas the $\alpha_{jk}(t)$ are random viewed from time point s where $s < t$, with the subtle condition of left continuity which leads to the notion of "predictability" described below. The idea of sequential standardization, the repeated subtraction of the mean, that leans on the evaluation of intensities, can only work when the mean exists. This requires a further technical property, that of being "adapted." We say

Definition 2.6 *A stochastic process $X(t)$ is said to be adapted to the filtration $\mathcal{F}_t$ if $X(t)$ is a random variable with respect to $\mathcal{F}_t$.*

Once again the definition is not particularly transparent to nonprobabilists and the reader need not be over-concerned since it will not be referred to here apart from in connection with the important concept of a predictable process. The basic idea is that the relevant quantities upon which we aim to use the tools of probability modeling should all be contained in $\mathcal{F}_t$. If any probability statement we wish to construct concerning $X(t)$ cannot be made using the set $\mathcal{F}_t$ but requires the set $\mathcal{F}_{t+u}$, where $u > 0$, then $X(t)$ is not adapted to $\mathcal{F}_t$. In our context just about all of the stochastic processes that are of interest to us are adapted and so this need not be a concern. A related property, of great importance, and which also will hold for all of those processes we focus attention on, is that of predictability. We have

Definition 2.7 *A real-valued stochastic process, $H(t)$, that is left continuous and adapted to the filtration $\mathcal{F}_t$ is called a predictable process.*

Since $H(t)$ is adapted to $\mathcal{F}_t$ it is a random variable with respect to $\mathcal{F}_t$. Since the process is left continuous it is also adapted to $\mathcal{F}_{t-}$. Therefore, whenever we condition upon $\mathcal{F}_{t-}$, $H(t)$ is simply a fixed and known constant. This is the real sense of the term "predictable" and, in practice, the property is a very useful one. It is frequently encountered in the probabilistic context upon which a great number of tests are constructed. Counting processes can be defined in many different ways and such a formulation allows for a great deal of flexibility. Suppose for instance that we have events of type 1 and events of type 2, indicated by $N_1(t)$ and $N_2(t)$ respectively. Then $N(t) = N_1(t) + N_2(t)$ counts the occurrences of events of either type. For this counting process we have

$$\alpha(t)dt = P(N(t) \text{ jumps in } [t, t + dt]|\mathcal{F}_{t-}),$$

i.e., the same as $P(N_1(t) \text{ or } N_2(t) \text{ jump in } [t, t+dt]|\mathcal{F}_{t-})$ and, if as is reasonable in the great majority of applications, where, we assume to be negligible the probability of seeing events occurring simultaneously compared to seeing them occur singly, then

$$\alpha(t)dt = E\{dN_1(t) + dN_2(t)|\mathcal{F}_{t-}\} = \alpha_1(t) + \alpha_2(t).$$

This highlights a nice linearity property of intensities, not shared by probabilities themselves. For example, if we consider a group of n subjects and n individual counting processes $N_i(t)$, then the intensity function, $\alpha(t)$, for the occurrence of an event, regardless of individual, is simply $\sum \alpha_i(t)$. This result does not require independence of the processes, only that we can consider as negligible the intensities we might associate with simultaneous events.

Another counting process of great interest in survival applications concerns competing risks. Suppose there are two types of event but that they cannot both be observed. The most common example of this is right censoring where, once the censoring event has occurred, it is no longer possible to make observations on $N_i(t)$. This is discussed more fully in the following chapters and we limit ourselves here to the observation that $N_i(t)$ depends on more than one variable. In the absence of further assumptions, we are not able to determine the intensity function, but if we are prepared to assume that the censoring mechanism is independent of the failure mechanism, i.e., that $\Pr(T_i > t | C_i > c) = \Pr(T_i > t)$, then a simple result is available.

Theorem 2.15 *Let the counting process, $N_i(t)$, depend on two independent and positive random variables, T_i and C_i such that $N_i(t) = I\{T_i \le t, T_i \le C_i\}$. Let $X_i = \min(T_i, C_i)$, $Y_i(t) = I(X_i \ge t)$; then $N_i(t)$ has intensity process*

$$\alpha_i(t)dt = Y_i(t)\lambda_i(t)dt. \qquad (2.39)$$

The counting process, $N_i(t)$, is one of great interest to us since the response variable in most studies will be of such a form, i.e., an observation when the event of interest occurs but an observation that is only possible when the censoring variable is greater than the failure variable. Also, when we study a heterogeneous group, our principal focus in this book, the theorem still holds in a modified form. Thus, if we can assume that $\Pr(T_i > t | C_i > c, Z = z) = \Pr(T_i > t | Z = z)$, we then have:

Theorem 2.16 *Let the counting processes, $N_i(t)$, depend on two independent and positive random variables, T_i and C_i, as well as Z such that*

$$N_i(t) = I\{T_i \le t, T_i \le C_i, Z = z\}. \qquad (2.40)$$

Then the intensity process for $N_i(t)$ can be written as $\alpha_i(t, z)dt = Y_i(t)\lambda_i(t, z)dt$.

The assumption needed for Theorem 2.16, known as the conditional independence assumption, is weaker than that needed for 2.15 in that the latter theorem contains the former as a special case. Note that the stochastic processes $Y_i(t)$ and $\alpha_i(t)$ are left continuous and adapted to $\mathcal{F}_t$. They are therefore predictable stochastic processes, which means that, given $\mathcal{F}_{t-}$, we treat $Y_i(t)$, $\alpha_i(t)$ and, assuming that $Z(t)$ is predictable, $\alpha_i(t, z)$ as fixed constants.

2.13 Excrcises and class projects

1. Use a simple sketch to informally demonstrate the mean value theorem.

2. Newton-Raphson iteration provides sequentially updated estimates to the solution to the equation $f(x_0) = 0$. At the nth step, we write $x_{n+1} = x_n - f(x_n)/f'(x_n)$ and claim that x_n converges (in the analytical sense) to x_0. Use the mean value theorem and, again, a simple sketch to show this. Intuitively, which conditions will lead to convergence and which ones can lead to failure of the algorithm.

3. Let $g(x)$ take the value 0 for $-\infty < x \leq 0 : 1/2$ for $0 < x \leq 1$; 1 for $1 < x \leq 2$: and 0 otherwise. Let $f(x) = x^2 + 2$. Evaluate the Riemann-Stieltjes integral of $f(x)$ with respect to $g(x)$ over the real linc.

4. Note that $\sum_{i=1}^{n} i = n(n+1)/2$. Describe a function such that a Riemann-Stieltjes integral of it is equal to $n(n+1)/2$. Viewing integration an an area under a curve, conclude that this integral converges to n^2 as n becomes large.

5. Suppose that in the Helly-Bray theorem for $\int h(x)dF_n(x)$, the function $h(x)$ is unbounded. Break the integral into components over the real line. For regions where $h(x)$ is bounded the theorem holds. For the other regions obtain conditions that would lead to the result holding generally.

6. Prove the probability integral transformation by finding the moment-generating function of the random variable $Y = F(X)$ where X has the continuous cumulative distribution function $F(x)$ and a moment-generating function that exists.

7. If X is a continuous random variable with probability density function $f(x) = 2(1-x), 0 < x < 1$, find that transformation $Y = \psi(X)$ such that the random variable Y has the uniform distribution over (0,2).

8. The order statistics for a random sample of size n from a discrete distribution are defined as in the continuous case except that now we have $X_{(1)} \leq X_{(2)} \leq \cdots \leq X_{(n)}$. Suppose a random sample of size 5 is

taken with replacement from the discrete distribution $f(x) = 1/6$ for $x = 1, 2, \ldots, 6$. Find the probability mass function of $X_{(1)}$, the smallest order statistic.

9. Ten points are chosen randomly and independently on the interval (0,1). Find (a) the probability that the point nearest 1 exceeds 0.8, (b) the number c such that the probability is 0.4 that the point nearest zero will exceed c.

10. Find the expected value of the largest order statistic in a random sample of size 3 from (a) the exponential distribution $f(x) = \exp(-x)$ for $x > 0$, (b) the standard normal distribution.

11. Find the probability that the range of a random sample of size n from the population $f(x) = 2e^{-2x}$ for $x \geq 0$ does not exceed the value 4.

12. Approximate the mean and variance of (a) the median of a sample of size 13 from a normal distribution with mean 2 and variance 9, (b) the fifth-order statistic of a random sample of size 15 from the standard exponential distribution.

13. Simulate 100 observations from a uniform distribution. Do the same for an exponential, Weibull and log-logistic distribution with different parameters. Next, generate normal and log-normal variates by summing a small number of uniform variates. Obtain histograms. Do the same for 5000 observations.

14. Obtain the histogram of 100 Weibull observations. Obtain the histogram of the logarithms of these observations. Compare this with the histogram obtained by the empirical transformation to normality.

15. Suppose that $T_1, \ldots, T_n$ are n exponential variates with parameter λ. Show that, under repeated sampling, the smallest of these also has an exponential distribution. Is the same true for the largest observation? Suppose we are only give the value of the smallest of n observations from an exponential distribution with parameter λ. How can this observation be used to estimate λ.

16. Suppose that X_i $i = 1, \ldots, n$ are independent exponential variates with parameter λ. Determine, via simple calculation, the variance of $\min(X_1, \ldots, X_n)$.

17. Having some knowledge of the survival distribution governing observations we are planning to study, how might we determine an interval of time to obtain with high probability a given number of failures? How should we proceed in the presence of censoring?

18. Derive the Bienaymé-Chebyshev inequality. Describe the advantages and drawbacks of using this inequality to construct confidence intervals in a general setting.

19. Suppose that the entropy described in Equation 2.14 depends on a parameter θ and is written $V_\theta(f, f)$. Consider $V_\alpha(f, f)$ as a function of α. Show that this function is maximized when $\alpha = \theta$.

20. Using the device of double expectation derive Equation 2.15. Why is this breakdown interpreted as one component corresponding to "signal" and one component corresponding to "noise."

21. Suppose that θ_n converges in probability to θ and that the variance of θ_n is given by $\psi(\theta)/n$. Using Equation 2.19, find a transformation of θ_n for which, at least approximately, the variance does not depend on θ.

22. Consider a stochastic process $X(t)$ on the interval $(2, 7)$ with the following properties: (a) $X(0) = 2$, (b) $X(t), t \in (2, 7)$ has increments such that (c), for each $t \in (2, 7)$ the distribution of $X(t)$ is Weibull with mean $2 + \lambda t^\gamma$. Can these increments be independent and stationary? Can the process be described using the known results of Brownian motion?

23. For Brownian motion, explain why the conditional distribution of $X(s)$ given $X(t)$ $(t > s)$ is normal with $E\{X(s)|X(t) = w\} = ws/t$ and $\mathrm{Var}\{X(s)|X(t) = w\} = s(t - s)/t$. Deduce the mean and the covariance process for the Brownian bridge.

24. The Ornstein-Uhlenbeck process can be thought of as transformed Brownian motion in which the variance has been standardized. Explain why this is the case.

25. Reread the subsection headed "Time-transformed Brownian motion" (Section 2.11) and conclude that the only essential characteristic underwriting the construction of Brownian motion is that of independent increments.

26. Find the value of $t \in (0, 1)$ for which the variance of a Brownian bridge is maximized.

27. Suppose that under H_0, $X(t)$ is Brownian motion. Under H_1, $X(t)$ is Brownian motion with drift, having drift parameter 2 as long as $X(t) < 1$ and drift parameter minus 2 otherwise. Describe likely paths for reflected Brownian motion under both H_0 and H_1. As a class exercise simulate ten paths under both hypotheses. Comment on the resulting figures.

Chapter 3

Background: General inference

3.1 Summary

We review the main theorems providing inference for sums of random variables. The theorem of de Moivre-Laplace is a well-known special case of the central limit theorem and helps provide the setting. Our main interest is on sums which can be considered to be composed of independent increments. The empirical distribution function $F_n(t)$ is readily seen to be a consistent estimator for $F(t)$ at all continuity points of $F(t)$. However, we can also view $F_n(t)$ as a constant number multiplying a sum of independent Bernoulli variates and this enables us to construct inference for $F(t)$ on the basis of $F_n(t)$. Such inference can then be extended to the more general context of estimating equations. Inference for counting processes and stochastic integrals is described since this is commonly used in this area and, additionally, shares a number of features with an approach based on empirical processes. The importance of estimating equations is stressed, in particular equations based on the method of moments and equations derived from the likelihood. Resampling techniques can also be of great value for problems in inference. Our final goal is the use of inferential tools to construct models and so the predictive power of a model is important. An approach to this question can be made via the idea of explained variation or that of explained randomness. Both are dealt with in later chapters. Here, since this does not appears to be well known, we present an outline of explained variation in general terms, i.e., without necessarily leaning on any specific model.

3.2   Motivation

The now classical approach to dealing with inference in survival time problems is via stochastic integrals and martingales. We recall this theory here together with the main results. Our main motivation though is to show that all of the practical problems we are ever likely to encounter can be attacked using very standard techniques, essentially the central limit theorem and simple variants of it. The statistics that we will derive can be seen quite easily to fall under the headings described below. The statistics described there will have known large sample distributions. We can then appeal immediately to known results from Brownian motion and other functions of Brownian motion. Using this approach to inference is reassuring since (1) the building blocks are elementary ones, well known to those who have followed introductory courses on inference (this is not the case, for instance, for the martingale central limit theorem) and (2) we obtain, as special cases, statistics that are currently widely used, the most notable examples being the partial likelihood score test and weighted log-rank statistics. However, we will obtain many more statistics, all of which can be seen to sit in a single solid framework and some of which, given a particular situation of interest, will suggest themselves as being potentially more suitable than others.

3.3   Limit theorems for sums of random variables

The majority of statistics of interest that arise in practical applications are directly or indirectly (e.g., after taking the logarithm to some base) expressible as sums of random variables. It is therefore of immense practical value that the distribution theory for such sums can, in a wide variety of cases, be approximated by normal distributions. Moreover, we can obtain some idea as to how well the approximation may be expected to behave. It is also possible to refine the approximation. In this section we review the main limit theorems applicable to sums of random variables.

Theorem of De Moivre-Laplace

Let $N_n = \sum_{i=1}^n X_i$ be the number of successes in n independent Bernoulli trials X_i, each trial having probability of success equal to p. Then

$$\{N_n - np\}/\sqrt{np(1-p)} \rightarrow \mathcal{N}(0,1)$$

where $\rightarrow$ means convergence in distribution. This is the oldest result of a central limit type and is the most well known special case of the more general result, just below, for sums of independent and identically distributed random variables.

Central limit theorem for i.i.d. variables

Let X_i, $i = 1, 2, \ldots$ be independent random variables having the same distribution $F(.)$. We assume that $\int u^2 dF(u) < \infty$. Let $\sigma^2 = \int u^2 dF(u) - \mu^2$ where $\mu = \int u dF(u)$. Let $\bar{x} = \int u dF_n(u)$ where $F_n(t) = n^{-1} \sum_{i=1}^{n} I$ $(T_i \leq t)$. Then the central limit theorem states that

$$\sigma \sqrt{n} (\bar{x} - \mu) \rightarrow \mathcal{N}(0,1).$$

Less formally we state that $\bar{x}$ converges to a normal distribution with mean μ and variance σ^2/n. This result is extremely useful and also quite general. For example, applying the mean value theorem, then for $g(\bar{x})$, where $g(x)$ is a differentiable function of x, we can see, using the same kind of informal statement, that $g(\bar{x})$ converges to a normal distribution with mean $g(\mu)$ and variance $\{g'(\mu)\}^2 \sigma^2/n$.

Central limit theorem for independent variables

For nonidentically distributed random variables the problem is very much more involved. This is because of the large number of potential situations that need be considered. The most succinct solution appeared as the condition described below. Let X_i, $i = 1, 2, \ldots$ be independent random variables having distributions $F_i(.)$. Let $\sigma_i^2 = \int u^2 dF_i(u) < \infty$ and $\mu_i = \int u dF_i(u)$. Let $B_n^2 = \sum \sigma_i^2$ and define $\int_\epsilon$ to be an integral over the real line such that $|t - \mu_i| > \epsilon B_n$. Introduce the following:

Condition 3.1 *For each $\epsilon > 0$, $\sum B_n^{-2} \int_\epsilon (t - \mu_i)^2 \rightarrow 0$, as $n \rightarrow \infty$ If this condition is satisfied then*

$$nB_n^{-1}(\bar{x} - n^{-1} \sum \mu_i) \rightarrow \mathcal{N}(0,1).$$

This condition is known as the Lindeberg condition. The statement is an "only if" statement. Less formally we say that $\bar{x}$ converges to a normal distribution with mean $\sum \mu_i/n$ and variance B_n^2/n^2. The condition is simply a way of formulating or expressing mathematically the need that the sum be composed of independent "relevant" contributions. If a single term or group of terms dominate the sum such that the remaining contributions are in some sense negligible, then our intuition may tell us it would not be reasonable to anticipate the central limit theorem to generally apply. This could happen in various ways, in particular if there is "too much" information in the tails of the distributions, i.e., the tails are too heavy or σ_i^2 diminishes with increasing i at too fast a rate. It follows from the Lindeberg condition that

$$B_n^{-2}\sigma_n^2 \to 0, \ \ B_n \to \infty \,, \text{as } n \to \infty.$$

It can be fairly easily shown that the condition below implies the Lindeberg condition and provides a more ready way of evaluating whether or not asymptotic normality obtains.

Condition 3.2 *The Lindeberg condition holds if, for $k > 2$,*

$$B_n^{-k} \sum \kappa_k \to 0 \ \ as \ n \to \infty.$$

Central limit theorem for dependent variables

Let $X_i \,, i = 1, 2, \ldots$ be a sequence of random variables having distributions $F_i(.)$. Let $\sigma_i^2 = \int u^2 dF_i(u) < \infty$ and $\mu_i = \int u dF_i(u)$. As before, let $B_n^2 = \sum \sigma_i^2$. Then, under certain conditions,

$$nB_n^{-1} \left(\bar{x} - n^{-1} \sum \mu_i \right) \to \mathcal{N}(0, 1).$$

As we might guess, the conditions in this case are much more involved and we need to use array notation in order to express the cross dependencies that are generated. If we take an extreme case we see immediately why the dependencies have to be carefully considered for, suppose $X_i = \alpha_{i-1}X_{i-1}$ where the α_i are nonzero deterministic coefficients such that $\sum_1^n \alpha_i \to 1$, then clearly X_n converges in distribution to X_1 which can be any chosen distribution. In rough terms, there needs to be enough independence between the variables for the result to hold. Describing what is meant by "enough" is important in certain contexts, time series analysis being an example, but, since it is not needed in this work, we do not spend any time on it here. A special case of nonidentical distributions, of value in survival analysis, is the following.

Central limit theorem for weighted sums of i.i.d. variables

Let X_i, $i = 1, 2, \ldots$ be independent random variables having the same distribution $F(.)$. Let $\sigma^2 = \int u^2 dF(u) < \infty$ and $\mu = \int u dF(u)$. Let a_i, $i = 1, \ldots, n$, be constants, $S_n = n^{-1/2} \sum_{i=1}^n a_i (X_i - \mu)$ and $\sigma_S^2(n) = \sigma^2 \sum_{i=1}^n a_i^2/n$, then $S_n/\sigma_S(n) \to \mathcal{N}(0, 1)$ where $\sigma_S(n) = \{\sigma_S^2(n)\}^{1/2}$, whenever the following condition holds;

Condition 3.3 *The coeficients a_i are constants and are such that*

$$\frac{\max |a_i|}{\sqrt{\sum_{j=1}^n a_j^2}} \to 0.$$

Many statistics arising in nonparametric theory come under this heading, e.g., linear sums of ranks. The condition is a particularly straightforward one to verify and leads us to conclude large sample normality for the great majority of the commonly used rank statistics in nonparametric theory. A related condition, which is sometimes of more immediate applicability, can be derived as a consequence of the above large sample result together with an application of Slutsky's theorem. Suppose, as before, that X_i, $i = 1, 2, \ldots$ are independent random variables having the same distribution $F(.)$, that $\sigma^2 = \int u^2 dF(u) < \infty$, $\mu = \int u dF(u)$ and that a_i, $i = 1, \ldots, n$, are constants. Again, letting $S_n = n^{-1/2} \sum_{i=1}^n a_i (X_i - \mu)$ and $\sigma_S^2(n) = \sigma^2 \sum_{i=1}^n a_i^2/n$, then $S_n \to \mathcal{N}(0, \sigma^2 \alpha^2)$ where:

Condition 3.4 *The mean of the constant coeficients a_i converges and*

$$\frac{1}{n} \sum_{j=1}^n a_j^2 \to \alpha^2, \quad 0 < \alpha^2 < \infty.$$

The condition is useful in that it will allow us to both conclude normality for the linear combination S_n and, at the same time, provide us with a variance for the linear combination. Weighted log-rank statistics and score statistics under non-proportional hazards models are close to coming under this heading. The weights in that case are not fixed in advance but, since the weights are typically bounded and converge to given quantities, it is relatively straightforward to put in the extra steps to obtain large sample normality in those cases too.

3.4   Functional Central Limit Theorem

If we limit our attention to sums of random variables, each of which is indexed by a value t lying between 0 and 1 (we lose no generality in practice by fixing an upper limit 1 rather than infinity), we can obtain many useful results. The important idea here is that of the order among the random variables indexed by t, since t will be a real number between 0 and 1. As always the sums of interest will be finite, sums of quantities evaluated at some finite number of time points on the interval (0,1), and, we will appeal to known results concerning the limiting continuous distributions, as the interval is "filled out," as a means to approximate the exact, but necessarily complicated, finite sample distributions. For this reason it is helpful to begin reasoning in terms of sums, indexed by a finite number of points, and consider what such sums look like as the number of points increases without limit.

Sums of i.i.d. variables on interval (0,1)

Imagine a process starting at the origin and making successive displacements $X_i, i = 1, \ldots n$, where the X_i are all independent. For every k, where $1 \leq k \leq n$, the total distance travelled from the origin can be represented by $U_k = \sum_{i \leq k} X_i$ (random walk). The simplest way of looking at such a process is to consider the interval (0,1) divided into n equal nonoverlapping intervals each of size $1/n$. This can only be achieved in one way. We make observations X_i, and therefore U_i, at the points $t = i/n$, $i = 1, \ldots, n$. The increments X_i are independent. We have $E(X_i) = 0$ and $\text{Var}(X_i) = \sigma^2 < \infty$. In consequence we see that $E(U_k) = 0$, that $E(U_k^2) = k\sigma^2$ and, in view of the central limit theorem, that $E(U_k^\ell) \to 0$, $\forall \ell$ odd. We make the process continuous by linearly interpolating between the points at which U_i, $i = 1, \ldots, n$ is defined. Note that there are much more general developments of the limiting process than we obtain here (Brownian motion) and that continuity can be demonstrated as a property of the limiting process. However, it seems easier to construct the process already having continuity as a property for finite situations. This avoids technical difficulties and, perhaps more importantly, helps illustrate why and how, in practice, we can construct processes that will look like Brownian motion. Indeed, not only will these processes look like Brownian motion, but their probabilistic behavior, of practical interest to us, can be accurately approximated by the known properties of Brownian motion. Finally, just as in the standardizations of the preceding sections, we need standardize the variance of our process. This we do by considering the sum

$$U_k^* = (\sigma\sqrt{n})^{-1}U_k = (\sigma\sqrt{n})^{-1}\sum_{i\leq k}X_i \,,$$

from which, letting $t = k/n$, we readily obtain the mean and the variance of U_k^* as

$$E(U_k^*) = (\sigma\sqrt{n})^{-1}\sum_{i\leq k}E(X_i) = 0\,; \quad \text{Var}\,(U_k^*) = (\sigma\sqrt{n})^{-2}k\sigma^2 = t. \quad (3.1)$$

Although this and the following section are particularly simple, the reader should make sure that he or she has a very solid understanding as to what is taking place. It underscores all the main ideas behind the methods of inference that are used. An example of such a process in which $\sigma^2 = 1$ and $n = 30$ is shown in Figure 3.1 As for $\text{Var}\,(U_k^*)$ we see in the same way that;

Theorem 3.1 *For $k < m$,* $\text{Cov}\,(U_k^*, U_m^*) = t$ *where $t = k/n$.*

The important thing to note is that the increments are independent, implying convergence to a Gaussian process. All we then need is the co-variance process. Figure 3.1 and Figure 3.2 represent approximations to Brownian motion in view of discreteness and the linear interpolation. The figures indicate two realizations from the above summed processes, and the reader is encouraged to carry out his or her own such simulations, an easy exercise, and yet invaluable in terms of building

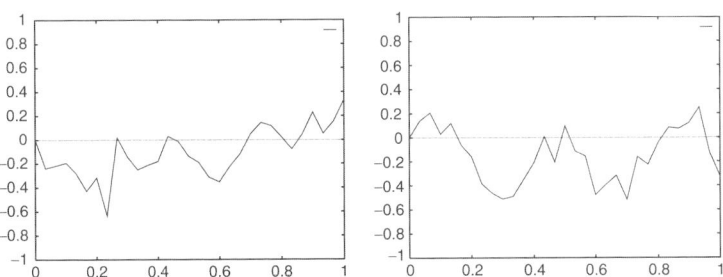

Figure 3.1: Two independent simulations of sums of 30 points on interval (0,1).

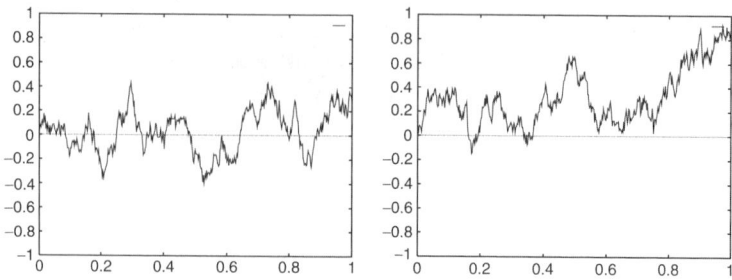

Figure 3.2: Two independent simulations of sums of 500 points on interval (0,1).

good intuition. An inspection of any small part of the curve (take, for example, the curve between 0.30 and 0.31 where the curve is based on less than 100 points), might easily be a continuous straight line, nothing at all like the limiting process, Brownian motion. But imagine, as very often is the case in applications, that we are only concerned about some simple aspect of the process, for instance, the greatest absolute distance travelled from the origin for the transformed process, tied down at $t = 1$. With as few as 30 observations our intuition would correctly lead us to believe that the distribution of this maximum will be accurately approximated by the same distribution, evaluated under the assumption of a Brownian bridge. Of course, such a statement can be made more technically precise via use, for example, of the law of the iterated logarithm or the use of Berry-Esseen bounds.

Sums of independent variables on (0,1)

The simple set-up of the previous section, for which the large sample theory, described above, is well established, can be readily extended to the non i.i.d. case. Let's begin by relaxing the assumption that the variances of the X_i do not depend on i. Suppose that as before $E(X_i) = 0$ and let $\text{Var}(X_i) = \sigma_i^2 < \infty$. Then clearly the process

$$U_k^* = (\sqrt{n})^{-1} \sum_{i \leq k} \sigma_i^{-1} X_i$$

will look like the process defined above. In particular, straightforward manipulation as above shows that $E(U_k^*) = 0$ and $\text{Cov}(U_k^*, U_m^*) = t$ where $k < m$ and $t = k/n$. We allow k to increase at the same rate, i.e., $k = nt$ where $0 < t < 1$. As $n \to \infty$ the number of possible values of t, $t \in (0,1)$ also increases without limit to the set of all rationals on this interval. We can also suppose that as $k, n \to \infty$; $k < n$, such that $k/n = t$ then σ_t^2 converges almost everywhere to some function $\sigma^2(t)$. To this set of rationals we can easily add the irrationals by, fixing t and, for every given n, choose k to produce the closest rational to t. We then allow n to increase without bound.

The functional central limit theorem states that the above process goes to a limit. The limiting process is defined on the real interval. Choosing any set of points $\{t_1, \ldots, t_k\}$, $(0 < t_i < 1, i = 1, \ldots, k)$ then the process $U_{t_1}^*, U_{t_2}^*, \ldots, U_{t_k}^*$ converges in distribution to the multivariate normal. As indicated above the covariance only depends on the distance between points so that $\text{Cov}\{U_s^*, U_t^*\} = s$; $s < t$. The basic idea is that the increments, making up the sum $U^*(t)$, get smaller and smaller as n increases. The increments have expectation equal to zero unless there is drift. Also, the way in which the increments become smaller with n is precisely of order $\sqrt{n}$. The variance therefore increases linearly with time out in the process. In practical applications, it is only necessary that the increments be independent and that these increments have a finite variance. It is then straightforward to carry out a time transformation to obtain the limiting process as an immediate consequence of the functional central limit theorem. The functional central limit theorem differs very little in essence to the usual central limit theorem, from which it derives. The key additional idea is that of sequential standardization. It is all very simple but, as we shall see, very powerful.

3.5 Empirical distribution function

The above results can be directly applied to the sample empirical distribution function $F_n(t)$, defined for a sample of size n (uncensored) to be the number of observations less than or equal to t divided by n, i.e., $F_n(t) = n^{-1} \sum_{i=1}^{n} I(T_i \leq t)$. For each t, and we may assume $F(t)$ to be a continuous function of t, we would hope that $F_n(t)$ converges to $F(t)$ in probability. This is easy to see but, in fact, we have stronger results, starting with the Glivenko-Cantelli theorem whereby

$$D_n = \sup_{0 \leq t \leq \infty} |F_n(t) - F(t)| = \sup_{0 \leq t \leq \infty} |S_n(t) - S(t)|$$

converges to zero with probability one and where $S_n(t) = 1 - F_n(t)$. This is analogous to the law of large numbers and, although important, is not all that informative. A central limit theorem can tell us much more and this obtains by noticing how $F_n(t)$ will simulate a process relating to the Brownian bridge. To see this it suffices to note that $nF_n(t)$, for each value of t, is a sum of independent Bernoulli variables. Therefore, for each t as $n \to \infty$, we have that $\sqrt{n}\{F_n(t) - F(t)\}$ converges to normal with mean zero and variance $F(t)\{1 - F(t)\}$. We have marginal normality. However, we can claim conditional normality in the same way, since, for each t and s $(s < t)$, $nF_n(t)$ given $nF_n(s)$, is also a sum of independent Bernoulli variables. Take k_1 and k_2 to be integers $(1 < k_1 < k_2 < n)$ such that k_1/n is the nearest rational smaller than or equal to s (i.e., $k_1 = \max j \, ; j \in \{1, \ldots, n, \}, j \le ns$) and k_2/n is the nearest rational smaller than or equal to t. Thus, k_1/n converges with probability one to s, k_2/n to t, and $k_2 - k_1$ increases without bound at the same rate as n. We then have:

Theorem 3.2 $\sqrt{n}\{F_n(t) - F(t)\}$ *is a Gaussian process with mean zero and covariance given by:*

$$\text{Cov}\left[\sqrt{n}\{F_n(s)\}, \sqrt{n}\{F_n(t)\}\right] = F(s)\{1 - F(t)\}. \tag{3.2}$$

It follows immediately that, for T uniform, the process $\sqrt{n}\{F_n(t) - t\}$ $(0 \le t \le 1)$, converges in distribution to the Brownian bridge. But note that, whatever the distribution of T, as long as it has a continuous distribution function, monotonic increasing transformations on T leave the distribution of $\sqrt{n}\{F_n(t) - F(t)\}$ unaltered. This means that we can use the Brownian bridge for inference quite generally. In particular, consider results of the Brownian bridge, such as the distribution of the supremum over the interval (0,1), that do not involve any particular value of t (and thereby $F(t)$). These results can be applied without modification to the process $\sqrt{n}\{F_n(t) - F(t)\}$ whether or not $F(t)$ is uniform. Among other useful results concerning the empirical distribution function we have:

Law of iterated logarithm

This law tells us something about the extreme deviations of the process. The following theorem (Serfling, page 62) provides the rate with n at which the largest absolute discrepancy between $F_n(t)$ and $F(t)$ is tending to zero.

Theorem 3.3 *With probability one,*

$$\lim_{n \to \infty} \frac{\sqrt{n} D_n}{(2 \log \log n)^{\frac{1}{2}}} = \sup\{F(t)(1 - F(t))\}^{\frac{1}{2}} \tag{3.3}$$

For the most common case in which $F(t)$ is continuous, we know that $\sup F(t)(1 - F(t))$ is equal to 0.5. As an illustration, for 50 subjects, we find that D_n is around 0.12. For 50 i.i.d. observations coming from some known or some hypothesized distribution, if the hypothesis is correct then we expect to see the greatest discrepancy between the empirical and the hypothesized distribution to be close to 0.12. Values far removed from that might then be indicative of either a rare event or that the assumed distribution is not correct. Other quantities, indicating how close to $F(t)$ we can anticipate $F_n(t)$ to be, are of interest, one in particular being;

$$C_n = n \int_0^\infty \{F_n(t) - F(t)\}^2 f(t) dt.$$

As for D_n the asymptotic distribution of C_n does not depend upon $F(t)$. For this case the law of the iterated logarithm is expressed as follows:

Theorem 3.4 *With probability one,*

$$\lim_{n \to \infty} \frac{C_n}{(2 \log \log n)^{\frac{1}{2}}} = \frac{1}{\pi^2}. \tag{3.4}$$

For D_n, inference can be based on the maximum of a Brownian bridge. In the case of C_n inference is less straightforward and is based on the following lemma;

Lemma 3.1 *Letting $\eta = \sum_{j=1}^\infty \chi_j^2 (\pi j)^{-2}$ where the χ_j^2 are independent chi-square variates then*

$$\lim_{n \to \infty} P(C_n \leq c) = P(\eta \leq c). \tag{3.5}$$

the results for both D_n and C_n are large sample ones but can nonetheless provide guidance when dealing with actual finite samples. Under assumed models it is usually possible to calculate the theoretical distribution of some quantity which can also be observed. We are then able to contrast the two and test the plausibility of given hypotheses.

3.6   Inference for martingales and stochastic integrals

The reader might look over Section 2.12 for the probability background behind martingales. A martingale $M = \{M(t) : t \geq 0\}$ is a stochastic process whose increment over an interval $(u, v]$, given the past up to and including time u, has expectation zero, i.e., $E\{M(v) - M(u)|\mathcal{F}_u\} = 0$ for all $0 \leq u < v < 1$. Equation 2.34 provides the essential idea for the discrete time case. We can rewrite the above defining property of martingales by taking the time instants u and v to be just before and just after the time instant t. Letting both v and u tend to t and u play the role of $t-$, we can write;

$$E\{dM(t)|\mathcal{F}_{t-}\} = 0. \tag{3.6}$$

Note that this is no more than a formal way of stating that, whatever the history $\mathcal{F}_t$ may be, given this history, expectations exist. If these expectations are not themselves equal to zero then we only need subtract the nonzero means to achieve this end. A counting process $N_i(t)$ is not of itself a martingale but note, for $0 \leq u < v \leq 1$, that $E\{N_i(v)|\mathcal{F}_u\} > E\{N_i(u)|\mathcal{F}_u\}$ and, as above, by taking the time instants u and v to be just before and just after the time instant t, letting v and u tend to t and u play the role of $t-$, we have

$$E\{dN_i(t)|\mathcal{F}_{t-}\} > 0. \tag{3.7}$$

A stochastic process $N_i(t)$ with the above property is known as a submartingale. Again, providing expectations are finite, it is only a matter of subtracting the sequentially calculated means in order to bring a submartingale under the martingale heading. This idea is made precise by the theorem of Doob-Meyer.

Doob-Meyer decomposition

For the submartingale $N_i(t)$, having associated intensity process $\alpha(t)$, we have from Equation 2.38 that $E\{dN(t)|\mathcal{F}_{t-}\} = \alpha(t)dt$. If we write $dM(t) = dN(t) - \alpha(t)dt$ then $E\{dM(t)|\mathcal{F}_{t-}\} = 0$. Thus $M(t)$ is a martingale. For the counting processes of interest to us we will always be able to integrate $\alpha(t)$ and we define $A(t) = \int_0^t \alpha(t)$. We can write

$$N_i(t) = M_i(t) + A_i(t). \tag{3.8}$$

Such a decomposition of a submartingale into the sum of a martingale and a predictable stochastic process, $A_i(t)$, is an example of a more general theorem for such decompositions known as the Doob-Meyer theorem. It can be applied to quite general submartingales, the precise conditions under which require measure-theoretic arguments. For the counting processes of interest to us in survival analysis the theorem always applies. The predictable process $A_i(t)$ is called the compensator of $N_i(t)$. In simple terms the compensator is used to make the means zero thereby producing the martingales that our theory needs.

The compensator $A_i(t)$

A counting process, $N_i(t)$, is simply a random variable indexed by t. This is the definition of a stochastic process so that $N_i(t)$ is, in particular, a stochastic process. For the majority of applications in survival analysis, $N_i(t)$ will count no further than one; at the outset, $N_i(0)$ takes the value zero and, subsequently, the value one for all times greater than or equal to that at which the event of interest occurs. But, generally, $N_i(t)$ may assume many, or all, integer values. Note that any sum of counting processes can be immediately seen to be a counting process in its own right. An illustrative example could be the number of goals scored during a soccer season by some team. Here, the indexing variable t counts the minutes from the beginning of the season. The expectation of $N_i(t)$ (which must exist given the physical constraints of the example) may vary in a complex way with t, certainly non-decreasing and with long plateau when it is not possible for a goal to be scored, for instance when no game is being played. At time $t = 0$, it might make sense to look forward to any future time t and to consider the expectation of $N_i(t)$.

As the season unfolds, at each t, depending on how the team performs, we may exceed, possibly greatly, or fall short of, the initial expectation of $N_i(t)$. As the team's performance is progressively revealed to us, the original expectations are of diminishing interest and it is clearly more useful to consider those conditional expectations in which we take account of the accumulated history at time point t. Working this out as we go along, we determine $A_i(t)$ so that $N_i(t) - A_i(t)$, given all that has happened up to time t, has zero expectation. When $\alpha_i(s)$ is the intensity function for $N_i(s)$, then

$$A_i(t) = \int_0^t \alpha_i(s)ds$$

and this important result is presented in Theorem 3.5 given immediately below.

Predictable variation process

Linear statistics of the form U described in Section 3.7, following standardization, are, not surprisingly, well approximated by standard normal variates. We will see this below using results for estimating equations and results for sums of independent, although not necessarily identical, random variables. The martingale central limit theorem can also be used in this context and, for all of our applications, it is possible to apply it in a simple form avoiding measure-theoretic arguments. Such an approach would then coincide with standard results for sums of independent random variables. In order to standardize U we will require an estimate of the variance as well as the mean. Unlike, say, Brownian motion or the Ornstein-Uhlenbeck processes mentioned in the previous chapter, where at time t we have a very simple expression for the variance, the variance of U can be complex and will clearly depend on $H(x)$. One way of addressing this question is through the use of the predictable variation process. We know from the above that:

$$E\{dN(t)|\mathcal{F}_{t-}\} = \alpha(t)dt\,, \quad E\{dM(t)|\mathcal{F}_{t-}\} = 0.$$

Conditional upon $\mathcal{F}_{t-}$, we can view the random variable $dN(t)$ as a Bernoulli $(0,1)$ having mean $\alpha(t)dt$ and variance given by $\alpha(t)dt$ $\{1 - \alpha(t)dt\}$. In contrast, the random variable $dM(t)$, conditional on $\mathcal{F}_{t-}$, has mean zero and the same variance. This follows since, given $\mathcal{F}_{t-}$, $\alpha(t)$ is fixed and known. As usual, all the equations are in an infinitesimal sense, the equal sign indicating a limiting value as $dt \to 0$. In this sense $\alpha^2(t)(dt)^2$ is negligible when compared to $\alpha(t)dt$ since the ratio of the first to the second goes to zero as t goes to zero. Thus, the incremental variances are simply the same as the means, i.e., $\alpha(t)dt$. This, of course, ties in exactly with the theory for Poisson counting processes.

Definition 3.1 *The predictable variation process of a martingale $M(t)$, denoted by $\langle M \rangle = \{\langle M \rangle(t) : t \geq 0\}$ is such that*

$$d\langle M \rangle(t) = E\{[dM(t)]^2|\mathcal{F}_{t-}\} = \text{Var}\{dM(t)|\mathcal{F}_{t-}\}. \qquad (3.9)$$

The use of pointed brackets has become standard notation here and, indeed, the process is often referred to as the pointed brackets process,

Note that $\langle M \rangle$ is clearly a stochastic process and that the process is predictable and nondecreasing. It can be thought of as the sum of conditional variances of the increments of M over small time intervals partitioning $[0, t]$, each conditional variance being taken given what has happened up to the beginning of the corresponding interval. We then have the following important result:

Theorem 3.5 *Let $M_i(t) = N_i(t) - A_i(t)$ where $A_i(t) = \int_0^t \alpha_i(s) ds$. Then*

$$\langle M_i \rangle(t) = A_i(t). \tag{3.10}$$

Corollary 3.1 *Define, for all t and $i \neq j$, the predictable covariation process, $\langle M_i, M_j \rangle$, of two martingales, M_i and M_j, analogously to the above. Then*

$$\langle M_i, M_j \rangle(t) = 0. \tag{3.11}$$

The corollary follows readily if, for $i \neq j$, the counting processes $N_i(t)$ and $N_j(t)$ can never jump simultaneously. In this case the product $dN_i(t)dN_j(t)$ is always equal to zero. Thus, the conditional covariance between $dN_i(t)$ and $dN_j(t)$ is $-\alpha_i(t)dt \cdot \alpha_j(t)dt$.

Stochastic integrals

The concept of a stochastic integral is very simple; essentially we take a Riemann-Stieltjes integral, from zero to time point t, of a function which, at the outset when $t = 0$ and looking forward, would be random. Examples of most immediate interest to us are: $N(t) = \int_0^t dN(s)$, $A(t) = \int_0^t dA(s)$ and $M(t) = \int_0^t dM(s)$. Of particular value are integrals of the form $\int_0^t H(s)dM(s)$ where $M(s)$ is a martingale and $H(s)$ a predictable function. By predictable we mean that if we know all the values of $H(s)$ for s less than t then we also know $H(t)$, and this value is the same as the limit of $H(s)$ as $s \to t$ for values of s less than t.

The martingale transform theorem provides a tool for carrying out inference in the survival context. Many statistics arising in practice will be of a form U described in the section on estimating equations just below. For these the following result will find immediate application:

Theorem 3.6 *Let M be a martingale and H a predictable stochastic process. Then M^* is also a martingale where it is defined by:*

$$M^*(t) = \int_0^t H(s)dM(s). \tag{3.12}$$

Corollary 3.2 *The predictable variation process of the stochastic process $M^*(t)$ can be written*

$$\langle M^* \rangle(t) = \int_0^t H(s)^2 d\langle M \rangle(s) = \int_0^t H^2(s)dA(s). \qquad (3.13)$$

There is a considerable theory for stochastic integrals, much of it developed in the econometric and financial statistical literature. For our purposes and for all the tests that have been developed in the framework of counting processes for proportional hazards models, the above theorem and corollary are all that is needed. A very considerable array of test procedures come directly under this heading. Many modern approaches to survival analysis lean on the theory of stochastic integrals in order to carry out inference. The main approaches here are sometimes different, based instead on the main theorem of proportional hazards regression, Donsker's theorem, and known results concerning Brownian motion and functions of Brownian motion. Behind both approaches are the ideas relating to the limits of sums of conditionally independent increments. In some situations the resulting statistics that we use are identical. Chapter 10 deals specifically with inference for the proportional hazards model based on the ideas described in this section.

Central limit theorem for martingale processes

The hard part of the work in obtaining a large sample result for stochastic integrals is to find a convergence in probability result for the variation process $\langle M \rangle(t)$. If this limit exists then we write it as $A(t)$. The result can be summarized in a theorem which depends on two conditions.

Theorem 3.7 *Suppose we have the following two conditions:*

1. *As n increases without bound, $\langle M \rangle(t)$ converges in probability to $A(t)$,*

2. *As n increases, the jumps in $M(t)$ tend to zero.*

Then, the martingale $M(t)$ converges to a Gaussian process with mean zero and variance $A(t)$.

Added conditions can make it easier to obtain the first one of these conditions and, as a result, there are a number of slightly different versions of these two criteria. The multivariate form has the same structure. In practical situations, we take $M(\infty)$ to be $\mathcal{N}(0, \sigma^2)$ where we estimate σ^2 by $\langle M \rangle(\infty)$.

Censoring and at-risk functions

The counting process structure does well on an intuitive level in dealing with the concept of censoring and the concept of being at risk. We will only observe the actual counts but we can imagine that the probability, more precisely the intensity when referring to infinitely small time periods, can change in complex ways through time. In particular, there may be time periods when, although a key event of interest may occur, we are unable to observe it because of some censoring phenomenon. As an example, allow $N(t)$ to count some event of interest and define $Y(t)$ to take the value zero when t lies in the semi-closed interval $(a, b]$, (when we are unable to observe the event of interest) and the value one otherwise. The counting process $N^*(t)$ where,

$$N^*(t) = \int_0^t Y(s) dN(s), \tag{3.14}$$

counts observable events. If the censoring does not modify the compensator, $A(t)$, of $N(t)$, then $N(t)$ and $N^*(t)$ have the same compensator. The difference, $M^*(t) = N^*(t) - A(t)$ would typically differ from the martingale $M(t)$ but would nonetheless still be a martingale in its own right. In addition, it is easily anticipated how we might go about tackling the much more complex situation in which the censoring would not be independent of the failure mechanism. Here, the compensators for $N^*(t)$ and $N(t)$ do not coincide. For this more complex case, we would need some model, $A^*(t)$, for the compensator of $N^*(t)$ in order that $M^*(t) = N^*(t) - A^*(t)$ would be a martingale.

The most common and the simplest form of the at-risk indicator $Y(t)$ is one where it assumes the value one at $t = 0$, retaining this value until censored or failed, beyond which time point it assumes the value zero. When dealing with n individuals, and n counting processes, we can write $\bar{N}(t) = \sum_{i=1}^n N_i(t)$ and use the at-risk indicator to denote the risk set. If $Y_i(t)$ refers to individual i, then $\bar{Y}(t) = \sum_{i=1}^n Y_i(t)$ is the risk set at time t. The compensator for $N_i(t)$ is $\alpha_i(t) = Y_i(t)\lambda_i(t)$, where $\lambda_i(t)$ is the hazard for subject i, written simply as $\lambda(t)$ in the case of i.i.d. replications. Then, the compensator, $\bar{A}(t)$, for $\bar{N}(t)$ is:

$$\bar{A}(t) = \int_0^t \{\textstyle\sum_{i=1}^n Y_i(s)\}\lambda(s)ds = \int_0^t \bar{Y}(s)\lambda(s)ds.$$

The intensity process for $\bar{N}(t)$ is then given by $\bar{Y}(t)\lambda(t)$. The multiplicative intensity model (Aalen 1978) has as its cornerstone the product of the fully observable quantity $\bar{Y}(t)$ and the hazard rate, $\lambda(t)$ which, typically, will involve unknown model parameters. In testing specific hypotheses we might fix some of these parameters at particular population values, most often the value zero.

Nonparametric statistics

The multiplicative intensity model just described and first recognized by Aalen (1978) allows a simple expression, and simple inference, for a large number of nonparametric statistics that have been used in survival analysis over the past half century. We return to these when looking at inference for the proportional hazards model based on counting processes and stochastic integrals in Chapter 10. Martingales are immediate candidates for forming an estimating equation with which inference can be made on unknown parameters in the model. In the next section we provide some general discussion on estimating equations. For our specific applications, these estimating equations will almost always present themselves in the form of a martingale.

3.7   Estimating equations

Most researchers, together with a large section of the general public, even if uncertain as to what the study of statistics entails, will be familiar with the concept, if not the expression itself, of the type $\bar{T} = n^{-1}\sum_{i=1}^n T_i$. The statistician may formulate this in somewhat more abstract terms, stating that; $\bar{T} = n^{-1}\sum_{i=1}^n T_i$ is a solution to the linear estimating equation for the parameter μ, the population mean of the random variable T, in terms of the n i.i.d. replicates of T. The estimating equation is simply $\mu - n^{-1}\sum_{i=1}^n T_i = 0$. This basic idea is very useful in view of the potential for immediate generalization.

The most useful approach to analyzing data is to postulate plausible models that may approximate some unknown, most likely very complex, mechanism generating the observations. These models involve unknown parameters and we use the observations, in conjunction with an estimating equation, to replace the unknown parameters by estimates. Deriving "good" estimating equations is a sizeable topic whose surface we only need to scratch here. We appeal to some general principles, the most common of which are very briefly recalled below, and note that, unfortunately, the nice simple form for the estimating equation for μ just above is more an exception than the rule. Estimating equations are mostly nonlinear and need to be solved by numerical algorithms. Nonetheless, an understanding of the linear case is more helpful than it may at first appear since solutions to the nonlinear case are achieved by local linearization (called also Newton-Raphson approximation) in the neighborhood of the solution. A fundamental result in the theory of estimation is described in the following theorem. Firstly, we define two important functions, $L(\theta)$ and $I(\theta)$ of the parameter θ by

$$L(\theta) = f(t_1, t_2, \ldots, t_n; \theta) \,; \qquad I(\theta) = -\partial^2 \log L(\theta)/\partial\theta^2. \quad (3.15)$$

We refer to $L(\theta)$ as the observed likelihood, or simply just the likelihood (note that, for $n = 1$, the expected log-likelihood is the negative of the entropy, also called the information). When the observations T_i, $i = 1, \ldots, n$, are independent and identically distributed then we can write $L(\theta) = \prod_{i=1}^{n} f(t_i; \theta)$ and $\log L(\theta) = \sum_{i=1}^{n} \log f(t_i; \theta)$. We refer to $I(\theta)$ as the information in the sample. Unfortunately the negative of the entropy is also called the information (the two are of course related, both quantifying precision is some sense). The risks of confusion are small given that the contexts are usually distinct. The function $I(\theta)$ is random because it depends on the data and reaches a maximum in the neighborhood of θ_0 since this is where the slope of the log likelihood is changing the most quickly.

Theorem 3.8 *For a statistic T we can write the following;*

$$\mathrm{Var}(T) \geq \{\partial E(T)/\partial\theta\}^2 / E\{I(\theta)\}.$$

This inequality is called the Cramer-Rao inequality (Cox and Hinkley 1974, page 254). When T is an unbiased estimate of θ then $\partial E(T)/\partial\theta = 1$ and $\mathrm{Var}(T) \geq 1/E\{I(\theta)\}$. The quantity $1/E\{I(\theta)\}$ is called the Cramer-Rao bound. Taking the variance as a measure of preciseness then, given unbiasedness, we prefer the estimator T that has the smallest variance. The Cramer-Rao bound provides the best that we can do in this sense and, below, we see that the maximum likelihood estimator achieves this for large samples, i.e., the variance of the maximum likelihood estimator becomes progressively closer to the bound as sample size increases without limit.

Basic equations

For a scalar parameter θ_0 we will take some function $U(\theta)$ that depends on the observations as well as θ. We then use $U(\theta)$ to obtain an estimate $\hat{\theta}$ of θ_0 via an estimating equation of the form

$$U(\hat{\theta}) = 0. \qquad (3.16)$$

This is too general to be of use and so we limit the class of possible choices of $U(\cdot)$. We require the first two moments of U to exist in which case, without loss of generality we can say

$$E\,U(\theta_0) = 0, \quad \operatorname{Var} U(\theta_0) = \sigma^2. \qquad (3.17)$$

Two widely used methods for constructing $U(\theta)$ are described below. It is quite common that U be expressible as a sum of independent and identically distributed contributions, U_i, each having a finite second moment. An immediate application of the central limit theorem then provides the large sample normality for $U(\theta_0)$. For independent but nonidentically distributed U_i, it is still usually not difficult to verify the Lindeburg condition and apply the central limit theorem for independent sums. Finally, in order for inference for U to carry over to $\hat{\theta}$, some further weak restrictions on U will be all we need. These require that U be monotone and continuous in θ and differentiable in some neighborhood of θ_0. This is less restrictive than it sounds. In practice it means that we can simply apply the mean value theorem (2.2) whereby:

Corollary 3.3 *For any $\epsilon > 0$, when $\hat{\theta}$ lies in an interval $(\theta_0 - \epsilon, \theta_0 + \epsilon)$ within which $U(\theta)$ is continuously differentiable, then there exists a real number $\xi \in (\theta_0 - \epsilon, \theta_0 + \epsilon)$ such that*

$$U(\hat{\theta}) = U(\theta_0) - (\hat{\theta} - \theta_0)I(\xi).$$

This expression is useful for the following reasons. A likelihood for θ will, with increasing sample size, look more and more normal. As a consequence, $I(\xi)$ will look more and more like a constant, depending only on sample size, and not ξ itself. This is useful since ξ is unknown. We approximate $I(\xi)$ by $I(\hat{\theta})$. We can then express $\hat{\theta}$ in terms of approximate constants and $U(\hat{\theta})$ whose distribution we can approximate by a normal distribution.

Finding equations

The guiding principle is always the same, that of replacing unknown parameters by values that minimize the distance between empirical (observed) quantities and their theoretical (model-based) equivalents. The large range of potential choices stem from two central observations: (1) there can be many different definitions of distance (indeed, the concept of distance is typically made wider than the usual mathematical one which stipulates that the distance between a and b must be the same as that between b and a) and (2) there may be a number of competing empirical and theoretical quantities to consider. To make this more concrete, consider a particular situation in which the mean is modelled by some parameter θ such that $E_\theta(T)$ is monotone in θ. Let's say that the true mean $E(T)$ corresponds to the value $\theta = \theta_0$. Then the mean squared error, variance about a hypothesized $E_\theta(T)$, let's say $\sigma^2(\theta)$, can be written as

$$\sigma^2(\theta) = E\{T - E_\theta(T)\}^2 = E\{T - E_\theta(T)\}^2 + \{E_\theta(T) - E_{\theta_0}(T)\}^2.$$

The value of θ that minimizes this expression is clearly θ_0. An estimating equation derives from minimizing the empirical equivalent of $\sigma^2(\theta)$. Minimum chi-squared estimates have a similar motivation. The idea is to bring, in some sense, via our choice of parameter value, the hypothesized model as close as possible to the data. Were we to index the distribution F by this parameter, calling this say $F_\theta(t)$, we could re-express the definition for $D_n(\theta)$ given earlier as

$$D_n(\theta) = \sup_{0 \leq t \leq \infty} |F_n(t) - F_\theta(t)|.$$

Minimizing $D_n(\theta)$ with respect to θ will often provide a good, although not necessarily very tractable, estimating equation. The same will apply to C_n and related expressions such as the Anderson-Darling statistic. We will see later that the so-called partial likelihood estimate for the proportional hazards model can be viewed as an estimate arising from an empirical process. It can also be seen as a method of moments estimate and closely relates to the maximum likelihood estimate. Indeed, these latter two methods of obtaining estimating equations are those most commonly used and, in particular, the ones given the closest attention in this work. It is quite common for different techniques, and even contending approaches from within the same technique, to lead to different estimators. It is not always easy to argue in favor of one over the others.

Other principles can sometimes provide guidance in practice, the principle of efficiency holding a strong place in this regard. The idea of efficiency is to minimize the sample size required to achieve any given precision or, equivalently, to find estimators having the smallest variance. However, since we are almost always in situations where our models are only approximately correct, and, on occasion, even quite far off, it is more useful to focus attention on other qualities of an estimator. How can it be interpreted when the data are generated by a mechanism much wider than that assumed by the model? How useful is it to us in our endeavor to build predictive models, even when the model is, at least to some extent, incorrectly specified. This is the reality of modeling data and efficiency, as an issue for us to be concerned with, does not take us very far. On the other hand, estimators that have demonstrably poor efficiency, when model assumptions are correct, are unlikely to redeem themselves in a broader context and so it would be a mistake to dismiss efficiency considerations altogether even though they are rather limited.

Method of moments

This very simple method derives immediately as an application of the Helly-Bray theorem (Theorem 2.3). The idea is to equate population moments to empirical ones obtained from the observed sample. Given that $\mu = \int x dF(x)$, the above example is a special case since, we can write $\bar{\mu} = \bar{x} = \int x dF_n(x)$. Properties of the estimate can be deduced from the well-known properties of $F_n(x)$ as an estimate of $F(x)$ (see Section 3.5). For the broad exponential class of distributions, the method of moments estimator, based on the first moment, coincides with the maximum likelihood estimator recalled below. In the survival context we will see that the so-called partial likelihood estimator can be viewed as a method of moments estimator. The main difficulty with method of moments estimators is that they are not uniquely defined for any given problem. For example, suppose we wish to aim to estimate the rate parameter λ from a series of observations, assumed to have been generated by a Poisson distribution. We can use either the empirical mean or the empirical variance as an estimate for λ. Typically they will not be the same. Indeed we can construct an infinite class of potential estimators as linear combinations of the two.

Maximum likelihood estimation

A minimum chi-square estimator for θ derives by minimizing an expression of variance. From Section 2.8 it may appear equally natural to minimize an estimate of the entropy as a function of θ, i.e., maximize the observed information as a function of θ. We write the information as $V(\theta)$ where

$$V(\theta) = E \log f(T; \theta).$$

Given the observations, $T_1, \ldots, T_n$, we replace the unknown function $V(\theta)$ by $\bar{V}(\theta) = n^{-1} \sum_{i=1}^{n} \log f(T_i, \theta)$. The maximization is easily accomplished when the parameter or parameter vector θ can only assume some finite number of discrete values. It is then sufficient to examine all the cases and select θ such that $\bar{V}(\theta)$ is maximized. For all the models under consideration here we can assume that $V(\theta)$ and $\bar{V}(\theta)$ are continuous smooth functions of θ. By smooth we mean that the first two derivatives, at least, exist for all values of θ. This is not at all a restrictive assumption and models that do not have such differentiability properties can nearly always be replaced by models that do via useful reparameterization. For instance, there are cases where a model, defined for all positive real θ, may break down at $\theta = 0$, the entropy not being differentiable at that point, whereas under the reparameterization $\theta = \exp(\alpha)$ for α defined over the whole real line, the problem disappears.

Two fundamental theorems and three corollaries enable us to appreciate the great utility of the maximum likelihood approach. All that we need are "suitable regularity conditions." We return to these immediately below. Assuming these conditions (a valid assumption for all the models in this book), we have a number of important results concerning $V(\theta)$ and consistent estimates of $V(\theta)$.

Theorem 3.9 *Viewed as a function of θ, $V(\theta)$ satisfies*

$$\left\{ \frac{\partial V(\theta)}{\partial \theta} \right\}_{\theta = \theta_0} = E \left\{ \frac{\partial \log f(T; \theta)}{\partial \theta} \right\}_{\theta = \theta_0} = 0. \qquad (3.18)$$

Note the switching of the operations, integration and differentiation, in the above equations. In many texts describing the likelihood method it is common to only focus on the second part of the equation. It helps understanding to also keep the first part of the equation in mind since this will enable us to establish the solid link between the information measure and likelihood. Having divided by sample size we should view the log-likelihood as an empirical estimate of $V(\theta)$. The law of large numbers alone would provide us with a convergence result but we can do better, in terms of fitting in with elementary results for estimating equations, by assuming some smoothness in $V(\theta)$ and a consistent estimator, $\bar{V}(\theta)$, as functions of θ. More precisely:

Corollary 3.4 *Let $\bar{V}(\theta)$ be a consistent estimate of $V(\theta)$. Then $\{\partial\bar{V}(\theta)/\partial\theta\}_{\theta=\theta_0}$ converges, with probability one, to zero.*

The expression "suitable regularity conditions" is not very transparent, and for those lacking a good grounding in analysis, or simply a bit rusty, it might be less than illuminating. We dwell on it for a moment since it appears frequently in the literature. We require continuity of the function $V(\theta)$, at least for general situations, in order to be able to claim that as $V(\theta)$ approaches $V(\theta_0)$ then θ approaches θ_0. This is nearly enough, although not quite. Differentiability is a stronger requirement since we can see that a function that is differentiable at some point must also be continuous at that same point. The converse is not so and can be seen immediately in a simple example, $y = |x|$, a function that is continuous everywhere but not differentiable at the origin. We need the differentiability condition in order to obtain the estimating equation and just a tiny bit more, the tiny bit more not being easily described but amounting to authorizing the switching of the processes of integration and differentiation. Such switching has to take place in order to be able to demonstrate the validity of the above theorem, and the one just below. All of this is summarized by the expression "suitable regularity conditions" and the reader need not worry about them since they will hold in all the practical cases of interest to us. The main result follows as a further corollary:

Corollary 3.5 *Suppose that $U(\alpha) = \{\partial\bar{V}(\theta)/\partial\theta\}_{\theta=\alpha}$ and that, for sample size n, $\hat{\theta}_n$ is the solution to the equation $U(\theta) = 0$. Then, if $\bar{V}(\theta)$ is consistent for $V(\theta)$, $\hat{\theta}_n$ converges with probability one to θ_0.*

This is almost obvious, and certainly very intuitive, but it provides a solid foundation to likelihood theory. The result is a strong and useful one, requiring only the so-called regularity conditions referred to above and that $\bar{V}(\theta)$ be consistent for $V(\theta)$. Independence of the observations or that the observations arise from random sampling is not appealed to or needed. However, when we do have independent and identically distributed observations then, not only do our operations become very straightforward, we also can appeal readily to central limit theorems, initially for the left-hand side of the estimating equation and, then, by extension, to a smooth function of the estimating equation. The smooth function of interest is of, course, $\hat{\theta}_n$ itself.

Corollary 3.6 *Suppose that $T_1, \ldots, T_n$ are independent identically distributed random variables having density $f(t; \theta_0)$. Then $\hat{\theta}_n$ converges with probability one to θ_0 where $\hat{\theta}_n$ is such that $U(\hat{\theta}_n) = 0$, where $U(\theta) = \sum_{i=1}^n U_i(\theta)$ and where $U_i(\alpha) = \{\partial \log f(T_i; \theta)/\partial \theta\}_{\theta=\alpha}$.*

We can say much more about $\hat{\theta}_n$. A central limit theorem result applies immediately to $U(\theta_0)$ and, via Slutsky's theorem, we can then also claim large sample normality for $U(\hat{\theta}_n)$. By expressing $\hat{\theta}_n$ as a smooth (not necessarily explicit) function of U we can also then claim large sample normality for $\hat{\theta}_n$. The fact that $U(\hat{\theta}_n)$ (having subtracted off the mean and divided by its standard deviation) will converge in distribution to a standard normal and that a smooth function of this, notably $\hat{\theta}_n$, will do the same, does not mean that their behavior can be considered to be equivalent. The result is a large sample one, i.e., as n tends to infinity, a concept that is not so easy to grasp, and, for finite samples, behavior will differ. Since U is a linear sum we may expect the large sample approximation to be more accurate, more quickly, than for $\hat{\theta}_n$ itself. Inference is often more accurate if we work directly with $U(\hat{\theta}_n)$, exploiting the monotonicity of $U(\cdot)$, and inverting intervals for $U(\hat{\theta}_n)$ into intervals for $\hat{\theta}_n$. In either case we need some expression for the variance and this can be obtained from the second important theorem:

Theorem 3.10 *Viewed as a function of θ, $V(\theta)$ satisfies:*

$$\left(\frac{\partial V(\theta)}{\partial \theta}\right)^2_{\theta=\theta_0} = \left(\frac{-\partial^2 V(\theta)}{\partial \theta^2}\right)_{\theta=\theta_0} = E\left(\frac{-\partial^2 \log f(T; \theta)}{\partial \theta^2}\right)_{\theta=\theta_0}. \quad (3.19)$$

As in the previous theorem, note the switching of the operations of integration (expectation) and differentiation. Since $E\, U(\theta_0) = E\{\partial \log f(T; \theta)/\partial \theta\}_{\theta=\theta_0} = 0$, then, from the above theorem, $\mathrm{Var}\, U(\theta_0) = E\{\partial^2 \log f(T; \theta)/\partial \theta^2\}_{\theta=\theta_0}$. In practical applications we approximate the variance expression by replacing θ_0 by its maximum likelihood estimate. It is also interesting to note that the above inequality will usually break down when the model is incorrectly specified and that, in some sense, the further away is the assumed model from that which actually generates the data, then the greater the discrepancy between the two quantities will be. This idea can be exploited to construct goodness-of-fit tests or to construct more robust estimators.

3.8 Inference using resampling techniques

Bootstrap resampling

The purpose of bootstrap resampling is twofold: (1) to obtain more accurate inference, in particular more accurate confidence intervals, than is available via the usual normal approximation, and (2) to facilitate inference for parameter estimators in complex situations. A broad discussion including several challenging applications is provided by Politis (1998). Here we will describe the basic ideas in so far as they are used for most problems arising in survival analysis. Consider the empirical distribution function $F_n(t)$ as an estimate for the unknown distribution function $F(t)$. The observations are T_1, T_2, ..., T_n. A parameter of interest, such as the mean, the median, some percentile, let's say θ, depends only on F. This dependence can be made more explicit by writing $\theta = \theta(F)$. The core idea of the bootstrap can be summarized via the simple expression $\tilde{\theta} = \theta(F_n)$ as an estimator for $\theta(F)$.

Taking infinitely many i.i.d. samples, each of size n, from F would provide us with the exact sampling properties of any estimator $\tilde{\theta} = \theta(F_n)$. If, instead of taking infinitely many samples, we were to take a very large number, say B, of samples, each sample again of size n from F, then this would provide us with accurate approximations to the sampling properties of $\tilde{\theta}$, the errors of the approximations diminishing to zero as B becomes infinitely large. Since F is not known we are unable to carry out such a prescription. However, we do have available our best possible estimator of $F(t)$, the empirical distribution function $F_n(t)$. The bootstrap idea is to sample from $F_n(t)$, which is known and available, instead of from $F(t)$ which, apart from theoretical investigations, is typically unknown and unavailable.

Empirical bootstrap distribution

The conceptual viewpoint of the bootstrap is to condition on the observed T_1, ..., T_n and its associated empirical cumulative distribution function $F_n(t)$, thereafter treating these quantities as though they were a population of interest, rather than a sample. From this "population" we can draw samples with replacement, each sample having size n. We repeat this whole process B times where B is a large number, typically in the thousands. Each sample is viewed as an i.i.d. sample from $F_n(t)$. The ith resample of size n can be written $T_{1i}^*, T_{2i}^*, ..., T_{ni}^*$ and has empirical distribution $F_n^{*i}(t)$. For any parameter of interest θ, the mean, median coefficient of variation for example, it is helpful to remind ourselves of the several quantities of interest, $\theta(F)$, $\theta(F_n)$, $\theta(F_n^{*i})$ and $F_B(\theta)$, the significance of each of these quantities needing a little explanation. First, $\theta(F)$ is simply the population quantity of interest. Second, $\theta(F_n)$ is this same quantity defined with respect to the empirical distribution of the data $T_1, ... , T_n$. Third, $\theta(F_n^{*i})$ is again the same quantity defined with respect to the ith empirical distribution of the resamples $T_{1i}^*, T_{2i}^*, ..., T_{ni}^*$. Finally, $F_B(\theta)$ is the bootstrap distribution of $\theta(F_n^{*i})$, i.e., the empirical distribution of $\theta(F_n^{*i})$ $(i = 1, \ldots, B)$.

To keep track of our asymptotic thinking we might note that, as $B \to \infty$, $\int u dF_B(u)$ converges in probability to $\theta(F_n)$ and, as $n \to \infty$, $\theta(F_n)$ converges in probability to $\theta(F)$. Thus, there is an important conceptual distinction between F_B and the other distribution functions. These latter concern the distribution of the original observations or resamples of these observations. F_B itself deals with the distribution of $\theta(F_n^{*i})$ $(i = 1, \ldots, n)$ and therefore, when our focus of interest changes from one parameter to another, from say θ_1 to θ_2 the function F_B will be generally quite different. This is not the case for F, F_n, and F_n^{*i} which are not affected by the particular parameter we are considering. Empirical quantities with respect to the bootstrap distribution, F_B are evaluated in a way entirely analogous to those evaluated with respect to F_n. For example,

$$\text{Var}\{\theta(F_n)\} = \sigma_B^2 = \int u^2 dF_B(u) - \left(\int u dF_B(u)\right)^2, \qquad (3.20)$$

where it is understood that the variance operator, $\text{Var}()$ is with respect to the distribution $F_B(t)$. Of greater interest in practice is the fact that $\text{Var}\{\theta(F_n)\}$, where $\text{Var}()$ is with respect to the distribution $F_B(t)\}$, can be used as an estimator of $\text{Var}\{\theta(F_n)\}$, where $\text{Var}(\cdot)$ is with respect to the distribution $F(t)$.

Bootstrap confidence intervals

For the normal distribution the standardized percentiles z_α are defined from $\Phi(z_\alpha) = \alpha$. Supposing that $\mathrm{Var}\,\theta(F_n) = \sigma^2$, the variance operator being taken with respect to F, the scaled percentiles Q_α are then given by $Q_\alpha = \sigma z_\alpha$, leading to a normal approximation for a confidence interval for θ as

$$I_{1-\alpha}(\theta) = \{\theta(F_n) - Q_{1-\alpha/2},\ \ \theta(F_n) - Q_{\alpha/2}\} \qquad (3.21)$$

which obtains from a rearrangement of the expression $\Pr\left[\sigma z_{\alpha/2} < \theta(F_n) - \theta < \sigma z_{1-\alpha/2}\right] = 1 - \alpha$. Since σ^2 is not generally available we would usually work with σ_B^2. Instead of using the normal approximation it is possible to define Q_α differently, directly from the observed bootstrap distribution F_B. We can define Q_α via the equation $F_B(Q_\alpha) = \alpha$. In view of the finiteness of B (and also n) this equation may not have an exact solution and we will, in practice, take the nearest point from $F_B^{-1}(\alpha)$ as Q_α. The values of $Q_{\alpha/2}$ and $Q_{1-\alpha/2}$ are then inserted into equation (3.21). Such intervals are referred to as bootstrap "root" intervals. The more common approach to constructing bootstrap confidence intervals is, however, slightly different and has something of a Bayesian or fiducial inference flavor to it. We simply tick off the percentiles, $Q_{\alpha/2}$ and $Q_{1-\alpha/2}$ and view the distribution F_B as our best estimate of a distribution for θ. The intervals are then written as

$$I_{1-\alpha}(\theta) = \{\theta(F_n) + Q_{\alpha/2},\ \ \theta(F_n) + Q_{1-\alpha/2}\}. \qquad (3.22)$$

These intervals are called percentile bootstrap confidence intervals. Whenever the distribution F_B is symmetric then the root intervals and the percentile intervals coincide since $Q_{\alpha/2} + Q_{1-\alpha/2} = 0$. In particular, they coincide if we make a normal approximation.

Accuracy of bootstrap confidence intervals

Using theoretical tools for investigating statistical distributions, the Edgeworth expansion in particular, it can be shown that the accuracy of the three types of interval described above is the same. The argument in favor of the bootstrap is then not compelling, apart from the fact that they can be constructed in cases where variance estimates may not be readily available. However, it is possible to improve on the accuracy of both the root and the percentile intervals. One simple, albeit slightly laborious, way to accomplish this is to consider studentized methods. By these we mean, in essence, that the variance in the "population" F_n is not considered fixed and known for each subsample but is estimated several, precisely B, times across the B bootstrap samples.

To get an intuitive feel for this it helps to recall the simple standard set-up we work with in the comparison of two estimated means, $\bar{X}_1$ and $\bar{X}_2$ where, when the variance σ^2 is known we use as test statistic, $(\bar{X}_1 - \bar{X}_2)/\sigma$, referring the result to standard normal tables. When σ^2 is unknown and replaced by its usual empirical estimate s^2 then, for moderate to large samples, we do the same thing. However, for small samples, in order to improve on the accuracy of inference, we appeal to the known distribution of $(\bar{X}_1 - \bar{X}_2)/s$ when the original observations follow a normal distribution. This distribution was worked out by Student and is his so-called Student's t distribution. For the i th bootstrap sample our estimate of the variance is

$$\text{Var}\,\{\theta(F_n^{*i})\} = \sigma_{*i}^2 = \int u^2 dF_n^{*i}(u) - \left(\int u dF_n^{*i}(u) \right)^2 \qquad (3.23)$$

and we then consider the standardized distribution of the quantity, $\theta(F_n^{*i})/\sigma_{*i}$. The essence of the studentized approach, having thus standardized, is to use the bootstrap sampling force to focus on the higher-order questions, those concerning bias and skewness in particular. Having, in some sense, spared our bootstrap resources from being dilapidated to an extent via estimation of the mean and variance, we can make real gains in accuracy when applying the resulting distributional results to the statistic of interest. For most day-to-day situations this is probably the best approach to take. It is computationally intensive (no longer a serious objection) but very simple conceptually. An alternative approach, with the same ultimate end in mind, is not to standardize the statistic but to make adjustments to the derived bootstrap distribution, the adjustments taking into account any bias and skewness. These lead to the so-called bias corrected, accelerated intervals (often written as BC_a intervals).

3.9 Explained variation

Although widely known and quoted for linear regression models, the concept of explained variation does not appear to be have been developed elsewhere as a concept in its own right, aside from ad hoc modifications for specific applications. In this section we present some general ideas that will be useful later on. The principal objective in studying the concept of explained variation is to derive suitable tools that can quantify the predictive value of any given model. Consider a pair of random variables (T, Z) defined on $\mathcal{T} \times \mathcal{Z}$. Let $F(t)$, $G(z)$, $F(t|z)$ and $G(z|t)$ be marginal and conditional distributions respectively. Using the device of double expectation it is almost immediate to show that

$$\text{Var}(T) = E_{\mathcal{Z}}\{\text{Var}_{\mathcal{T}}(T|Z)\} + \text{Var}_{\mathcal{Z}}\{E_{\mathcal{T}}(T|Z)\}, \qquad (3.24)$$

where the subscripts indicate the order of conditioning. It is common practice to drop these subscripts, leaving the order in which expectations and variances are taken to be implicit. We can view $\sigma_T^2(Z) = E_{\mathcal{Z}}\{\text{Var}_{\mathcal{T}}(T|Z)\}$ to represent the residual noise, i.e., the average variance after having taken into account the variable Z. The greater the dispersion of $E_{\mathcal{T}}(T|Z)$ across the distribution of Z, then the stronger is the signal. We can then call $\text{Var}_{\mathcal{Z}}\{E_{\mathcal{T}}(T|Z)\}$ the signal. So that we can write the decomposition as

$$\text{Var}(T) = \text{residual noise} + \text{signal},$$

the noise not depending on Z for the classical linear model but more generally needing to be averaged over the distribution of Z. We then have an obvious definition for Ω^2,

$$\Omega_T^2(Z) = \frac{\text{signal}}{\text{signal+residual noise}} = \frac{\text{Var}(T) - \sigma_T^2(Z)}{\text{Var}(T)} \qquad (3.25)$$

as the amount of variation in T explained by conditioning upon Z. We use the notation $\Omega_T^2(Z)$ where the subscript indicates the variable that we are trying to explain, the argument Z then indicating the variable being used to explain the variability in T. The quantity $\Omega_Z^2(T)$ is defined by interchanging the symbols and, interestingly, in the case of a bivariate normal pair (T, Z) it is easily shown that $\Omega_T^2(Z)$ and $\Omega_Z^2(T)$ are the same quantity. These definitions make no distributional assumptions on Z and T and we can deduce the following properties;

Properties of Ω^2

1. When there is no reduction in variance by conditioning upon Z then $\text{Var}(T) = E\{\text{Var}(T|Z)\}$ and $\Omega^2 = 0$.

2. If, given $Z = z$, T assumes some given value with probability one, then $\text{Var}(T|Z) = 0$ and $\Omega^2 = 1$.

3. Intermediary values of Ω^2 provide an ordering of predictive strength for a normal model directly in terms of symmetric prediction intervals and for other situations by virtue of the Chebyshev inequality. The greater the reduction in variance, i.e., the more of the overall variance that can be explained, then the greater the predictability. A situation where $\Omega^2 = 0.4$ has, in this precise sense, explained twice as much of the variance as a situation in which $\Omega^2 = 0.2$. This is, of course, a very important, and quite basic property. It is not shared by a number of suggested measures of "explained variation" in the literature in which, not only are we not able to claim that a value of 0.4 represents twice as much explained variance as a value 0.2, but, for certain proposed coefficients, it is not even clear that the latter corresponds to lesser predictive power than the former.

As an illustration, take T to be a Bernoulli variate with parameter 0.5, $\Pr(Z = 1) = 0.5$, where $E(T|Z = 0) = 0.4$ and $E(T|Z = 1) = 0.6$. We find $\Omega^2 = .04$. In this situation a knowledge of Z, although predictive, tells us relatively little about the probability of seeing a 0 or a 1 for T. This corresponds to our intuition. If, instead of considering individual observations on T, we consider groups of 5, the outcomes now being the successes from 0 to 5, then we find $\Omega^2 = .45$. In this case a knowledge of Z is much more informative and, indeed, as the group size becomes larger the information conveyed by Z increases, Ω^2 tending to one as groups size becomes large without bound. Thus, knowing the value of Z does not help much in predicting binary T. We can do much better in predicting T here when T is binomial having a range of counts as outcomes.

Empirical estimates of Ω^2

In order to evaluate Ω^2 in Equation 3.25, apart from $\text{Var}(T)$ we need

$$E\{\text{Var}(T|Z)\} = \int_T \int_Z \left\{ t - \int_T t dF(t|z) \right\}^2 dF(t|z) dG(z). \quad (3.26)$$

Consistent estimates for $\Omega_T^2(Z)$ follow if we can consistently estimate $F(t|z)$ and $G(z)$. Given $(t_i, z_i; i = 1, \ldots, n)$, (i.i.d.) it is only then necessary to replace $F(t)$, $G(z)$ and $F(t|z)$ by the empirical estimates $F_n(t)$, $G_n(z)$ and $F_n(t|z)$ to obtain an estimate; let's call it R^2. By virtue of the Helly-Bray theorem R^2 will provide a consistent estimate of Ω^2. Note also that a straightforward application of Bayes formula enables us to use an alternate expression for $E\{\text{Var}(T|Z)\}$ since

$$E\{\text{Var}(T|Z)\} = \int_T \int_{\mathcal{Z}} \left\{ t - \frac{\int_T ug(z|u)dF(u)}{\int_T g(z|u)dF(u)} \right\}^2 dG(z|t)dF(t). \quad (3.27)$$

Consistent estimates for $\Omega_T^2(Z)$ follow if we can consistently estimate the conditional distribution $G(z|t)$ and the marginal distribution $F(t)$. In certain contexts, proportional hazards regression being one of them, these latter conditional distributions are very much easier to estimate.

Explained variation given a stochastic process $Z(t)$

The question here is the interpretation of $E\{\text{Var}(T|Z(t))\}$ when $Z(t)$ is a stochastic process. Recall that, by a stochastic process $Z(t)$, we indicate a possibly infinite collection of random variables, each indexed by t. However, T itself can be taken as being random and, in the special case where the support of T coincides with the set of indices to $Z(t)$ we can refer to the bivariate pair (T, Z). This follows immediately by virtue of the existence of the marginal distribution of T and the conditional distribution of Z given t. The marginal distribution of Z, obtained by integrating out t can be denoted either by Z or by $Z(T)$, the latter notation being conceptually useful in indicating a value from Z as arising first by selecting a value from T and then, given that $T = t$, we observe some value $Z(t)$. The distribution $F(t|z)$ can be understood via $F(t|z) = F(t|z, T > 0)$.

Rank invariance

Recalling that $\text{Var}(Z) = E\{\text{Var}(Z|T)\} + \text{Var}\{E(Z|T)\}$, we could have equally well used an alternative definition for the population parameter of explained variation in which we reverse the roles played by Z and T. This corresponds to studying the variation in Z at fixed time points $T = t$ and we write,

$$\Omega_Z^2(T) = \frac{\text{Var}(Z) - E\{\text{Var}(Z|T)\}}{\text{Var}(Z)} = \frac{\text{Var}\{E(Z|T)\}}{\text{Var}(Z)}. \quad (3.28)$$

We can make the following observations. The two definitions coincide for (T, Z) bivariate normal and, for many other models, are likely to be close in practice. The idea of "close" can be formalized via an appeal to Berry-Esseen bounds or to bivariate Edgeworth expansions in given situations. More importantly, there is a qualitative distinction between the two definitions in that, the first provides an Ω^2 that remains invariant to monotonic transformations on Z, whereas the second provides an Ω^2 that remains invariant to monotonic transformations on T. The measure of dependency Ω^2 then applies to all pairs $(T, \psi(Z))$ or $(\phi(T), Z)$, respectively, for all increasing transformations $\psi(\cdot)$ and $\phi(\cdot)$. The above formulation is then more general than that of the commonly employed Pearson coefficient, which is not invariant to monotonic transformations on either of the variables. Although, it is possible to construct a nonparametric product moment, Pearson-type, estimate of the correlation coefficient using the Fisher-Yates scores in place of the original observations. This coefficient is invariant to monotonic transformation on either Z or T.

Multivariate and partial coefficients

For the case in which Z is a vector the extension is obvious and the basic formula is essentially the same. Thus

$$\text{Var}(T) = E\{\text{Var}(T|Z)\} + \text{Var}\{E(T|Z)\}$$

which, apart from the fact that we are now viewing Z as a vector, has exactly the same form as before. We can write $E\{\text{Var}(T|Z)\}$ as $\sigma_T^2(Z_1, \ldots, Z_p)$ and, again, refer to this as the residual noise. Adding variables to the conditioning will reduce the residual noise in a very well-defined, and useful, way, described below. The signal is then $\text{Var}_Z\{E_T(T|Z)\}$ and we define;

$$\Omega_T^2(Z_1, \ldots, Z_p) = \frac{\text{Var}(T) - \sigma_T^2(Z_1, \ldots, Z_p)}{\text{Var}(T)} = \frac{\text{Var}\{E(T|Z)\}}{\text{Var}(T)}. \quad (3.29)$$

This is then the multivariate coefficient and, conceptually, the step up from the simple coefficient is all but immediate. The multivariate coefficient quantifies the amount of variation that we can explain by simultaneously taking into account several potential explanatory variables. A different, although related and equally important problem in the multivariate setting, is the idea of a partial coefficient. A partial coefficient quantifies the amount of variability explained after having, in a well-defined way, taken account of the effects of a third variable or, possibly, a vector of variables. Consider then the vector (T, Z_1, Z_2). We are interested in a partial measure of strength of association between T and Z_2 after having taken into account any effects of Z_1. For Z_1 not random and fixed at some particular value, say $Z_1 = z_1$ we have the classic breakdown:

$$\text{Var}(T|Z_1 = z_1) = E\{\text{Var}(T|Z_2, Z_1 = z_1)\} + \text{Var}\{E(T|Z_2, Z_1 = z_1)\}.$$

Clearly this breakdown can be carried out for any other value of Z_1 and, indeed, can be undertaken for all values of Z_1 over its domain of definition. Taking expectations with respect to the distribution of Z_1 we have

$$E\,\text{Var}(T|Z_1) = EE\{\text{Var}(T|Z_2, Z_1)\} + E\,\text{Var}\{E(T|Z_2, Z_1)\},$$

from which a natural definition of partial explained variation arises as

$$\begin{aligned}\Omega_T^2(Z_2|Z_1) &= \frac{E\,\text{Var}(T|Z_1) - EE\{\text{Var}(T|Z_2, Z_1)\}}{E\,\text{Var}(T|Z_1)} \\ &= \frac{E\,\text{Var}\{E(T|Z_2, Z_1)\}}{E\,\text{Var}(T|Z_1)}.\end{aligned}$$

The variance breakdown is carried out at all possible values of Z_1, considered, thereby, as fixed. In this concrete sense we have taken account of Z_1 in evaluating the variation of T explained by Z_2. The coefficient is described as a partial coefficient. Keep in mind that, as before, all of this development takes place in terms of random variables alone. No appeal, so far, is made to any specific model. The expressions hold generally. For the multivariate normal model the relationship between partial correlation and multiple correlation is well known. However, without appealing to any model, we can establish useful results linking partial coefficients to multiple coefficients.

Variance decomposition

Suppose that we have two variables of interest; Z_1 and Z_2. Considering at first only Z_1 we can write the usual decomposition as

$$\text{Var}(T) = \sigma^2(Z_1) + \text{Var}_{Z_1}\{E_T(T|Z_1)\}$$

and, from the above result for the partial coefficient, we can further decompose the noise $\sigma^2(Z_1)$ writing in the formula, instead of $\sigma^2(Z_1)$, its decomposition so that

$$\text{Var}(T) = \sigma^2(Z_1, Z_2) + E_{Z_1}\text{Var}_{Z_2}\{E_T(T|Z_1, Z_2) + \text{Var}_{Z_1}\{E_T(T|Z_1)\}\}$$

where the new residual noise, $\sigma^2(Z_1, Z_2)$ is equal to $EE\{\text{Var}(T|Z_1, Z_2)\}$. In words, by introducing a further variable Z_2 into the decomposition, the old signal provided by Z_1 alone remains unchanged. The noise is reduced in such a way that the old noise now equates to the added signal, provided by Z_2 after having already accounted for Z_1, together with a new component that corresponds to the new noise. We can summarize this by

$$\text{Var}(T) = \text{new residual noise} + \text{added signal} + \text{old signal}.$$

We can then continue this process each time we wish to add a further variable to those that have already been accounted for. Any new contribution is orthogonal in the sense that the previous signals remain unchanged. Only the residual noise is affected. Furthermore, unless the signal corresponding to an added variable is exactly zero, then the amount of explained variance must necessarily increase. For the case of p variables, $Z_1, \ldots, Z_p$, straightforward manipulation then leads to

$$\text{Var}(T) = \sigma^2(Z_1, \ldots, Z_p) + V(Z_1, \ldots, Z_p)$$

in which

$$
\begin{aligned}
V(Z_1, \ldots, Z_p) = \; & \text{Var}_{Z_1}\{E_T(T|Z_1)\} \\
& + E_{Z_1}\text{Var}_{Z_2}\{E_T(T|Z_1, Z_2)\} \\
& + E_{Z_1}E_{Z_2}\text{Var}_{Z_3}\{E_T(T|Z_1, Z_2, Z_3)\} + \cdots \\
& + E_{Z_1}E_{Z_2}\cdots E_{Z_p}\text{Var}_{Z_p}\{E_T(T|Z_1, \ldots, Z_p)\}. \quad (3.30)
\end{aligned}
$$

This decomposition provides the main result supporting the usefulness of the concept of explained variation in multivariate problems. Interpretation is clear. Note, however, that there are several ways in which the breakdown can be carried out. The interpretation of the equation is that, once we have accounted for certain variables, then further increases in explained variance can only arise if the residual noise can be further decomposed into, once again, a new noise component, together with a new signal provided by information orthogonal to that we already have.

Multivariate normal case

The special case of the multivariate normal model provides yet further insight since there turn out to be a number of interesting and useful algebraic identities relating the partial and the multiple coefficients. In particular we have

$$
\begin{aligned}
1 - \Omega_T^2(Z_1, Z_2, \dots, Z_p) = {} & \{1 - \Omega_T^2(Z_p | Z_1, \dots, Z_{p-1})\} \\
& \times \{1 - \Omega_T^2(Z_{p-1} | Z_1, \dots, Z_{p-2})\} \times \cdots \\
& \times \{1 - \Omega_T^2(Z_2 | Z_1)\} \times \{1 - \Omega_T^2(Z_1)\}. \quad (3.31)
\end{aligned}
$$

In this expression it becomes very clear that the multiple coefficient must take on values at least as great as those of lower order. It is also apparent that, if $\Omega_T^2(Z_2 | Z_1)$ is zero, then $\Omega_T^2(Z_2, Z_1) = \Omega_T^2(Z_1)$, a result which is also true outside the bivariate normal model. For the linear combination $\eta = \beta' Z$, since $F(t | Z = z)$ and $F(t | \eta = \beta' z)$ are the same, all the needed information is contained in η. Thus, for multidimensional Z, $\Omega_T^2(Z) = \Omega_T^2(\eta)$, the predictive power of Z being summarized by η, the actual values assumed by the components of Z itself having no impact unless they modify η. Although this does not hold generally, for regression models in which the predictor is of the form η, a sensible definition for the multivariate coefficient, possibly easier to manipulate than that provided above, would be to replace the multidimensional Z by $\hat{\eta}$ for some consistent estimate of η and then proceed as in the bivariate case.

General expression for explained variation

In a bid to obtain suitable measures of explained variation for specific contexts, proportional hazards being an example, some authors have focused attention on some specific property of the normal model and then generalized it. This may or may not produce coefficients of predictability with good properties and it seems necessary to provide a more fundamental generalization of the basic definition. If the more fundamental generalization is successful, then not only should we be able to recover all the known situations as special cases, it should also be clear how to proceed when faced with new situations. With this in mind consider the following definition: take subsets of $\mathcal{Z}$, $\mathcal{A}(z)$ and $\mathcal{B}(z)$, where $\mathcal{A}(z) \subseteq \mathcal{B}(z)$. We then define Ω^2 by

$$
1 - \Omega^2 = \frac{E\{E[T - E(T | \mathcal{A}(Z))]^2\}}{E\{E[T - E(T | \mathcal{B}(Z))]^2\}}. \quad (3.32)
$$

The choices $\mathcal{A}(z) = \{z\}$ and $\mathcal{B}(z) = \mathcal{Z}$ in Equation 3.32, corresponding to the most common case enables Equation 3.32 to reduce to

Equation 3.25. Other choices can be helpful. For example, it is known, and indeed has sometimes been advanced as a criticism of Ω^2, that the coefficient depends on restrictions on $\mathcal{Z}$, this being made very clear via the explicit appearance of $\mathcal{B}(z)$ in the above formula. It may also be of interest to consider other definitions of $\mathcal{A}(z)$, allowing for say grouping (e.g., Cox and Wermuth 1992). Thus, in any given context, when discussing Ω^2, the amount of explained variation, it can be useful to keep in mind both how it is we are trying to explain the variation as well as just how much variation there is to explain.

Our reasons for being interested in the more general definition (3.32) have more to do with difficulties in establishing appropriate reference sets in the presence of right censoring. This inevitably involves conditioning on the risk sets and a suitable choice of $\mathcal{B}$ enables this to be carried out. Consider a bivariate regression model in which the special value $\beta = 0$ corresponds to absence of association between T and Z. This will often reduce itself to the equivalent statement that $z \in \mathcal{B}(z)$ where $\mathcal{B}(z)$ is the whole support of Z. The population value β_0 will often correspond to $z \in \mathcal{A}(z)$ where $\mathcal{A}(z) = \{z\}$. We can then write; as a definition for $\Omega_T^2(Z)$

$$\Omega^2 = \frac{E\{\mathrm{Var}(Z(t)|T=t; \beta=0)\} - E\{\mathrm{Var}(Z(t)|T=t; \beta)\}}{E\{\mathrm{Var}(Z(t)|T=t; \beta=0)\}}.$$

which we can rewrite as;

$$\Omega^2 = 1 - \frac{\int E_\beta\{[Z(t) - E_\beta(Z(t)|t)]^2|t\}dF(t)}{\int E_\beta\{[Z(t) - E_0(Z(t)|t)]^2|t\}dF(t)}. \tag{3.33}$$

This expression can be immediately adopted in the regression setting. Expectations derived from our chosen models and appropriate estimators arise naturally by replacing β by a consistent estimate. We return to this later in the text.

3.10 Exercises and class projects

1. Suppose that $\sum_{i=1}^{\infty} a_i = 1$. Also suppose that there exist random variable X_i such that $X_i = a_i X_1$. Show that $\sum X_i$, standardized by mean and variance, would not generally tend to a normal distribution. Comment.

2. Describe in simple terms what is quantified by the Lindeburg condition.

3. For a weighted sum of independent identically distributed random variables, suppose that each variable is standard uniform and the weights a_i are defined as $a_i = 10^i$. Will a central limit result hold and, if not, why not.

4. As a class project, construct graphs based on summing (1) 20 uniform random variates, (2) 100 uniform random variates and (3) 10000 uniform random variates. Standardize the x axis to lie between 0 and 1 and the y axis such that the mean is equal to zero and the variance increases linearly with x. Replicate each graph ten times. What conclusions can be drawn from the figures, in particular the influence of the number of variates summed?

5. Repeat the above class exercise, replacing the uniform distribution by (1) the log-logistic distribution with different means and variances, (2) the exponential distribution and (3) the normal distribution. Again replicate each graph ten times. Comment on your findings.

6. Consider two hypotheses for the sequence of observations; $X_1, \ldots, X_n$ in which $\mu_i = E(X_i)$; $H_1 : \mu_i = 0, \forall i$, against $H_1 : \mu_i = b^2 i$ for some $b \neq 0$. Construct different tests based on Brownian motion that would enable us to test H_0 versus H_1. Discuss the relative merits of the different tests.

7. Simulate 20 values from a standard exponential distribution and evaluate the greatest absolute distance between $F_n(t)$ and $F(t)$. Repeat this 1000 times, store the 1000 values of D_{20}, and then plot a histogram of these values. Add to the histogram the distribution of D_{20} using the Brownian bridge approximation.

8. For a sample size of 220, provide an approximate calculation of how large you would anticipate the greatest discrepancy between $F_n(t)$ and $F(t)$ to be.

9. Let $T_1, \ldots, T_n$ be i.i.d. observations from an exponential distribution with mean θ. Obtain an estimating equation for θ in terms of $E(T_i)$. Obtain an estimating equation for θ in terms of Var (T_i). Which of the two equations, if any, would be the most preferable. Give reasons.

10. Consider the following bivariate situation. We have a random variable, T, which is continuous and a binary variable Z, which take the

values 0 or 1. Given that $T = 0$, $\Pr(Z = 0) = 0.5$. Given that $Z = 0$, the distribution of T is exponential with mean equal to 1. Given that $Z = 1$ the distribution of T is exponential with mean equal to $\exp(\beta)$. Given n pairs of observations (T_i, Z_i) our purpose is to estimate the parameter β. Considering the values of Z as being fixed, obtain an estimating equation based on the observations T_i $(i = 1, \ldots, n)$. Secondly, we could view the random aspect of the experiment differently and now take the observations T_1 to T_n as being fixed. Derive the conditional distribution of Z given $T = T_i$. Use this to obtain a different estimating equation. Discuss the relative merits and disadvantages of the two sets of estimating equations.

11. In the next chapter we discuss the idea of right censoring where, for certain of the observations T_i, the exact value is not known. All that we can say for sure is that it is greater than some censoring time. How might the discussion of the previous exercise on the two types of estimating equations arising from reversing the conditioning variable have a bearing on this.

12. Consider the pair of independent random variables (Y_i, X_i), $i = 1, \ldots, n$. The null hypothesis is that $Y_i = \phi(X_i) + \epsilon_i$ where ϵ_i is an error term independent of the pair (Y_i, X_i) and where $\phi(u)$ is a nondecreasing function of u. Describe how you would carry out tests based on D_n and C_n of Section 3.5.

13. By investigating different classes of functions ϕ, describe the relative advantages and disadvantages of tests based upon C_n rather than D_n.

Chapter 4

Background: Survival analysis

4.1 Summary

We recall some elementary definitions concerning probability distributions, putting an emphasis toward one minus the usual cumulative distribution function, i.e., the survival function. This is also sometimes called the survivorship function. The closely related hazard function has, traditionally, been the most popular function around which to construct models. For multistate models it can be helpful to work with intensity functions, rather than hazard functions since these allow the possibility of moving in and out of states. This is facilitated by the very important function, $Y(t)$, the "at-risk" indicator. A number of special parametric cases of proportional hazards models are presented. The issue of censoring and the different kinds of censoring is discussed. The "at-risk" indicator $Y_i(w, t)$, taking the value one when the subject i is at risk of making a transition of a certain kind, indicated by w, makes it particularly simple to address more complex issues in survival such as repeated events, competing risks, and multistate modelling. We consider some tractable parametric models, the exponential model in particular.

4.2 Motivation

Survival time T will be a positive random variable, typically right skewed and with a non-negligible probability of sampling large values, far above the mean. The fact that an ordering, $T_1 > T_2$, corresponds to a solid physical interpretation has led some authors to consider that time is somehow different from other continuous random variables, reminiscent of discussion among early twentieth century physicists about the nature of time "flowing inexorably in and of itself." These characteristics are sometimes put forward as a reason for considering techniques other than the classic techniques of linear regression. From a purely statistical viewpoint, this reasoning is incorrect. Elementary transformations fix the skewness problems which, in consequence, reveal themselves as quite superficial. Nor is there any worthwhile, statistical, distinction between time and, say, height or weight. The reason for considering particular techniques, outside of the classical ones of linear regression, is the presence of censoring. In early work censoring came to be viewed as a nuisance feature of the data collection, hampering our efforts to study the main relationships of interest. A great breakthrough occurred when this feature of the data, the censoring, was modelled by the "at-risk" function. Almost immediately it became clear that all sorts of much more involved problems; competing risks, repeated events, correlated outcomes, could all be handled with almost no extra work. Careful use of the "at-risk" indicator was all that would be required. At the heart then of survival analysis is the idea of being at risk for some event of interest taking place in a short time frame (for theoretical study this short time will be made arbitrarily small). Transition rates are then very natural quantities to consider. In epidemiology these ideas have been well rooted for a half-century where age-dependent rates of disease incidence have been the main objects under investigation.

4.3 Basic tools

Time and risk

The insurance example in the introduction highlights an obvious, but important, issue. If driver A, on average, has a higher daily risk than driver B, then his mean time to be involved in an accident will be shorter. Conversely, if driver B has a longer mean time to accident, then he has, on average, a lower daily risk. For many examples we may tend to have in mind the variable time and how it is affected by other variables. But we can think equally well in terms of risk over short time periods, a viewpoint that we will see generalizes more readily to be able to deal with complicated situations. The connection between time and risk is outlined more formally below.

Hazard and related functions

The purpose here is to continue the introduction of preliminary notions and some basic concepts. Before discussing data and estimation we consider the problem in its most simplified form as that of the study of the pair of random variables (T, Z), T being the response variable "survival" of principal interest and Z an associated "explanatory" variable. There would be little difficulty in applying the host of techniques from linear regression to attacking this problem were it not for the presence of a "censoring" variable C. The particularity of C is that, when observed, i.e., $C = c$, we are no longer able to observe values of T for which $T > c$. Also, in most cases, when T is observed, we are no longer able to observe C. Nonetheless an observation on one tells us something about the other, in particular that it must assume some greater value.

Although the joint distribution of (T, Z) can be of interest, we are particularly interested in the conditional distribution of T given Z. First let us consider T alone. The probability density function of T is defined as

$$f(t) = \lim_{\Delta t \to 0^+} \frac{1}{\Delta t} \Pr(t < T < t + \Delta t), \qquad (4.1)$$

where $\lim_{\Delta t \to 0^+}$ means that Δt goes to 0 only through positive values. We define as usual $F(t) = \int_0^t f(u)du$. The survivorship function is written as $S(t) = 1 - F(t)$. If we view the density as the unconditional failure rate, we can define a conditional failure rate as being the same quantity after having accounted for the fact that the individual has already survived until the time point t. We call this $\lambda(t)$ and we define

$$\lambda(t) = \lim_{\Delta t \to 0^+} \frac{1}{\Delta t} \Pr(t < T < t + \Delta t | T > t). \qquad (4.2)$$

It helps understanding to contrast equation (4.2) and (4.1) and we can see that $\lambda(t)$ and $f(t)$ are closely related quantities. In a sense the function $f(t)$ for all values of t is seen from the standpoint of an observer sitting at $T = 0$, whereas, for the function $\lambda(t)$, the observer moves along with time looking at the same quantity but viewed from the position $T = t$. Analogous to a density, conditioned by some event, we can define

$$\lambda(t|C > t) = \lim_{\Delta t \to 0^+} \frac{1}{\Delta t} \Pr(t < T < t + \Delta t | T > t, C > t). \qquad (4.3)$$

The conditioning event $C > t$ is of great interest since, in practical investigations, all our observations at time t have necessarily been conditioned by the event. All associated probabilities are also necessarily conditional. But note that, under an independent censoring mechanism, $\lambda(t|C > t) = \lambda(t)$. This result underlies the great importance of certain assumptions, in this case that of independence between C and T. The conditional failure rate, $\lambda(t)$, is also sometimes referred to as the hazard function, the force of mortality, the instantaneous failure rate or the age-specific failure rate. If we consider a small interval then $\lambda(t) \times \Delta t$ closely approximates the probability of failing in a small interval for those aged t, the approximation improving as Δt goes to zero. If units are one year then these are yearly death rates. The cumulative hazard function is also of interest and this is defined as $\Lambda(t) = \int_0^t \lambda(u)du$. For continuous $\lambda(t)$, using elementary calculus we can see that:

$$\lambda(t) = f(t)/S(t), \ \ S(t) = \exp\{-\Lambda(t)\}, \ \ f(t) = \lambda(t)\exp\{-\Lambda(t)\}.$$

Although mathematically equivalent, we may prefer to focus attention on one function rather than another. The survival function, $S(t)$, is the function displaying most clearly the information the majority of applied workers are seeking. The hazard function, $\lambda(t)$, of central concern in much theoretical work, provides the most telling visual representation of time effects. An important function, of theoretical and practical interest, is the conditional survivorship function,

$$S(t, u) = \Pr(T > t | T > u) = \exp\{\Lambda(u) - \Lambda(t)\}, \ \ (u < t).$$

From this it is clear that $S(t, u) = S(t)/S(t)$ and that $S(u, u) = 1$ so that it is as though the process had been restarted at time $t = u$. Other quantities that may be of interest in some particular contexts are the mean residual lifetime, $m(t)$, and the mean time lived in the interval $[0, t]$, $\mu(t)$, defined as

$$m(t) = E(T - t | T \geq t), \quad \mu(t) = \int_0^t S(u)du. \qquad (4.4)$$

Like the hazard itself, these functions provide a more direct reflection on the impact of having survived until time t. The mean residual life-time provides a very interpretable measure of how much more time we can expect to survive, given that we have already reached the time-point t. This can be useful in actuarial applications. The mean time lived in the interval $[0, t]$ is not so readily interpretable, requiring a lit-tle more thought (it is not the same as the expected lifetime given that $T < t$). It has one strong advantage in that it can be readily estimated from right censored data in which, without additional assumptions, we may not even be able to estimate the mean itself. The functions $m(t)$ and $\mu(t)$ are mathematically equivalent to one another as well as the three described above and, for example, a straightforward integration by parts shows that $m(t) = S^{-1}(t) \int_t^\infty S(u)du$ and that $\mu(\infty) = E(T)$. If needed, it follows that the survivorship function can be expressed in terms of the mean residual lifetime by

$$S(t) = m^{-1}(t)m(0) \exp\left(-\int_0^t m^{-1}(u)du\right).$$

We may wish to model directly in terms of $m(t)$, allowing this func-tion to depend on some vector of parameters θ. If the expression for $m(t)$ is not too intractable then, using $f(t) = -S'(t)$ and the above relationship between $m(t)$ and $S(t)$, we can write down a likelihood for estimation purposes in the situation of independent censoring. An in-teresting and insightful relationship (see for instance the Kaplan-Meier estimator) between $S(t)$ and $S(t, u)$ follows from considering some dis-crete number of time points of interest. Thus, for any partition of the time axis, $0 = a_0 < a_1 <, \ldots, a_n = \infty$, we see that

$$S(a_j) = S(a_{j-1})S(a_j, a_{j-1}) = \prod_{\ell \leq j} S(a_\ell, a_{\ell-1}).$$

The implication of this is that the survival function $S(t)$ can always be viewed as the product of a sequence of conditional survival functions, $S(t, u)$. Although more cumbersome, a theory could equally well be constructed for the discrete case whereby $f(t_i) = \Pr(T = t_i)$ and $S(t_i) = \sum_{\ell \geq i} f(t_\ell)$. We do not explore this here.

Intensity functions and compartment models

Modern treatment of survival analysis tends to focus more on inten-sity than hazard functions. This leads to great flexibility, enabling, for

Figure 4.1: A simple alive/dead transition model.

example, the construction of simple models to address questions in complex situations such as repeated events (Andersen and Gill 1982). We believe that both concepts can be useful and we will move back and forth between them according to the application. Intensity functions find their setting in the framework of stochastic processes where the random nature of T is suppressed, t being taken simply as an index to some stochastic process. The *counting process* $N(t)$, takes the value 0 at $t = 0$, remaining at this same value until some time point, say $T = u$, at which the event under study occurs and then $N(t) = 1$ $(t \geq u)$. We can then define, in an infinitesimal sense, i.e., the equality only holds precisely in the limit as dt goes to zero through positive values

$$\Pr\{(N(t) - N(t - dt) = 1|\mathcal{F}_{t-dt})\} = \alpha(t)dt \qquad (4.5)$$

where $\mathcal{F}_{t-dt}$, written as $\mathcal{F}_{t-}$ when we allow $dt > 0$ to be arbitrarily close to zero, is the accumulated information, on all processes under consideration, observed up until time $t - dt$. The observed set $\mathcal{F}_{t-}$ is referred to as the history at time t. The set is necessarily non decreasing in size as t increases, translating the fact that more is being observed or becoming known about the process. The Kolmogorov axioms of probability, in particular sigma additivity, may not hold for certain noncountable infinite sets. For this reason probabilists take great care, and use considerable mathematical sophistication, to ensure, in broad terms, that the size of the set $\mathcal{F}_{t-}$ does not increase too quickly with t. The idea is to ensure that we remain within the Kolmogorov axiomatic framework, in particular that we do not violate sigma additivity. Much of these concerns have spilled over into the applied statistical literature where they do not have their place. No difficulties will arise in applications, with the possible exception of theoretical physics, and the practitioner, unfamiliar with measure theory, ought not be deterred from applying the techniques of stochastic processes simply because he or she lacks a firm grasp of concepts such as filtrations. It is hard to imagine an application in which a lack of understanding of the term "filtration" could have led to error. On the other hand, the more accessible notions of history, stochastic process, and conditioning sets are central and of great importance both to understanding and to deriving creative structures around which applied problems can be solved. Viewing t as an index to a stochastic process rather than simply the realization of a random variable T, and defining the intensity process $\alpha(t)$ as above, will enable great flexibility and the possibility to model events dynamically as they unfold.

At risk functions $Y(t)$, $Y(w,t)$ and multistate models

The simplest case we can consider occurs when following a randomly chosen subject through time. The information in $\mathcal{F}_{t-}$ tells us whether or not the event has yet occurred and if the subject is still at risk i.e., the set $\mathcal{F}_{t-}$ is providing the same information as an observation on the function $Y(t)$ where we take $Y(t)$ to be left continuous, assuming the value one until the occurrence of an event, or removal from observation, at which time it assumes the value zero. If the simple fact of not having been removed from the study, the event $(C > t)$ is independent of the event $(t < T < t + dt)$, then conditioning on $Y(t) = 1$ is the same as conditioning on $T > t$. Referring then to Equation (4.2) it is clear that if $Y(t) = 0$ then $\alpha(t) = 0$ and, if $Y(t) = 1$ then $\alpha(t) = \lambda(t)$. Putting these two results together we have

$$\alpha(t) = Y(t)\lambda(t). \tag{4.6}$$

This relation is important in that, under the above condition, referred to as the independent censoring condition, the link between the intensity function and the hazard function is clear. Note that the intensity function is random since Y is random when looking forward in time. Having reached some time point, t say, then $\alpha(t)$ is fixed and known since the function $Y(u)$, $0 < u < t$ is known and $Y(t)$ is left continuous.

We call $Y(\cdot)$ the "at risk" function (left continuous specifically so that at time t the intensity function $\alpha(t)$ is not random). The idea generalizes readily and in order to cover a wide range of situations we also allow Y to have an argument w where w takes integer values counting the possible changes of state. For the ith subject in any study we will typically define $Y_i(w,t)$ to take the value 1 if this subject, at time t, is at risk of making a transition of type w, and 0 otherwise. Figure 4.2 summarizes a situation in which there are four states of interest, an absorbing state, death, and three states from which an individual is able to make a transition into the death state. Transitions

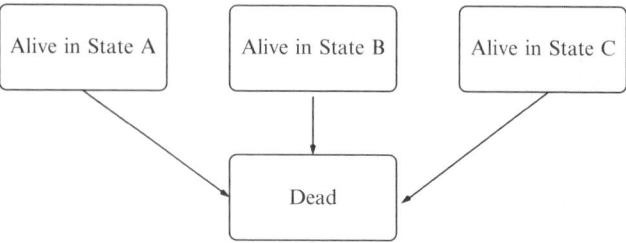

Figure 4.2: A simple compartment model with an absorbing state.

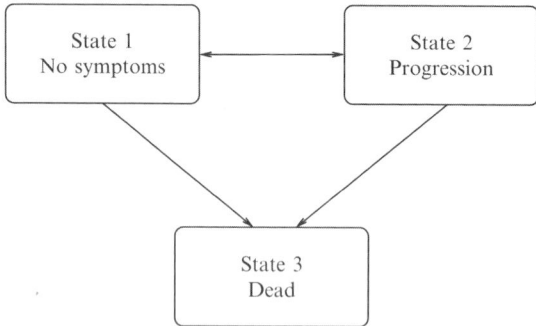

Figure 4.3: A simple compartment model with a single absorbing state.

among the three nondeath states themselves are not allowed. Later we will consider different ways of modeling such a situation, depending upon further assumptions we may wish or not wish to make.

In Figure 4.3 there is one absorbing state, the death state, and two non absorbing states between which an individual can make transitions. We can define $w = 1$ to indicate transitions from state 1 to state 2, $w = 2$ to indicate transitions from state 2 to state 1, $w = 3$ to indicate transitions from state 1 to state 3 and, finally, $w = 4$ to indicate transitions from state 2 to state 3. Note that such an enumeration only deals with whether or not a subject is at risk for making the transition, the transition probabilities (intensities) themselves could depend on the path taken to get to the current state. We can then appreciate why it can be helpful to frame certain questions in terms of compartment models, intensity functions and the risk function. Rather complex situations can be dealt with quite straightforwardly, the figures illustrating simple cases where we can use the argument w in $Y_i(w, t)$ to indicate, at any t, which kinds of transition any given subject i is available to make. In Figure 4.4 there are two absorbing states, one of

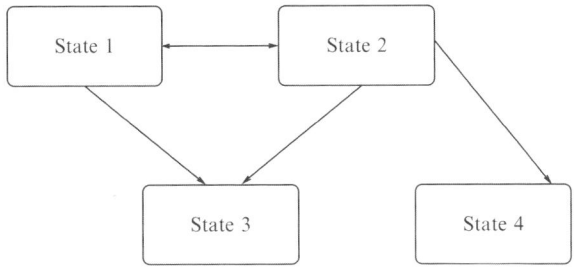

Figure 4.4: A complex compartment model with two absorbing states.

which can only be reached from state 2. The transition rate between
state 2 and state 4 may or may not depend on the number of times
a subject moves between states 1 and 2. Allowing for transitions between states greatly adds to the flexibility of any model so that, in
Figure 4.2, although the explanatory variable (state) has three levels,
the model is, in principle, much simpler than that described in Figure
4.3 where the explanatory variable can assume only two states.

At-risk indicator $Y(w,t)$ and repeated events

Some studies have the particularity that an occurrence of the event of
interest does not remove the subject from further observation. Additional events, of the same or of different types, may happen. An example is benign breast disease, potentially followed by malignant disease.
A patient may have several incidences of benign breast disease at different intervals of time. Following any one of these incidences, or even
before such an incidence takes place the subject may become incident
for malignant disease. If our interest is essentially focussed on the incidence of malignant disease then we would treat the time-dependent
history of benign breast disease as a potential explanatory variable for
incidence of malignant disease. However, we may also be interested in
modelling directly the repeated incidence of benign breast disease in
its own right. Clearly a patient can only be at risk of having a third
incident of benign breast disease if she has already suffered two earlier
incidents. We can model the rate of incidence for the jth occurrence
of benign disease as,

$$\alpha_j(t) = Y(j,t)\lambda_j(t - t_{j-1}), \tag{4.7}$$

where $t_0 = 0$ and t_j is the observed occurrence of the jth event.
Different options may be considered for modeling $\lambda_j(t)$. Usually there

will be at least one covariate, Z, indicating two distinct prognostic groups, possibly established on the basis of different treatments. The model will involve coefficients multiplying Z and thereby quantifying treatment affect. Allowing these coefficients to also depend upon j provides the broadest generality and is equivalent to analyzing separate studies for each of the occurrences. Stronger modeling, imposing greater structure, might assume that the coefficients do not depend upon j, in which case the information provided by a subject having three incident cases is comparable to that of three independent subjects each providing information on a single incident. So-called marginal models have been proposed in this context. Here, it would be as though the subject, after an event, starts the clock from zero and, aside from covariate information, is deemed to be in the same position as another subject who has just entered the study without having yet suffered a single event. A lot of information would appear to be thereby gained but the set-up seems rather artificial and implausible. Starting the clock from zero, after each event, is sensible but it is more realistic to assume that the underlying hazard rates, i.e., those not adjusted by covariate information, would change with the number of prior incidents. In other words the most sensible model would condition on this information allowing the baseline hazard rate to change according to the number of events counted so far.

4.4   Some potential models

Simple exponential

The simple exponential model is fully specified by a single parameter λ. The hazard function, viewed as a function of time, does not in fact depend upon time so that $\lambda(t) = \lambda$. By simple calculation we find that $\Pr(T > t) = \exp(-\lambda t)$. Note that $E(T) = 1/\lambda$ and, indeed, the exponential model is often parameterized directly in terms of the mean $\theta = E(T) = 1/\lambda$. Also $\mathrm{Var}(t) = 1/\lambda^2$. This model expresses the physical phenomenon of no aging or wearing out since, by elementary calculations, we obtain $S(t + u, u) = S(t)$; the probability of surviving a further t units of time, having already survived until time u, is the same as that associated with surviving the initial t units of time. The property is sometimes referred to as the lack of memory property of the exponential model.

For practical application the exponential model may suggest itself in view of its simplicity or sometimes when the constant hazard assumption appears realistic. A good example is that of a light bulb which may only fail following a sudden surge in voltage. The fact that no such surge has yet occurred may provide no information about the chances for such a surge to take place in the next given time period. If T has an exponential distribution with parameter λ then λT has the so-called standard exponential distribution, i.e., mean and variance are equal to one.

Recall that for a random variable Y having normal distribution $\mathcal{N}(\mu, \sigma^2)$ it is useful to think in terms of a simple linear model $Y = \mu + \sigma\epsilon$, where ϵ has the standard distribution $\mathcal{N}(0, 1)$. As implied above, scale changes for the exponential model lead to a model still within the exponential class. However, this is no longer so for location changes so that, unlike the normal model in which linear transformations lead to other normal models, a linear formulation for the exponential model is necessarily less straightforward. It is nonetheless of interest to consider the closest analogous structure and we can write

$$Y = \log T = \alpha + bW, \qquad (4.8)$$

where W has the standard extreme value density $f(w) = \exp\{w - \exp(w)\}$. When $\alpha = 0$ we recover an exponential model for T with parameter b, values other than zero for α pushing the variable T out of the restricted exponential class into the broader Weibull class discussed below.

Proportional hazards exponential

In anticipation of the central topic of this book (that of heterogeneity among the subjects under study) imagine that we have two groups, indicated by a binary variable $Z = 0$ or $Z = 1$. For $Z = 0$ the subjects follow an exponential law with parameter λ_0. For $Z = 1$ the subjects follow an exponential law with parameter λ_1. It is clear that for the hazard functions there exists real β ($= \log \lambda_1 - \log \lambda_0$) such that

$$\lambda(t|Z) = \lambda(t|Z = 0) \exp(\beta Z) = \lambda_0 \exp(\beta Z). \qquad (4.9)$$

The important point to note here is that the ratio of the hazards, $\lambda(t|Z = 1)/\lambda(t|Z = 0)$ does not involve t. It also follows that $S(t|Z = 1) = S(t|Z = 0)^\alpha$ where $\alpha = \exp(\beta)$. The survival curves are power transformations of one another. This is an appealing parameterization since, unlike a linear parameterization, whatever the true value of β, the constraints that we impose upon $S(t|Z = 1)$ and $S(t|Z = 0)$ in order to be well-defined probabilities, i.e., remaining between 0 and 1, are always respected. Such a model is called a proportional hazards model. For three groups we can employ two indicator variables, Z_1 and Z_2, such that, for group 1 in which the hazard rate is equal to λ_0, $Z_1 = 0$ and $Z_2 = 0$, for group 2, $Z_1 = 1$ and $Z_2 = 0$ whereas for group 3, $Z_1 = 0$ and $Z_2 = 1$. We can then write;

$$\lambda(t|Z) = \lambda_0 \exp(\beta_1 Z_1 + \beta_2 Z_2), \tag{4.10}$$

where $\lambda_0 = \lambda(t|Z_1 = Z_2 = 0)$. It is worthwhile bringing the reader's attention to just where the constraints of the model express themselves here. They concern the hazard rates for all groups, which are assumed to be constant. Given this constraint there are no further constraints concerning the relationship between the groups. Suppose, though, that we were to consider a further group, group 4, defined by $Z_1 = 1$ and $Z_2 = 1$. In order to add a fourth group without introducing a further binary coding variable Z_3, we introduce the constraint that the hazard for group 4 is simply expressed in terms of the hazards for groups 2 and 3. Such assumptions are commonly made in routine data analysis but, nonetheless, ought come under critical scrutiny. We return to this issue in later chapters. The extension to many groups follows in the same way. For this we take Z to be a p dimensional vector of indicator variables and β a vector of parameters having the same dimension as Z, the product βZ in Equation 4.9 now implying an inner product, i.e., $\beta Z = \sum_{i=1}^{p} \beta_i Z_i$. In this case the proportional hazards exponential model (4.9) implies that every group follows some simple exponential law, a consequence being that the survivorship function for any group can be expressed as a power transformation of any other group. Once again, it is important to keep in mind just which assumptions are being made, the potential impact of such assumptions on conclusions, and techniques for bringing under scrutiny these assumptions. The proportional hazards constraint then appears as a very natural one in which we ensure that the probabilities $S(t|z)$ and subsequent estimates always remain between 0 and 1. A linear shift added to $S(t|0)$ would not allow for this. We do nonetheless have a linear shift although on a different, and thereby more appropriate, scale and we can write

$$\log - \log S(t|Z) = \log - \log S(t|0) + \sum_{i=1}^{p} \beta_i Z_i.$$

This formulation is the same as the proportional hazards formulation. Noting that $-\log S(T|Z = z)$ is an exponential variate some authors prefer to write a model down as a linear expression in the transformed random variable itself with an exponential error term. This then provides a different link to the more standard linear models we are familiar with.

Piecewise exponential

The lack of flexibility of the exponential model will often rule it out as a potential candidate for application. Many other models, only one or two of which are mentioned here, are more tractable, a property stemming from the inclusion of at least one additional parameter. Even so, it is possible to maintain the advantages of the exponential model's simplicity while simultaneously gaining in flexibility. One way to achieve this is to construct a partition of the time axis $0 = a_0 < a_1 < \ldots < a_k = \infty$. Within the jth interval $(a_{j-1}, a_j), (j = 1, \ldots, k)$ the hazard function is given by $\lambda(t) = \lambda_j$. We can imagine that this may provide quite a satisfactory approximation to a more involved smoothly changing hazard model in which the hazard function changes through time. We use $S(t) = \exp\{-\Lambda(t)\}$ to obtain the survival function where

$$\Lambda(t) = \sum_{j=1}^{k} I(t \geq a_j)\lambda_j(a_j - a_{j-1})$$

$$+ \sum_{j=1}^{k} I(a_{j-1} \leq t < a_j)\lambda_j(t - a_{j-1}). \qquad (4.11)$$

Properties such as the lack of memory property of the simple exponential have analogues here by restricting ourselves to remaining within an interval. Another attractive property of the simple exponential is that the calculations are straightforward and can be done by hand and, again, there are ready analogues for the piecewise case. Although the ready availability of sophisticated computer packages tends to eliminate the need for hand calculation, it is still useful to be able to work by hand if for no other purposes than those of teaching. Students gain invaluable insight by doing these kind of calculations the long way.

Proportional hazards piecewise exponential

In the same way as for the simple exponential model, for two groups, indicated by a binary variable $Z = 0$ or $Z = 1$, each having constant piecewise rates on the same intervals, it is clear that there exists β_j such that, for $t \in [a_{j-1}, a_j)$,

$$\lambda(t|Z) = \lambda(t|Z = 0)\exp(\beta_j Z) = \lambda_0(t)\exp\{\beta(t)Z\}, \qquad (4.12)$$

where we now have a function $\beta(t) = \sum_{j=1}^{k} \beta_j I(a_{j-1} \leq t < a_j)$. This can be described as a nonproportional hazards model and, if, under a further restriction that $\beta(t)$ is a constant function of time, i.e., $\beta_1 = \beta_2 = \cdots = \beta_k = \beta$, then, as for the simple exponential model, we have $S(t|Z = 1) = S(t|Z = 0)^\alpha$ where $\alpha = \exp(\beta)$ and, once again, such a model is called a proportional hazards model. The model can once more be described in terms of a linear translation on $\log - \log S(t|z)$.

Weibull model

Another way to generalize the exponential model to a wider class is to consider a power transformation of the random variable T. For any positive γ, if the distribution of T^γ is exponential with parameter λ, then the distribution of T itself is said to follow a Weibull model whereby

$$f(t) = \lambda\gamma(\lambda t)^{\gamma-1}\exp\{-(\lambda t)^\gamma\}$$

and $S(t) = \exp -(\lambda t)^\gamma$. The hazard function follows immediately from this and we see, as expected, that when $\gamma = 1$ an exponential model with parameter λ is recovered. It is of interest to trace out the possible forms of the hazard function for any given λ. It is monotonic, increasing for values of γ greater than 1 and decreasing for values less than 1. This property, if believed to be reflected in some given physical situation, may suggest the appropriateness of the model for that same situation. An example might be the time taken to fall over for a novice roller blade enthusiast - the initial hazard may be high, initially decreasing somewhat rapidly as learning sets in and thereafter continuing to decrease to zero, albeit more slowly.

The Weibull model, containing the exponential model as a special case, is an obvious candidate structure for framing questions of the sort - is the hazard decreasing to zero or is it remaining at some constant level? A null hypothesis would express this as $H_0 : \gamma = 1$.

Straightforward integration shows that $E(T^r) = \lambda^{-r}\Gamma(1 + r/\gamma)$ where $\Gamma(\cdot)$ is the gamma function,

$$\Gamma(p) = \int_0^\infty u^{p-1}e^{-u}du \quad p > 0.$$

For p integer $\Gamma(p) = (p-1)!$ The mean and the variance are $\lambda^{-1}\Gamma(1 + 1/\gamma)$ and $\lambda^{-2}\Gamma(1 + 2/\gamma) - E^2$, respectively. The Weibull model can be motivated from the theory of statistics of extremes. The distribution coincides with the limiting distribution of the smallest of a collection of random variables, under broad conditions on the random variables in question (Kalbfleisch and Prentice 1980, page 48).

Proportional hazards Weibull

Once again, for two groups indicated by a binary variable $Z = 0$ or $Z = 1$, sharing a common γ but different values of λ, then there exists a β such that $\lambda(t|Z)/\lambda(t|Z = 0) = \exp(\beta Z)$. Since, as above, the right-hand side of the equation does not depend on t, then we have a proportional hazards model. This situation and the other two described above are the only common parametric models that come under the heading proportional hazards models by simply expressing the logarithm of the location parameter linearly in terms of the covariates. The situation for more than two groups follows as before. Consider however a model such as

$$\lambda(t|Z) = \lambda\gamma(\lambda t)^{\gamma - 1}\exp(\beta Z), \tag{4.13}$$

in which Z indicates three groups by assuming the values $Z = 1, 2, 3$.

Unlike the model just above in which three groups were represented by two distinct binary covariates, Z_1 and Z_2, we have only one covariate. In the context of estimation and a given set of data we will almost invariably achieve greater precision in our estimates when there are less parameters to estimate. We would then appear to gain by using such a model. As always though, any such gain comes at a price and the price here is that we have made much stronger assumptions. We are assuming that the signed "distance" between groups 1 and 2, as measured by the logarithm of the hazard, is the same as the signed distance between groups 2 and 3. If this is not the case in reality then we are estimating some sort of compromise, the exact nature of which is determined by our estimating equations. In an extreme case in which the distances are the same but the signs are opposite we might erroneously conclude that there is no effect at all. At the risk of being repetitive, it cannot be stressed too much just how important it is to identify the assumptions we are making and how they may influence our conclusions. Here the assumptions concern both the parametric form of the underlying risk as well as the nature of how the different groups are related. Allowing a shape parameter γ to be other than one provides a more flexible model for the underlying risk than that furnished by the simple exponential model. The choice of covariate coding, on the other hand, is more restrictive than the earlier choice. All of this needs to be studied in applications. An interesting point is that, for the three group case defined as above, the "underlying" hazard, $\lambda(t|Z = 0) = \lambda\gamma(\lambda t)^{\gamma-1}$ does not correspond to the hazard for any of the three groups under study. It is common in practice to consider a recoding of Z, a simple one being $Z - \bar{Z}$, so that the underlying hazard will correspond to some kind of average across the groups. For the case just outlined, another simple recoding is to rewrite Z as $Z - 2$, in which case the underlying hazard corresponds to the middle group, the other two groups having hazard rates lower and greater than this, respectively.

Log-minus-log transformation

As a first step to constructing a model for $S(t|Z)$ we may think of a linear shift, based upon the value of Z, the amount of the shift to be estimated from data. However, the function $S(t|Z)$ is constrained, becoming severely restricted for both $t = 0$ and for large t where it approaches one and zero respectively. Any model would need accommodate these natural constraints. It is usually easiest to do this by eliminating the constraints themselves during the initial steps of model construction. Thus, $\log S(t|Z) = -\Lambda(t)$ is a better starting point for modeling, weakening the hold the constraints have on us. However, $\log - \log S(t|Z) = \log \Lambda(t)$ is better still. This is because $\log \Lambda(t)$ can take any value between $-\infty$ and $+\infty$, whereas $\Lambda(t)$ itself is constrained to be positive. The transformation $\log - \log S(t|Z)$ is widely used and is called the log-minus-log transformation. The above cases of the exponential and Weibull proportional hazards models, as already seen, fall readily under this heading.

Other models

The exponential, piecewise exponential and Weibull models are of particular interest to us because they are especially simple and of the proportional hazards form. Nonetheless there are many other models which have found use in practical applications. Some are directly related to the above, such as the extreme value model in which

$$S(t) = \exp\left(-\exp\left(\frac{t - \mu}{\sigma}\right)\right),$$

since, if T is Weibull, then $\log T$ is extreme value with $\sigma = 1/\gamma$ and $\mu = \log \lambda$. These models may also be simple when viewed from some particular angle. For instance, if $M(s)$ is the moment-generating function for the extreme value density then we can readily see that $M(s) = \Gamma(1 + s)$. A distribution, closely related to the extreme value distribution (see Johnson and Johnson 1980), and which has found wide application in actuarial work is the Gompertz where

$$S(t) = \exp\left(\beta\alpha^{-1}(1 - e^{\alpha t})\right).$$

The hazard rates for these distributions increase with time, and, for actuarial work, in which time corresponds to age, such a constraint makes sense for studying disease occurrence or death. The normal distribution is not a natural candidate in view of the tendency for survival data to exhibit large skewness, not forgetting that times themselves are constrained to be positive. The log normal distribution has seen some use but is most often replaced by the log-logistic, similar in shape apart from the extreme tails, and much easier to work with. The form is particularly simple for this model and we have

$$S(t) = (1 + (\alpha t)^{\gamma})^{-1}.$$

For two groups, sharing a common γ but different values of α it is interesting to note that the hazard ratio declines monotonically with time t to its asymptotic value of one. Such a model may be appropriate when considering group effects which gradually wane as we move away from some initial time point.

Parametric proportional hazards models

In principle, for any parametric form, the above providing just a very few examples, we can make a straightforward extension to two or more groups via a proportional hazards representation. For example, if the survivorship functions of two groups are $S(t|Z = 1)$ and $S(t|Z = 0)$ then we can introduce the parameter α to model one group as a power transform of the other. Rewriting α to include Z via $\alpha = \exp(\beta Z)$ then we have an expression involving the regressors,

$$\log - \log S(t|Z) = \log - \log S(t|Z = 0) + \beta Z. \qquad (4.14)$$

All parameters, including β, can be estimated using standard techniques, maximum likelihood in particular, the only restriction being that we require some conditions on the censoring variable C. In practice, standard techniques are rarely used, most likely as a consequence of the attractive proposal of Cox (1972) whereby we can estimate β without having to consider the form of $S(t|Z = 1)$ or $S(t|Z = 0)$. As attractive as the Cox approach is though, we should not overlook the fact that, in exchange for generality concerning the possible parametric forms of functions of interest, such as $S(t|Z)$, making inferences on these population quantities becomes that much more involved. Parametric proportional hazards models may be an area that merits renewed interest in applications.

4.5   Censoring

The most important particularity of survival data is the presence of censoring. Other aspects such as the positivity and skewness of the main random variable under study, time T, and other complex situations such as repeated measures or random effects, are not of themselves reasons for seeking methods other than linear regression. Using transformations and paying careful attention to the structure of the error, linear models are perfectly adequate for dealing with almost any situation in which censoring does not arise. It is the censoring that forces us to consider other techniques. Censoring can arise in different ways.

We typically view the censoring as a nuisance feature of the data, and not of direct interest in its own right, essentially something that hinders us from estimating what it is we would like to estimate. In order for our endeavors to succeed we have to make some assumptions about the nature of the censoring mechanism. The assumptions may often be motivated by convenience, in which case it is necessary to give consideration as to how well grounded the assumptions appear to be as well as to how robust are the procedures to departures from any such assumptions. In other cases the assumptions may appear natural given the physical context of interest, a common case being the uniform recruitment into a clinical trial over some predetermined time interval. When the study closes patients for whom the outcome of interest has not been observed are censored at study close and until that point occurs it may be reasonable to assume that patients are included in the study at a steady rate.

It is helpful to think of a randomly chosen subject being associated with a pair of random variables (T, C), an observation on one of the pair impeding observation on the other, while at the same time indicating that the unobserved member of the pair must be greater than the observed member. This idea is made more succinct by saying that only the random variable $X = \min(T, C)$ can be fully observed. Clearly $\Pr(X > x) = \Pr(T > x, C > x)$ and we describe censoring as being independent whenever

$$\Pr(X > x) = \Pr(T > x, C > x) = \Pr(T > x) \Pr(C > x). \quad (4.15)$$

Type I censoring

Such censoring most often occurs in industrial or animal experimentation. Items or animals are put on test and observed until failure. The study is stopped at some time T^*. If any subject does not fail it will have observed survival time at least equal to T^*. The censoring times for all those individuals being censored is then equal to T^*. Equation (4.15) is satisfied and so this is a special case of independent censoring, although not very interesting since all subjects, from any random sample, have the same censoring time.

Type II censoring

The proportion of censoring is determined in advance. So if we wish to study 100 individuals and observed half of them as failures we determine the number of failures to be 50. Again all censored observations have the same value T^* although, in this case, this value is not known in advance. This is another special case of independent censoring.

Type III censoring

In a clinical trial patients enter randomly. A model for entry is often assumed to be uniform over a fixed study period, anywhere from a few months to several years but determined in advance. Survival time is the time from entry until the event of interest. Subjects can be censored because (1) the end of the study period is reached, (2) they are lost to follow-up (3) the subject fails due to something unrelated to the event of interest. This is called random censoring. So, unlike for *Type I* or *Type II* censoring, for a random sample $C_1, \ldots, C_n$, the C_i could all be distinct.

For a random sample of pairs (T_i, C_i), $i = 1, \ldots, n$, we are only able to observe $X_i = \min(T_i, C_i)$. A fundamental result in this context was discovered by Tsiatis (1975). The result says that, for such data, we are unable to estimate the joint distribution of the pair (T, C). Only the marginal distributions can be estimated under the independent censoring assumption, the assumption itself not being testable from such data. It is common then to make the assumption of independent censoring, sometimes referred to as non informative censoring, by stipulating that

$$\Pr(X_i > x) = \Pr(T_i > x, C_i > x) = \Pr(T_i > x)\Pr(C_i > x). \quad (4.16)$$

The assumption is strong but not entirely arbitrary. For the example of the clinical trial with a fixed closing date for recruitment it seems reasonable to take the length of time from entry up until this date as not being associated with the mechanism generating the failures. For loss to follow-up due to an automobile accident or due to leaving the area, again the assumption may be reasonable, or, at least, a good first approximation to a much more complex, unknown, and almost certainly unknowable, reality.

Informative censoring

When censoring is informative, which we can take to be the negation of non-informative, then it is no longer possible to estimate the main quantities of interest without explicitly introducing some model for the censoring. The number of potential models relating C and T is infinite and, in the absence of special knowledge, it can be helpful to postulate some simple relationship between the two, the proportional hazards model itself having been used in this context (Koziol and Green 1976, Slud and Rubinstein 1983). Obvious examples might be surrogate endpoints in the study of the evolution of AIDS following treatment, where, for falling CD4 cell counts, below a certain point patients can be withdrawn from study. Censoring here is clearly informative. This will be the case whenever the fact of removing a subject, yet to experience the event of interest, from study implies a change in risk. Informative censoring is necessarily more involved than non informative censoring and we have to resort to more elaborate models for the censoring itself in order to make progress. If, as might be the case for a clinical trial where the only form of censoring would be the termination of the study, we know for each subject, in advance, their censoring time C, we might then postulate that

$$\log - \log S(t) = \log - \log(S(t|C < t) + \beta I(C > t).$$

This would be a proportional hazards model for a dependent censoring mechanism. More generally we would not know C in advance of making observations on T, but we could write down a similar model in terms of intensity functions, viewing the censoring indicator as a predictable stochastic process. For the purposes of estimation we may require empirical quantities indicating how the risk changes once censoring is observed, and for this we need to be able to compare rates between those censored at some point and those who are not. Mostly, once censoring has occurred, it is no longer possible to observe the main event under study so that, for data of this nature, we are not able to estimate parameters of interest without further assumptions. These assumptions are usually that the censoring is independent of the failure process or that it is conditionally independent given covariate values. The paper of Tsiatis (1975) demonstrates this intuitive observation formally.

Marginal and conditionally independent censoring

When considering many groups, defined by some covariate value Z, there are essentially two types of independence commonly needed. The stronger assumption is that of marginal independence in which the variables T, C, and Z are pairwise independent. The censoring distribution for C is the same for different values of Z. A weaker assumption that is often made, is that of conditional independence. Here, the pair (T, C) are independent given Z. In other words, for each possible value of Z, the pair (T, C) is independent, but the censoring distribution C can be different for different values of Z.

Finite censoring support

Many mathematical issues simplify immediately when the failure variable T is continuous, as we generally suppose, but that the censoring variable is restricted to having support on some finite subset. We can imagine that censoring times are only allowed to take place on the set $\{a_0, a_1, \ldots, a_k\}$. This is not a practical restriction since we can make the division (a_j, a_{j-1}) as fine as we wish. We will frequently need to consider the empirical distribution function and analogues (Kaplan-Meier estimate, Nelson-Aalen estimate) in the presence of censoring. If we adopt this particular censoring set-up of finite censoring support, then generalization from the empirical distribution function to an analogue incorporating censoring is very straightforward. We consider this in greater detail when we discuss the estimation of marginal survival.

4.6 Competing risks as a particular type of censoring

Recalling the "at-risk" indicator function, $Y_i(w, t)$, which takes the value one if, at time t, the i th subject is at risk of making a transition of type w, and is zero otherwise, we can imagine a simple situation in which w takes only one of two values. Calling these $w = 1$ and $w = 2$, consider a constraint whereby $Y_i(1, t) = Y_i(2, t)$. In words, if the ith subject is at risk of one kind of transition, then he or she is also at risk of the other kind. If the subject is no longer at risk then this means that they are not at risk for either kind of transition. Thus, if a subject suffers an event of type $w = 1$ then he is no longer considered at risk of suffering an event of type $w = 2$, and conversely.

This is the situation of so-called competing risks. As long as the subject is at risk, then either of the event types can occur. Once one type of event has occurred, then it is no longer possible to observe an occurrence of an event of the other type. Such a construction fits in immediately with the above models for survival involving censoring. If at time $t = t_1$ an event of type $w = 1$ takes place, then, as far as events of type $w = 2$ are concerned, the subject is simply censored at $t = t_1$. In Figure 4.5 a subject may be at risk of death from stroke or at risk from either stroke or cirrhosis of the liver. Once one of the types of death has occurred, then the other type of event can no longer be observed. We will assume that the subject is censored at this point, in as much as our attention focuses on the second type of event, and the above discussion on the different censoring models applies in the same way. We will need make some assumptions, most often that of

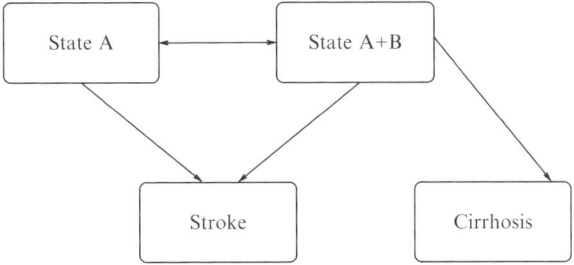

Figure 4.5: A situation of competing risks for subjects in states A+B.

independent censoring or that of independent censoring conditional on covariate information in order to make progress.

4.7 Exercises and class projects

1. Using the definition for $\lambda(t) = f(t)/S(t)$, show that $S(t) = \exp\{-\Lambda(t)$ and that $f(t) = \lambda(t)\exp\{-\Lambda(t)$.

2. For a Weibull variate with parameters λ and k, derive an expression for the conditional survivorship function $S(t + u, u)$. How does this function vary with t for fixed u? With u for fixed t?

3. Use numerical integration to calculate the mean residual lifetime $m(t)$ and the mean time lived in the interval $[0, t]$, $\mu(t)$ for the Weibull with parameters 2 and 1.5. With parameters 2 and 0.7. Plot these as functions of time t.

4. Consider two groups each of which follows a Weibull distribution, i.e., $f(t) = \lambda\gamma(\lambda t)^{\gamma-1}\exp\{-(\lambda t)^{\gamma}\}$. For the first group, $\lambda = \lambda_1$, $\gamma = \gamma_1$. For the second, $\lambda = \lambda_2$, $\gamma = \gamma_2$. Under which conditions will this situations be described by proportional hazards?

5. Undertake a numerical and graphical study of the conditional survivorship function, $S(t + u, u)$, for the Weibull model, the extreme value model, the Gompertz model and the log-logistic model. What conclusions can be drawn from this?

6. Repeat the previous class project, focusing this time on the mean residual lifetime. Again what conclusions can be drawn from the graphs.

7. Consider a disease with three states of gravity (state 1, state 2 and state 3), the severity corresponding to the size of the number. State 4 corresponds to death and is assumed to follow state 3. New treatments offer the hope of prolonged survival. The first treatment, if it is effective, is anticipated to slow down the rate of transition from state 2 to state 3. Write down a compartmental model and a survival model, involving a treatment indicator, for this situation. A second treatment, if effective, is anticipated to slow down all transition rates. Write down the model for this. Write down the relevant null and alternative hypotheses for the two situations.

8. Consider a nondegenerative disease with several states; $1, 2, \ldots,$ counting the occurrence of these together with a disease state indicating a progression to something more serious, e.g., benign and malignant tumors or episodes of mild asthma with the possibility of progression to a more serious respiratory ailment. Write down possible models for this and how you might formulate tests of hypotheses of interest under varying assumptions on the role of the less serious states.

9. Suppose we have data; $T_1, \ldots, T_n$, from a Weibull distribution in which the shape parameter γ is known to be equal to 1.3. Use the delta-method to find an estimate for the variance of the estimated median (transform to a standard form).

10. For a proportional hazards Weibull model describe the relationship between the respective medians.

11. Investigate the function $S(t, u)$ for different parametric models described in this chapter. Draw conclusions from the form of this two-dimensional function and suggest how we might make use of these properties in order to choose suitable parametric models when faced with actual data.

12. Consider two possible structures for a parametric proportional hazards model;

$$\log S(t|Z) = \log\{S[t|E(Z)]\}\exp(\beta Z)$$
$$\log S(t|Z) = \log\{ES[t|Z]\}\exp(\beta Z).$$

How do the interpretations differ and what difficulties are likely to be encountered in fitting either of the models?

13. Consider a clinical trial comparing two treatments in which patients enter sequentially. Identify situations in which an assumption of an independent censoring mechanism may seem a little shaky.

14. On the basis of a single data set, fit the exponential, the Weibull, the Gompertz and the log-normal models. On the basis of each model estimate the mean survival. On the basis of each model estimate the 90th percentile. What conclusions would you draw from this.

15. Suppose our focus of interest is on the median. Can you write down a model directly in terms of the median. Would there be any advantage/drawback to modeling in this way rather than modeling the hazard and then obtaining the median via transformations of the hazard function?

Chapter 5

Marginal survival

5.1 Summary

In this chapter we examine in some detail estimates of marginal survival. Some assumptions on the censoring mechanism are needed in order to make progress. Attention is paid to the exponential and piece-wise exponential models. The exponential model, fully characterized by its mean, can appear over restrictive. However, via the probability integral transform and empirical estimates of marginal survival considered in this chapter, it can be used in more general situations. The piecewise exponential is seen, in some sense, to lie somewhere between the simple exponential and the empirical estimate. Particular attention is paid to empirical processes and how the Kaplan-Meier estimator, very commonly employed in survival-type problems, can be seen to be a natural generalization of the empirical distribution function. In the presence of parametric assumptions it is also straightforward to derive suitable estimating equations. The equations for the exponential model are particularly simple.

5.2 Motivation

Our interest is mostly in the survival function $S(t)$. Later we will focus on how $S(t)$, written as $S(t|Z)$, depends on covariates Z. Even though such studies of dependence are more readily structured around the hazard function $\lambda(t|Z)$, the most interpretable quantity we often would like to be able to say something about is the survival function itself. In order to distinguish the study of the influence of Z on $S(t|Z)$

from the less ambitious goal of studying $S(t)$, we refer to the former as conditional survival and the latter as marginal survival.

Since we will almost always have in mind some subset Z from the set of all possible covariates, and some distribution for this subset, we should remind ourselves that, although Z has been "integrated out" of the quantity $S(t)$, the distribution of Z does impact $S(t)$. Different experimental designs will generally correspond to different $S(t)$. Marginal survival, $S(t)$, corresponds to two situations: (i) the subjects are considered as *i.i.d.* replicates from a single population or (ii) the subjects can be distinct, from many, and potentially an infinite number of populations, each population being indexed by a value of some covariate Z. It may also be that we have no information on the covariates Z that might distinguish these populations. In case (ii), $S(t)$ is an average over these several populations, not necessarily representing any particular population of interest in itself. It is important to appreciate that, in the absence of distributional assumptions, and the absence of observable Z, it is not possible, on the basis of data, to distinguish case (i) from case (ii). The homogeneous case then corresponds to either case (i) or case (ii) and it is not generally useful to speculate on which of the cases we might be dealing with. They are not, in the absence of observable Z, identifiable from data. We refer to $S(t|Z)$ as the conditional survival function given the covariate Z. This whole area, the central focus of this work, is studied in the following chapter. First, we need consider the simpler case of a single homogeneous group.

5.3   Maximum likelihood estimation

The reader may wish to first recall Section 3.7 on estimating equations in general and maximum likelihood estimation in particular. Let us suppose that the survival distribution can be completely specified via some parametric model, the parameter vector being, say, θ. We take θ to be a scalar in most cases in order to facilitate the presentation. The higher-dimensional generalization is, in most cases, very straightforward. The data will consist of the n pairs (x_i, δ_i); $i = 1, \ldots, n$. We assume an independent censoring mechanism. This leads to the important theorem:

Theorem 5.1 *Under an independent censoring mechanism the log-likelihood can be written* $\log L(\theta) = \sum_{i=1}^{n} \log L_i(\theta)$ *where*

$$\log L_i(\theta) = \delta_i \log f(x_i; \theta) + (1 - \delta_i) \log S(x_i; \theta) \tag{5.1}$$

This covers the majority of cases in which parametric models are used. Later, when we focus on conditional survival involving covariates Z, rather than marginal survival, the same arguments follow through. In this latter case the common assumption, leading to an analogous expression for the log-likelihood, is that of conditional independence of the pair (T, C) given Z.

Estimating equation

The maximum likelihood estimate obtains as the value of θ, denoted $\hat{\theta}$, which maximizes $L(\theta)$ over the parameter space. Such a value also maximizes $\log L(\theta)$ (by monotonicity) and, in the usual case where $\log L(\theta)$ is a continuous function of θ this value is then the solution to the estimating equation $U(\theta) = 0$ where; $U(\theta) = \partial \log L(\theta)/\partial\theta = \sum_i \partial \log L_i(\theta)/\partial\theta$. Next, notice that at the true value of θ, denoted θ_0, we have $\mathrm{Var}\{U(\theta_0)\} = EU^2(\theta_0) = EI(\theta_0)$ where

$$I(\theta) = \sum_{i=1}^{n} I_i(\theta) = -\partial^2 \log L(\theta)/\partial\theta^2 = -\sum_{i=1}^{n} \partial^2 \log L_i(\theta)/\partial\theta^2.$$

As for likelihood in general, some care is needed in thinking about the meaning of these expressions and the fact that the operators $E(\cdot)$ and $\mathrm{Var}(\cdot)$ are taken with respect to the distribution of the pairs (x_i, δ_i) but with θ_0 fixed. The score equation is $U(\hat{\theta}) = 0$ and the large sample variance is approximated by $\mathrm{Var}(\hat{\theta}) \approx 1/I(\hat{\theta})$. It is usually preferable to base calculations on $I(\hat{\theta})$ rather than $EI(\hat{\theta})$, the former being, in any event, a consistent estimate of the latter (after standardizing by $1/n$). The expectation itself would be complicated to evaluate, involving the distribution of the censoring, and unlikely, in view of the study by Efron and Hinkley (1978) to be rewarded by more accurate inference. Newton-Raphson iteration is set up from

$$\hat{\theta}_{j+1} = \hat{\theta}_j + I(\hat{\theta}_j)^{-1}U(\hat{\theta}_j), \quad j \geq 1, \tag{5.2}$$

where $\hat{\theta}_1$ is some starting value, often zero, to the iterative cycle. The Newton-Raphson formula arises as an immediate application of the mean value theorem (Equation 2.2). The iteration is brought to a halt once we achieve some desired level of precision.

Large sample inference can be based on any one of the three tests based on the likelihood function; the score test, the likelihood ratio test or the Wald test. For the score test there is no need to estimate the unknown parameters. Many well-established tests can be derived in this way. In exponential families, also the so-called curved exponential families (Efron 1975), such tests reduce to contrasting some observed value to its expected value under the model. Good confidence intervals (see Cox and Hinkley 1974, page 212) can be constructed from "good" tests. For the exponential family class of distributions the likelihood ratio forms a uniformly most powerful test and, as such, qualifies as a "good" test in the sense of Cox and Hinkley. The other tests are asymptotically equivalent so that confidence intervals based on the above test procedures will agree as sample size increases. Also we can use such intervals for other quantities of interest such as the survivorship function which depends on these unknown parameters.

Estimating the survival function

We can estimate the survival function as $S(t; \hat{\theta})$. If Θ_α provides a $100(1 - \alpha)\%$ confidence region for the vector θ then we can obtain a $100(1 - \alpha)\%$ confidence region for $S(t; \theta)$ in the following way. For each t let

$$S_\alpha^+(t; \hat{\theta}) = \sup_{\theta \in \Theta_\alpha} S(t; \theta), \quad S_\alpha^-(t; \hat{\theta}) = \inf_{\theta \in \Theta_\alpha} S(t; \theta), \tag{5.3}$$

then $S_\alpha^+(t; \hat{\theta})$ and $S_\alpha^-(t; \hat{\theta})$ form the endpoints of the $100(1 - \alpha)\%$ confidence interval for $S(t; \theta)$. Such a quantity may not be so easy to calculate in general, simulating from Θ_α or subdividing the space being an effective way to approximate the interval. Some situations nonetheless simplify such as the following example, for scalar θ, based on the exponential model in which $S(t; \theta)$ is monotonic in θ. For such cases it is only necessary to invert any interval for θ to obtain an interval with the same coverage properties for $S(t; \theta)$.

Exponential survival

For this model we only need estimate a single parameter, λ which will then determine the whole survival curve. Referring to Equation (5.1) in which, for $\delta_i = 1$, the contribution to the likelihood is, $f(x_i; \lambda) = \lambda \exp(-\lambda x_i)$ and, for $\delta_i = 0$, the contribution is, $S(x_i; \lambda) = \exp(-\lambda x_i)$. Equation (5.1) then becomes:

$$\log L(\lambda) = k \log \lambda - \lambda \sum_{j=1}^n x_j, \tag{5.4}$$

where $k = \sum_{i=1}^{n} N_i(\infty)$. Differentiating this and equating with zero we find that $\hat{\lambda} = k / \sum_{j=1}^{n} x_j$. Differentiating a second time we obtain $I(\lambda) = k/\lambda^2$. Note, by conditioning upon the observed number of failures k, that $EI(\lambda) = I(\lambda)$, the observed information coinciding with the expected Fisher information, a property of exponential families, but which we are not generally able to recover in the presence of censoring.

An ancillary argument would nonetheless treat k as being fixed and this is what we will do as a general principle in the presence of censoring, the observed information providing the quantity of interest. Some discussion of this is given by Efron and Hinkley (1978) and Barndorff Nielsen and Cox (1999). We can now write down an estimate of the large sample variance which, interestingly, only depends on the number of observed failures. Thus, in order to correctly estimate the average, it is necessary to take into account the total time on study for both the failures and those observations that result in censoring. On the other hand, given this estimate of the average, the precision we will associate with this only depends on the observed number of failures. This is an important observation and will be made again in the more general stochastic process framework.

Multivariate setting

In the majority of applications, the parameter θ will be a vector of dimension p. The notation becomes heavier but otherwise everything is pretty much the same. The estimating equation, $U(\theta) = 0$ then corresponds to a system of p estimating equations and $I(\theta)$ is a $p \times p$ symmetric matrix in which the (q, r) th element is given by $-\partial^2 \log L(\theta)/\partial\theta_q\partial\theta_r$ where θ_q and θ_r are elements of the vector θ. Also, the system of Newton-Raphson iteration can be applied to each one of the components of θ so that we base our calculations on solving the set of equations:

$$\hat{\theta}_{j+1,q} = \hat{\theta}_{j,q} + I(\hat{\theta}_j)^{-1}U(\hat{\theta}_j), \quad q = 1, \ldots, p; \quad j \geq 1, \qquad (5.5)$$

where, in this case, $\hat{\theta}_1$ is a vector of starting values to the iterative cycle, again most often zero.

Example 5.1 *For the Freireich data we calculate for the 6-MP group* $\hat{\lambda} = 9/359 = 0.025$. *For the placebo group we obtain* $\hat{\lambda} = 21/182 = 0.115$. *Furthermore in the 6-MP group we have* Var $(\hat{\lambda}) = 9/$

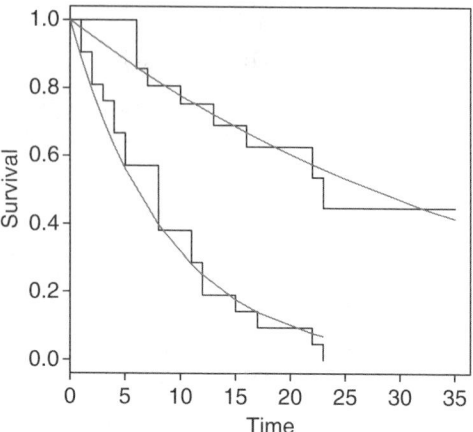

Figure 5.1: Exponential fitted curves to two-group Freireich data.

$(359)^2 = 0.000070$ *whereas for the placebo group we have* $\text{Var}(\hat{\lambda}) = 21/(182)^2 = 0.0006$.

The non-parametric empirical estimate (described below) agrees well with curves based on the exponential model and this is illustrated in Figure 5.1. Inference can be based on an appeal to the usual large sample theory. In this particular case, however, we can proceed in a direct way by recognizing that, for the case of no censoring $k = n$, the sample size and $\sum_{j=1}^{n} T_j$ is a sum of n independent random variables each exponential with parameter λ. We can therefore treat $n/\hat{\lambda}$ as a gamma variate with parameters (λ, n). When there is censoring, in view of the consistency of $\hat{\lambda}$, we can take $k/\hat{\lambda}$ as a gamma variate with parameters (λ, k), when $k < n$. This is not an exact result, since it hinges on a large sample approximation, but it may provide greater accuracy than the large sample normal approximation.

In order to be able to use standard tables we can multiply each term of the sum by 2λ since this then produces a sum of n exponential variates, each with variance 2. Such a distribution is a gamma $(2, n)$, equivalent to a chi-square distribution with $2n$ degrees of freedom (Hastings and Peacock 1975, page 46). Taking the range of values of $2k\lambda/\hat{\lambda}$ to be between $\chi_{\alpha/2}$ and $\chi_{1-\alpha/2}$ gives a $100(1 - \alpha)\%$ confidence interval for λ. For the Freireich data we find 95% $CI = (0.0115, 0.0439)$. Once we have intervals for λ, we immediately have intervals with the same coverage properties for the survivorship function, this being a monotonic function of λ. Denoting the upper and lower limits of the $100(1-\alpha)\%$ confidence interval $S_\alpha^+(t;\hat\lambda)$ and $S_\alpha^-(t;\hat\lambda)$ respectively, we have:

$$S_\alpha^+(t;\hat\lambda)=\exp\left\{-\left(\frac{\hat\lambda\chi_{\alpha/2}}{2k}\right)t\right\}, \quad S_\alpha^-(t;\hat\lambda)=\exp\left\{-\left(\frac{\hat\lambda\chi_{1-\alpha/2}}{2k}\right)t\right\}.$$

An alternative approximation, based on large sample normality, and also very simple in form, can be written as:

$$S_\alpha^+(t;\hat\lambda)\approx\exp\left\{-\left(\frac{\hat\lambda}{\sqrt k}\right)(\sqrt k - z_{1-\alpha/2})t\right\}, \tag{5.6}$$

where, this time, the corresponding expression for $S_\alpha^-(t;\hat\lambda)$ obtains by replacing $z_{1-\alpha/2}$ by $z_{\alpha/2}$.

Piecewise exponential model

We can achieve considerably greater flexibility than the standard exponential model by constructing a partition of the time axis $0 = a_0 < a_1 < \ldots < a_k = \infty$. Within the jth interval $(a_{j-1}, a_j), (j = 1, \ldots, k)$ the hazard function is given by $\lambda(t) = \lambda_j$. Using Equations (4.11) and (5.1) and equating first derivatives to zero we find

$$\hat\lambda_j = k_j/\sum_{\ell:x_\ell>a_{j-1}} \{(x_\ell - a_{j-1})I(x_\ell < a_j) + (a_j - a_{j-1})I(x_\ell \geq a_j)\}.$$

Differentiating a second time we find $I(\lambda_1, ..., \lambda_k)$ to be block diagonal where the (j, j)th element is given by $I_{jj} = k_j/\lambda_j^2$. Verify. In view of the orthogonality of the parameter estimates, we can construct a $100(1-\alpha)\%$ simultaneous confidence interval for $\lambda_1, \ldots, \lambda_k$ by choosing a sequence of α_j such that $1 - \prod_j(1 - \alpha_j) = \alpha$. An estimate of the survivorship function derives from $\hat\Lambda(t)$ in which we define $\hat S(t;\hat\Lambda) = \exp\{-\hat\Lambda(t)\}$ and where

$$\hat\Lambda(t) = \sum_{j:a_j\leq t} \hat\lambda_j(a_j - a_{j-1}) + \sum_\ell \hat\lambda_\ell(t - a_{\ell-1})I(a_{\ell-1} \leq t < a_\ell). \tag{5.7}$$

Confidence intervals for this function can be based on Equation (5.3). We can view the simple exponential survival model as being at one extreme of the parametric spectrum, leaning as it does on a single parameter, the mean. It turns out that we can view the piecewise exponential model, with a division so fine that only single failures occur in any interval, as being at the other end of the parametric spectrum, i.e., a nonparametric estimate. Such an estimate corresponds to that obtained from the empirical distribution function. This is discussed below.

Other parametric models

The exponential and piecewise exponential models hold a special place in the survival literature for a number of reasons. The models are simple in form, simple to understand, have a clear physical property that we can interpret (lack of memory property) and the parameters can be estimated so easily that analysis based on such models clearly falls under the heading of desirable procedures as defined by Student; ones that can be calculated on the back of a railway ticket while awaiting the train. The piecewise model also allows quite considerable flexibility.

Given these facts, it is hard to justify the use of other parametric models unless motivated by some compelling physical argument. Cox (1958) used the Weibull model in analyzing the strengths of wool, but the model was not just pulled out of the air. Physical considerations and the fact that the Weibull distribution obtains as the limiting case of the minimum of a collection of random variables provide a powerful case in its favor. Another physical case might be the sum of exponential variates, each having the same parameter, since this can be seen to be a gamma variate. Generally we may have no good reason to believe some model over most others is likely to provide a good fit to data. In these cases, given the generally good performance of empirical estimates, it may be preferable to base our analyses on these.

5.4   Empirical estimate (no censoring)

Empirical or nonparametric estimates, i.e., those making no model assumptions, can be most readily understood in the context of some finite division of the time axis. Theoretical results stem from continuous-time results, the transition from the discrete to the continuous, as a consequence of sample size n going to infinity, presenting no conceptual difficulty. Unlike the discussion on the piecewise exponential model in which we most often anticipate having very few intervals, rarely more than three or four, the idea being to pool resources (estimating power) per interval, while keeping within the bounds of serious model violation, for empirical estimates we do the opposite.

We imagine a fine division of the real line in which, given real data, the vast majority of the intervals are likely to contain no observations. Consider the time interval to be fixed, divided equally into k non overlapping intervals $(a_{j-1}, a_j]$, $j = 1, \ldots, k$, the notation "(" indicating that the interval is open on the left, and "]" closed on the right, i.e., a_{j-1} does not belong to the interval but a_j does. We have that $\Pr(T \geq a_j) = \Pr(T > a_j) = S(a_j)$ and that $\Pr(T > a_j | T > a_{j-1}) = S(a_j, a_{j-1})$. Recall from Section 4.3 that

$$S(a_j) = S(a_{j-1})S(a_j, a_{j-1}) = \prod_{\ell \leq j} S(a_\ell, a_{\ell-1}).$$

For each $t = a_\ell$, $(\ell > 0)$, the empirical estimate of $S(t)$ based on a sample of size n, and denoted $S_n(t)$, is simply the observed number of observations that are greater than t. For a random sample of observations T_i, $(i = 1, \ldots, n)$ we use the indicator variable $I(\cdot)$ to describe whether or not the subject i survives beyond point t, i.e., for $t = a_j$,

$$S_n(a_j) = \frac{1}{n} \sum_{i=1}^{n} I(T_i > a_j) = \prod_{\ell=1}^{j} S_n(a_\ell, a_{\ell-1}) \qquad (5.8)$$

in which the empirical $S_n(a_\ell, a_{\ell-1})$ is defined in an entirely analogous way to $S_n(a_\ell)$, i.e.,

$$S_n(a_\ell, a_{\ell-1}) = \frac{1}{n_{\ell-1}} \sum_{i=1}^{n} I(T_i > a_\ell), \quad n_\ell = \sum_{i=1}^{n} I(T_i \geq a_\ell). \quad (5.9)$$

It is readily seen, and instructive for understanding the Kaplan-Meier estimate of the next section, that, if no failure is observed in $(a_{\ell-1}, a_\ell]$ then $S_n(a_\ell, a_{\ell-1}) = 1$, whereas for an observed failure, $S_n(a_\ell, a_{\ell-1}) = (n_{\ell-1} - 1)/n_{\ell-1}$. The empirical distribution has been well studied and, in particular, we have:

Lemma 5.1 *For any fixed value of t, $S_n(t)$ is asymptotically normal with mean and variance given by;*

$$E\{S_n(t)\} = S(t); \quad \text{Var}\{S_n(t)\} = F(t)S(t)/n. \qquad (5.10)$$

We should understand the operators E and Var to refer to expectations over a set of repetitions, the number of repetitions increasing without bound, and with n and t fixed. As an estimator of $S(t)$, the function

$S_n(t)$ is then unbiased. The variance, as we might anticipate, is the variance of a binomial variate with parameters n and $S(t)$. These two moments, together with a normal approximation of DeMoivre-Laplace (see Section 3.3) enable the calculation of approximate confidence intervals. The result, which is well known (Glivenko-Cantelli), enables us to carry out inference for $S(t)$ on the basis of $S_n(t)$ for any given point t. Very often we will be interested in the whole function $S_n(t)$, over all values of t and, in this case, it is more helpful to adopt the view of $S_n(t)$ as a stochastic process. In this regard we have

Theorem 5.2 $\sqrt{n}\{S_n(t) - S(t)\}$ *is a Gaussian process with mean zero and covariance given by*

$$\mathrm{Cov}\left[\sqrt{n}\{S_n(s)\}, \sqrt{n}\{S_n(t)\}\right] = F(s)\{1 - F(t)\}. \qquad (5.11)$$

An important consequence of the theorem arises when the distribution of T is uniform, for, in this case, $F(s) = s$ and all the conditions are met for the process $\sqrt{n}\{F_n(t) - t\}$ ($0 \leq t \leq 1$) to converge in distribution to the Brownian bridge. Finally these results apply more generally than just to the uniform case for, as long as T has a continuous distribution, there exists a unique monotonic transformation from T to the uniform, such a transformation not impacting $\sqrt{n}\{F_n(t) - F(t)\}$ itself. In particular this enables us to use the result of Section 2.11 to make inference for an arbitrary continuous cumulative distribution function, $F(t)$, whereby, for $W_n(t) = \sqrt{n}\{F_n(t) - F(t)\}$,

$$\Pr\left\{\sup_t |W_n(t)| \leq D\right\} \to 1 - 2\sum_{k=1}^{\infty} (-1)^{k+1} \exp(-2k^2 D^2), \quad D \geq 0. \quad (5.12)$$

Most often we are interested in events occurring with small probability in which case a good approximation obtains by only taking the first term of the sum, i.e., $k = 1$. Under this approximation $|\sqrt{n}\{F_n(t) - F(t)\}|$ will be greater than about 1.5 less than 5% of the time. This is a simple and effective working rule.

5.5   Empirical estimate (with censoring)

First, recall the model for finite censoring support, described in Section 4.5. Such a model adds no additional practical restriction but helps us to avoid much mathematical complexity, leading to a simple and useful theorem (Theorem 5.3). The theorem enables us to circumvent the difficulty arising from the fact that we are unable to obtain $\sum_{i=1}^{n} I(T_i \geq a_j)$ as in Equation 5.8. This is because at a_j we cannot ascertain whether or not earlier observed values of X_i ($\delta_i = 0$), arising from the pair (T_i, C_i) in which $C_i = X_i$, are such that $T_i > a_j$. Since $X_i = \min(T_i, C_i)$ we do not observe the actual value of T_i when $X_i = C_i$. The trick is to notice in the right hand-side of (5.8) that we are able to obtain the empirical estimates, $G_n(a_\ell, a_{\ell-1})$ for $G(a_\ell, a_{\ell-1}) = \Pr(X \geq a_\ell | X > a_{\ell-1})$. But, unlike Equation (5.8)

$$G_n(a_j) = \frac{1}{n} \sum_{i=1}^{n} I(X_i \geq a_j) \neq \prod_{\ell \leq j} G_n(a_\ell, a_{\ell-1}). \qquad (5.13)$$

This is because of the non zero masses being associated to the times at which the censorings occur. Nonetheless, the rest follows through readily, although, unlike (5.9), we now define $n_\ell = \sum_{i=1}^{n} I(X_i \geq a_\ell)$, noting that this definition contains (5.9) as a special case when there is no censoring.

Kaplan-Meier estimate

The distribution of $X = \min(T, C)$ is mostly of indirect interest but turns out to be important in view of the following theorem and corollary.

Theorem 5.3 *Under an independent censoring mechanism,*

$$G(a_\ell, a_{\ell-1}) = S(a_\ell, a_{\ell-1}). \qquad (5.14)$$

Corollary 5.1 *Under an independent censoring mechanism, a consistent estimator of $S(t)$ at $t = a_j$ is given by $\hat{S}(t) = \prod_{\ell \leq j} G_n(a_\ell, a_{\ell-1})$.*

Corollary 5.2 *In the absence of censoring $\hat{S}(t) = S_n(t)$.*

The estimator of the corollary is known as the Kaplan-Meier estimator. In the light of our particular set-up, notably the use of the independent censoring mechanism having finite support (Section 4.5), it can be argued that our estimate is only available at distinct values of $t = a_j$. But this is not a practical restriction since our finite division can be as fine as we wish, the restriction amounting to limiting accuracy to some given number of decimal places. One way or another it is not possible to

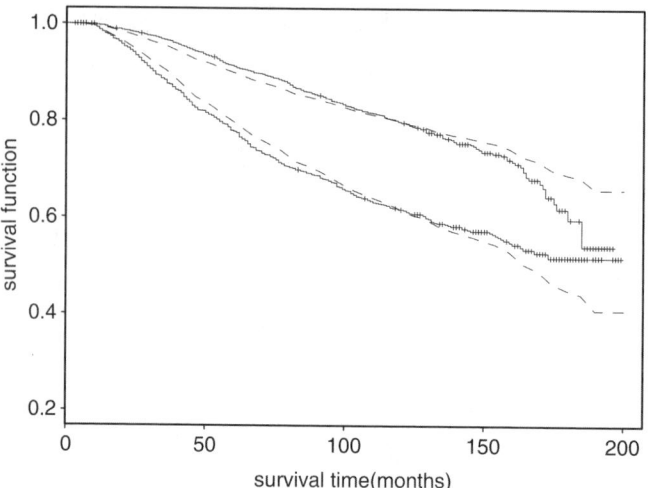

Figure 5.2: Individual Kaplan-Meier survival estimates for two groups.

completely avoid some mathematical fine points, our preference being
to use the mechanism of Section 4.5.

In practice we talk about the Kaplan-Meier estimate at time point
t, the underlying support of C and T rarely being a central concern.
The one exception to this arises when the range of C is less than that
for T. A single sampled T_i which is greater than the greatest value that
can be taken by C_i will necessarily appear as a censored observation.
In this case the observed empirical Kaplan-Meier estimate is never
able to reach zero. The true distribution $F(t)$ cannot be estimated
at values greater than the upper limit for the support of C. This is
intuitively clear. Again, in practice, this ought not be a real concern.
If we have no information on something (in this case a certain upper
region of the survival curve), it is most likely too optimistic to hope
to be able to carry out useful estimation there, unless, of course, we
make additional parametric assumptions, which amount to saying that
information obtained over some range can tell us a lot about what is
taking place elsewhere, and certainly more than just the provision of
bounds on $F(t)$ for t greater than the upper limit of C.

Remarks on the Kaplan-Meier curve

The Kaplan-Meier (1958) product-limit estimator provides a nonparametric estimate for the survival function $S(t)$ under the independent

censorship assumption. It is a generalization of the commonly known empirical distribution function to the case of censoring since, in the absence of censoring, the two coincide. Expressing the estimate $\hat{S}(t)$ in terms of the fine division $(a_\ell, a_{\ell-1})$, $\ell = 1, \ldots, N$, is conceptually useful. However, since intervals not containing events produce no change in $\hat{S}(t)$ it is, for the purposes of practical calculation, only necessary to consider evaluation at the distinct observed failure times $t_1 < t_2 < \cdots < t_k$. All divisions of the time interval $(a_\ell, a_{\ell-1})$, $\ell = 1, \ldots, N$, for different N, lead to the same estimate $\hat{S}(t)$, provided that the set of observed failure points is contained within the set $\{a_\ell ; \ell = 1, \ldots, N\}$. A minimal division of the time axis arises by taking the set $\{a_\ell\}$ to be the same as the set of the distinct observed failure times. So, for practical purposes $a_j = t_j$, $j = 1, \ldots, k$. and the Kaplan-Meier estimate can be defined as

$$\hat{S}(t) = \prod_{j:t_j < t} \frac{n_j - d_j}{n_j} = \prod_{j:t_j < t} 1 - \frac{d_j}{n_j}. \tag{5.15}$$

Note that $\hat{S}(t)$ is a left-continuous step function that equals 1 at $t = 0$ and drops immediately after each failure time t_j. The estimate does not change at censoring times. When a censoring time and a failure time t_j are recorded as equal, the convention is that censoring times are adjusted an infinitesimal amount to the right so that the censoring time is considered to be infinitesimally larger that t_j. Any subjects censored at time t_j are therefore included in the risk set of size n_j, as are those that fail at t_j. This convention is sensible because a subject censored at time t_j almost certainly survives beyond t_j. Note also that when the last observation is a censoring time rather than a failure time, the KM estimate is taken as being defined only up to this last observation.

Continuous version of Kaplan-Meier curve

The Kaplan-Meier curve is so useful and so commonly employed that we reserve the most standard notation for this estimate, i.e., $\hat{S}(t)$. For the majority of applications this is all we need. However, there are some situations in which it is useful to be able to invert the function in order to estimate $S^{-1}(t)$. Now, the usual Kaplan-Meier estimate takes a constant value between adjacent failures, t_{j-1} and t_j and so, in general, we are unable to invert this function in a unique way. This is not a serious concern and any intuitive solution such as taking $\hat{S}(t)$ to be $S(t_{j-1})$ or $S(t_j)$ for all $t \in (t_{j-1}, t_j)$ would be fine. Rather than this we suggest a simple linear interpolation. We then define the continuous (and invertible since strictly decreasing) function $\bar{S}(t)$ to be given by

$$\bar{S}(t) = \hat{S}(t_{j-1}) + \left(\frac{t - t_{j-1}}{t_j - t_{j-1}} \right) \left\{ \hat{S}(t_j) - \hat{S}(t_{j-1}) \right\} ; \quad t \in (t_{j-1}, t_j)$$

(5.16)

Note that at the distinct failure times t_j the two estimates, $\hat{S}(t)$ and $\bar{S}(t)$ coincide. An example of where we make an appeal to $\bar{S}(t)$ is illustrated in the two-group exponential model where a transformation to exponentiality for one group is then applied to the other group.

Precision of Kaplan-Meier estimate

Recalling the results of Sections 2.3 and 2.9 we can derive a Taylor series approximation to a function of random variables wherever the function of expectation converges in probability to the expectation of the function. An immediate application leads to the following theorem.

Theorem 5.4 *For each* $t = a_\ell$, *the estimate* $\hat{S}(t)$ *is asymptotically normal with mean* $S(t)$ *and variance:*

$$\text{Var}\, S(a_\ell) \approx S(a_\ell)^2 \sum_{m \leq \ell} \sum_{m \leq \ell} \frac{d_m}{n_m (n_m - d_m)}.$$

(5.17)

The above expression for the variance of $\hat{S}(t)$ is known as Greenwood's formula. Breslow and Crowley (1974) in a detailed large sample study of the Kaplan-Meier estimator obtained a result asymptotically equivalent to the Greenwood formula, making a slight correction to overestimation of the variation in the estimated survival probability. The formula's simplicity, however, made it the most commonly used when computing the variance of the Kaplan-Meier estimate and related quantities. We also have:

Corollary 5.3 *When there is no censoring, the approximation* $\text{Var}\hat{S}(t)$ *from theorem 5.4 reduces to the usual binomial variance estimate* $\hat{S}(t)\{1 - \hat{S}(t)\}/n$.

The usual use to which we put such variance estimates is in obtaining approximate confidence intervals. Thus, using the large sample normality of $\hat{S}(t)$, adding and subtracting to this $z_{1-\alpha/2}$ (the $1 - \alpha/2$ quantile from the standard normal distribution) multiplied by the square root of the variance estimate, provides approximate $100(1-\alpha)\%$ confidence intervals for $\hat{S}(t)$. As mentioned before the constraints on $\hat{S}(t)$, lying between 0 and 1, will impact the operating characteristics of such intervals, in particular it may not be realistic, unless sample sizes are large, to limit attention to symmetric intervals around $\hat{S}(t)$. Borgan and Liestøl (1990) investigate some potential transformations, especially the log-minus-log transformation discussed in Section 4.4, leading to;

Corollary 5.4 *Let* $w(\alpha) = \text{Var}^{1/2} \hat{S}(t) z_{1-\alpha/2}/\hat{S}(t) \log \hat{S}(t)$. *For each* $t = a_\ell$, *a* $100(1-\alpha)\%$ *confidence intervals for* $\hat{S}(t)$ *can be approximated by*

$$\{\hat{S}(t)^{\exp -w(\alpha)}, \ \hat{S}(t)^{\exp w(\alpha)}\} \tag{5.18}$$

The same arguments which led to Greenwood's formula also lead to approximate variance expressions for alternative transformations of the survivorship function. In particular we have;

Corollary 5.5 *For each* $t = a_\ell$ *the estimate* $\log \hat{S}(t)$ *is asymptotically normal with asymptotic mean* $\log S(t)$ *and variance*

$$\text{Var} \log S(a_\ell) \approx \sum_{m \leq \ell} \sum_{m \leq \ell} \frac{d_m}{n_m(n_m - d_m)} \tag{5.19}$$

Corollary 5.6 *For each* $t = a_\ell$ *the estimate* $\log \hat{S}(t)/\{1 - \hat{S}(t)\}$ *is asymptotically normal with asymptotic mean* $\log S(t)/\{1 - S(t)\}$ *and variance*

$$\text{Var} \log \left\{ \frac{S(a_\ell)}{1 - S(a_\ell)} \right\} \approx \{1 - S(a_\ell)\}^{-2} \sum_{m \leq \ell} \sum_{m \leq \ell} \frac{d_m}{n_m(n_m - d_m)}. \tag{5.20}$$

Confidence intervals calculated using any of the above results will be of help in practice. Following some point estimate, obtained from $\hat{S}(t)$ at some given t, these intervals are useful enough to quantify the statistical precision that we wish to associate with the estimate. All of the variance estimates involve a comparable degree of complexity of calculation so that choice is to some extent a question of taste. Nonetheless, intervals based on the log-minus-log or the logit transformation will behave better for smaller samples, and guarantee that the endpoints of the intervals themselves stay within the interval (0,1). This is not so

for the Greenwood formula, the main argument in its favor being that
it has been around the longest and is the most well known. For moderate to large sample sizes, and for $\hat{S}(t)$ not too close to 0 or 1, all the
intervals will, for practical purposes, coincide.

Kaplan-Meier curve and redistribution to the right algorithm

Another way to look at the KM estimate, which turns out to be conceptually useful and of value to later developments, is to focus on the
increments, or step size, of the function at points where it changes. If we
denote t_j+ the time instant immediately after t_j, then $\hat{S}(t_j) - \hat{S}(t_j+)$
is the stepsize, or jump, of the KM curve at time t_j. From Equation
5.15 we see that

$$\hat{S}(t_j+) = \hat{S}(t_j) \cdot \frac{n_j - d_j}{n_j},$$

so the stepsize is $\hat{S}(t_j) \cdot d_j/n_j$. That is to say, when the total "leftover"
probability mass is $\hat{S}(t_j)$, each observed failure gets one-n_jth of it,
where $n_j = \sum_{i=1}^{n} Y_i(t_j)$ is the number of subjects at risk at time t_j.
In the absence of censoring this corresponds exactly to the way in
which the empirical estimate behaves. When there is censoring, then
one way of looking at a censored observation is to consider that the
mass that would have been associated with it is simply reallocated to
all of those observations still remaining in the risk set (hence the term
"redistribution to the right.")

Kaplan-Meier estimates of median and mean survival

Since the support of T is only on the positive real line the distributions
we deal with are asymmetrical. In consequence it is more common
to take as summary measures simple functions of different quantiles,
most often the median or interquartile range, rather than the mean
and variance. Nonetheless, the mean is of interest and in special cases,
such as the exponential distribution, relates directly to the median
via a constant scaling factor. Also, of course, there is no compelling
reason not to work with symmetric distributions defined for say $\log T$
and then transform back to T, although this is not very commonly
done. From Section 4.3 we can estimate the expected life time over
some given interval $[0, t]$ as

$$\hat{\mu}(t) = \int_0^t \hat{S}(u)du, \qquad (5.21)$$

the mean itself being then estimated by $\hat{\mu}(\infty)$. However, the theory comes a little unstuck here since, not only must we restrict the time scale to be within the range determined by the largest observation, the empirical distribution itself will not correspond to a probability distribution whenever $\hat{F}(t) = 1 - \hat{S}(t)$ fails to reach one. In practice then it makes more sense to consider mean life time $\hat{\mu}(t)$ over intervals $[0, t]$, acknowledging that t needs to be kept within the range of our observations. The following result provides the required inference for $\hat{\mu}(t)$;

Lemma 5.2 *For large samples,* $\hat{\mu}(t)$ *can be approximated by a normal distribution with* $E\hat{\mu}(t) = \mu(t)$ *and*

$$\text{Var } \hat{\mu}(t) \approx \sum_{m \leq \ell} \sum_{m \leq \ell} \frac{\{\hat{\mu}(t) - \hat{\mu}(a_m)\}^2 d_m}{n_m(n_m - d_m)}. \tag{5.22}$$

In view of the consistency of the Kaplan-Meier estimate it is, in principle, straightforward to obtain estimates for any desired quantile, or function of the quantiles, such as the median or interquartile range. However, again, we can be limited by the observations and, if $\hat{S}(t)$, for the largest observed survival time, is not less than p then we are not able to obtain point estimates of quantiles corresponding to such values, although we could obtain interval estimates. We define the pth quantile ξ_p to satisfy $F(\xi_p) = p$. The Kaplan-Meier function, or indeed the uncensored empirical distribution function itself, being a step function, is not invertible. To overcome this, we define the estimate $\hat{\xi}_p$ to be the smallest value of t such that $\hat{F}(\xi_p) \geq p$. In order to make inferences for $\hat{\xi}_p$ we can appeal to some basic results from Section 2.7 to obtain:

Lemma 5.3 *For large samples,* $\hat{\xi}_p$ *approaches a normal distribution with* $E\hat{\xi}_p = \xi_p$ *and*

$$\text{Var } \hat{\xi}_p \approx \sum_{m \leq \ell} \sum_{m \leq \ell} \frac{d_m(1 - p^2)f^{-2}(\hat{\xi}_p)}{n_m(n_m - d_m)}. \tag{5.23}$$

It is difficult to use the above result in practice in view of the presence of the density $f(\cdot)$ in the expression. Smoothing techniques and the methods of density estimation can be used to make progress here but our recommendation would be to use a more direct, albeit more heavy, approach, constructing intervals based on sequences of hypothesis tests. In principle at least the programming of these is straightforward.

Nelson-Aalen estimate of survival

An alternative approach to estimating the empirical survival distribution, adapted to accommodate censoring, and, essentially, equivalent to the Kaplan-Meier estimate arises from considerations of the basic formulae in Section 4.3. Recalling that $S(t) = \exp\{-\Lambda(t)\}$ and that $\Lambda(t) = \int_0^t \lambda(u)du$ we can consider empirical estimates of $\Lambda(t)$. The integral can be approximated by a Riemann sum so that for $t = a_j$

$$\sum_{\ell \leq j} \lambda(a_\ell)(a_\ell - a_{\ell-1}) \rightarrow \int_0^{a_j} \lambda(u)du \tag{5.24}$$

as $a_\ell - a_{\ell-1}$ goes to zero. Now, applying a local linearization to 4.2 we obtain

$$\lambda(a_\ell)(a_\ell - a_{\ell-1}) \approx P(a_\ell < T < a_{\ell-1}|T > a_{\ell-1}) = 1 - S(a_\ell, a_{\ell-1}).$$

Applying theorem 5.3 and then, first replacing $S(a_\ell, a_{\ell-1})$ by $G(a_\ell, a_{\ell-1})$, second replacing $G(a_\ell, a_{\ell-1})$ by $G_n(a_\ell, a_{\ell-1})$ i.e., d_ℓ/n_ℓ, we obtain, at $t = a_j$, $\tilde{\Lambda}(a_j) = \sum_{\ell=1}^j d_\ell/n_\ell$ as a consistent estimator for $\Lambda(t)$. The resulting estimator

$$\tilde{S}(t) = \exp\{-\tilde{\Lambda}(t)\} \tag{5.25}$$

is called the Nelson-Aalen estimate of survival. Recalling the Taylor series expansion, $\exp(x) = 1 + x + x^2/2! + \cdots$, for small values of x we have $\exp(-x) = 1 - x + O(x^2)$, the error of the approximation being strictly less than $x^2/2$ since the series is convergent with alternating sign. Applying this approximation to $\tilde{S}(t)$ we recover the Kaplan-Meier estimate described above. In fact we can use this idea to obtain:

Lemma 5.4 *Under the Breslow-Crowley conditions, $|\tilde{S}(t) - \hat{S}(t)|$ converges almost surely to zero.*

In view of the lemma, large sample results for the Nelson-Aalen estimate can be deduced from those already obtained for the Kaplan-Meier estimate. This is the main reason that there is relatively little study of the Nelson-Aalen estimate in its own right. We can exploit the wealth of results for the Kaplan-Meier estimate that are already available to us. Indeed, in most practical finite sample applications, the level of agreement is also very high and the use of one estimator rather than the other is really more a question of taste than any theoretical advantage. In some ways the Nelson-Aalen estimate appears very natural in the survival setting, and it would be nice to see it used more in practice.

Model verification using empirical estimate

A natural approach to model assessment, i.e., whether or not some parametric model appears as a reasonable choice for the observed data, is to contrast the empirical estimates to those leaning on the model assumptions, the role of the data being reduced to that of providing estimates for any unknown parameters. We do not propose tackling the broad issues of goodness of fit until later but notice that, if the assumed parametric form is reasonable, then $\hat{S}(t)$ and $S_k(t)$ should broadly agree. For the Weibull model, for example, we know that $S(t) = \exp\{-(\lambda t)^p\}$. Therefore, a plot, from the Kaplan-Meier, or Nelson estimate, of $\log\{-\log \hat{S}(t)\}$ against $\log t$ should be linear with intercept equal to $p \log \lambda$ and slope equal to p. For all the other parametric models it is possible to devise similar constructions. A visual impression of the adequacy of any postulated model is obtained, alongside the possibility of obtaining simple parameter estimates for the unknown parameters. Such estimates are typically less efficient than maximum likelihood estimates so, in a more thorough analysis, we may wish to use them either as a rough guide or as a first approximation in some iterative scheme. In practice it is often the case that quite different parametric assumptions, unless particularly restrictive like that for the exponential model, will produce very similar survival curves. Important differences between competing parametrizations tend to manifest themselves mostly in the tails of the distribution where there may be few observations. As a goodness of fit tool then these procedures are not usually very powerful.

Nonparametric exponential analysis

Referring to Section 2.5 and Theorem 2.8, we have the important result that, for any continuous positive random variable T, with distribution function $F(t)$, the variate $\Lambda(T) = \int_0^T f(u)/[1 - F(u)]du$ has a standard exponential distribution. As a consequence, if we consider the empirical survivorship function, $\hat{S}(t)$, then we can take the observations $-\log \hat{S}(X_i)$ as arising from a standard exponential distribution. All of the simple results that are available to us when data are generated by an exponential distribution can be used. In particular, if we wish to compare the means of two distributions, both subject to censoring, then we can transform one of them to standard exponential via its empirical survival function, then use this same transformation on the other group. The simple results for contrasting two censored exponential samples can then be applied even though, at least initially, the data arose from samples generated by some other mechanism.

5.6   Exercises and class projects

1. Write down the estimating equations for a Weibull model based on maximum likelihood. Write down the estimating equations for a Weibull model based on the mean and variance.

2. Consider the following observations; 1, 3, 4, 4, 5, 6^*, 6, 7, 9^*, 16 where a $*$ indicates a censored observation. Fit a Weibull model to these observations based on (i) maximum likelihood, (ii) method of moments.

3. For the data of the previous question, calculate and plot the survivorship function. Calculate an approximate 90% confidence interval for $S(4)$. Do the same for $S(7)$.

4. Describe how you might calculate a simultaneous 90% confidence interval for $S(4)$ and $S(7)$ together.

5. Compare the variance expression for $S(7)$ with that approximated by the binomial formula based on $\hat{S}(7)$ and 8 failure times.

6. Take 100 bootstrap samples, fit the Weibull model to each one separately and estimate $S(4)$ and $S(7)$. Calculate empirical variances based on the 100 sample estimates. How do these compare with those calculated on the basis of large sample theory.

7. In the previous question, rather than fit the model, we could calculate empirical estimates of $S(4)$ and $S(7)$. Describe possible difficulties with this approach.

8. For small samples generated via an exponential distribution with unknown mean, discuss the relative merits of the different possible confidence interval approximations for the survival function. Describe a study you might set up in order to make a recommendation regarding the "best" confidence interval to work with. How do you understand "best" in this context.

9. Simulate 100 uncensored observations from a standard exponential distribution. Calculate the empirical survival function $S_n(t)$. Choose two points, s and t and repeat the whole process 100 times obtaining 100 pairs of $S_{100}(s)$ and $S_{100}(t)$. Calculate the empirical covariance between $S_{100}(s)$ and $S_{100}(t)$. Use different values of s and t to suggest the validity of Equation 5.2.

10. Explain why the result of Equation 5.12 for $W_n(t) = \sqrt{n}\{F_n(t) - F(t)\}$, when $F(t)$ is uniform continues to hold for any other continuous distribution.

11. Explain the importance of Theorem 5.3 and how it is used in order to obtain consistent estimates of survival in the presence of an independent censoring mechanism.

12. Simulate 200 uncensored observations from the log-normal distribution. Suppose that we had been led to believe that the observations had been generated from a Weibull law. Carry out graphical procedures to challenge the validity of the assumption. Repeat the exercise under the supposition that the observations had been generated via a log-logistic model.

13. For the 200 observations of the previous question, introduce an independent censoring mechanism so that approximately half of the observations are censored. Calculate the logarithm of the Kaplan-Meier and Nelson-Aalen estimates and plot one against the other. Fit a least squares line to the plot and comment on the values of the slope.

14. Use the results of Lemma 5.2 to show that $\hat{\mu}(t)$ is consistent for $\mu(t)$.

15. Show that when there is no censoring, the Greenwood estimate of the variance of the Kaplan-Meier estimate reduces to the usual variance estimate for the empirical distribution function. Conclude from this that confidence intervals based on the Greenwood estimate of variance are only valid at a single given time point, t, and would not provide bounds for the whole Kaplan-Meier curve.

16. Carry out a study on the coverage properties based on $\hat{S}(t)$, $\log \hat{S}(t)$ and $\log \hat{S}(t)/\{1-\hat{S}(t)\}$. Describe what you anticipate to be the relative merits of the different functions.

17. Malani (1995) outlined an operationally simple approach for estimating survival in the presence of a dependent censoring mechanism. The method requires that the dependency be captured via some explanatory variable. Appealing to the redistribution to the right algorithm, each censored observation has its remaining mass redistributed. However, unlike the simple version of the algorithm, Malani proposes to only redistribute among subjects in the risk set sharing the same covariate value as the subject censored at that point. Give an intuitive explanation as to why this would work.

18. Following the idea of Malani (1995) suppose, in the presence of dependent censoring, we obtained a Nelson-Aalen estimate of survival for each level of the covariate. Subsequently, appealing to the law of total probability, we estimate marginal survival by a linear combination of these several estimates. Comment on such an estimate and contrast it with that of Malani.

19. Recall that the uncensored Kaplan-Meier estimator, i.e., the usual empirical estimate, is unbiased. This is no longer generally so for the Kaplan-Meier estimate. Can you construct a situation in which the estimate of Equation 5.16 would exhibit less bias than the Kaplan-Meier estimate?

20. Using data from a cancer registry, show how you could make use of the piecewise exponential model to obtain conditional survival estimates for $S(T > t + s|T > s)$.

Chapter 6

Regression models and subject heterogeneity

6.1 Summary

We consider several models that describe survival in the presence of observable covariates, these covariates measuring subject heterogeneity. The most general situation can be described by a model with a parameter of high, possibly unbounded, dimension. Proportional hazards models, partially proportional hazards models (O'Quigley and Stare 2002), stratified models or models with frailties or random coefficients all arise as special cases of this model (O'Quigley and Xu 2000). One useful parameterization (O'Quigley and Pessione 1991, O'Quigley and Prentice 1991) can be described as a non proportional hazards model with intercept. Changepoint models are a particular form of a non proportional hazards model with intercept (O'Quigley and Natarajan 2004). Any model can be viewed as a special case of the general model, lying somewhere on a conceptual scale between this general model and the most parametric extreme, which would be the simple exponential model. Models can be placed on this scale according to the extent of model constraints and, for example, a random effects model would lie strictly between a stratified model and the simple exponential model. Relative risk models used in epidemiology come under these headings. For relative risk models the time component is usually taken to be age and great generalization, e.g., period or cohort analysis is readily accomplished. Time-dependent covariates, $Z(t)$, in combination with the at-risk indicator, $Y(t)$, can be used to describe states. Multistate models in which subjects can move in and out of different states, or into an absorbing state such as death, can then be analyzed using the same methodology.

6.2   Motivation

The presence of subject heterogeneity, summarized by risk factors Z, known or suspected of being related to $S(t)$, is our central concern. The previous chapter dealt with the issue of marginal survival, i.e., survival ignoring any indicator of heterogeneity and which treats the data in hand as though the observations came from a single popula-tion. In Figure 6.1 there are two groups. This can be described by two distinct Kaplan-Meier curves or, possibly, two independently calcu-lated fitted parametric curves. If, however, the curves are related, then each estimate provides information not only about its own population curve but also about the other group's population curve. The curve estimates would not be independent. Exploiting such dependence can lead to considerable gains in our estimating power. The agreement between an approach modeling dependence and one ignoring it can be more or less strong and, in Figure 6.1, agreement is good apart from observations beyond 150 months where a proportional hazards assumption may not hold very well. Returning to the simplest case, we can imagine a compartmental model describing the occurrence of

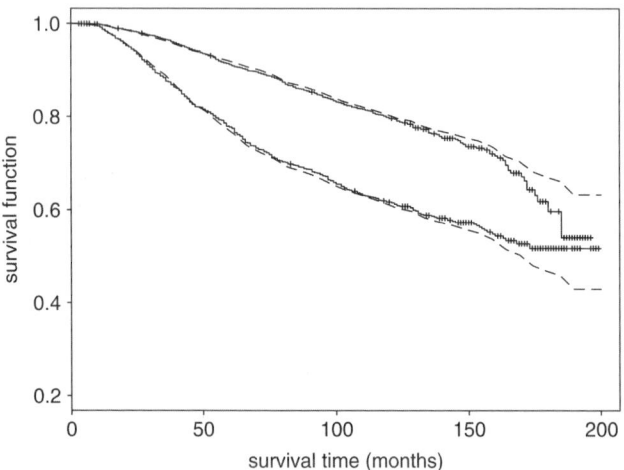

Figure 6.1: Kaplan-Meier survival curves and PH model curves for two groups defined by a binary covariate. Dashed lines represent PH estimates.

deaths independently of group status in which all individuals are assumed to have the same hazard rates. As pointed out in the previous chapter, the main interest then is in the survival function $S(t)$ when the Z are either unobservable or being ignored. Here we study the conditional survival function given the covariates Z and we write this as $S(t|Z)$. In the more complex situations (multicompartment models, time-dependent Z) it may be difficult, or even impossible, to given an interpretation to $S(t)$ as an average over conditional distributions, but the idea of conditioning is still central although we may not take it beyond that of the probability of a change of state conditional upon the current state as well as the relevant covariate history which led to being in that state.

The goal here is to consider models with varying degrees of flexibility applied to the summary of n subjects each with an associated covariate vector Z of dimension p. The most flexible models will be able to fully describe any data at hand but, as a price for their flexibility, little reduction in dimension from the $n \times p$ data matrix we begin with. Such models will have small bias in prediction compared with large sampling errors. The most rigid models can allow for striking reductions in dimension. Their consequent impact on prediction will be associated with much smaller sampling errors. However, as a price for such gains, the biases in prediction can be large. The models we finally work with will lie between these two extremes. Their choice then depends on an artful balance between the two conflicting characteristics.

6.3 General or nonproportional hazards model

In the most straightforward cases we can express the conditional dependence of survival upon fixed covariates in terms of the hazard function. A general expression for the hazard function given the value of the covariate Z is given by:

Figure 6.2: A simple alive/dead transition model. At time t the only information being used is whether the subject is dead or alive. Covariate information (eg. group status) is not used.

$$\lambda(t|Z) = \lambda_0(t) \exp\{\beta(t)Z\}, \tag{6.1}$$

where $\lambda(t|\cdot)$ is the conditional hazard function, $\lambda_0(t)$ the baseline hazard corresponding to $Z = 0$, and $\beta(t)$ a time-varying regression effect. Whenever Z has dimension greater than one we view $\beta(t)Z$ as an inner product in which $\beta(t)$ has the same dimension as Z so that $\beta(t)Z = \beta_1(t)Z_1+, \cdots , +\beta_p(t)Z_p$.

Recalling the discussion of Chapter 5, we are only interested in situations where observations on Z can be made in the course of any study. In Equation 6.1 Z is not allowed to depend upon time. If we also disallow the possibility of continuous covariates, which, in practice, we can approximate as accurately as we wish via high dimensional Z together with $\beta(t)$ of the same dimension, we see that model (6.1) is completely general and, as such, not really a model. It is instead a representation, or re-expression, of a very general reality, an expression that is convenient and which provides a framework to understanding many of the models described in this chapter. At the cost of losing the interpretation of a hazard function, we can immediately generalize (6.1) to

$$\lambda(t|Z) = \lambda_0(t) \exp\{\beta(t)Z(t)\}. \tag{6.2}$$

As long as we do not view $Z(t)$ as random, i.e., the whole time path of $Z(t)$ is known at $t = 0$, then a hazard function interpretation for $\lambda(t|Z)$ is maintained. Otherwise we lose the hazard function interpretation, since this requires knowledge of the whole function at the origin $t = 0$, i.e., the function is a deterministic and not a random one. In some ways this loss is of importance in that the equivalence of the hazard function, the survival function, and the density function means that we can easily move from one to another. However, when $Z(t)$ is random, we can reason in terms of intensity functions and compartmental models, a structure that enables us to deal with a wide variety of applied problems. The parameter $\beta(t)$ is of infinite dimension and therefore the model would not be useful without some restrictions upon $\beta(t)$.

6.4   Proportional hazards model

Corresponding to the truth or reality under scrutiny, we can view Equation (6.2) as being an extreme point on a large scale which calibrates model complexity. The opposite extreme point on this scale might have been the simple exponential model, although we will start with a restriction that is less extreme, specifically the proportional hazards model in which $\beta(t) = \beta$ so that;

$$\lambda(t|Z) = \lambda_0(t) \exp\{\beta Z(t)\}. \tag{6.3}$$

Putting restrictions on $\beta(t)$ can be done in many ways, and the whole art of statistical modeling, not only for survival data, is in the search for useful restrictions upon the parameterization of the problem in hand. Our interpretation of the word "useful" depends very much on the given particular context.

Just where different models find themselves on the infinite scale between Equation 6.3 and Equation 6.2 and how they can be ordered is a very important concept we need master if we are to be successful at the modeling process, a process which amounts to feeling our way up this scale (relaxing constraints) or down this scale (adding constraints), guided by the various techniques at our disposal. From the outset it is important to understand that the goal is not one of establishing some unknown hidden truth. We already have this, expressed via the model described in Equation (6.1). The goal is to find a much smaller, more restrictive model, which, for practical purposes is close enough or which is good enough to address those questions that we have in mind; for example, deciding whether or not there is an effect of treatment on survival once we have accounted for known prognostic factors which may not be equally distributed across the groups we are comparing. For such purposes, no model to date has seen more use than the Cox regression model.

6.5 The Cox regression model

In tackling the problem of subject heterogeneity, Cox's (1972) proportional hazards regression model has enjoyed outstanding success, a success, it could be claimed, matching that of classic multilinear regression itself. The model has given rise to considerable theoretical work and continues to provoke methodological advances. Research and development into the model and the model's offspring have become so extensive that we cannot here hope to cover the whole field, even at the time of writing. We aim nonetheless to highlight what seem to be the essential ideas and we begin with a recollection of the seminal paper of D.R. Cox, presented at a meeting of the Royal Statistical Society in London, England, March 8, 1972.

Regression models and life tables (D.R. Cox 1972)

After summarizing earlier work on the life table (Kaplan and Meier 1958, Chiang 1968), Professor Cox introduced his, now famous, model postulating a simplified form for the relationship between the hazard function $\lambda(t)$, at time t and the value of an associated fixed covariate Z. As its name suggests, the proportional hazards model assumes that the hazard functions among subjects with different covariates are proportional to one another. The hazard function can then be written:

$$\lambda(t|Z) = \lambda_0(t) \exp\{\beta Z\}, \tag{6.4}$$

where $\lambda_0(t)$ is a fixed "baseline" hazard function, and β is a relative risk parameter to be estimated. Whenever $Z = 0$ has a concrete interpretation (which we can always obtain by recoding) then so does the baseline hazard $\lambda_0(t)$ since, in this case, $\lambda(t|Z = 0) = \lambda_0(t)$. As mentioned just above, when Z is a vector of covariates, then the model is the same, although with the scalar product βZ interpreted as an inner product. It is common to replace the expression βZ by $\beta' Z$ where β and Z are $p \times 1$ vectors, and $a'b$ denotes the inner product of vectors a and b. Usually, though, we will not distinguish notationally between the scalar and the vector inner product since the former is just a special case of the latter. We write them both as βZ. Again we can interpret $\lambda_0(t)$ as being the hazard corresponding to the group for which the vector Z is identically zero.

The model is described as a multiplicative model, i.e., a model in which factors related to the survival time have a multiplicative effect on the hazard function. An illustration in which two binary variables are used to summarize the effects of four groups is shown in Figure 6.3. As pointed out by Cox, the function (βZ) can be replaced by any function of β and Z, the positivity of $\exp(\cdot)$ guaranteeing that, for any hazard function $\lambda_0(t)$, and any Z, we can always maintain a hazard function interpretation for $\lambda(t|Z)$. Indeed it is not necessary to restrict ourselves to $\exp(\cdot)$, and we may wish to work with other functions $R(\cdot)$, although care is required to ensure that $R(\cdot)$ remains positive over the range of values of β and Z of interest. Figure 6.3 represents the case of two binary covariables indicating four distinct groups (in the figure we take the logarithm of $\lambda(t)$) and the important thing to observe is that the distance between any two groups on this particular scale, i.e., in terms of the log-hazards, does not change through time. In view of the relation between the hazard function and the survival function,

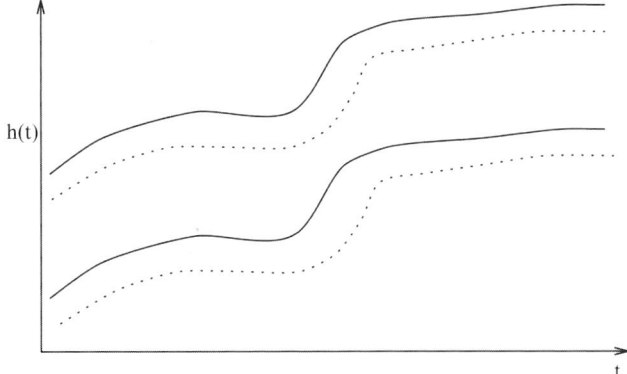

Figure 6.3: Proportional hazards with two binary covariates indicating 4 groups. Log-hazard rate written as $h(t) = \log \lambda(t)$.

there is an equivalent form of Equation 6.4 in terms of the survival function. Defining $S_0(t)$ to be the baseline survival function; that is, the survival function corresponding to $S(t|Z = 0)$, then, for scalar or vector Z, we have that,

$$S(t|Z) = \{S_0(t)\}^{\exp(\beta Z)}. \qquad (6.5)$$

When the covariate is a single binary variable indicating, for example, treatment groups, the model simply says that the survival function of one group is a power transformation of the other, thereby making an important connection to the class of Lehmann alternatives (Lehmann 1953).

Cox took the view that "parametrization of the dependence on Z is required so that our conclusions about that dependence are expressed concisely," adding that any choice "needs examination in the light of the data." "So far as secondary features of the system are concerned ... it is sensible to make a minimum of assumptions." This view led to focusing on inference that allowed $\lambda_0(t)$ to remain arbitrary. The resulting procedures are nonparametric with respect to t in that inference is invariant to any increasing monotonic transformation of t, but parametric in as much as concerns Z. For this reason the model is often referred to as Cox's semi-parametric model. Let's keep in mind, however, that it is the adopted inferential procedures that are semiparametric rather than the model itself. Although, of course, use of the term $\lambda_0(t)$ in the model, in which $\lambda_0(t)$ is not specified, implies use of procedures that will work for all allowable functions $\lambda_0(t)$.

Having recalled to the reader how inference could be carried out following some added assumptions on $\lambda_0(t)$, the most common assumptions being that $\lambda_0(t)$ is constant, that $\lambda_0(t)$ is a piecewise constant function, or that $\lambda_0(t)$ is equal to t^γ for some γ, Cox presented his innovatory likelihood expression for inference, an expression that subsequently became known as a partial likelihood (Cox 1975). We look more closely at these inferential questions in later chapters. First note that the quantity $\lambda_0(t)$ does not appear in the expression for partial likelihood given by

$$L(\beta) = \prod_{i=1}^{n} \left\{ \frac{\exp(\beta Z_i)}{\sum_{j=1}^{n} Y_j(X_i) \exp(\beta Z_j)} \right\}^{\delta_i}, \qquad (6.6)$$

and, in consequence, $\lambda_0(t)$ can remain arbitrary. Secondly, note that each term in the product is the conditional probability that at time X_i of an observed failure, it is precisely individual i who is selected to fail, given all the individuals at risk and given that one failure would occur. Taking the logarithm in Equation 6.6 and its derivative with respect to β, we obtain the score function which, upon setting equal to zero, can generally be solved without difficulty using the Newton-Raphson method, to obtain the maximum partial likelihood estimate $\hat{\beta}$ of β. We will discuss more deeply the function $U(\beta)$ under the various approaches to inference. We can see already that it has the same form as that encountered in the standard linear regression situation where the observations are contrasted to some kind of weighted mean. The exact nature of this mean is described later. Also, even though the expression

$$U(\beta) = \sum_{i=1}^{n} \delta_i \left\{ Z_i - \frac{\sum_{j=1}^{n} Y_j(X_i) Z_j \exp(\beta Z_j)}{\sum_{j=1}^{n} Y_j(X_i) \exp(\beta Z_j)} \right\} \qquad (6.7)$$

looks slightly involved, we might hope that the discrepancies between the Z_i and the weighted mean, clearly some kind of residual, would be uncorrelated, at least for large samples, since the Z_i themselves are uncorrelated.

All of this turns out to be so and makes it relatively easy to carry out appropriate inference. The simplest and most common approach to inference is to treat $\hat{\beta}$ as asymptotically normally distributed with mean β and large sample variance $I(\hat{\beta})^{-1}$, where $I(\beta)$, called the information in view of the analogy with classical likelihood, is minus the second derivative of $L(\beta)$ with respect to β, i.e., letting

$$I_i(\beta) = \frac{\sum_{j=1}^{n} Y_j(X_i) Z_j^2 \exp(\beta Z_j)}{\sum_{j=1}^{n} Y_j(X_i) \exp(\beta Z_j)} - \left\{ \frac{\sum_{j=1}^{n} Y_j(X_i) Z_j \exp(\beta Z_j)}{\sum_{j=1}^{n} Y_j(X_i) \exp(\beta Z_j)} \right\}^2, \quad (6.8)$$

then $I(\beta) = \sum_{i=1}^{n} \delta_i I_i(\beta)$. Inferences can also be based on likelihood ratio methods. A third possibility, which is sometimes convenient, is to base tests on the score $U(\beta)$, which in large samples can be considered to be normally distributed with mean zero and variance $I(\beta)$. Multivariate extensions are completely natural, with the score being a vector and I an information matrix.

Early applications of the model

The first success of the model was in its use for the two-sample problem, i.e., testing the null hypothesis of no difference in the underlying true survival curves for two groups. In this case Cox (1972) showed that the test statistic $U(0)/\sqrt{I(0)}$ is formally identical to a test, later known under the heading of the log-rank test, obtained by setting up at each failure point a 2×2 contingency table, group against failed/survived, and combining the many 2×2 tables. As in a standard analysis of a single such contingency table we use the marginal frequencies to obtain estimates of expected rates under the null hypothesis of no effect. Assuming, as we usually do here, no ties we can obtain a table such as described in Table 6.1 in which, at time $t = X_i$ the observed failure occurs in group A and there are $n_A(t)$ and $n_B(t)$ individuals at risk in the respective groups.

The observed rates and the expected rates are simply summed across the distinct failure points, each of which gives rise to its own contingency table where the margins are obtained from the available risk sets at that time. From the above, if $Z_i = 1$ when subject i is in group A and zero otherwise, then elementary calculation gives that,

$$U(0) = \sum_{i=1}^{n} \delta_i \{ Z_i - \pi(X_i) \}, \quad I(0) = \sum_{i=1}^{n} \delta_i \pi(X_i) \{ 1 - \pi(X_i) \}$$

Time point $t = X_i$	Group A	Group B	Totals
Number of failures	1	0	1
Number not failing	$n_A(t) - 1$	$n_B(t)$	$n_A(t) + n_B(t) - 1$
Total at risk	$n_A(t)$	$n_B(t)$	$n_A(t) + n_B(t)$

Table 6.1: 2×2 table at failure point $t = X_i$ for group A and group B.

where $\pi(t) = n_A(t)/\{n_A(t)+n_B(t)\}$. The statistic U then contrasts the observations with their expectations under the null hypothesis of no effect. This expectation is simply the probability of choosing, from the subjects at risk, a subject from group A. The variance expression is the well-known expression for a Bernoulli variable. Readers interested in a deeper insight into this test should also consult (Cochran 1954, Mantel and Haenzel 1959, Mantel 1963, Peto and Peto 1972). As pointed out by Cox, "whereas the test in the contingency table situation is, at least in principle, exact, the test here is only asymptotic ..."

However, the real advantage of Cox's approach was that while contributing significantly toward a deeper understanding of the log-rank and related tests, it opened up the way for more involved situations; additional covariates, continuous covariates, random effects and, perhaps surprisingly, in view of the attribute "proportional hazards," a way to tackle problems involving time varying effects or time dependent covariates. Cox illustrated his model via an application to the now famous Freireich data (Freireich et al. 1963) describing a clinical trial in leukemia in which a new treatment was compared to a placebo. Treating the two groups independently and estimating either survivorship function using a Kaplan-Meier curve gave good agreement with the survivorship estimates derived from the Cox model. Such a result can also, of course, be anticipated by taking a $\log(-\log)$ transform of the Kaplan-Meier estimates and noting that they relate to one another via a simple shift. This shift exhibits only the weakest, if any, dependence on time itself.

Multivariate applications

Recovering the usual two-group log rank statistic as a special case of a test based on model (6.4) is reassuring. In fact, exactly the same approach extends to the several group comparison (Breslow 1972). More importantly, model (6.4) provides the framework for considering the multivariate problem from its many angles; global comparisons of course but also more involved conditional comparisons in which certain effects are controlled for while others are tested. We look at this in more detail below under the heading "Modeling multivariate problems." The partially proportional hazards model (in particular the stratified model) were to appear later to Cox's original work of 1972 and provide great flexibility in addressing regression problems in a multivariate context.

Discussion of Professor Cox's paper

Professor Cox's paper represented an important step forward in dealing with survival problems for heterogeneous populations and a nonnegligible subset of a whole generation of academic biostatisticians has spent over a quarter of a century, keeping up and clarifying the many ideas originally outlined in Cox's 1972 paper. The discussion continues but, already, back in 1972 a group drawn from among the most eminent statisticians of the time, made a collective contribution to the new developments in a discussion that turned out to be almost as significant as the paper itself.

The issue which, arguably, gave rise to the most fertile exchanges concerned the partial likelihood, not yet named as such and referred to by Cox as a conditional likelihood. Kalbfleisch and Prentice took issue with Cox's naming of the likelihood used for inference as a "conditional" likelihood. They pointed out that the likelihood expression is not obtainable as a quantity proportional to a probability after having conditioned on some event. Conditioning was indeed taking place in the construction of the likelihood expression but in a sequential manner, a dynamic updating whose inferential home would later be seen to lie more naturally within the context of stochastic processes, indexed by time, rather than regular likelihoods, whether marginal or conditional.

The years following this discussion gave rise to a number of papers investigating the nature of the "conditional" likelihood proposed in Cox's original paper. Given the striking success of the model, together with the suggested likelihood expression, in reproducing and taking further a wide range of statistics then in use, most researchers agreed that Cox's proposal was correct. They remained uncertain, though, as to how to justify the likelihood itself. This thinking culminated in several major contributions; those of Cox (1975), Prentice and Kalbfleisch (1975), Aalen (1979) and Andersen and Gill (1982), firmly establishing the likelihood expression of Cox. In our later chapter on inference we discuss some of the issues raised in those contributions. It turned out that Cox was correct, not just on the appropriateness of his proposed likelihood expression but also in describing it as a "conditional" likelihood, this description being the source of all the debate.

Not unlike other major scientific thinkers of the twentieth century, Cox showed quite remarkable insight and although his likelihood derivation may not have been conditional, in the sense of taking as observed some single statistic upon which we condition before proceeding, his likelihood is not only very much a conditional one but also it conditions in just the right way. Not in the most straightforward sense whereby all the conditioning is done in one go, but in the sense of sequentially conditioning through time. Cox's "conditional" likelihood is now called a "partial" likelihood although, as an inferential tool in its own right, i.e., as a tool for inference independent of the choice of any particular model the partial likelihood is not as useful a concept as believed by many. We return to this in the chapter on inference.

Professor Downton of the University of Birmingham and Professor Peto of the University of Oxford pointed out the connection to rank test procedures. Although the formulation of Cox allowed the user to investigate more complex structures, many existing set-ups, framed in terms of tests based on the ranks, could be obtained directly from the use of the Cox likelihood. The simplest example was the sign test for the median. Using permutation arguments, other tests of interest in the multivariate setting could be obtained, in particular tests analogous to the Friedman test and the Kruskal-Wallis test. Richard Peto referred to some of his own work with Julian Peto. Their work demonstrated the asymptotic efficiency of the log-rank test and that, for the two-group problem and for Lehmann alternatives, this test was locally most powerful. Since the log-rank test coincides with a score test based on Cox's likelihood, Peto argued that Cox's method necessarily inherits the same properties.

Professor Bartholomew of the University of Kent considered a lognormal model in current use and postulated its extension to the regression situation by writing down the likelihood. Such an analysis, being fully parametric, represents an alternative approach since the structure is not nested in a proportional hazards one. Bartholomew made an insightful observation that allowing for some dependence of the explanatory variable Z on t can enable the lognormal model and a proportional hazards model to better approximate each another. This is indeed true and allows for a whole development of a class of non proportional hazards models where Z is a function of time and within which the proportional hazards model arises as a special case.

Professors Oakes and Breslow discussed the equivalence between a saturated piecewise exponential model and the proportional hazards model. By a saturated piecewise exponential model we mean one allowing for constant hazard rates between adjacent failures. The model is data dependent in that it does not specify in advance time regions of constant hazard but will allow these to be determined by the observed failures. From an inferential standpoint, in particular making use of likelihood theory, we may expect to run into some difficulties. This is because the number of parameters of the model (number of constant hazard rates) increases at the same rate as the effective sample size (number of observed failure times). However, the approach does nonetheless work, although justification requires the use of techniques other than standard likelihood. A simple estimate of the hazard rate, the cumulative hazard rate, and the survivorship function are then available. When $\beta = 0$ the estimate of the cumulative hazard rate coincides with that of Nelson (1969).

Professor Lindley of University College London writes down the full likelihood which involves $\lambda_0(t)$ and points out that, since terms involving $\lambda_0(t)$ do not factor out we cannot justify Cox's conditional likelihood. If we take $\lambda_0(t)$ as an unknown nuisance parameter having some prior distribution, then we can integrate the full likelihood with respect to this in order to obtain a marginal likelihood (this would be different to the marginal likelihood of ranks studied later by Kalbfleisch and Prentice 1973). Lindley argues that the impact of censoring is greater for the Cox likelihood than for this likelihood which is then to be preferred. The author of this text confesses to not fully understanding Lindley's argument and there is some slight confusion there since, either due to a typo or to a subtlety that escapes me, Lindley calls the Cox likelihood a "marginal likelihood" and what I am referring to as a marginal likelihood, an "integrated likelihood." We do, of course, integrate a full likelihood to obtain a marginal likelihood, but it seems as though Professor Lindley was making other, finer, distinctions which are best understood by those in the Bayesian school. His concern on the impact of censoring is echoed by Mr. P. Glassborow of British Rail underlining the strength behind the independent censoring assumption, an assumption which would not be reasonable in many practical cases.

Professor Zelen, a pioneer in the area of regression analysis of survival data, pointed out important relationships in tests of regression effect in the proportional hazards model and tests of homogeneity of the odds ratio in the study of several contingency tables. Dr. John Gart of the National Cancer Institute also underlined parallels between contingency table analysis and Cox regression. These ideas were to be developed extensively in later papers by Ross Prentice and Norman

Breslow in which the focus switched from classical survival analysis
to studies in epidemiology. The connection to epidemiological applications was already alluded to in the discussion of the Cox paper by Drs.
Meshalkin and Kagan of the World Health Organization. Finally, algorithms for carrying out an analysis based on the Cox model became
quickly available thanks to two further important contributions to the
discussion of Cox's paper. Richard Peto obtained accurate approximations to the likelihood in the presence of ties, obviating the need
for computationally intensive permutation algorithms, and Susannah
Howard showed how to program efficiently by exploiting the nested
property of the risk sets in reversed time.

Historical background to Cox's paper

Alternative hypotheses to a null which assumes that two probabilities
are equal, such as in Equation (6.5), taking the form of a simple power
transformation, have a long history in statistical modeling. Such alternatives which, in the special case where the probabilities in question
are survival functions, are known as Lehmann alternatives (Lehmann
1953). Lehmann alternatives are natural in that, under the restriction
that the power term is positive, always achievable by reparameteriz-
ing the power term to be of an exponential form; then, whatever the
actual parameter estimates, the resulting probability estimates satisfy
the laws of probability. In particular, they remain in the interval (0,1).
Linear expressions for probabilities are less natural although, at least
prior to the discovery of the logistic and Cox models, possibly more
familiar. Feigl and Zelen (1965) postulated a linear regression for the
location parameter, λ_0, of an exponential law. In this case the location
parameter and the (constant) hazard coincide so that the model could
be written;

$$\lambda(t|Z) = \lambda_0 \exp\{\beta Z\}. \tag{6.9}$$

In Feigl and Zelen their model was not written exactly this way, expressed as $\lambda = \alpha + \beta Z$. However, since λ is constant, the two expressions are equivalent and highlight the link to Cox's more general
formulation. Feigl and Zelen only considered the case of uncensored
data. Zippin and Armitage (1966) used a modeling approach, essentially the same as that of Feigl and Zelen, although allowing for the
possibility of censoring. This was achieved by an assumption of independence between the censoring mechanism and the failure mechanism

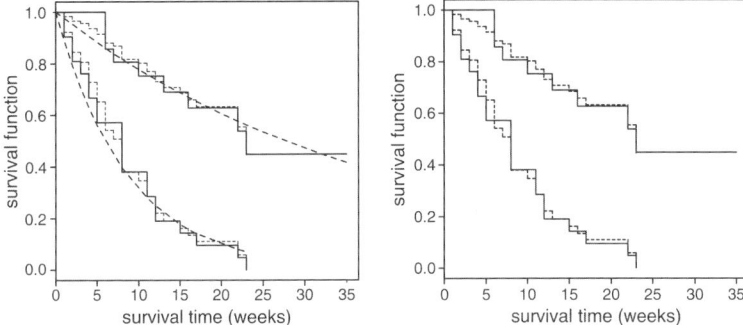

Figure 6.4: Kaplan-Meier curves and model based curves for Freireich data. Dashed lines represent model based estimates; exponential model (left), Cox model (right).

enabling an expression for the full likelihood to be obtained. Further discussion on these ideas can be found in Myers, Hankey and Mantel (1973) and Brown (1975). The estimates of the survival function for the different groups in the Freireich study, based on a simple exponential model or a Cox model, are shown in Figure 6.4. For these data the level of agreement between the two approaches appears to be high. This early work on the exponential model certainly helped anticipate the more general development of Cox and, for many more straightforward comparisons, such as the one illustrated by the Freireich data, it is perhaps unfortunate that the exponential model has been relegated to a historical role alone and is rarely, if ever, used in current practical analysis of similar data.

6.6 Modeling multivariate problems

The strength of the Cox model lies in its ability to describe and characterize involved multivariate situations. Crucial issues concern the adequacy of fit of the model, how to make predictions based on the model, and how strong is the model's predictive capability. These are considered in detail later. Here, in the following sections and in the chapter on inference we consider how the model can be used as a tool to formulate questions of interest to us in the multivariate setting. The simplest case is that of a single binary covariate Z taking the values zero and one. The zero might indicate a group of patients undergoing a standard therapy, whereas the group for which $Z = 1$ could be undergoing some experimental therapy. Model 6.4 then indicates the hazard rate for the standard group to be $\lambda_0(t)$ and for the experimental group to be $\lambda_0(t)\exp(\beta)$. Testing whether or not the new therapy has any effect on survival translates as testing the hypothesis $H_0 : \beta = 0$. If β is less than zero then the hazard rate for the experimental therapy is less than that for the standard therapy at all times and is such that the arithmetic difference between the respective logarithms of the hazards is of magnitude β. Suppose the problem is slightly more complex and we have two new experimental therapies. We can write;

$$\lambda(t|Z) = \lambda_0(t)\exp\{\beta_1 Z_1 + \beta_2 Z_2\}$$

and obtain Table 6.2. As we shall see the two covariate problem is very much more complex than the case of a single covariate. Not only do we need to consider the effect of each individual treatment on the hazard rate for the standard therapy but we also need to consider the effect of each treatment in the presence or absence of the other as well as the combined effect of both treatments together. The particular model form in which we express any relationships will typically imply assumptions on those relationships and an important task is to bring under scrutiny (goodness of fit) the soundess of any assumptions.

It is also worth noting that if we are to assume that a two-dimensional covariate proportional hazards model hold exactly, then, integrating over one of the covariates to obtain a one dimensional model will not result (apart from in very particular circumstances) in a lower-dimensional proportional hazards model. The lower dimensional model would be in a much more involved non proportional hazards form. This observation also holds when adding a covariate to a one-dimensional proportional hazards model, a finding that compels us, in realistic modeling situations, to only ever consider the model as an approximation.

By extension the case of several covariates becomes rapidly very complicated. If, informally, we were to define complexity as *the number*

Treatment group	Z_1	Z_2	Log of group effect
Standard therapy	0	0	0
Experimental therapy 1	1	0	β_1
Experimental therapy 2	0	1	β_2

Table 6.2: Effects for two treatment groups

of things you have to worry about, then we could, even more informally, state an important theorem.

Theorem 6.1 (Theorem of complexity) *The complexity of any problem grows exponentially with the number of covariates in the equation.*

Obviously such a theorem cannot hold in any precise mathematical sense without the need to add conditions and restrictions such that its simple take-home message would be lost. For instance, if each added covariate was a simple constant multiple of the previous one, then there would really be no added complexity. But, in some broad sense, the theorem does hold and to convince ourselves of this we can return to the case of two covariates. Simple combinatorial arguments show that the number of possible hypotheses of potential interest is increasing exponentially. But it is more complex than that. Suppose we test the hypothesis $H_0 : \beta_1 = \beta_2 = 0$. This translates the clinical null hypothesis: neither of the experimental therapies impacts survival against the alternative, $H_1 : \exists \beta_i \neq 0$, $i = 1, 2$. This is almost, yet not exactly, the same as simply regrouping the two experimental treatments together and reformulating the problem in terms of a single binary variable.

Next we might consider testing the null hypothesis $H_0 : \beta_1 = 0$ against the alternative hypothesis $H_1 : \beta_1 \neq 0$. Such a test focuses only on the first experimental treatment, but does not, as we might at first imagine, lump together both the second experimental treatment and the standard treatment. This test makes no statement about β_2 and so this could indeed take the value zero (in which case the standard and the second experimental therapy are taken to be the same) or any other value in which case, detecting a nonzero value for β_1 translates as saying that this therapy has an effect different to the standard regardless of the effect of the second experimental therapy. Clearly this is different from lumping together the second experimental therapy with the standard and testing the two together against the first experimental therapy. In such a case, should the effect of the first experimental therapy lie somewhere between that of the standard and the second, then, plausibly, we might fail to detect a nonzero β_1 even though there exist real differences between the standard and the first therapy.

All of this discussion can be repeated, writing β_1 in the place of β_2. Already, we can see that there are many angles from which to consider an equation such as the above. These angles, or ways of expressing the scientific question, will impact the way of setting up the statistical hypotheses. In turn, these impact our inferences.

Another example would be testing the above null hypothesis H_0 : $\beta_1 = \beta_2 = 0$ against an alternative $H_1 : 0 < \beta_1 < \beta_2$ instead of that initially considered (i.e., $H_1 : \exists \beta_i \neq 0$, $i = 1, 2$). The tests, and their power properties, would not typically be the same. We might consider recoding the problem, as in Equation 6.10, so that testing H_0 : $\beta_1 = 0$ against $H_1 : \beta_1 \neq 0$ corresponds to testing for an effect in either group. Given this effect we can test H_0 : $\beta_2 = 0$ against $H_1 : \beta_2 \neq 0$ which will answer the question as to whether, given that their exists a treatment effect, it is the same for both of the experimental treatments:

$$
\begin{aligned}
\lambda(t|Z) &= \lambda_0(t) \exp\{\beta_1 Z_1 + (\beta_1 + \beta_2)Z_2\} \\
&= \lambda_0(t) \exp\{\beta_1(Z_1 + Z_2) + \beta_2 Z_2\}.
\end{aligned}
\tag{6.10}
$$

Note that fitting the above models needs no new procedures or software for example, since both cases come under the standard heading. In the first equation all we do is write $\alpha_1 = \beta_1$ and $\alpha_2 = \beta_1 + \beta_2$. In the second we simply redefine the covariates themselves. The equivalence expressed in the above equation is important. It implies two things. Firstly, that this previous question concerning differential treatment effects can be re-expressed in a standard way enabling us to use existing structures, and computer programs. Secondly, since the effects in our models express themselves via products of the form βZ, any recoding of β can be artificially carried out by re-coding Z and vice versa. This turns out to be an important property and anticipates the fact that a non proportional hazards model $\beta(t)Z$ can be re-expressed as a time-dependent proportional hazards model $\beta Z(t)$. Hence the very broad sweep of proportional hazards models.

It is easy to see how the above considerations, applied to a situation in which we have $p > 2$ covariates, become very involved. Suppose we have four ordered levels of some risk factor. We can re-code these levels using three binary covariates as in Table 6.3; For this model we can, again, write the hazard function in terms of these binary coding variables, noting that, as before, there are different ways of expressing this. In standard form we write

$$
\lambda(t|Z) = \lambda_0(t) \exp\{\beta_1 Z_1 + \beta_2 Z_2 + \beta_3 Z_3\}
$$

so that the hazard rate for those exposed to the risk factor at level i, $i = 1, \dots, 4$, is given by $\lambda_0(t) \exp(\beta_i)$ where we take $\beta_0 = 0$. Our

Risk factor	Z_1	Z_2	Z_3	Log of risk factor effect
Level 1	0	0	0	0
Level 2	1	0	0	β_1
Level 3	0	1	0	β_2
Level 4	0	0	1	β_3

Table 6.3: Coding for four ordered levels of a risk factor.

interest may be more on the incremental nature of the risk as we increase through the levels of exposure to the risk factor. The above model can be written equivalently as

$$\lambda(t|Z) = \lambda_0(t)\exp\{\beta_1 Z_1 + (\beta_1 + \beta_2)Z_2 + (\beta_1 + \beta_2 + \beta_3)Z_3\}$$
$$= \lambda_0(t)\exp\{\beta_1(Z_1 + Z_2 + Z_3) + \beta_2(Z_2 + Z_3) + \beta_3 Z_3\} \quad (6.11)$$

so that our interpretation of the β_i is in terms of increase in risk. The coefficient β_1 in this formulation corresponds to an overall effect, common to all levels above the lowest. The coefficient β_2 corresponds to the amount by which the log-hazard rate for the second level differs from that at the first. Here then, a value of β_2 equal to zero does not mean that there is no effect at level 2, simply that the effect is no greater than that already quantified at level 1. The same arguments follow for levels 3 and 4.

Writing the model in these different ways is not changing the basic model. It changes the interpretation that we can give to the different coefficients. The equivalent expression shown in Equation 6.11 for example means that we can carefully employ combinations of the covariates in order to use existing software. But we can also consider the original coding of the covariates Z. Suppose that, instead of the coding given in Table (6.3), we use the coding given in Table 6.4. This provides an equivalent description of the four levels. As we move up the levels, changing from level i to level $i + 1$, the log hazard is increased by β_i.

Let's imagine a situation, taken from Table 6.4, in which $\beta_1 = \beta_2 = \beta_3$. Real situations may not give rise to strict equalities but may well provide good first approximations. The hazards at each level can now be written very simply as $\lambda_0(t)\exp(j\beta_1)$ for $j = 0, 1, 2, 3$, and this is described in Table (6.5). Taking $\beta_1 = \beta$, we are then able to write a model for this situation as; $\lambda(t|Z) = \lambda_0(t)\exp(\beta Z)$, in which the covariate Z, describing group level, takes the values 0 to 3. This model

Risk factor	Z_1	Z_2	Z_3	Log of risk factor effect
Level 1	0	0	0	0
Level 2	1	0	0	β_1
Level 3	1	1	0	$\beta_1 + \beta_2$
Level 4	1	1	1	$\beta_1 + \beta_2 + \beta_3$

Table 6.4: Coding for four ordered levels of a risk factor.

Risk factor	Z	Log of risk factor effect
Level 1	0	0
Level 2	1	β
Level 3	2	2β
Level 4	3	3β

Table 6.5: Coding for four ordered levels of a risk factor.

has a considerable advantage over the previous one, describing the same situation of four levels, in that only a single coefficient appears in the model as opposed to three. We will use our data to estimate just a single parameter. The gain is clear. The cost, however, is much less so, and is investigated more thoroughly in the chapters on prediction (explained variation, explained randomness) and goodness of fit. If the fit is good, i.e., the assumed linearity is reasonable, then we would certainly prefer the latter model to the former. If we are unsure we may prefer to make less assumptions and use the extra flexibility afforded by a model which includes three binary covariates rather than a single linear covariate. In real data analytic situations we are likely to find ourselves somewhere between the two, using the tools of fit and predictability to guide us.

Returning once more to Table 6.4 we can see that the same idea prevails for the β_i not all assuming the same values. A situation in which four ordered levels is described by three binary covariates could be recoded so that we only have a single covariate Z, together with a single coefficient β. Next, suppose that in the model; $\lambda(t|Z) = \lambda_0(t) \exp(\beta Z)$, Z not only takes the ordered values, 0, 1, 2 and 3 but also all of those in between. In a clinical study this might correspond to some prognostic indicator, such as blood pressure or blood cholesterol, recorded continuously and re-scaled to lie between 0 and 3.

Including the value of Z, as a continuous covariate, in the model amounts to making very strong assumptions. It supposes that the log hazard increases by the same amount for every given increase in Z, so that the relative risk associated with $\Delta = z_2 - z_1$ is the same for all values of z_1 between 0 and $3 - \Delta$. Let's make things a little more involved. Suppose we have the same continuous covariate, this time let's call it Z_1, together with a single binary covariate Z_2 indicating one of two groups. We can write

$$\lambda(t|Z_1, Z_2) = \lambda_0(t) \exp(\beta_1 Z_1 + \beta_2 Z_2).$$

Such a model supposes that a given change in exposure Z_1 results in a given change in risk, as just described, but that, furthermore, this resulting change is the same at both levels of the discrete binary covariate Z_2. This may be so but such strong assumptions must be brought under scrutiny. Given the ready availability of software, it is not at all uncommon for data analysts to simply "throw in" all of the variables of interest, both discrete and continuous, without considering potential transformations or recoding, turn the handle, and then try to make sense of the resulting coefficient estimates together with their standard errors. Such an exercise will rarely be fruitful. In this respect it is preferable to write one's own computer programs when possible or to use available software such as the R package, which tends to accompany the user through model development. Packages that present a "complete" one-off black box analysis based on a single model are unlikely to provide much insight into the nature of the mechanisms generating the data at hand.

The user is advised to exercise great care when including continuous covariates in a model. We can view a continuous covariate as equivalent to an infinite dimensional vector of indicator variables so that, in accordance with our informal theorem of complexity, the number of things we need worry about is effectively infinite. Let us not however overstate things, and it is of course useful to model continuous covariates. But be wary. Also consider the model

$$\lambda(t|Z) = \lambda_0(t) \exp(\beta_1 Z + \beta_2 Z^2).$$

If Z is binary then $Z^2 = Z$ and there is no purpose to the second term in the equation. If Z is ordinal or continuous then the effect of Z is quadratic rather than linear. And, adding yet higher-order terms enables us, at least in principle, to model other nonlinear functions. In practice, in order to carry out the analysis, we would use existing tools by simply introducing a second variable Z_2 defined by $Z_2 = Z^2$; an important observation in that the linear representation of the covariate can be relaxed with relatively little effort. For example, suppose that the log-relative risk is expressed via some smooth function $\psi(z)$ of a continuous covariate z. Writing the model

$$\lambda(t|Z) = \lambda_0(t) \exp\{\beta\psi(Z)\}$$

supposes that we know the functional form of the relative risk, at least up to the constant multiple β. Then, a power series approximation to this would allow us to write $\psi(Z) = \sum \beta_j Z^j$ in which any constant term β_0 is absorbed into $\lambda_0(t)$. We then introduce the covariates $Z_j = Z^j$ to bring the model into its standard form.

6.7 Partially proportional hazards models

In the case of a single binary variable, model (6.2) and model (6.4) represent the two extremes of the modeling options open to us. Under model (6.2) there would be no model constraint and any consequent estimation techniques would amount to dealing with each level of the variable independently. Under model (6.4) we make a strong assumption about the nature of the relative hazards, an assumption that allows us to completely share information between the two levels. There exists an important class of models lying between these extremes and, in order to describe this class, let us now imagine a more complex situation; that of three groups, A, B and C, identified by a vector Z of binary covariates; $Z = (Z_2, Z_3)$. This is summarized in Table 6.6. We are mainly interested in a treatment indicator Z_1, mindful of the fact that the groups themselves may have very different survival probabilities. Under model (6.4) we have

$$\lambda(t|Z) = \lambda_0(t) \exp\{\beta_1 Z_1 + \beta_2 Z_2 + \beta_3 Z_3\}. \tag{6.12}$$

	Z_2	Z_3	Log of group effect
Group A	0	0	0
Group B	1	0	β_2
Group C	1	1	$\beta_2 + \beta_3$

Table 6.6: Coding for three groups.

Our assumptions are becoming stronger in that not only are we modeling the treatment affect via β_1 but also the group effects via β_2 and β_3. Expressing this problem in complete generality, i.e., in terms of model (6.2), we write

$$\lambda(t|Z) = \lambda_0(t) \exp\{\beta_1(t)Z_1 + \beta_2(t)Z_2 + \beta_3(t)Z_3\}. \qquad (6.13)$$

Unlike the simple case of a single binary variable where our model choices were between the two extremes of model (6.2) and model (6.4), as the situation becomes more complex, we have open to us the possibility of a large number of intermediary models. These are models that make assumptions lying between model (6.2) and model (6.4) and, following O'Quigley and Stare (2002) we call them partially proportional hazards models. A model in between (6.12) and (6.13) is

$$\lambda(t|Z) = \lambda_0(t) \exp\{\beta_1 Z_1 + \beta_2(t)Z_2 + \beta_3(t)Z_3\}. \qquad (6.14)$$

This model is of quite some interest in that the strongly modeled part of the equation concerns Z_1, possibly the major focus of our study. Figure 6.5 illustrates a simple situation. The only way to leave any state is to die, the probabilities of making this transition varying from state to state and the rates of transition themselves depending on time. Below, under the heading time-dependent covariates, we consider the case where it is possible to move within states. Here it will be possible to move from a low-risk state to a high-risk state, to move from either to the death state, but to also, without having made the transition to the absorbing state, death, to move back from high-risk to low-risk.

Stratified models

Coming under the heading of a partially proportional hazards model is the class of models known as stratified models. In the same way

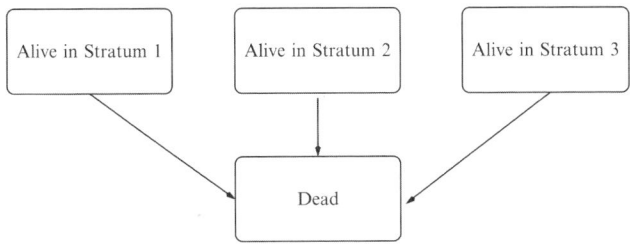

Figure 6.5: A stratified model with transitions only to death state.

these models can be considered as being situated between the two extremes of Equation 6.2 and Equation 6.3 and have been discussed by Kalbfleisch and Prentice (1980) among others. Before outlining why stratified models are simply partially proportional hazards models we recall the usual expression for the stratified model as;

$$\lambda(t|Z(t), w) = \lambda_{0w}(t) \exp\{\beta Z(t)\}, \tag{6.15}$$

where w takes integer values $1, \ldots, m$. If the coefficient β were allowed to depend on each stratum, indicated by w, say $\beta(w)$, then this would exactly correspond to a situation in which we consider each stratum completely independent, i.e., we have independent models for each stratum. This would be nothing more than w separate, independent, proportional hazards models. The estimation of $\beta(w)$ for one model has no impact on the estimation of $\beta(w)$ for another. If we take β to be common to the different strata, which is of course the whole purpose of the stratified model, then, using data, whatever we learn about one stratum tell us something about the others. They are no longer independent of one another. Stratified models are necessarily broader than (6.3), lying, in the precise sense described below, between this model and the non proportional hazards model (6.2). To see this, consider a restricted case of model (6.2) in which we have two binary covariates $Z_1(t)$ and $Z_2(t)$. We put the restriction on the coefficient β_2, constrained to be constant in time. The model is then

$$\lambda\{t|Z_1(t), Z_2(t)\} = \lambda_0(t) \exp\{\beta_1(t)Z_1(t) + \beta_2 Z_2(t)\}, \tag{6.16}$$

a model clearly lying, in a well-defined way, between models (6.3) and (6.2). It follows that

$$\lambda\{t|Z_1(t) = 0, Z_2(t)\} = \lambda_0(t) \exp\{\beta_2 Z_2(t)\}$$

and

$$\lambda\{t|Z_1(t) = 1, Z_2(t)\} = \lambda_0^*(t) \exp\{\beta_2 Z_2(t)\},$$

where $\lambda_0^*(t) = \lambda_0(t)e^{\beta_1(t)}$. Recoding the binary $Z_1(t)$ to take the values 1 and 2, and rewriting $\lambda_0^*(t) = \lambda_{02}(t)$, $\lambda_0(t) = \lambda_{01}(t)$ we recover the stratified PH model (6.15) for $Z_2(t)$. The argument is easily seen to be reversible and readily extended to higher dimensions so we can conclude an equivalence between the stratified model and the partially proportional hazards model in which some of the $\beta(t)$ are constrained to be constant. We can exploit this idea in the goodness of fit or the model construction context. If a PH model holds as a good approximation, then the main effect of Z_2 say, quantified by β_2, would be similar over different stratifications of Z_1 and remain so when these stratifications are re-expressed as a PH component to a two covariate model. Otherwise the indication is that $\beta_1(t)$ should be allowed to depend on t. The predictability of any model is studied later under the headings of explained variation and explained randomness and it is of interest to compare the predictability of a stratified model and an un-stratified one. For instance, we might ask ourselves just how strong is the predictive strength of Z_2 after having accounted for Z_1. Since we can account for the effects of Z_1 either by stratification or by its inclusion in a single PH model we may obtain different results. Possible discrepancies tell us something about our model choice.

The relation between the hazard function and the survival function follows as a straightforward extension of (6.5). Specifically, we have

$$S(t|Z) = \sum_{w} \phi(w)\{S_{0w}(t)\}^{\exp(\beta Z)}, \qquad (6.17)$$

where $S_{0w}(t)$ is the corresponding baseline survival function in stratum w and $\phi(w)$ is the probability of coming from that particular stratum. This is then slightly more involved than the nonstratified case in which, for two groups the model expressed the survival function of one group as a power transformation of the other. Nonetheless the connection to the class of Lehmann alternatives is still there although somewhat weaker. For the stratified model, once again the quantity $\lambda_{0w}(t)$ does not appear in the expression for the partial likelihood given now by

$$L(\beta) = \prod_{i=1}^{n} \left\{ \frac{\exp(\beta Z_i)}{\sum_{j=1}^{n} Y_j\{w_i(X_i), X_i\} \exp(\beta Z_j)} \right\}^{\delta_i} \qquad (6.18)$$

and, in consequence, once again, $\lambda_{0w}(t)$ can remain arbitrary. Note also that each term in the product is the conditional probability that at time X_i of an observed failure, it is precisely individual i who is selected to fail, given all the individuals at risk from stratum w and that one failure from this stratum occurs.

The notation $w_i(t)$ indicates the stratum in which the subject i is found at time t. Although we mostly consider $w_i(t)$ which do not depend on time, i.e., the stratum is fixed at the outset and thereafter remains the same, it is almost immediate to generalize this idea to time dependency and we can anticipate the later section on time-dependent covariates where the risk indicator $Y_j\{w_i(t), t\}$ is not just a function taking the value one until it drops at some point to zero, but can change between zero and one with time, as the subject moves from one stratum to another. For now the function $Y_j\{w_i(t), t\}$ will be zero unless the subject is at risk of failure from stratum w_i, i.e., the same stratum in which the subject i is to be found. Taking the logarithm in (6.18) and derivative with respect to β, we obtain the score function

$$U(\beta) = \sum_{i=1}^{n} \delta_i \left\{ Z_i - \frac{\sum_{j=1}^{n} Y_j\{w_i(X_i), X_i\} Z_j \exp(\beta Z_j)}{\sum_{j=1}^{n} Y_j\{w_i(X_i), X_i\} \exp(\beta Z_j)} \right\}, \quad (6.19)$$

which, upon setting equal to zero, can generally be solved without difficulty using standard numerical routines, to obtain the maximum partial likelihood estimate $\hat{\beta}$ of β. The parameter β then is assumed to be common across the different strata.

Inferences about β are made by treating $\hat{\beta}$ as asymptotically normally distributed with mean β and variance $I(\hat{\beta})^{-1}$, where, now, $I(\beta)$ is given by $I(\beta) = \sum_{i=1}^{n} \delta_i I_i(\beta)$. In this case each I_i is, as before, obtained as the derivative of each component to the score statistic $U(\beta)$. For the stratified score this is

$$I_i = \frac{\sum_{j=1}^{n} Y_j\{w_i(X_i), X_i\} Z_j^2 \exp(\beta Z_j)}{\sum_{j=1}^{n} Y_j\{w_i(X_i), X_i\} \exp(\beta Z_j)}$$
$$- \left\{ \frac{\sum_{j=1}^{n} Y_j\{w_i(X_i), X_i\} Z_j \exp(\beta Z_j)}{\sum_{j=1}^{n} Y_j\{w_i(X_i), X_i\} \exp(\beta Z_j)} \right\}^2.$$

The central notion of the risk set is once more clear from the above expressions and we most usefully view the score function as contrasting the observed covariates at each distinct failure time with the means of those at risk from the same stratum. A further way of looking at the score function is to see it as having put the individual contributions on a linear scale. We simply add them up within a stratum and then, across the strata, it only remains to add up the different sums. Once again, inferences can also be based on likelihood ratio methods or on the score $U(\beta)$, which in large samples can be considered to be normally distributed with mean zero and variance $I(\beta)$. Multivariate extensions follow as before. For the stratified model the only important distinction impacting the calculation of $U(\beta)$ and $I_i(\beta)$ is that the sums are carried out over each stratum separately and then combined at the end. The indicator $Y_j\{w_i(X_i)\}$ enables this to be carried out in a simpler way as indicated by the equation.

Random effects and frailty models

Also coming under the heading of partially proportional hazards model are the classes of models, which include random effects. When the effects concern a single individual such models have been given the heading frailty models (Vaupel 1979) since, for an individual identified by w, we can write $\lambda_{0w}(t) = \alpha_w \lambda_0(t)$ implying a common underlying hazard $\lambda_0(t)$ adjusted to each individual by a factor, the individual's *frailty*, unrelated to the effects of any other covariates that are quantified by the regression coefficients. The individual effects are then quantified by the α_w.

Although of some conceptual interest, such models are indistinguishable from models with time-dependent regression effects and therefore, unless there is some compelling reason to believe (in the absence of frailties) that a proportional hazards model would hold, it seems more useful to consider departures from proportional hazards in terms of model (6.2). On the other hand, random effects models, as commonly described by Equation (6.20) in which the α_w identify a potentially large number of different groups, are interesting and potentially of use. We express these as

$$\lambda(t|Z(t), w) = \alpha_w \lambda_0(t) \exp\{\beta Z(t)\}. \tag{6.20}$$

These models are also partially parametric in that some effects are allowed not to follow a proportional hazards constraint. However, unlike the stratified models described above, restrictions are imposed. The most useful view of a random effects model is to see it as a stratified model with some structure imposed upon the strata. A random effects model is usually written

$$\lambda(t|Z(t), w) = \lambda_0(t) \exp\{\beta Z(t) + w\}, \tag{6.21}$$

in which we take w as having been sampled from some distribution $G(w; \theta)$. Practically there will only be a finite number of distinct values of w, however large. For any value w we can rewrite $\lambda_0(t)e^w = \lambda_{0w}(t)$ and recover model (6.15). For the right hand side of this equation, and as we might understand from (6.15), we suppose w to take the values 1,2, ... The values on the left-hand side, being generated from

$G(\cdot)$ would generally not be integers but this is an insignificant notational issue and not one involving concepts. Consider the equation
to hold. It implies that the random effects model is a stratified model
in which added structure is placed on the strata. In view of Equation
6.16 and the arguments following this equation we can view a random
effects model equivalently as in Equation 6.2 where, not only are PH
restrictions imposed on some of the components of $\beta(t)$, but the time
dependency of the other components is subject to constraints. These
latter contraints, although weaker than imposing constancy of effect,
are all the stronger as the distribution of $G(w; \theta)$ is concentrated.

Structure of random effects models

Consider firstly the model of Equation (6.3). Suppose we have one main
variable, possibly a treatment variable of interest, coded by $Z_1 = 0$ for
group A and $Z_1 = 1$ for group B. The second variable, say a center
variable, which may or may not have prognostic importance and for
which we may wish to control for possible imbalance is denoted Z_2.
A strong modeling approach would include both binary terms in the
model so that the relationship between the hazard functions is as
described in Figure 6.6. If our main focus is on the effect of treatment,
believed to be comparable from one center to another, even though
the effects of the centers themselves are not absent, it makes sense to
stratify. This means that we do not attempt to model the effects of
the centers but, instead, remove any such potential effects from our

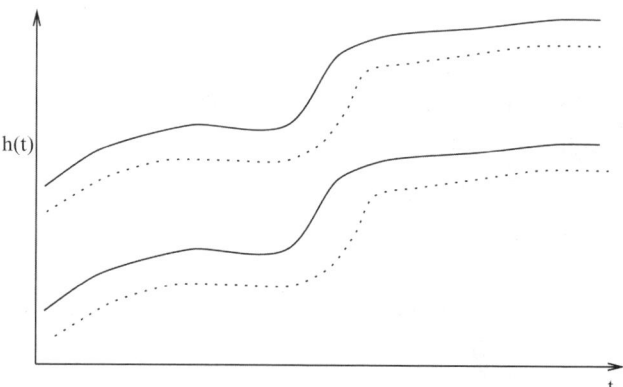

Figure 6.6: PH model with binary covariates denoting center and treatment groups.

analysis. This is nice in that it allows for rather greater generality than that illustrated in Figure 6.6. We maintain an assumption of constant treatment effect but the center effects can be arbitrary. This is illustrated in Figure 6.7. The illustration makes it clear that, under the assumption, a weaker one than that implied by Equation 6.3, we can estimate the treatment effect whilst ignoring center effects. A study of these figures is important to understanding what takes place when we impose a random effects model as in Equation (6.21). For many centers, Figure 6.8, rather than having two curves per center, parallel but otherwise arbitrary, we have a family of parallel curves. We no

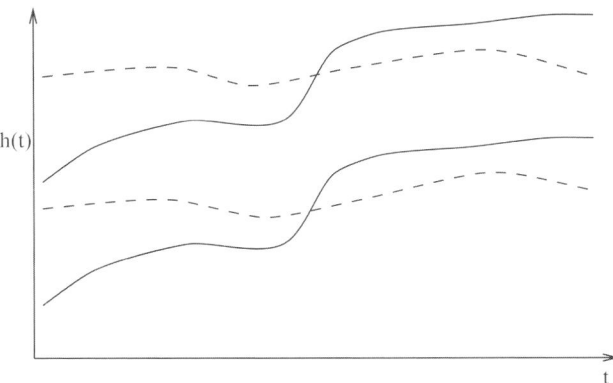

Figure 6.7: An outline sketch of a stratified PH model. Main variable in two strata: stratum 1, i.e., center 1 given by dotted line; stratum 2 by continuous line.

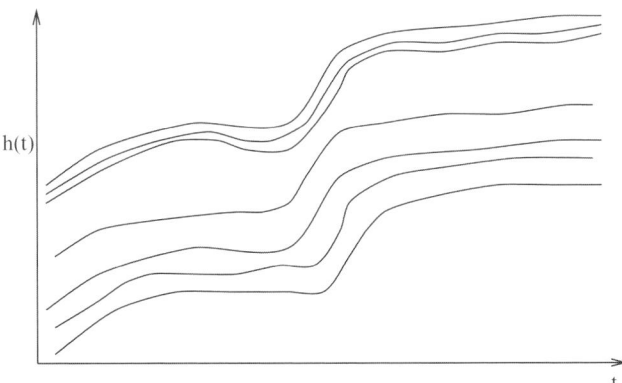

Figure 6.8: An outline sketch of a PH model with centers as random effects.

longer are able to say anything about the distance between any given centers, as we could for the model of Equation 6.3, a so-called fixed effects model, but the distribution of the distances between centers is something we aim to quantify. This is summarized by the distribution $G(w; \theta)$ and our inferences are then partly directed at θ.

Random effects models versus stratified models

The stratified model is making weaker assumptions than the random effects model. This follows since the random effects model is just a special case of a stratified model in which some structure is imposed upon the differences between strata. The stratified model not only leaves any distribution of differences between strata unspecified, but it also makes no assumption about the form of any given stratum. Whenever the stratified model is valid, then so also is the random effects model, the converse not being the case.

It may then be argued that we are making quite a strong assumption when we impose this added structure upon the stratified model. In exchange we would hope to make non-negligible inferential gains, i.e., greater precision of our estimates of errors for the parameters of main interest, the treatment parameters. In practice gains tend to be small for most situations and give relatively little reward for the extra effort made. Since any such gains are only obtainable under the assumption that the chosen random effects model actually generates the data, actual gains in practice are likely to be yet smaller and, of course, possibly negative when our additional model assumptions are incorrect. A situation where gains for the random effects model may be of importance is one where a non-negligeable subset of the data include strata containing only a single subject. In such a case simple stratification would lose information on those subjects. A random effects model, assuming the approximation to be sufficiently accurate, enables us to recover such information.

Efficiency of random effects models

Most of our discussion here focuses on different possible representations of the infinitely complex reality we are hoping to model. Our purpose in modeling is, ultimately, to draw simple, at least clear-cut, inferences. The question of inference no longer concerns the general but rather the specific data set we have at hand. If our main concern is on estimating risk functions then the question becomes, to what extent do we gain by including in our inferential setup the presence of random effect terms. Since our main objective is estimation and quantification of regression parameters enabling us to say something about the risk factors under study, the idea behind the inclusion of additional random effect terms is to make more precise this estimation and quantification.

As already argued above the inclusion of individual random effects (frailties) is of no practical interest and simply amounts to expressing the idea, albeit in an indirect way, of model inadequacy (O'Quigley and Stare 2002). We therefore assume that we are dealing with groups, some of which, but not all, may only include a isolated individual. We know that a partial likelihood analysis, stratified by group, is estimating the same regression parameter. Inference is based on the stratified score statistic. We contrast the observed covariate value with its estimated expectation under the model. Different model assumptions will impact this estimated expectation and it is here that any efficiency gains can be made. For a stratified model, these estimated expectations may be with respect to relatively small risk sets. A random effects model on the other hand, via the inclusion of a different w per group, will estimate the relevant expectations over the whole risk set and not just that relative to the group defined by the covariate value.

Comparisons for the stratified model are made with respect to the relatively few subjects of the group risk sets. This may lead us to believe that much information could be recovered were we able to make the comparison, as does the alternative random effects analysis, with respect to the whole risk set. Unfortunately this is not quite so because each contribution to the score statistic involves a difference between an observation on a covariate and its expectation under the model and the "noise" in the expectation estimate is of lower order that the covariate observations themselves. There is not all that much to be gained by improving the precision of the expectation estimate.

In other words, using the whole of the risk set or just a small sample from it will provide similar results. This idea of risk set sampling has been studied in epidemiology and it can be readily seen that the efficiency of estimates based on risk set samples of size k, rather than the whole risk set, is of the order

$$\frac{k}{k+1}\left\{1+\sum_{j=1}^{n}\frac{1}{n(n-j+1)}\right\}. \tag{6.22}$$

This function increases very slowly to one but, with as few as four subjects on average in each risk set comparison, we have already achieved 80% efficiency. With nine subjects this figure is close to 90%. Real efficiency will be higher for two reasons: (1) the above assumes that the estimate based on the full risk set is without error, (2) in our context we are assuming that each random effect w is observed precisely.

Added to this is the fact that, since the stronger assumptions of the random effects model must necessarily depart to some degree from the truth, it is by no means clear that there is much room to make any kind of significant gains. As an aside, it is of interest to note that, since we do not gain much by considering the whole of the risk set as opposed to a small sample from it, the converse must also hold, i.e., we do not lose very much by working with small samples rather than the whole of the risk set. In certain studies, there may be great economical savings made by only using covariate information, in particular when time dependent, from a subset of the full risk set.

Table 6.7 was taken from O'Quigley and Stare (2002). The table was constructed from simulated failure times where the random effects model was taken to be exactly correct. Data were generated from this model in which the gamma frailty had a mean and variance equal to one. The regression coefficient of interest was exactly equal to 1.0. Three situations were considered; 100 strata each of size 5, 250 strata each of size 2 and 25 strata each of size 20. The take-home message from the table is that, in these cases for random effects models, not much is to be gained in terms of efficiency. Any biases appear negligible and the mean of the point estimates for both random effects and stratified models, while differing notably from a crude model ignoring model inadequacy, are effectively indistinguishable. As we would expect there is a gain for the variance of estimates based on the random effects model but, even for highly stratified data (100×5), any gain is very small. Indeed for the extreme case of 250 strata, each of size 2, surely the worst situation for the stratified model, it is difficult to become enthusiastic over the comparative performance of the random effects model.

	100×5	250×2	25×20
Ignoring effect	0.52 (0.16)	0.51 (0.16)	0.54 (0.16)
Random effect model	1.03 (0.19)	0.99 (0.22)	1.01 (0.17)
Stratified model	1.03 (0.22)	1.02 (0.33)	1.01 (0.18)

Table 6.7: Simulations for three models under different groupings.

We might conclude that we only require around 80% of the comparative sample size needed for estimating relative risk based on the stratified model. But, such a conclusion, leaning entirely on the assumption that we know not only the class of distributions from which the random effects come but also the exact value of the population parameters, suggests, in practice, that the hoped for gain, in this most hopeful of cases, is more likely to be greater than the 80% indicated by our calculations. The only real situation that can be clearly disadvantageous to the stratified model is one where a non-negligible subset of the strata are seen to only contain a single observation. For such cases, and assuming a random effects model to provide an adequate fit, information from states with a single observation (which would be lost by a stratified analysis) can be recovered by a random effects analysis.

6.8 Non proportional hazards model with intercept

Recalling the general model, i.e., the non proportional hazards model for which there is no restriction on $\beta(t)$, note that we can re-express this so that the function $\beta(t)$ is written as a constant term, the intercept, plus some function of time multiplied by a constant coefficient. Writing this as

$$\lambda(t|Z) = \lambda_0(t) \exp\{[\beta_0 + \theta Q(t)]Z\}, \tag{6.23}$$

we can describe the term β_0 as the intercept and $Q(t)$ as reflecting the nature of the time dependency. The coefficient θ will simply scale this dependency and we may often be interested in testing the particular value, $\theta = 0$, since this value corresponds to a hypothesis of proportional hazards. Fixing the function $Q(t)$ to be of some special functional form allows us to obtain tests of proportionality against alternatives of a particular nature. Linear or quadratic decline in the log-relative risk, change-point, and crossing hazard situations are all then easily accommodated by this simple formulation. Tests of goodness of fit of the proportional hazards assumption can be then be constructed which may be optimal for certain kinds of departures.

Although not always needed it can sometimes be helpful to divide the time axis into r nonoverlapping intervals, $B_1, \ldots, B_r$ in an ordered sequence beginning at the origin. In a data-driven situation these intervals may be chosen so as to have a comparable number of events in each interval or so as not to have too few events in any given interval. Defined on these intervals is a vector, also of dimension r, of some known or estimable functions of time, not involving the parameters of interest, β. This is denoted $Q(t) = \{Q_1(t), \ldots, Q_r(t)\}$ This model is then written in the form,

$$\lambda(t|Z) = \lambda_0(t) \exp\{[\beta + \theta Q(t)]Z\}, \tag{6.24}$$

where θ is a vector of dimension r. Thus, $\theta Q(t)$ (here the usual inner product) has the same dimension as β, i.e., one. In order to investigate the time dependency of particular covariates in the case of multivariate Z we would have β of dimension greater than one, in which case $Q(t)$ and θ are best expressed in matrix notation (O'Quigley and Pessione 1989).

Here, as through most of this text, we concentrate on the univariate case since the added complexity of the multivariate notation does not bring any added light to the concepts being discussed. Also, for the majority of the cases of interest, $r = 1$ and θ becomes a simple scalar. We will often have in mind some particular form for the time-dependent regression coefficient $Q(t)$, common examples being a linear slope (Cox 1972), an exponential slope corresponding to rapidly declining effects (Gore et al. 1984) or some function related to the marginal distribution, $F(t)$ (Breslow, Edler and Berger 1984). In practice we may be able to estimate this function of $F(t)$ with the help of consistent estimates of $F(t)$ itself, in particular the Kaplan-Meier estimate. The non proportional hazards model with intercept is of particular use in questions of goodness of fit of the proportional hazards model pitted against specific alternatives. These specific alternatives can be quantified by appropriate forms of the function $Q(t)$. We could also test a joint null hypothesis $H_0 : \beta = \theta = 0$ corresponding to no effect, against an alternative H_1, either θ or β nonzero. This leads to a test with the ability to detect non proportional hazards, as well as proportional hazards departures to the null hypothesis of no effect. We could also test a null hypothesis $H_0 : \theta = 0$ against $H_1 : \theta \neq 0$, leaving β itself unspecified. This would then provide a goodness-of-fit test of the proportional hazards assumption. We return to these issues later on when we investigate in greater detail how these models give rise to simple goodness of fit tests.

Changepoint models

A simple special case of a non proportional hazards model with an intercept is that of a changepoint model. O'Quigley and Pessione (1991), O'Quigley (1994), and O'Quigley and Natarajan (2004) develop such models whereby we take the function $Q(t)$ to be defined by, $Q(t) = I(t \leq \gamma) - I(t > \gamma)$ with γ an unknown changepoint. This function $Q(t)$ depends upon γ but otherwise does not depend upon the unknown regression coefficients and comes under the above heading of a non proportional hazards model with an intercept. For the purposes of a particular structure for a goodness of fit test we can choose the intercept to be equal to some fixed value, often zero (O'Quigley and Pessione 1991). The model is then

$$\lambda(t|Z) = \lambda_0(t) \exp\{[\beta + \alpha Q(t)]Z(t)\}. \tag{6.25}$$

The parameter α is simply providing a scaling (possibly of value zero) to the time dependency as quantified by the function $Q(t)$. The chosen form of $Q(t)$, itself fixed and not a parameter, determines the way in which effects change through time; for instance whether they decline exponentially to zero, whether they decline less rapidly or any other way in which effects might potentially change through time.

Inference for the changepoint model is not straightforward and in the series of chapters dealing with approaches to inference one chapter is devoted specifically to changepoint models. Note that were γ to be known, then inference would come under the usual headings with no additional difficulty. The changepoint model expressed by Equation (6.25) deals with the regression effect changing through time and putting the model under the heading of a non proportional hazards model. A related, although entirely different model, is one which arises as a simplification of a proportional model with a continuous covariate and the idea is to replace the continuous covariate by a discrete classification.

The classification problem itself fall into two categories. If we are convinced of the presence of effects and simply wish to derive the most predictive classification into, say, two groups, then the methods using explained randomness or explained variation will achieve this goal. If, on the other hand, we wish to test a null hypothesis of absence of effect, and, in so doing, wish to consider all possible classifications based on a family of potential cutpoints of the continuous covariate, then special techniques of inference are required. We return to this in Chapter 12.

6.9   Time-dependent covariates

In all of the above models we can make a simple change by writing the covariate Z as $Z(t)$, allowing the covariate to assume different values at different time points. Our model then becomes

$$\lambda(t|Z(t)) = \lambda_0(t)\exp\{\beta(t)Z(t)\} \tag{6.26}$$

and allows situations such as those described in Figure 6.9 to be addressed. As we change states the intensity function changes. This enables us to immediately introduce further refinement into a simple alive/dead model whereby we can suppose one or more intermediary states. A subject can move across states thereby allowing prognosis to improve or to worsen, the rates of these changes themselves depending upon other factors. The state death is described as an absorbing state and so we can move into this state but, once there, we cannot move out of it again.

Mostly we will work with the proportional hazard restriction on the above model so that

$$\lambda(t|Z(t)) = \lambda_0(t)\exp\{\beta Z(t)\}, \tag{6.27}$$

Such a simple, albeit very much more sophisticated, model than our earlier one describes a broad range of realistic situations. We will see that models with time-dependent covariates do not raise particular difficulties, either computationally or from the viewpoint of interpretation, when we deal with inference. This will be clear from the main

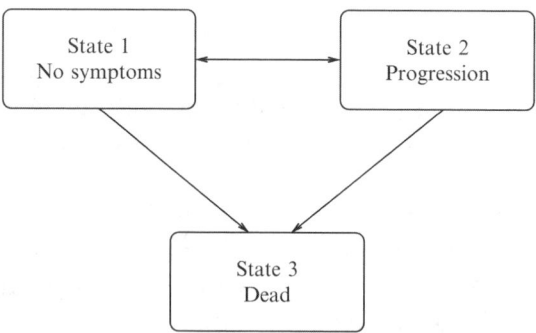

Figure 6.9: Compartment model where ability to move between states other than death state can be characterized by time dependent indicator covariates $Z(t)$.

theorem of proportional hazards regression (Section 7.4). The model simply says that the effect of the covariate remains constant, i.e., the regression coefficient remains constant, but that the covariate, or state, can itself change with time. Models with time-dependent covariates can also be used as a purely artificial construction in order to be able to express non proportional hazards models in a proportional hazards form. That is not our main purpose here, however, and we are assuming that $Z(t)$ does correspond to some real physical measurement which can be obtained through time.

We can also imagine a slightly more involved situation than the above. Suppose that the covariate Z remains fixed, but that a second covariate, known to influence survival, also needs to be accounted for. Furthermore this second covariate is time dependent. We could, of course, simply use the above model extended to the case of two covariates. This is straightforward, apart from the fact that, as previously underlined by the complexity theorem, care is needed. If, however, we do not wish to model the effects of this second covariate, either because it is only of indirect concern or because its effects might be hard to model, then we could appeal to a stratified model. We write;

$$\lambda(t|Z(t), w(t)) = \lambda_{0w(t)}(t) \exp\{\beta Z(t)\}, \qquad (6.28)$$

where, as for the non time-dependent case, $w(t)$ takes integer values $1,..., m$ indicating status. The subject can move in and out of the m strata as time proceeds. Two examples illustrate this. Consider a new treatment to reduce the incidence of breast cancer. An important time-dependent covariate would be the number of previous incidents of benign disease. In the context of inference, the above model simply means that, as far as treatment is concerned, the new treatment and the standard are only ever contrasted within patients having the same previous history. These contrasts are then summarized in final estimates and possibly tests. Any patient works her way through the various states, being unable to return to a previous state. The states themselves are not modeled. A second example might be a sociological study on the incidence of job loss and how it relates to covariates of main interest such as training, computer skills etc. Here, a stratification variable would be the type of work or industry in which the individual finds him or herself. Unlike the previous example a subject can move between states and return to previously occupied states.

Time-dependent covariates describing states can be used in the same way for transition models in which there is more than one

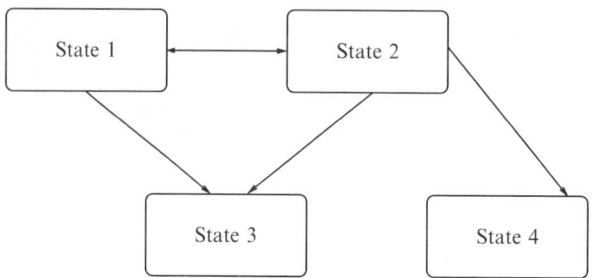

Figure 6.10: Compartment model with 2 absorbing "death" states.

absorbing "death" state. Many different kinds of situations can be constructed, these situations being well described by compartment models with arrows indicating the nature of the transitions that are possible (Figure 6.10). For compartment models with time-dependent covariates there is a need for some thought when our interest focuses on the survival function. The term external covariate is used to describe any covariate $Z(t)$ such that, at $t = 0$, for all other $t > 0$, we know the value of $Z(t)$. The paths can be described as deterministic. In the great majority of the problems that we face this is not the case and a more realistic way of describing the situation is to consider the covariate path $Z(t)$ to be random. Also open to us as a modeling possibility, when some covariate $Z_1(t)$ is of secondary interest assuming a finite number of possible states, is to use the at risk function $Y(s,t)$. This restricts our summations to those subjects in state s as described above for stratified models.

6.10   Time-dependent covariates
and non proportional hazards models

A non proportional hazard model with a single constant covariate Z is written

$$\lambda(t|Z) = \lambda_0(t) \exp\{\beta(t)Z\}. \tag{6.29}$$

The multivariate extension is immediate and, in keeping with our convention of only dealing with the univariate problem whenever possible, we focus our attention on the simple product $\beta(t)Z$ at some given point in time t. If we define $\beta_0 = \beta(0)$ we can rewrite this product as $\beta_0 Z(t)$ where $Z(t) = Z\beta(t)/\beta_0$. We could take any other time point t'

and, once again, we observe that we can re-write the product $\beta(t')Z$ as $\beta_0 Z(t')$ where $Z(t') = Z\beta(t')/\beta_0$. This equivalence is then true for any, and all, values of t. Thus, a non proportional hazards model with a constant covariate can be re-expressed, equivalently, as a simple proportional hazards model with a time-dependent covariate.

It is almost immediate, and perhaps worth carrying out as an exercise, to show that we can reverse these steps to conclude also that any model with time dependent covariates can be expressed in an equivalent form as a non proportional hazards model. In conclusion, for every non proportional hazards model there exists an equivalent proportional hazards model with time-dependent covariates. Indeed, it also clear that this argument can be extended. For, if we have the model; $\lambda(t|Z) = \lambda_0(t)\exp\{\beta(t)Z(t)\}$, then, via a re-expression of the model using $\beta_0 Z^*(t)$ where $Z^*(t) = Z(t)\beta(t)/\beta_0$, we can construct a proportional hazards model with a time-dependent regression effect from a model which began with both time-dependent regression effects as well as time changing regression coefficient.

This equivalence is a formal one and does not of itself provide any new angle on model development. It may be exploited nonetheless in theoretical investigation or used as a means to enable the structuring of a particular problem. For example, many available softwares, as well as user written code, will cater for time-dependent covariables. This facility can then be made use of should we wish to study particular types of non proportional hazards models.

6.11 Proportional hazards models in epidemiology

For arbitrary random variables X and Y with joint density $f(x,y)$, conditional densities $g(x|y)$ and $h(y|x)$, marginal densities $v(x)$ and $w(y)$, we know that

$$f(x,y) = g(x|y)w(y) = h(y|x)v(x),$$

so that, in the context of postulating a model for the pair (X,Y), we see that there are two natural potential characterizations. Recalling the discussion from Section 4.3 note that, for survival studies, our interest in the binary pair (T,Z), time and covariate, can be seen equivalently from the viewpoint of the conditional distribution of time given the

covariate, along with the marginal distribution of the covariate, or from
the viewpoint of the conditional distribution of the covariate given
time, along with the marginal distribution of time. This equivalence
we exploit in setting up inference where, even though the physical
problem concerns time given the covariate, our analysis describes the
distribution of the covariate given time.

In epidemiological studies the variable time T is typically taken to
be age. Calendar time and time elapsed from some origin may also be
used but, mostly, the purpose is to control for age in any comparisons
we wish to make. Usually we will consider rates of incidence of some
disease within small age groups or possibly, via the use of models, for a
large range of values of age. Unlike the relatively artificial construction
of survival analysis which exploits the equivalent ways of expressing
joint distributions, in epidemiological studies our interest naturally
falls on the rates of incidence for different values of Z given fixed
values of age T. It is not then surprising that the estimating equations
we work with turn out to be essentially the same for the two situations.

The main theorem of proportional hazards regression (Section 7.4)
applies more immediately in epidemiology than in survival type studies. We return to this in the chapter on inference. One important distinction, although already well catered for by use of our "at risk"
indicator variables, is that for epidemiological studies the subjects in
different risk sets are often distinct subjects. Even so, as we will see,
the form of the equations is the same, and software which allows an
analysis of survival data will also allow an analysis of certain problems
in epidemiology.

For a binary outcome, indicated by $Y = 1$ or $Y = 0$, and a binary
risk or exposure factor, $Z = 1$ or $Z = 0$, the relative risk is defined as
the ratio of the probabilities $P(Y = 1|Z = 1)/P(Y = 1|Z = 0)$ and
the, related, odds ratio ψ as

$$\psi = \frac{P(Y = 1|Z = 1)P(Y = 0|Z = 0)}{P(Y = 1|Z = 0)P(Y = 0|Z = 1)}.$$

In the above and in what follows, in order for the notation not to
become too cluttered, we write $\Pr(A) = P(A)$. Under a "rare disease
assumption," i.e., when $P(Y = 0|Z = 0)$ and $P(Y = 0|Z = 1)$ are close
to 1, then the odds ratio and relative risk approximate one another.
One reason for being interested in the odds ratio, as a measure of the

impact of different levels of the covariate (risk factor) Z follows from the easily obtained identity

$$\frac{P(Y=1|Z=1)P(Y=0|Z=0)}{P(Y=1|Z=0)P(Y=0|Z=1)} = \frac{P(Z=1|Y=1)P(Z=0|Y=0)}{P(Z=1|Y=0)P(Z=0|Y=1)}.$$
(6.30)

Thus, the impact of different levels of the risk factor Z can equally well be estimated by studying groups defined on the basis of this same risk factor and their corresponding incidence rates of $Y = 1$. This provides the rationale for the case-control study in which, in order to estimate ψ, we make our observations on Z over fixed groups of cases and controls (distribution of Y fixed), rather than the more natural, but practically difficult if not impossible, approach of making our observations on Y for a fixed distribution of Z. Assumptions and various subtleties are involved. The subject is vast and we will not dig too deeply into this. The points we wish to underline in this section are those that establish the link between epidemiological modeling and proportional hazards regression.

Series of 2×2 tables

The most elementary presentation of data arising from either a prospective study (distribution of Z fixed) or a case-control study (distribution of Y fixed) is in the form of a 2×2 contingency table in which the counts of the number of observations are expressed. Estimated probabilities, or proportions of interest are readily calculated. In Table 6.8, $a_{1*} = a_{11} + a_{12}$, $a_{2*} = a_{21} + a_{22}$, $a_{*1} = a_{11} + a_{21}$, $a_{*2} = a_{12} + a_{22}$ and $a_{**} = a_{1*} + a_{2*} = a_{*1} + a_{*2}$. For prospective studies the proportions a_{11}/a_{*1} and a_{12}/a_{*2} estimate the probabilities of being a case $(Y = 1)$ for both exposure groups while, for case-control studies, the proportions a_{11}/a_{1*} and a_{21}/a_{2*} estimate the

	$Z = 1$	$Z = 0$	totals
$Y = 1$	a_{11}	a_{12}	a_{1*}
$Y = 0$	a_{21}	a_{22}	a_{2*}
Totals	a_{*1}	a_{*2}	a_{**}

Table 6.8: Basic 2×2 table for cases $(Y = 1)$ and controls $(Y = 0)$.

probabilities of exhibiting the risk or exposure factor $(Z = 1)$ for both cases and controls. For both types of studies we can estimate ψ by the ratio $(a_{11}a_{22})/(a_{21}a_{12})$, which is also the numerator of the usual chi-squared test for equality of the two probabilities. If we reject the null hypothesis of the equality of the two probabilities we may wish to say something about how different they are based on the data from the table. As explained below, under the heading "Logistic regression," quantifying the difference between two proportions is not best done via the most obvious, and simple, arithmetic difference. There is room for more than one approach, the simple arithmetic difference being perfectly acceptable when sample sizes are large enough to be able to use the De Moivre-Laplace approximation (Section 3.3) but, more generally, the most logical in our context is to express everything in terms of the odds ratio. We can then exploit the following theorem;

Theorem 6.2 *Taking all the marginal totals as fixed, the conditional distribution of a_{11} is written*

$$P(a|a_{1*}, a_{2*}, a_{*1}, a_{*2}) = \binom{a_{1*}}{a}\binom{a_{2*}}{a_{*1}-a}\psi^a \bigg/ \sum_u \binom{a_{1*}}{u}\binom{a_{2*}}{a_{*1}-u}\psi^u,$$

the sum over u being over all integers compatible with the marginal totals. The conditionality principle appears once more, in this instance in the form of fixed margins. The appropriateness of such conditioning, as in other cases, can be open to discussion. But again, insightful conditioning has greatly simplified the inferential structure. Following conditioning of the margins, it is only necessary to study the distribution of any one entry in the 2×2 table, the other entries being then determined. It is usual to study the distribution of a_{11}. A nonlinear estimating equation can be based on $a_{11} - E(a_{11})$, expectation obtained from Theorem 6.2, and from which we can estimate ψ and associate a variance term with the estimator. The nonlinearity of the estimating equation, the only approximate normality of the estimator, and the involved form of variance expressions has led to much work in the methodological epidemiology literature; improving the approximations, obtaining greater robustness and so on. However, all of this can be dealt with in the context of a proportional hazards (conditional logistic) regression model. Since it would seem more satisfactory to work with a single structure rather than deal with problems on a case-by-case basis the recommendation is to work with proportional and non proportional hazards models. Not only does a model enable us

Table i	$Z = 1$	$Z = 0$	Totals
$Y = 1$	$a_{11}(i)$	$a_{12}(i)$	$a_{1*}(i)$
$Y = 0$	$a_{21}(i)$	$a_{22}(i)$	$a_{2*}(i)$
Totals	$a_{*1}(i)$	$a_{*2}(i)$	$a_{**}(i)$

Table 6.9: 2×2 table for ith age group of cases and controls.

to more succinctly express the several assumptions which we may be making it offers, more readily, well established ways of investigating the validity of any such assumptions. In addition the framework for studying questions such as explained variation, explained randomness and partial measures of these is clear and requires no new work.

The "rare disease" assumption, allowing the odds ratio and relative risk to approximate one another, is not necessary in general. However, the assumption can be made to hold quite easily and is therefore not restrictive. To do this we construct fine strata, within which the probabilities $P(Y = 0|Z = 0)$ and $P(Y = 0|Z = 1)$ can be taken to be close to 1. For each stratum, or table, we have a 2×2 table as in Table 6.9, indexed by i. Each table provides an estimate of relative risk at that stratum level and, assuming that the relative risk itself does not depend upon this stratum, although the actual probabilities themselves composing the relative risk definition may themselves depend upon strata, then the problem is putting all these estimates of the same thing into a single expression. The most common such expression for this purpose is the Mantel-Haenszel estimate of relative risk.

Mantel-Haenszel estimate of relative risk

The, now famous, Mantel-Haenszel estimate of relative risk was described by Mantel and Haenszel (1959) and is particularly simple to calculate. Referring to the entries of observed counts in Table 6.10, if we first define for the i th subtable $R_i = a_{11}(i)a_{22}(i)/a_{**}(i)$ and $S_i = a_{12}(i)a_{21}(i)/a_{**}(i)$, then the Mantel-Haenszel summary relative risk estimate across the tables is given by $\hat{\psi}_{MH} = \sum_i R_i / \sum_i S_i$. Breslow (1996) makes the following useful observations concerning $\hat{\psi}_{MH}$ and $\hat{\beta}_{MH} = \hat{\psi}_{MH}$. First, $E(R_i) = \psi_i E(S_i)$ where the true odds ratio in the ith table is given by ψ_i. When all of these odds ratios coincide then $\hat{\psi}_{MH}$ is the solution to the unbiased estimating equation; $R - \psi S = 0$, where $R = \sum_i R_i$ and $S = \sum_i S_i$. Under an assumption of

Table i	$Z = 1$	$Z = 0$	Totals	Table i	$Z = 1$	$Z = 0$	Totals
$Y = 1$	$a_{11}(i)$	$a_{12}(i)$	$a_{1*}(i)$	$Y = 1$	$e_{11}(i)$	$e_{12}(i)$	$e_{1*}(i)$
$Y = 0$	$a_{21}(i)$	$a_{22}(i)$	$a_{2*}(i)$	$Y = 0$	$e_{21}(i)$	$e_{22}(i)$	$e_{2*}(i)$
Totals	$a_{*1}(i)$	$a_{*2}(i)$	$a_{**}(i)$	Totals	$e_{*1}(i)$	$e_{*2}(i)$	$e_{**}(i)$

Table 6.10: 2×2 table for ith age group of cases and controls. Left-hand table: observed counts. Right hand table: expected counts.

binomial sampling, Breslow shows that the variances of the individual contributions to the estimating equation are such that the quantity $2a_{**}^2(i)\mathrm{Var}\,(R_i - \psi S_i)$ can be equated to;

$$E\left\{[a_{11}(i)a_{22}(i)+\psi a_{12}(i)a_{21}(i)]\,[a_{11}(i) + a_{22}(i) + \psi\,(a_{12}(i) + a_{21}(i))]\right\},$$

from which, by a simple application of the delta method we can obtain estimates of the variance of $\hat{\psi}_{MH}$.

Logistic regression

Without any loss in generality we can express the two probabilities of interest, $P(Y = 1|Z = 1)$ and $P(Y = 1|Z = 0)$ as simple power transforms of one another. This follows, since, whatever the true values of these probabilities, there exists some positive number α such that $P(Y = 1|Z = 1) = P(Y = 1|Z = 0)^\alpha$. The parameter α is constrained to be positive in order that the probabilities themselves remain between 0 and 1. To eliminate any potential dangers that may arise, particularly in the estimation context where, even though the true value of α is positive, the estimate itself may not be, a good strategy is to re-express this parameter as $\alpha = \exp(\beta)$. We then have

$$\log\log P(Y = 1|Z = 1) = \log\log P(Y = 1|Z = 0) + \beta. \qquad (6.31)$$

The parameter β can then be interpreted as a linear shift in the log-log transformation of the probabilities, and can take any value between $-\infty$ and ∞, the inverse transformations being one-to-one and guaranteed to lie in the interval (0,1). An alternative model to the above is

$$\mathrm{logit}\,P(Y = 1|Z = 1) = \mathrm{logit}\,P(Y = 1|Z = 0) + \beta. \qquad (6.32)$$

where the logit transformation, again one-to-one, is defined by $\mathrm{logit}\,\theta = \log\{\theta/(1-\theta)\}$. Although a natural model, the model of Equation 6.31 is not usually preferred to that of Equation 6.32, motivated in an analogous way (i.e., avoiding constraints) but having a slight advantage from the viewpoint of interpretation. This is because the parameter β is the logarithm of the odds ratio, i.e., $\beta = \log \psi$.

In the light of the equivalence of the the odds for disease given the risk factor and the odds for the risk factor given the disease, as expressed in Equation 6.30, we conclude immediately that, equivalent to the above model involving β, expressed in Equation 6.32, we have a model expressing the conditional probability of Z given Y and using the same β. This highlights an important feature of proportional hazards modeling whereby we focus attention on the conditional distribution of the covariates given an event yet, when thinking of the applied physical problem behind the analysis, we would think more naturally in terms of the conditional distribution of the event given the covariates. The essential point is that the unknown regression parameter, β, of interest to us is the same for either situation so that in place of Equation (6.32), we can write

$$\text{logit } P(Z = 1|Y = 1) = \text{logit } P(Z = 1|Y = 0) + \beta. \qquad (6.33)$$

Since the groups are indicated by a binary Z, we can exploit this in order to obtain the more concise notation, now common for such models, whereby

$$\text{logit } P(Y = 1|Z) = \text{logit } P(Y = 1|Z = 0) + \beta Z. \qquad (6.34)$$

As we have tried, in as much as is possible throughout this text, to restrict attention to a single explanatory variable, this is once more the case here. Extension to multiple explanatory variables, or risk factors, is immediate and, apart from the notation becoming more cumbersome, there are no other concepts to which to give thought. We write the model down, as above in Equation (6.34), and use several binary factors Z (Z now a vector) to describe the different group levels. The coefficients β (β now a vector) then allow the overall odds ratio to be modeled or, allows the modeling of partial odds ratios whereby certain risk factors are included in the model, and our interest focuses on those remaining after having taken account of those already included. The above model can also be written in the form

$$\frac{P(Y = 1|Z)}{1 - P(Y = 1|Z)} = \exp(\beta_0 + \beta Z), \qquad (6.35)$$

where $\beta_0 = \text{logit } P(Y = 1|Z = 0)$. Maintaining an analogy with the usual linear model we can interpret β_0 as an intercept, simply a function of the risk for a "baseline" group defined by $Z = 0$.

Assigning the value $Z = 0$ to some group and thereby giving that group baseline status is, naturally, quite arbitrary and there is nothing special about the baseline group apart from the fact that we define it as such. We are at liberty to make other choices and, in all events, the only quantities of real interest to us are relative ones. In giving thought to the different modeling possibilities that arise when dealing with a multivariate Z, the exact same kind of considerations, already described via several tables in the section on modeling multivariate problems will guide us (see Section 6.6 and those immediately following it). Rather than repeat or reformulate those ideas again here, the reader, interested in these aspects of epidemiological modeling, is advised to go over those earlier sections. Indeed, without a solid understanding as to why we choose to work with a particular model rather than another, and as to what the different models imply concerning the complex inter-relationships between the underlying probabilities, it is not really possible to carry out successful modeling in epidemiology.

Stratified and conditional logistic regression

In the above model, and Z being multivariate, we may wish to include alongside the main factors under study, known risk factors, and particularly risk factors such as age, or period effects, for which we would like to control. Often age alone is the strongest factor and its effect can be such that the associated errors of estimation in quantifying its impact can drown the effect of weaker risk factors. One possibility in controlling for such factors, S, it to appeal to the idea of stratification. This means that analysis is carried out at each level of S and, within a level, we make the same set of assumptions concerning the principle factors under study. We write

$$\frac{P(Y = 1|Z, S)}{1 - P(Y = 1|Z, S)} = \exp(\beta_0 + \beta Z), \qquad (6.36)$$

where, in the same way as before, $\beta_0 = \text{logit } P(Y = 1|Z = 0, S)$. The important aspect of a stratified model is that the levels of S only appear in the left-hand side of the equation.

We might conclude that this is the same model as the previous one but it is not quite and, in later discussions on inference, we see that it does impact the way the likelihood is written. In the simpler cases, in as far as β is concerned, the stratified model is exactly equivalent to a regular logistic model if we include in the regression function indicator variables, of dimension one less than the number of strata. However, when the number of strata is large, use of the stratified model enables us to bypass estimation of the stratum-level effects. If these are not of real interest then this may be useful in that it can result in gains in estimating efficiency even though the underlying models may be equivalent. In a rough intuitive sense we are spending the available estimating power on the estimation of many less parameters, thereby increasing the precision of each one. This underlines an important point in that the question of stratification is more to do with inference than the setting up of the model itself.

This last remark is even more true when we speak of conditional logistic regression. The model will look almost the same as the unconditional one but the process of inference will be quite different. Suppose we have a large number of strata, very often in this context defined by age. A full model would be as in Equation (6.35), including in addition to the risk factor vector Z, a vector parameter of indicator variables of dimension one less than the number of strata. Within each age group, for the sake of argument let's say age group i, we have the simple logistic model. However, rather than write down the likelihood in terms of the products $P(Y = 1|Z)$ and $P(Y = 0|Z)$ we consider a different probability upon which to construct the likelihood, namely the probability that the event of interest, the outcome or case in other words, occurred on an individual (in particular the very individual for whom the event *did* occur, given that one event occurred among the set $S\{i\}$ of the $a_{**}(i)$ cases and controls. Denoting Z_i to be the risk factor for the case, corresponding to the age group i, then this probability is simply; $\exp(\beta Z_i)/\sum I[j \in S\{i\}]\exp(\beta Z_j)$. The likelihood is then the product of such terms across the number of different age groups for which a case was selected. If we carefully define the "at-risk" indicator $Y(t)$ where t now represents age, we can write the conditional likelihood as

$$L(\beta) = \prod_{i=1}^{n} \left\{ \frac{\exp(\beta Z_i)}{\sum_{j=1}^{n} Y_j(X_i)\exp(\beta Z_j)} \right\}^{\delta_i}. \tag{6.37}$$

Here we take the at-risk indicator function to be zero unless, for the subject j, X_j has the same age, or is among the same age group as that given by X_i. In this case the at-risk indicator $Y_j(X_i)$ takes the value one. To begin with, we assume that there is only a single case per age group, that the ages are distinct between age groups, and that, for individual i, the indicator δ_i takes the value one if this individual is a case. Use of the δ_i would enable us to include in an analysis sets of controls for which there was no case. This would be of no value in the simplest case but, generalizing the ideas along exactly the same lines as for standard proportional hazards models, we could easily work with indicators $Y(t)$ taking the value one for all values less than t and becoming zero if the subject becomes incident or is removed from the study. A subject is then able to make contributions to the likelihood at different values of t, i.e., at different ages, and appears therefore in different sets of controls. Indeed, the use of the risk indicator $Y(t)$ can be generalized readily to other complex situations.

One example is to allow it to depend on two time variables, for example, an age and a cohort effect, denoting this as $Y(t, u)$. Comparisons are then made between individuals having the same age and cohort status. Another useful generalization of $Y(t)$ is where individuals go on and off risk, either because they leave the risk set for a given period or, possibly, because their status cannot be ascertained. Judicious use of the at-risk indicator Y makes it possible then to analyze many types of data that, at first glance, would seem quite intractable. This can be of particular value in longitudinal studies involving time-dependent measurements where, in order to carry out unmodified analysis we would need, at each observed failure time, the time dependent covariate values for all subjects at risk. These would not typically all be available. A solution based on interpolation, assuming that measurements do not behave too erratically, is often employed. Alternatively we can allow for subjects for whom, at an event time, no reliable measurement is available, to simply temporarily leave the risk set, returning later when measurements have been made.

The striking thing to note about the above conditional likelihood is that it coincides with the expression for the partial likelihood given earlier in the chapter. This is no real coincidence of course and the main theorem of proportional hazards regression (Section 7.4), described in the following chapter, applies equally well here. For this we need one more concept, described later, and that is the idea of sampling from the risk set. The difference between the $Y(t)$ in a classical survival study, where it is equal to one as long as the subject is under study and then drops to zero, as opposed to the $Y(t)$ in the simple epidemiological application in which it is zero most of time, taking the value one when indicating the appropriate age group, is a small one. It can be equated with having taken a small random sample from a conceptually much larger group followed since time (age) is zero. On the basis of the above conditional likelihood we obtain the estimating equation

$$U(\beta) = \sum_{i=1}^{n} \delta_i \left\{ Z_i - \frac{\sum_{j=1}^{n} Y_j(X_i) Z_j \exp(\beta Z_j)}{\sum_{j=1}^{n} Y_j(X_i) \exp(\beta Z_j)} \right\}, \quad (6.38)$$

which we equate to zero in order to estimate β. The equation contrasts the same quantities written down in Table 6.10 in which the expectations are taken with respect to the model. The estimating equations are then essentially the same as those given in Table 6.10 for the Mantel-Haenszel estimator. Furthermore, taking the second derivative of the expression for the log-likelihood, we have that $I(\beta) = \sum_{i=1}^{n} \delta_i I_i(\beta)$ where

$$I_i(\beta) = \frac{\sum_{j=1}^{n} Y_j(X_i) Z_j^2 \exp(\beta Z_j)}{\sum_{j=1}^{n} Y_j(X_i) \exp(\beta Z_j)} - \left\{ \frac{\sum_{j=1}^{n} Y_j(X_i) Z_j \exp(\beta Z_j)}{\sum_{j=1}^{n} Y_j(X_i) \exp(\beta Z_j)} \right\}^2, \quad (6.39)$$

then $I(\beta) = \sum_{i=1}^{n} \delta_i I_i(\beta)$. Inferences can then be carried out on the basis of these expressions. In fact, once we have established the link between the applied problem in epidemiology and its description via a proportional hazards model, we can then appeal to those model-building techniques (explained variation, explained randomness, goodness of fit, conditional survivorship function etc.) which we use for applications in time to event analysis. In this context the building of models in epidemiology is no less important, and no less delicate, than the building of models in clinical research.

6.12 Exercises and class projects

1. One of the early points of discussion on Cox's 1972 paper was how to deal with tied data. Look up the Cox paper and write down the various different ways that Cox and the contributors to the discussion suggested that tied data be handled. Explain the advantages and disadvantages to each approach.

2. One suggestion for dealing with tied data, not in that discussion, is to simply break the ties via some random split mechanism. What are the advantages and drawbacks to such an approach?

3. As an alternative to the proportional hazards model consider the two models (i) $S(t|Z) = S_0(t) + \beta Z$, and (ii) $\text{logit} S(t|Z) = \text{logit} S_0(t) + \beta Z$. Discuss the relative advantages and drawbacks of all three models.

4. Show that the relation; $S(t|Z) = \{S_0(t)\}^{\exp(\beta Z)}$ implies the Cox model and vice versa.

5. Suppose that we have two groups and that a proportional hazards model is believed to apply. Suppose also that we know for one of the groups that the hazard rate is a linear function of time, and equal to zero at the origin. Given data from such a situation, suggest different ways in which it can be analyzed and the possible advantages and disadvantages of the various approaches.

6. Explain in what sense the components of Equation 6.7 and equation (6.8) can be viewed as an equation for the mean and an equation for the variance.

7. Using equations (6.7) and (6.8) work out the calculations explicitly for the two-group case, i.e., the case in which there are $n_1(t)$ subjects at risk from group 1 at time t and $n_2(t)$ from group 2.

8. Suppose that we have available software able to analyze a proportional hazards model with a time-dependent covariate $Z(t)$. Suppose that, for the problem in hand the covariate, Z, does not depend on time. However, the regression effect $\beta(t)$ is known to decline as an exponential function of time. How would you proceed?

9. Suppose we fit a proportional hazards model, using some standard software, to a continuous covariate Z defined on the interval (1,4). Unknown to us our model assumption is incorrect and the model applies exactly to $\log Z$ instead. What effect does this have on our parameter estimate?

10. Consider an experiment in which there are eight levels of treatment. The levels are ordered. The null hypothesis is that there is no treatment effect. The alternative is that there exists a non-null effect increasing with level until it reaches one of the levels, say level j, after which the remaining levels all have the same effect as level j. How would you test for this?

11. Write down the joint likelihood for the underlying hazard rate and the regression parameter β for the two-group case in which we assume the saturated piecewise exponential model. Use this likelihood

to recover the partial likelihood estimate for β. Obtain an estimate of the survivorship function for both groups.

12. For the previous question derive an approximate large sample confidence interval for the estimate of the survivorship function for both groups in cases: (i) where the parameter β is exactly known, (ii) where the parameter is replaced by an estimate with approximate large sample variance σ^2.

13. Carry out a large sample simulation for a model with two binary variables. Each study is balanced with a total of 100 subjects. Choose $\beta_1 = \beta_2 = 1.5$ and simulate binary Z_1 and Z_2 to be uncorrelated. Show the distribution of $\hat{\beta}_1$ in two cases: (i) where the model used includes Z_2, (ii) where the model used includes only Z_1. Comment on the distributions, in particular the mean value of $\hat{\beta}_1$ in either case.

14. In the previous exercise, rather than include in the model Z_2, use Z_2 as a variable of stratification. Repeat the simulation in this case for the stratified model. Comment on your findings.

15. Consider the following regression situation. We have one-dimensional covariates Z, sampled from a density $g(z)$. Given z we have a proportional hazards model for the hazard rates. Suppose that, in addition, we are in a position to know exactly the marginal survivorship function $S(t) = \int S(t|z)g(z)dz$. How can we use this information to obtain a more precise analysis of data generated under the PH model with Z randomly sampled from $g(z)$?

16. Suppose we have two groups defined by the indicator variable $Z = \{0,1\}$. In this example, unlike the previous in which we know the marginal survival, we know the survivorship function $S_0(t)$ for one of the groups. How can this information be incorporated into a two-group comparison in which survival for both groups is described by a proportional hazards model? Use a likelihood approach.

17. Use known results for the exponential regression model in order to construct an alternative analysis to that of the previous question based upon likelihood.

18. A simple test in the two-group case for absence of effects is to calculate the area between the two empirical survival curves. We can

evaluate the null distribution by permuting the labels corresponding to the group assignment indicator Z. Carry out this analysis for the Freireich data and obtain a p-value. How does this compare with that obtained from an analysis based on the assumption of an exponential model and that based on partial likelihood?

19. Carry out a study, i.e., the advantages, drawbacks and potentially restrictive assumptions, of the test of the previous example. How does this test compare with the score test based on the proportional hazards model?

20. Obtain a plot of the likelihood function for the Freireich data. Using simple numerical integration routines, standardize the area under the curve to be equal to one.

21. For the previous question, treat the curve as a density. Use the mean as an estimate of the unknown β. Use the upper and lower 2.5% percentiles as limits to a 95% confidence interval. Compare these results with those obtained using large sample theory.

22. Suppose we have six ordered treatment groups indicated by $Z = 1, \ldots, 6$. For all values of $Z \leq \ell$ the hazards are the same. For $Z > \ell$ the hazards are again the same and either the same as those for $Z \leq \ell$ or all strictly greater than for $Z \leq \ell$. The value of ℓ is not known. How would you model and set up tests in this situation?

23. Consider an epidemiological application in which workers may be exposed to some carcinogen during periods in which they work in some particular environment. When not working in that particular environment their risk falls back to the same as that for the reference population. Describe this situation via a proportional hazards model with time-dependent effects. How do you suggest modifying such a model if the risk from exposure rather than falling back to the reference group once exposure is removed is believed to be cumulative?

24. Write down a conditional logistic model in which we adjust for both age and cohort effects where cohorts are grouped by intervals of births from 1930-35, 1936-40, 1940-45, etc. For such a model is it possible to answer the question: was there a peak in relative risk during the nineteen sixties?

Chapter 7

Inference: Estimating equations

7.1 Summary

The results of this chapter and, for the most, all of the succeeding chapters, are based on an elementary and central theorem. We call this theorem the main theorem of proportional hazards regression. Its development is essentially that of O'Quigley (2003) which generalizes earlier results of Schoenfeld (1980), O'Quigley and Flandre (1994) and Xu and O'Quigley (2000). The theorem has several immediate corollaries and we can use these to write down estimating equations upon which we can then construct suitable inferential procedures for our models. While a particular choice of estimating equation can result in high efficiency when model assumptions are correct or close to being correct, other equations may be less efficient but still provide estimates which can be interpreted when model assumptions are incorrect. For example, when the regression function $\beta(t)$ might vary with time we are able to construct an estimating equation, the solution of which provides an estimate of β, in the case where $\beta(t)$ is a constant β, and $E\{\beta(T)\}$, the average effect, in the case where $\beta(t)$ changes through time. It is worth underlining that the usual partial likelihood estimate fails to achieve this.

7.2   Motivation

The earlier chapter on marginal survival is important in its own right and we lean on the results of that chapter throughout this work. We need keep in mind the idea of marginal survival for two reasons: (1) it provides a natural backdrop to the ideas of conditional survival and (2), together with the conditional distribution of the covariate given $T = t$, we are able to consider the joint distribution of covariate and survival time T. Conditional survival, where we investigate the conditional distribution of survival given different potential covariate configurations, as well as possibly time elapsed, is a central concern. More generally we are interested in survival distributions corresponding to transitions from one state to another, conditional on being in some particular state or of having mapped out some particular covariate path. The machinery that will enable us to obtain insight into these conditional distributions is that of proportional hazards regression.

When we consider any data at hand as having arisen from some experiment the most common framework for characterizing the joint distribution of the covariate Z and survival T is one where the distribution of Z is fixed and known, and the conditional survivorship distribution the subject of our inferential endeavors. In fact, as underlined in the main theorem of proportional hazards regression, just below, it is more useful to characterize the joint distribution of Z and T via the conditional distribution of Z given $T = t$ and the marginal distribution of T. This is one of the reasons why, in the previous chapter, we dealt with the marginal distribution of T. We can construct estimating equations based on these ideas and from these build simple tests or make more general inferences.

One of the most intriguing aspects of the Cox model concerns estimation of the regression parameter β while ignoring any precise specification of $\lambda_0(t)$. Otherwise, under a conditional independent censoring mechanism and a specified functional form for the underlying hazard $\lambda_0(t)$, likelihood methods, at least in principle, are straightforward. But mostly we prefer to relax assumptions concerning $\lambda_0(t)$, possibly considering it to be entirely unknown, and construct inference for β that remains invariant to any change in $\lambda_0(t)$. Any such changes can be made to correspond to monotonic increasing transformations on T, in which case we can take inference procedures to be rank invariant. This follows since monotonic increasing transformations on the observed times X_i will not affect the rank ordering.

7.3 The observations

Our data will consist of the observations $(Z_i(t), Y_i(t), (t \leq X_i), X_i \, ; \, i = 1 \ldots n)$. The Z_i are the covariates (possibly time dependent), the $X_i = \min(T_i, C_i)$, the observed survival which is the smallest of the censoring time and the actual survival time and the $Y_i(t)$ are time-dependent indicators taking the value one as long as the ith subject is at risk at time t and zero otherwise. For the sake of large sample constructions we make $Y_i(t)$ to be left continuous. At some level we will be making an assumption of independence, an assumption that can be challenged via the data themselves, but that is often left unchallenged, the physical context providing the main guide. Mostly, we think of independence as existing across the indices $i \, (i = 1, \ldots, n)$, i.e., the triplets $\{Z_i(t), Y_i(t), X_i \, ; \, i = 1, \ldots, n\}$. It is helpful to our notational construction to have:

Definition 7.1 *Let $\mathcal{Z}(t)$ be a data-based step function of t, everywhere equal to zero except at the points X_i, $i = 1, \ldots, n$, at which the function takes the value $Z_i(X_i)$. We assume that $|Z_i|$ is bounded, if not the definition is readily broadened.*

The reason for this definition is to unify notation. Our practical interest will be on sums of quantities such as $Z_i(X_i)$ with i ranging from 1 to n. Using the Stieltjes integral, we will be able to write such sums as integrals with respect to an empirical process. In view of the Helly-Bray theorem (Section 2.3) this makes it easier to gain an intuitive grasp on the population structure behind the various statistics of interest. Both T and C are assumed to have supports on some finite interval, the first of which is denoted $\mathcal{T}$. The time-dependent covariate $Z(\cdot)$ is assumed to be a left continuous stochastic process and, for notational simplicity, is taken to be of dimension one whenever possible. Let $F(t) = \Pr(T < t)$, $D(t) = \Pr(C < t)$ and $H(t) = F(t)\{1 - D(t)\} - \int_0^t F(u) dD(u)$.

For each subject i we observe $X_i = \min(T_i, C_i)$, and $\delta_i = I(T_i \leq C_i)$ so that δ_i takes the value one if the ith subject corresponds to a failure and is zero if the subject corresponds to a censored observation. A more general situation allows a subject to be dynamically censored in that he or she can move in and out of the risk set. To do this we define the "at-risk" indicator $Y_i(t)$ where $Y_i(t) = I(X_i \geq t)$. The events on the ith individual are counted by $N_i(t) = I\{T_i \leq t, T_i \leq C_i\}$ and $\bar{N}(t) = \sum_1^n N_i(t)$ counts the number of events before t. It is also helpful to be able to refer to the total number of observed failures $k = \bar{N}\{\sup t : t \in \mathcal{T}\}$, and the inverse function $\bar{N}^{-1}(\cdot)$, where $\bar{N}^{-1}(\ell) = \{\inf t : t \in \mathcal{T}, \bar{N}(t) = \ell\}$, the smallest time by which a given number of events ℓ have occurred. Consistent estimators of $F(t)$ and $H(t)$ are indicated by hats, the examples here being the Kaplan-Meier estimator for $1 - F(t)$ and $\hat{H}(t) = n^{-1}\bar{N}(t)$.

Some other sums of observations will frequently occur. In order to obtain an angle on empirical moments under the model, Andersen and Gill (1982) define

$$S^{(r)}(\beta,t) = n^{-1}\sum_{i=1}^{n} Y_i(t)e^{\beta Z_i(t)}Z_i(t)^r, \quad s^{(r)}(\beta,t) = ES^{(r)}(\beta,t),$$

for $r = 0, 1, 2$, where the expectations are taken with respect to the true distribution of $(T, C, Z(\cdot))$. Define also

$$V(\beta,t) = \frac{S^{(2)}(\beta,t)}{S^{(0)}(\beta,t)} - \frac{S^{(1)}(\beta,t)^2}{S^{(0)}(\beta,t)^2}, \quad v(\beta,t) = \frac{s^{(2)}(\beta,t)}{s^{(0)}(\beta,t)} - \frac{s^{(1)}(\beta,t)^2}{s^{(0)}(\beta,t)^2}. \quad (7.1)$$

The Andersen and Gill notation is now classic in this context. Their notation lends itself more readily to large sample theory based upon martingales and stochastic integrals. We will frequently keep this notation in mind although our approaches to inference do not appeal to special central limit theorems (the martingale central limit theorem in particular) and, as a result, our notation is typically lighter. The required conditions for the Andersen and Gill theory to apply are sightly broader although this advantage is more of a theoretical than a practical one. For their results, as well as ours, the censorship is restricted in such a way that, for large samples, there remains information on F in the tails. The conditional means and the conditional variances, $\mathcal{E}_{\beta(t)}(Z|t)\,\mathcal{V}_{\beta(t)}(Z|t)$, introduced immediately below, are related to the above via $V(\beta,t) \equiv \mathcal{V}_\beta(Z|t)$ and $S^{(1)}(\beta,t)/S^{(0)}(\beta,t) \equiv \mathcal{E}_\beta(Z|t)$. In the counting process framework of Andersen and Gill (1982), we imagine n as remaining fixed and the asymptotic results obtaining as a result of asymptotic theory for n-dimensional counting processes, in which we understand the expectation operator E to be with respect to infinitely many repetitions of the process. Subsequently we allow n to increase without bound. For the quantities $\mathcal{E}_{\beta(t)}(Z^k|t)$ we take the E operator to be these same quantities when n becomes infinitely large.

7.4 Main theorem

A simple theorem underpins all of the key results discussed in this book (testing the presence of regression effect, estimating average regression effect under non-proportional hazards, quantifying predictability via the conditional survivorship function as well as via summary indices such as explained randomness and explained variation, assessing fit, contrasting competing models etc). In view of all these several applications the theorem then appears to be quite fundamental and, as such, it seems appropriate to refer to it as the main theorem of proportional hazards regression.

We most often view time as providing the set of indices to certain stochastic processes, so that, for example, we consider $Z(t)$ to be a random variable having different distributions for different t. Also, the failure time variable T can be viewed as a non-negative random variable with distribution $F(t)$ and, whenever the set of indices t to the stochastic process coincide with the support for T, then not only can we talk about the random variables $Z(t)$ for which the distribution corresponds to $P(Z \leq z|T = t)$ but also marginal quantities such as the random variable $Z(T)$ having distribution $G(z) = P(Z \leq z)$. An important result concerning the conditional distribution of $Z(t)$ given $T = t$ follows. First we need the following definitions:

Definition 7.2 *The discrete probabilities $\pi_i(\beta(t), t)$ are given by*

$$\pi_i(\beta(t), t) = \frac{Y_i(t) \exp\{\beta(t) Z_i(t)\}}{\sum_{j=1}^{n} Y_j(t) \exp\{\beta(t) Z_j(t)\}}. \tag{7.2}$$

The $\pi_i(\beta(t), t)$ are easily seen to be bona fide probabilities (for all real values of $\beta(t)$) since $\pi_i \geq 0$ and $\sum_i \pi_i = 1$. Note that this continues to hold for values of $\beta(t)$ different to those generating the data, and even when the model is incorrectly specified. As a consequence, replacing β by $\hat{\beta}$ results in a probability distribution that is still valid but different to the true one. Means and variances with respect to this distribution maintain their interpretation as means and variances.

Under the proportional hazards assumption, i.e., the constraint $\beta(t) = \beta$, the product of the π's over the observed failure times gives the partial likelihood (Cox 1972, 1975). When $\beta = 0$, $\pi_i(0, t)$ is the empirical distribution that assigns equal weight to each sample subject in the risk set. Based on the $\pi_i(\beta(t), t)$ we have:

Definition 7.3 *Conditional moments of Z with respect to $\pi_i(\beta(t), t)$ are given by*

$$\mathcal{E}_{\beta(t)}(Z^k|t) = \sum_{i=1}^{n} Z_i^k(t)\pi_i(\beta(t), t), \quad k = 1, 2, \ldots, . \qquad (7.3)$$

These two definitions are all that we need in order to set about building the structures upon which inference is based. This is particularly so when we are able to assume an independent censoring mechanism, although the weaker assumption of a conditionally independent censoring mechanism (see Chapter 4) will mostly cause no conceptual difficulties; simply a slightly more burdensome notation. Another, somewhat natural, definition will also be appealed to on occasion and this concerns unconditional expectations.

Definition 7.4 *Marginal moments of Z with respect to the bivariate distribution characterized by $\pi_i(\beta(t), t)$ and $F(t)$ are given by*

$$\mathcal{E}_{\beta(t)}(Z^k) = \int \mathcal{E}_{\beta(t)}(Z^k|t)dF(t), \quad k = 1, 2, \ldots, . \qquad (7.4)$$

Recall that for arbitrary random variables A and B, assuming expectation to be defined, we have the result of double expectation whereby $E(A) = EE(A|B)$. This is the motivation behind the above definition. Once again, these expectations are to be interpreted as population quantities in as much as $\beta(t)$ and $F(t)$ are taken to be known. They can also, of course, be viewed as sample-based quantities since n is finite and the $Y_i(t)$ are random until time point t. At the end of the study the paths of all the $Y_i(t)$ are known and we are, to use a common expression, "conditioning on the data." The art of inference, and its understanding, stem, to a great extent, from knowing which aspects of an experiment to view as random (given that once the experiment is over there is not really anything truly random). Also which distributions are relevant and these can change so that, here for example, we should think carefully about the meaning of the expectation operators E and $\mathcal{E}$ in its particular context. These expectations are still well defined, but with respect to different distributions; when replacing β by $\hat{\beta}$, when replacing F by F_n and $\hat{F}$, and when allowing n to go to infinity. The quantity ϕ of the following definition is not of any essential interest, featuring in the main theorem but disappearing afterwards.

Definition 7.5 *In order to distinguish conditionally independent censoring from independent censoring we define $\phi(z,t)$ where*

$$\phi(z^*, t) = \frac{\int P(C \geq t|z)g(z)dz}{P(C \geq t|z^*)}.$$

Note that when censoring does not depend upon z then $\phi(z,t)$ will depend upon neither z nor t and is, in fact, equal to one. Otherwise, under a conditionally independent censoring assumption, we can consistently estimate $\phi(z,t)$ and we call this $\hat{\phi}(z,t)$. The following theorem is presented in O'Quigley (2003).

Theorem 7.1 *Under model (6.2) and assuming $\beta(t)$ known, the conditional distribution function of $Z(t)$ given $T = t$ is consistently estimated by*

$$\hat{P}\{Z(t) \leq z|T = t\} = \frac{\sum_{z_i \leq z} Y_i(t) \exp\{\beta(t)z_i(t)\}\hat{\phi}(z_i, t)}{\sum_{j=1}^{n} Y_j(t) \exp\{\beta(t)z_j(t)\}\hat{\phi}(z_j, t)}. \qquad (7.5)$$

The theorem, which we refer to as the main theorem of proportional hazards regression, has many important consequences including:

Corollary 7.1 *Under model (6.2) and an independent censorship, assuming $\beta(t)$ known, the conditional distribution function of $Z(t)$ given $T = t$ is consistently estimated by*

$$\hat{P}(Z(t) \leq z|T = t) = \sum_{j=1}^{n} \pi_j(\beta(t), t)I(Z_j(t) \leq z). \qquad (7.6)$$

The observation we would like to make here is that we can fully describe a random variable indexed by t, i.e., a stochastic process. All of our inference will follow from this. In essence, we first fix t and then we fix our attention on the conditional distribution of Z given that $T = t$ and models which enable us to characterize this distribution. Indeed, under the broader censoring definition of conditional independence, common in the survival context, we can still make the same basic observation. In this case we condition upon something more complex that just $T = t$ but the actual random outcome that we condition upon is of less importance than the simple fact that we are able to described sets of conditional distributions all indexed by t, i.e., a stochastic process indexed by t. Specifically:

Corollary 7.2 *For a conditionally independent censoring mechanism we have*

$$\hat{P}(Z(t) \leq z | T = t, C > t) = \sum_{j=1}^{n} \pi_j(\beta(t), t) I(Z_j(t) \leq z). \qquad (7.7)$$

Whether we condition on the event $T = t$ or the event $(T = t, C > t)$, we identify a random variable indexed by t. This is all we need to construct appropriate stochastic processes (functions of $Z(t)$) enabling inference. Again simple applications of Slutsky's theorem shows that the result still holds for $\beta(t)$ replaced by any consistent estimate. In particular, when the hypothesis of proportionality of risks is correct, the result holds for the estimate $\hat{\beta}$. The following two corollaries follow immediately from those just above and form the basis to the main tests we construct. For integer k we have:

Corollary 7.3 $\mathcal{E}_{\hat{\beta}(t)}(Z^k | t)$ *provides a consistent estimate of $E_{\beta(t)}(Z^k (t) | t)$, under model (6.2). In particular $\mathcal{E}_{\hat{\beta}}(Z^k | t)$ provides a consistent estimate of $E_{\beta}(Z^k(t) | t)$, under the model expressed by Equation 6.3.*

Furthermore, once again working under the model, we consider:

Definition 7.6 $\mathcal{V}_{\beta(t)}(Z | t) = \mathcal{E}_{\beta(t)}(Z^2 | t) - \mathcal{E}_{\beta(t)}^2(Z | t).$

In practical data analysis the quantity $\beta(t)$ may be replaced by a value constrained by some hypothesis or an estimate. The quantity $\mathcal{V}_{\beta(t)}(Z | t)$ can be viewed as a conditional variance which may vary little with t, in a way analogously to the residual variance in linear regression which, under classic assumptions, remains constant with different levels of the independent variable. Since $\mathcal{V}_{\beta(t)}(Z | t)$ may change with t, even if not a lot, it is of interest to consider some average quantity and so we also introduce:

Definition 7.7 $E\mathcal{V}_{\beta(t)}(Z) = \int \mathcal{V}_{\beta(t)}(Z | t) dF(t).$

These sample-based variances relate to population variances via the following corollary;

Corollary 7.4 *Under model (6.3), $\text{Var}(Z | t)$ is consistently estimated by $\mathcal{V}_{\hat{\beta}}(Z | t)$. $E\text{Var}(Z | t)$ is consistently estimated by $E\mathcal{V}_{\hat{\beta}}(Z | t)$. In addition, $\int \mathcal{V}_{\hat{\beta}}(Z | t) d\hat{F}(t)$ is consistent for $E\text{Var}(Z | t)$.*

These quantities are all useful in our construction. Interpretation requires some care. For example, although $E\,\mathcal{V}_{\hat{\beta}}(Z|t)$ is, in some sense, a marginal quantity, it is not the marginal variance of Z since we have neglected the variance of $E_{\beta(t)}(Z(t)|t)$ with respect to the distribution of T. The easiest case to interpret is the one where we have an independent censoring mechanism (Equation 7.6). However, we do not need to be very concerned about any interpretation difficulty, arising for instance in Equation 7.7 where the censoring time appears in the expression, since, in this or the simpler case, all that matters to us is that our observations can be considered as arising from some process, indexed by t and, for this process, we are able, under, as usual, some model assumptions, to consistently estimate the mean and the variance of the quantities that we observe. It is also useful to note another natural relation between $\mathcal{V}_\beta(Z|t)$ and $\mathcal{E}_\beta(Z|t)$ since

$$\mathcal{V}_\beta(Z|t) = \partial\,\mathcal{E}_\beta(Z|t)/\partial\beta.$$

This relation is readily verified for fixed β. In the case of time-dependent $\beta(t)$ then, at each given value of t, it is again clear that the same relation holds. The result constitutes one of the building blocks in the overall inferential construction and, under weak conditions, for example Z being bounded, then it also follows that

$$\int \mathcal{V}_\beta(Z|t) = \int \partial\mathcal{E}_\beta(Z|t)/\partial\beta = \partial\left\{\int \mathcal{E}_\beta(Z|t)\right\}/\partial\beta.$$

Throughout the rest of this book we will see just why the main theorem is so fundamental. Essentially all the information we need, for almost any conceivable statistical goal, arising from considerations of any of the models considered, is contained in the joint probabilities $\pi_i(\beta(t), t)$ of the fundamental definition 7.2. We are often interested, in the multivariate setting for example, in the evaluation of the effects of some factor while having controlled for others. This can be immediately accommodated. Specifically, taking Z to be of some dimension greater than one (β being of the same dimension) and writing $Z^T = (Z_1^T, Z_2^T)$ and $Z_i^T = (Z_{1i}^T, Z_{2i}^T)$ then, summing over the multivariate probabilities, we have two obvious extensions to Corollaries 7.1 and 7.2.

Corollary 7.5 *Under model (6.2) and an independent censorship, assuming $\beta(t)$ known, the conditional distribution function of $Z_2(t)$ given $T = t$ is consistently estimated by*

$$\hat{P}(Z_2(t) \leq z | T = t) = \sum_{j=1}^{n} \pi_j(\beta(t), t) I(Z_{2j}(t) \leq z). \qquad (7.8)$$

The corollary enables component wise inference. We can consider the components of the vector Z_i individually. Also we could study some functions of the components, usually say a simple linear combination of the components such as the prognostic index. Note also that:

Corollary 7.6 *For a conditionally independent censoring mechanism we have*

$$\hat{P}(Z_2(t) \leq z | T = t, C > t) = \sum_{j=1}^{n} \pi_j(\beta(t), t) I(Z_{2j}(t) \leq z), \qquad (7.9)$$

where in Definition 7.2 for $\pi_j(\beta(t), t)$ we take $\beta(t) Z_j(t)$ to be an inner product, which we may prefer to write as $\beta(t)^T Z_j(t)$ and where $Z_j(t)$ are the observed values of the vector $Z(t)$ for the jth subject. Also, by $Z_2(t) \leq z$ we mean that all of the scalar components of $Z_2(t)$ are less than or equal to the corresponding scalar components of z. As for the corollaries and definitions following Corollaries 7.1 and 7.2 they have obvious equivalents in the multivariate setting and so we can readily write down expressions for expectations, variances and covariances as well as their corresponding estimates.

Moments for stratified models

Firstly we recall from the previous chapter that the stratified model is simply a partially proportional hazards model in which some of the components of $\beta(t)$ remain unspecified while the other components are constant terms. The definition for the stratified model was

$$\lambda(t | Z(t), s) = \lambda_{0s}(t) \exp\{\beta(t) Z(t)\},$$

where s takes integer values $1, \ldots, m$. In view of the equivalence between stratified models and partially proportional hazards models described in the previous chapter, the main theorem and its corollaries apply immediately. However, in light of the special importance of stratified models, as proportional hazards models with relaxed assumptions, it will be helpful to our development to devote a few words to this case. Analogous to the above definition for $\pi_i(\beta(t), t)$, and using the, possibly time-dependent, stratum indicator $s(t)$ we now define these probabilities via:

Definition 7.8 *For the stratified model, having strata* $s = 1, \ldots, m$ *the discrete probabilities* $\pi_i(\beta(t), t)$ *are now given by*

$$\pi_i(\beta(t), t) = \frac{Y_i\{s(t), t\} \exp\{\beta(t) Z_i(t)\}}{\sum_{j=1}^{n} Y_j\{s(t), t\} \exp\{\beta(t) Z_j(t)\}}. \tag{7.10}$$

When there is a single stratum then this definition coincides with the earlier one and, indeed, we use the same $\pi_i(\beta(t), t)$ for both situations, since it is only used indirectly and there is no risk of confusion. Under equation (6.3), i.e. the constraint $\beta(t) = \beta$, the product of the π's over the observed failure times gives the so-called stratified partial likelihood (Kalbfleisch and Prentice 1980). The series of above definitions for the non-stratified model, in particular Definition 7.2, theorems, corollaries, all carry over in an obvious way to the stratified model and we do not propose any additional notation. It is usually clear from the context although it is worth making some remarks. Firstly, we have no direct interest in the distribution of Z given t (note that this distribution depends on the distribution of Z given $T > 0$, a distribution which corresponds to our design and is quite arbitrary).

We will exploit the main theorem in order to make inferences on β and, in the stratified case, we would also condition upon the strata from which transitions can be made. In practice, we contrast the observations $Z_i(X_i)$, made at time point X_i at which an event occurs ($\delta_i = 1$) with those subjects at risk of the same event. The "at risk" indicator, $Y(s(t), t)$, makes this very simple to express. We can use $Y(s(t), t)$ to single out appropriate groups for comparison. This formalizes a standard technique in epidemiology whereby the groups for comparison may be matched by not just age but by other variables. Such variables have then been controlled for and eliminated from the analysis. Their own specific effects can be quite general and we are not in a position to estimate them. Apparently very complex situations, such as subjects moving in and out of risk categories, can be easily modeled by the use of these indicator variables.

Moments for other relative risk models

Instead of Equation 6.2 some authors have suggested a more general form for the hazard function whereby

$$\lambda(t|Z) = \lambda_0(t) R\{\beta(t) Z\}, \tag{7.11}$$

and where, mostly, $\beta(t)$ is not time-varying, being equal to some unknown constant. The most common choices for the function $R(r)$ are $\exp(r)$, in which case we recover the usual model, and $1+r$ which leads to the so-called additive model. Since both $\lambda(t|Z)$ and λ_0 are necessarily positive we would generally need constraints on the function $R(r)$. In practice this can be a little bothersome and is, among several other good reasons, a cause for favoring the multiplicative risk model $\exp(r)$ over the additive risk model $1+r$. If we replace our earlier definition for $\pi_i(\beta(t),t)$ by:

Definition 7.9 *The discrete probabilities* $\pi_i(\beta(t),t)$ *are given by;*

$$\pi_i(\beta(t),t) = \frac{Y_i(t)R\{\beta(t)Z_i(t)\}}{\sum_{j=1}^n Y_j(t)R\{\beta(t)Z_j(t)\}}, \qquad (7.12)$$

then all of the above definitions, theorems, and corollaries have immediate analogues and we do not write them out explicitly. Apart from one interesting exception, which we look at more closely in the chapters dealing with inference, there are no particular considerations we need concern ourselves over if we choose $R(r) = 1 + r$ rather than $R(r) = \exp(r)$. Note also that if we allow the regression functions, $\beta(t)$, to depend arbitrarily upon time then, given either model, the other model exists with a different function of $\beta(t)$. The only real reason for preferring one model over another would be due to parsimony; for example, we might find in some given situation that in the case of the additive model the regression function $\beta(t)$ is in fact constant unlike the multiplicative model where it may depend on time. But otherwise both functions may depend, at least to some extent, on time and then the multiplicative model ought be preferred since it is the more natural. We say the more natural because the positivity constraint is automatically satisfied.

Transformed covariate models

For some transformation ψ of the covariate we can postulate a model of the form;

$$\lambda(t|Z(t)) = \lambda_0(t)\exp\{\beta(t)\psi[Z(t)]\}. \qquad (7.13)$$

All of the calculations proceed as above and no real new concept is involved. Such models can be considered in the case of continuous covariates, Z, which may be sufficiently asymmetric, implying very great changes of risk at the high or low values, to be unlikely to provide a satisfactory fit. Taking logarithms, or curbing the more extreme values via a defined plateau, or some other such transformation will produce models of potentially wider applicability. Note that this is a different approach to working with, say,

$$\lambda(t|Z(t)) = \lambda_0(t) \exp\{\beta(t)Z(t)\},$$

and using the main theorem, in conjunction with estimating equations described here below and basing inference upon the observations $\psi Z(X_i)$ and their expectations under this model. In this latter case we employ ψ in the estimating equation as a means to obtain greater robustness or to reduce sensitivity to large observations. In the former case the model itself is different and would lead to different estimates of survival probabilities.

Our discussion so far has turned around the hazard function. However, it is equally straightforward to work with intensity functions and these allow for increased generality, especially when tackling complex time-dependent effects. O'Brien (1978) introduced the logit-rank test for survival data when investigating the effect of a continuous covariate on survival time. His purpose was to construct a test that was rank invariant with respect to both time and the covariate itself. O'Quigley and Prentice (1991) showed how a broad class of rank invariant procedures can be developed within the framework of proportional hazards models. The O'Brien logit-rank procedure was a special case of this class. In these cases we work with intensity rather than hazard functions. Suppose then that $\lambda_i(t)$ indicates an intensity function for the ith subject at time t. A proportional hazards model for this intensity function can be written

$$\lambda_i(t) = Y_i(t)\lambda_0(t) \exp\{\beta Z_i(t)\},$$

where $Y_i(t)$ indicates whether or not the ith subject is at risk at time t, $\lambda_0(t)$ the usual "baseline" hazard function and $Z_i(t)$ is a constructed covariate for the ith subject at time t. Typically, $Z_i(t)$ in the estimating equation is defined as a function of measurements on the ith subject alone, but it can be defined more generally as $Z_i(t) = \psi_i(t, \mathcal{F}_t)$ for ψ some function of $\mathcal{F}_t$, the collective failure, censoring and covariate information prior to time t on the entire study group. The examples in O'Quigley and Prentice (1991) included the rank of the the subject's covariate at X_i and transformations on this such as the normal order statistics. This represents a departure from most regression situations because the value used in the estimating equation depends not only on what has been observed on the particular individual but also upon what has been observed on other relevant subsets of individuals.

Misspecified models

For multinormal linear regression involving p regressors we can eliminate from consideration some of these and focus our attention on models involving the remaining regressors strictly less than p. We could eliminate these by simple integration, thereby obtaining marginal distributions. Under the usual assumptions of multiple linear regression the resulting lower dimensional model remains a multinormal one. As an example, in the simple case of a two dimensional covariate normal model, both the marginal models involving only one of the two covariates are normal models. However, for non-linear models this result would only be expected to hold under quite unusual circumstances. Generally, for non-linear models, and specifically proportional hazards models, the result will not hold so that if the model is assumed true for a covariate vector of dimension p, then, for any submodel, of dimension less that p, the model will not hold exactly. A corollary to this is that no model of dimension greater that p could exactly follow a proportional hazards prescription if we claim that the model holds precisely for some given p covariates.

These observations led some authors to claim that "forgotten" or "overlooked" variables would inevitably lead to misleading results. Such a claim implies that *all* analyses based on proportional hazards models are misleading and since, to say the least, such a conclusion is unhelpful we offer a different perspective. This says that *all* practical models are only ever approximately correct. In other words, the model is always making a simplifying assumption, necessarily overlooking potential effects as well as including others which may impact the proportionality of those key variables of interest. Our task then focuses on interpreting our estimates when our model cannot be exactly true. In terms of analysing real data, it makes much more sense to take as our underlying working assumption that the model is, to a greater or lesser degree, misspecified.

A model can be misspecified in one of two clear ways; the first is that the covariate form is not correctly expressed and the second is that the regression coefficient is not constant through time. An example of the first would be that the true model holds for $\log Z$ but that, not knowing this, we include Z in the model. An example of the second might have $\beta(t)$ declining through time rather than remaining constant.

It has been argued that the careful use of residual techniques can indicate which kind of model failure may be present. This is not so. Whenever a poor fit could be due to either cause it is readily seen that a misspecified covariate form can be represented correctly via a time-dependent effect. In some sense the two kinds of misspecification are unidentifiable. We can fix the model by working either with the covariate form or the regression coefficient $\beta(t)$. Of course, in certain cases, a discrete binary covariate describing two groups, for example, there can only be one cause of model failure - the time dependency of the regression coefficient. This is because the binary coding imposes no restriction of itself since all possible codings are equivalent.

The important issue is then the interpretation of an estimate, say $\hat{\beta}$ under a proportional hazards assumption when, in reality, the data are generated under the broader non-proportional hazards model with regression coefficient function $\beta(t)$. This is not a straightforward endeavor and the great majority of the currently used procedures, including those proposed in the widely distributed R, SAS, STATA and S-Plus packages, produce estimates which cannot be interpreted unless there is no censoring. To study this question we first define $\mu = \int \beta(t)dF(t)$, which is an average of $\beta(T)$ with respect to the distribution $F(t)$. It is also of interest to consider the approximation

$$\hat{P}(Z(t) \leq z | T = t, C > t) \approx \sum_{j=1}^{n} \pi_j(\mu, t) I(Z_j(t) \leq z) \qquad (7.14)$$

and, for the case of a model making the stronger assumption of an independent censoring mechanism as opposed to a conditionally independent censoring mechanism given the covariate, we have

$$\hat{P}(Z(t) \leq z | T = t) \approx \sum_{j=1}^{n} \pi_j(\mu, t) I(Z_j(t) \leq z). \qquad (7.15)$$

For small samples it will be unrealistic to hope to obtain reliable estimates of $\beta(t)$ for all of t so that, often, we take an estimate of some summary measure, in particular μ. It is in fact possible to construct an estimating equation which provides an estimate of μ without estimating $\beta(t)$ (Xu and O'Quigley 1998) and it is very important to stress that, unless there is no censoring, the usual estimating equation which leads to the partial likelihood estimate does not accomplish this. In fact, the partial likelihood estimate turns out to be equivalent to obtaining the solution of an estimating equation based on $H(t)$ (see Section 7.3) and using $\hat{H}(t)$ as an estimate whereas, to consistently estimate μ, it is necessary to work with some consistent estimate of $F(t)$, in particular the Kaplan-Meier estimate.

Some thought needs be given to the issues arising when our estimating equation is based on certain assumptions (in particular, a proportional hazards assumption), whereas the data themselves can be considered to have been generated by something broader (in particular, a non proportional hazards model). To this purpose we firstly consider a definition that will allow us to anticipate just what is being estimated when the data are generated by model (6.2) and we are working with model (6.3). This is contained in the definition for β^* just below.

Let's keep in mind the widely held belief that the partial likelihood estimate obtained when using a proportional hazards model in a situation where the data are generated by a broader model must correspond to some kind of average effect. It does correspond to something (as always) but nothing very useful and not something we can hopefully interpret as an average effect. This is considered in the following sections. Firstly we need:

Definition 7.10 *Let β^* be the constant value satisfying*

$$\int_T \mathcal{E}_{\beta^*}(Z|t)dF(t) = \int_T \mathcal{E}_{\beta(t)}(Z|t)dF(t). \qquad (7.16)$$

The definition enables us to make sense out of using estimates based on (6.3) when the data are in fact generated by (6.2). Since we can view T as being random, whenever $\beta(t)$ is not constant, we can think of having sampled from $\beta(T)$. The right-hand side of the above equation is then a double expectation and β^*, occurring in the left-hand side of the equation, is the best fitting value under the constraint that $\beta(t) = \beta$. We can show the existence and uniqueness of solutions to Equation (7.16) (Xu and O'Quigley 1998). More importantly, β^* can be shown to have the following three properties: (i) under model (6.3) $\beta^* = \beta$; (ii) under a subclass of the broad class of models known as the Harrington-Fleming models, we have an exact result in that $\beta^* = \int_T \beta(t)dF(t)$; and (iii) for very general situations we can write that

$\beta^* \approx \int_{\mathcal{T}} \beta(t) dF(t)$, an approximation which is in fact very accurate. Estimates of β^* are discussed in (Xu and O'Quigley 1998, Xu and O'Quigley 2000) and, in the light of the foregoing, we can take these as estimates of μ.

Theorem 7.1 and its corollaries provide the ingredients necessary to constructing a number of relevant stochastic processes, in particular functions of Brownian motion. We will be able to construct a process that will look like simple Brownian motion under the chosen model and with given parameter values. We can then consider what this process will look like when, instead of those null values, the data are generated by a model from the same class but with different parameter values. First we consider the estimating equations that can be readily constructed as a result of the preceding theory.

7.5   The estimating equations

The above setting helps us anticipate the properties of the estimators we will be using. First, recall our definition of $\mathcal{Z}(t)$ as a step function of t with discontinuities at the points X_i, $i = 1, ..., n$, at which the function takes the value $Z_i(X_i)$. Next, consider $F_n(t)$, the empirical marginal distribution function of T. Note that $F_n(t)$ coincides with the Kaplan-Meier estimate of $F(t)$ in the absence of censoring. When there is no censoring, a sensible estimating equation (which we will see also arises as the derivative of a log likelihood, as well as the log partial likelihood) is

$$U_1(\beta) = \int \{\mathcal{Z}(t) - \mathcal{E}_\beta(Z|t)\} dF_n(t) = 0. \qquad (7.17)$$

The above integral is simply the difference of two sums, the first the empirical mean without reference to any model and the second the average of model-based means. It makes intuitive sense as an estimating equation and the only reason for writing the sum in the less immediate form as an integral is that it helps understand the large sample theory when $F_n(t) \xrightarrow{\text{P}} F(t)$. Each component in the above sum includes the size of the increment, $1/n$, a quantity that can then be taken outside of the summation (or integral) as a constant factor. Since the right-hand side of the equation is identically equal to zero, the incremental size $1/n$ can be canceled, enabling us to rewrite the equation as

$$U_2(\beta) = \int \{\mathcal{Z}(t) - \mathcal{E}_\beta(Z|t)\}d\bar{N}(t) = 0. \qquad (7.18)$$

It is this expression where the integral is taken with respect to increments $d\bar{N}(t)$, rather than with respect to $dF_n(t)$ that is the more classic representation in this context. The expression equates $U_2(\beta)$ in terms of the counting processes $N_i(t)$. These processes, unlike the empirical distribution function, are available in the presence of censoring. It is the above equation that is used to define the partial likelihood estimator, since, unless the censoring is completely absent, the quantity $U_1(\beta)$ is not defined.

A natural question would be the following: suppose two observers were to undertake an experiment to estimate β. A certain percentage of observations remain unobservable to the first observer as a result of an independent censoring mechanism but are available to the second observer. The first observer uses Equation 7.18 to estimate β, whereas the second observer uses Equation 7.17. Will the two estimates agree? By "agree" we mean, under large sample theory, will they converge to the same quantity. We might hope that they would; at least if we are to be able to usefully interpret estimates obtained from Equation 7.18. Unfortunately though (especially since Equation 7.18 is so widely used), the estimates do not typically agree. Table 7.1 below indicates just how severe the disagreement might be. However, the form of $U_1(\beta)$ remains very much of interest and, before discussing the properties of

Table 7.1: Comparison of β^*, $\int \beta(t)dF(t)$, and the estimates $\tilde{\beta}$ and $\hat{\beta}_{PL}$

β_1	β_2	t_0	% censored	β^*	$\int \beta(t)dF(t)$	$\tilde{\beta}$	$\hat{\beta}_{PL}$
1	0	0.1	0%	0.156	0.157	0.155 (0.089)	0.155 (0.089)
			17%	0.156	0.157	0.158 (0.099)	0.189 (0.099)
			34%	0.156	0.157	0.160 (0.111)	0.239 (0.111)
			50%	0.156	0.157	0.148 (0.140)	0.309 (0.130)
			67%	0.156	0.157	0.148 (0.186)	0.475 (0.161)
			76%	0.156	0.157	0.161 (0.265)	0.654 (0.188)
3	0	0.05	0%	0.721	0.750	0.716 (0.097)	0.716 (0.097)
			15%	0.721	0.750	0.720 (0.106)	0.844 (0.107)
			30%	0.721	0.750	0.725 (0.117)	1.025 (0.119)
			45%	0.721	0.750	0.716 (0.139)	1.294 (0.133)
			60%	0.721	0.750	0.716 (0.181)	1.789 (0.168)
			67%	0.721	0.750	0.739 (0.255)	2.247 (0.195)

the above equations let us consider a third estimating equation which we write as

$$U_3(\beta) = \int \{\mathcal{Z}(t) - \mathcal{E}_\beta(Z|t)\}d\hat{F}(t) = 0. \qquad (7.19)$$

Note that, upon defining the stochastic process $W(t) = \hat{S}(t)\{\sum_{i=1}^n Y_i(t)\}^{-1}$ we can rewrite (7.19) in the usual counting process terminology as

$$U_3(\beta) = \int W(t)\{\mathcal{Z}(t) - \mathcal{E}_\beta(Z|t)\}d\bar{N}(t) = 0.$$

For practical calculation note that $W(X_i) = \hat{F}(X_i+) - \hat{F}(X_i)$ at each observed failure time X_i, i.e., the jump in the KM curve. When there is no censoring, then clearly

$$U_1(\beta) = U_2(\beta) = U_3(\beta).$$

More generally $U_1(\beta)$ may not be available and solutions to $U_2(\beta) = 0$ and $U_3(\beta) = 0$ do not coincide or converge to the same population counterparts even under independent censoring. They would only ever converge to the same quantities under the unrealistic assumption that the data are exactly generated by a proportional hazards model. As argued in the previous section we can assume that this never really holds in practical situations.

Many other possibilities could be used instead of $U_3(\beta)$, ones in which other consistent estimates of $F(t)$ are used in place of $\hat{F}(t)$, for example, the Nelson-Aalen estimator or, indeed, any parametric estimate for marginal survival. If we were to take the route of parametric estimates of marginal survival, we would need to be a little cautious since these estimates could also contain information on the parameter β which is our central focus. However, we could invoke a conditional argument, i.e., take the marginal survival estimate as fixed and known at its observed value or argue that the information contained is so weak that it can be ignored. Although we have not studied any of these we would anticipate the desirable properties described below to still hold. Stronger modelling assumptions are also possible (Moeschberger and Klein 1985, Klein et al. 1990).

Note also that the left-hand side of the equation is a special case of the weighted scores under the proportional hazards model (Harrington and Fleming 1982, Lin 1991, Newton and Raftery 1994). However those weighted scores were not proposed with the non-proportional hazards model in mind, and the particular choice of $W(\cdot)$ used here was not considered in those papers. Indeed other choices for the weights will lead to estimators closer to the partial likelihood itself, in the sense that under a non-proportional hazards model and in the presence of censoring, the broader class of weighted estimates will not converge to quantities that remain unaffected by an independent censoring mechanism. On the other hand, the estimating equation based on U_3 is in the same spirit as the approximate likelihood of Oakes (1986) for censored data and the M-estimate of Zhou (1992) for censored linear models. Hjort (1992) also mentioned the use of the reciprocal of the Kaplan-Meier estimate of the censoring distribution as weights in parametric survival models, and these weights are the same as $W(\cdot)$ defined here. For the random effects model - a special case of this is the stratified model which, in turn, can be expressed in the form (6.2) - we can see, even when we know that (6.3) is severely misspecified, that we can still obtain estimates of meaningful quantities. The average effect resulting from the estimating equation U_3 is clearly of interest.

For the stratified model, $\mathcal{Z}(X_i)$ is contrasted with its expectation $\mathcal{E}_\beta(Z|X_i, s)$. Here, the inclusion of s is used to indicate that if Z_i belongs to stratum s then the reference risk set for $\mathcal{E}_\beta(Z|X_i, s)$ is restricted to members of this same stratum. Note that for time-dependent $s(t)$ the risk set is dynamic, subjects entering and leaving the set as they become at risk. The usual estimating equation for stratified models is again of the form $U(\beta)$ and, for the same reasons as recalled above and described more fully in Xu and O'Quigley (1998) we might prefer to use

$$U_s(\beta) = \int \{\mathcal{Z}(t) - \mathcal{E}_\beta(Z|t, s)\}d\hat{F}(t) = 0 \,. \tag{7.20}$$

Even weaker assumptions (not taking the marginal $F(t)$ to be common across strata) can be made and, at present, this is a topic that remains to be studied.

Zeros of estimating equations

Referring back to Section 7.4 we can immediately deduce that the zeros of the estimating equations provide consistent estimates of β under the model. Below we consider zeros of the estimating equations when the model is incorrectly specified. This is important since, in practice, we can assume this to be the case. Most theoretical developments proceed under the assumptions that the model is correct. We would have that $\hat{\beta}$ where $U_2(\hat{\beta}) = 0$ is consistent for β. Also $\tilde{\beta}$ where $U_3(\tilde{\beta}) = 0$ is consistent for a parameter of interest, namely the average effect. From the mean value theorem we write

$$U_2(\hat{\beta}) = U_2(\beta_0) + (\hat{\beta} - \beta_0)\left\{\frac{\partial U_2(\beta)}{\partial \beta}\right\}_{\beta = \xi},$$

where ξ lies strictly on the interior of the interval with endpoints β_0 and $\hat{\beta}$. Now $U_2(\hat{\beta}) = 0$ and $U_2'(\xi) = \sum_{i=1}^{n}\delta_i\mathcal{V}_\xi(Z|X_i)$ so that $\text{Var}(\hat{\beta}) \approx 1/\sum_{i=1}^{n}\delta_i\text{Var}(Z|X_i)$. This is the Cramer-Rao bound and so the estimate is a good one. Although the sums are of variables that we can take to be independent they are not identically distributed. Showing large sample normality requires verification of the Lindeburgh condition but, if awkward, this is not difficult. All the necessary ingredients are then available for inference. However, as our recommended approach, we adopt a different viewpoint based on the functional central limit theorem rather than a central limit theorem for independent variables. This is outlined in some detail in the following chapter.

Large sample properties of solutions to estimating equations

The reason for considering estimating equations other than (7.18) is because of large sample properties. Without loss of generality, for any multivariate categorical situation, a non-proportional hazards model (Equation 6.2) can be taken to generate the observations. Suppose that for this more general situation we fit the best available model, in particular the proportional hazards model (Equation 6.3). In fact, this is what always takes place when fitting the Cox model to data. It will be helpful to have the following definition:

Definition 7.11 *The average conditional variance $A(\beta)$ is defined as;*

$$A(\beta) = \int_0^\infty \left\{E_\beta(Z^2|t) - E_\beta^2(Z|t)\right\} dF(t).$$

Note that the averaging does not produce the marginal variance for that we would need to include a further term which measures the variance of the conditional expectations. Under the conditions on the censoring of Breslow and Crowley (1974), essentially requiring that, for each t, as n increases, the information increases at the same rate, then $nW(t)$ converges in probability to $w(t)$. Under these same conditions, recall that the probability limit as $n \to \infty$ of $\mathcal{E}_\beta(Z|t)$ under model (6.2) is $E_\beta(Z|t)$, that of $\mathcal{E}_\beta(Z^2|t)$ is $E_\beta(Z^2|t)$ and that of $\mathcal{V}_\beta(Z|t)$ is $V_\beta(Z|t)$. The population conditional expectation and variance, whether the model is correct or not, are denoted by $E(Z|t)$ and $V(Z|t)$, respectively. We have an important result due to Struthers and Kalbfleisch (1986).

Theorem 7.2 *Under model 6.2 the estimator $\hat{\beta}$, such that $U_2(\hat{\beta}) = 0$, converges in probability to the constant β_{PL}, where β_{PL} is the unique solution to the equation*

$$\int_0^\infty w^{-1}(t) \left\{ E(Z|t) - E_\beta(Z|t) \right\} dF(t) = 0, \qquad (7.21)$$

provided that $A(\beta_{PL})$ is strictly greater than zero.

Should the data be generated by model (6.3) then $\beta_{PL} = \beta$, but otherwise the value of β_{PL} would depend upon the censoring mechanism in view of its dependence on $w(t)$. Simulation results below on the estimation of average effect show a very strong dependence of β_{PL} on an independent censoring mechanism. Of course, under the unrealistic assumption that the data are exactly generated by the model, then, for every value of t, the above integrand is identically zero, thereby eliminating any effect of $w(t)$. In such situations the partial likelihood estimator is more efficient and we must anticipate losing efficiency should we use the estimating equation $U_3(\beta)$ rather than the estimating equation $U_2(\beta)$.

Viewing the censoring mechanism as a nuisance feature of the data we might ask the following question: were it possible to remove the censoring then to which population value do we converge? We would like an estimating equation that, in the presence of an independent censoring mechanism, produces an estimate that converges to the same quantity we would have converged to had there been no censoring. The above estimating equation (7.19) has this property. This is summarized in the following theorem of Xu and O'Quigley (1998), which is an application of Theorem 3.2 in Lin (1991).

Theorem 7.3 *Under model 6.2 the estimator $\tilde{\beta}$, such that $U_3(\tilde{\beta}) = 0$, converges in probability to the constant β^*, where β^* is the unique solution to the equation*

$$\int_0^\infty \left\{ E(Z|t) - E_\beta(Z|t) \right\} dF(t) = 0, \qquad (7.22)$$

provided that $A(\beta^*)$ is strictly greater than zero.

None of the ingredients in the above equation depends on the censoring mechanism. In consequence the solution itself, $\beta = \beta^*$, is not influenced by the censoring. Thus the value we estimate in the absence of censoring, β^*, is the same as the value we estimate when there is censoring. A visual inspection of equations (7.21) and (7.22) suffices to reveal why we argue in favor of (7.19) as a more suitable estimating equation than (7.18) in the presence of non proportional hazard effects. Furthermore, the solution to (7.19) can be given a strong interpretation in terms of average effects. We return to this in more detail, but we can already state a compelling argument for the broader interpretability of β^*.

7.6 Consistency and asymptotic normality of $\tilde{\beta}$

We have that $\mathcal{E}_\beta(Z|t) = S^{(1)}(\beta, t)/S^{(0)}(\beta, t)$, and that $W(t) = \hat{S}(t)/\{nS^{(0)}(0, t)\}$. Under an independent censoring mechanism, $s^{(1)}(\beta(t), t)/s^{(0)}(\beta(t), t) = E\{Z(t)|T = t\}$, and $s^{(1)}(\beta, t)/s^{(0)}(\beta, t)$ is what we get when we impose a constant β through time in place of $\beta(t)$, both of which do not involve the censoring distribution. In addition $v(t) = v(\beta(t), t) = \text{Var}\{Z(t)|T = t\}$. We take it that $nW(t)$ converges in probability to a non-negative bounded function $w(t)$ uniformly in t. Then we have $w(t) = S(t)/s^{(0)}(0, t)$. Using the same essential approach as that of Andersen and Gill (1982) it is seen, under the model and an independent censoring mechanism, that the marginal distribution function of T can be written

$$F(t) = \int_0^t w(t)s^{(0)}(\beta(t), t)\lambda_0(t)dt. \tag{7.23}$$

Theorem 7.4 *Under the non-proportional hazards model and an independent censorship the estimator $\tilde{\beta}$ converges in probability to the constant β^*, where β^* is the unique solution to the equation*

$$\int_0^\infty \left\{ \frac{s^{(1)}(\beta(t), t)}{s^{(0)}(\beta(t), t)} - \frac{s^{(1)}(\beta, t)}{s^{(0)}(\beta, t)} \right\} dF(t) = 0, \tag{7.24}$$

provided that $\int_0^\infty v(\beta^*, t)dF(t) > 0$.

It is clear that equation (7.24) does not involve censoring. Neither then does the solution to the equation, β^*. As a contrast the maximum partial likelihood estimator $\hat{\beta}_{PL}$ from the estimating equation $U_2 = 0$ converges to the solution of the equation

$$\int_0^\infty \left\{ \frac{s^{(1)}(\beta(t), t)}{s^{(0)}(\beta(t), t)} - \frac{s^{(1)}(\beta, t)}{s^{(0)}(\beta, t)} \right\} s^{(0)}(\beta(t), t)\lambda_0(t)dt = 0. \qquad (7.25)$$

This result was obtained by Struthers and Kalbfleisch (1986). Should the data be generated by the proportional hazards model, then the solutions of (7.24) and (7.25) are both equal to the true regression parameter β. In general, however, these solutions will be different, the solution to (7.25) depending on the unknown censoring mechanism through the factor $s^{(0)}(\beta(t), t)$. The simulation results of Table 7.1 serve to underline this fact in a striking way. The estimate $\tilde{\beta}$ can be shown to be asymptotically normal with mean zero and variance that can be written down. The expression for the variance is nonetheless complicated and is not reproduced here since it is not used. Instead we base inference on functions of Brownian motion as described in the next chapter.

7.7 Interpretation for β^* as average effect

The solution β^* to the large sample equivalent to the estimating equation $U_3(\beta)$, i.e., Equation 7.24 can be viewed as an average regression effect. In the equation $s^{(1)}(\beta(t), t)/s^{(0)}(\beta(t), t) = E\{Z(t)|T = t\}$, and $s^{(1)}(\beta^*, t)/s^{(0)}(\beta^*, t)$ results when $\beta(t)$ is restricted to be a constant; the difference between these two is zero when integrated out with respect to the marginal distribution of failure time. Suppose, for instance, that $\beta(t)$ decreases over time, then earlier on $\beta(t) > \beta^*$ and $s^{(1)}(\beta(t), t)/s^{(0)}(\beta(t), t) > s^{(1)}(\beta^*, t)/s^{(0)}(\beta^*, t)$; whereas later we would have the opposite effect whereby $\beta(t) < \beta^*$ and $s^{(1)}(\beta(t), t)/s^{(0)}(\beta(t), t) < s^{(1)}(\beta^*, t)/s^{(0)}(\beta^*, t)$. We can write, $v(\beta, t) = \partial/\partial\beta\{s^{(1)}(\beta, t)/s^{(0)}(\beta, t)\}$ and, applying a first-order Taylor series approximation to the integrand of (7.24), we have

$$\int_0^\infty v(t)\{\beta(t) - \beta^*\}dF(t) \approx 0, \qquad (7.26)$$

where $v(t) = v(\beta(t), t) = \text{Var}\{Z(t)|T = t\}$. Therefore

$$\beta^* \approx \frac{\int_0^\infty v(t)\beta(t)dF(t)}{\int_0^\infty v(t)dF(t)} \qquad (7.27)$$

is a weighted average of $\beta(t)$ over time. According to Equation 7.27 more weights are given to those $\beta(t)$'s where the marginal distribution of T is concentrated, which simply means that, on average, we anticipate there being more individuals subjected to those particular levels of $\beta(t)$. The approximation of Equation 7.27 also has an interesting connection with Murphy and Sen (1991), where they show that if we divide the time domain into disjoint intervals and estimate a constant β on each interval, in the limit as $n \to \infty$ and the intervals become finer at a certain rate, the resulting $\hat{\beta}(t)$ estimates $\beta(t)$ consistently. In their large sample studies, they used a (deterministic) piecewise constant parameter $\bar{\beta}(t)$, which is equivalent to Equation 7.27 restricted to individual intervals. They showed that $\bar{\beta}(t)$ is the best approximation to $\hat{\beta}(t)$, in the sense that the integrated squared difference $\int \{\hat{\beta}(t) - \bar{\beta}(t)\}^2 dt \to 0$ in probability as $n \to \infty$, at a faster rate than any other choice of such piecewise constant parameters. In Equation (7.27) if $v(t)$, the conditional variance of $Z(t)$, changes relatively little with time apart from for large t, when the size of the risk sets becomes very small, we can make the approximation $v(t) \equiv c$ and it follows that

$$\beta^* \approx \int_0^\infty \beta(t)dF(t) = E\{\beta(T)\}. \qquad (7.28)$$

In practice, $v(t)$ will often be approximately constant, an observation supported by our own practical experience as well as with simulated data sets. For a comparison of two groups coded as 0 and 1, the conditional variance is of the form $p(1-p)$ for some $0 < p < 1$, and this changes relatively little provided that, throughout the study, p and $1-p$ are not too close to zero. The approximate constancy of this conditional variance is used in the sample size calculation for two-group comparisons (Kim and Tsiatis 1990). In fact, we only require the weaker condition that $\text{Cov}(v(T), \beta(T)) = 0$ to obtain Equation 7.28, a constant $v(t)$ being a special case of this. Even when this weaker condition does not hold exactly, $\int \beta(t)dF(t)$ will still be close to β^*.

Xu and O'Quigley (1998) carried out simulations to study the approximation of $\int \beta(t)dF(t)$ to β^*. Some of those findings are shown in Table 7.1 and these are typical of the findings from a wide variety of other situations. The results are indeed striking. It is also most likely true that it is not well known just how strong is the dependence of the partial likelihood estimator on an independent censoring mechanism when the data are generated by a non-proportional hazards model. Since, in practical data analysis, such a situation will almost always hold, we ought be rather more circumspect about the usual estimators furnished by standard software.

In the table the data are simulated from a simple two-step time-varying regression coefficients model, with baseline hazard $\lambda_0(t) = 1$, $\beta(t) = \beta_1$ when $t < t_0$ and β_2 otherwise. The covariate Z is distributed as Uniform(0,1). At time t_0 a certain percentage of subjects at risk are censored. The value $\hat{\beta}_{PL}$ is the partial likelihood estimate when we fit a proportional hazards model to the data. Table 7.1 summarizes the results of 200 simulations with sample size of 1600. We see that $\int \beta(t)dF(t)$ is always close to β^*, for the values of β that we might see in practice. The most important observation to be made from the table is the strong dependence of $\hat{\beta}_{PL}$ on an independent censoring mechanism, the value to which it converges changing substantially as censoring increases. The censoring mechanism here was chosen to emphasize the difference between $\hat{\beta}_{PL}$ and $\tilde{\beta}$, since $\tilde{\beta}$ puts (asymptotically) the correct weights on the observations before and after t_0. In other cases the effect of censoring may be weaker. Nonetheless, it is important to be aware of the behavior of the partial likelihood estimator under independent censoring and non-proportional hazards and the subsequent difficulties in interpreting the partial likelihood estimate in general situations.

The bracketed figures in Table 7.1 give the standard errors of the estimates from the simulations. From these we can conclude that any gains in efficiency of the partial likelihood estimate can be very quickly lost to biases due to censoring. When there is no censoring the estimators are the same. As censoring increases we see differences in the standard errors of the estimates, the partial likelihood estimate being more efficient; but we also see differences in the biases. Typically, these latter differences are at least an order of magnitude greater.

7.8   Exercises and class projects

1. Show that, under an independent censoring mechanism, $\hat{H}(t)$, as defined in Section 7.3, provides a consistent estimate of $H(t)$.

2. Show that the variance expression, $V(\beta, t)$, using the Andersen and Gill notation (see Section 7.3) is the same as $\mathcal{V}_\beta(Z|t)$ using the notation of Section 7.4. Explain why $\text{Var}(Z|t)$ is consistently estimated by $\mathcal{V}_{\hat{\beta}}(Z|t)$ but that $\text{Var}(Z|t)$ is not generally equal to $v(\beta, t)$.

3. For the general model, suppose that $\beta(t)$ is linear so that $\beta(t) = \alpha_0 + \beta t$. Show that $\mathcal{E}_{\beta(t)}(Z^k|t)$ does not depend upon α_0.

4. Sketch an outline of a proof that $\text{Var}(Z|t)$ is consistently estimated by $\mathcal{V}_{\hat{\beta}}(Z|t)$ and that $E\,\text{Var}(Z|t)$ is consistently estimated by $E\,\mathcal{V}_{\hat{\beta}}(Z|t)$.

5. As for the previous question, indicate why $\int \mathcal{V}_{\hat{\beta}}(Z|t)d\hat{F}(t)$ would be consistent for $E\,\text{Var}(Z|t)$.

6. Show that $\mathcal{V}_\beta(Z|t) = \partial \mathcal{E}_\beta(Z|t)/\partial\beta$ and identify the conditions for the relationship; $\int \mathcal{V}_\beta(Z|t) = \int \partial \mathcal{E}_\beta(Z|t)/\partial\beta = \partial \{\int \mathcal{E}_\beta(Z|t)\}/\partial\beta$ to hold.

7. Consider some parametric non proportional hazards model (see Chapter 4), in which the conditional density of T given $Z = z$ is expressed as $f(t|z)$. Suppose the marginal distribution of Z is $G(z)$. Write down estimating equations for the unknown parameters based on the observations Z_i at the failure times X_i.

8. Use some data set to fit the proportional hazards model. Estimate the parameter β on the basis of estimating equations for the observations Z_i^2 rather than Z_i. Derive another estimate based on estimating equations for $\sqrt{Z_i}$. Compare the estimates.

9. Write down a set of estimating equations based on the observations, Z_i^p, $p > 0$, $i = 1, \ldots, n$. Index the estimate $\hat{\beta}$ by p, i.e., $\hat{\beta}(p)$. For a given data set, plot $\hat{\beta}(p)$ as a function of p.

10. Use analytical or heuristic arguments to described the expected behavior of $\hat{\beta}(p)$ as a function of p under (1) data generated under a proportional hazards model, (2) data generated under a non-proportional hazards model where the effect declines monotonically with time.

11. Consider a proportional hazards model in which we also know that the marginal survival is governed by a distribution $F(t; \theta)$ where θ is

not known. Suppose that it is relatively straightforward to estimate θ, by maximum likelihood or by some graphical technique. Following this we base an estimating equation for the unknown regression coefficient, β, on $U(\beta|\hat{\theta}) = \int \{\mathcal{Z}(t) - \mathcal{E}_\beta(Z|t)\} dF(t; \hat{\theta})$. Comment on this approach and on the properties you anticipate it conferring on the estimate $\hat{\beta}$.

12. Use the approach of the preceding question on some data set by (1) approximating the marginal distribution by an exponential distribution, (2) approximating the marginal distribution by a log-normal distribution.

13. Using again the approach of the previous two questions show that, if the proportional hazards models is correctly specified then the estimate $\hat{\beta}$ based on $F(t; \theta)$ is consistent whether or not the marginal model $F(t; \theta)$ is correctly specified.

14. Supposing that the function $\beta(t)$ is linear so that $\beta(t) = \alpha_0 + \beta t$. Show how to estimate the function $\beta(t)$ in this simple case. Note that we can use this model to base a test of the proportional hazards assumption via a hypothesis test that $H_0 : \beta = 0, \alpha_0 \neq 0$ (Cox 1972).

15. Investigate the assertion that it is not anticipated for $v(t)$, the conditional variance of $Z(t)$, to change much with time. Use the model-based estimates of $v(t)$ and different data sets to study this question informally.

16. In epidemiological studies of breast cancer it has been observed that the tumor grade is not well modeled on the basis of a proportional hazards assumption. A model allowing a monotonic decline in the regression coefficient $\beta(t)$ provides a better fit to observed data. On the basis of observations some epidemiologists have argued that the disease is more aggressive (higher grade) in younger women. Can you think of other explanations for this observed phenomenon?

Chapter 8

Inference: Functions of Brownian motion

8.1 Summary

Inference based on likelihood and counting processes is treated in the following two chapters and, here, we outline our recommended approach to inference based on functions of Brownian motion. Commonly used tests, in particular those based on partial likelihood, arise as special cases. The basic theory is made possible by virtue of the main theorem and its consequences, described in the previous chapter. The theory itself requires no particular extra effort and is quite classical, being based on a simple application of Donsker's theorem. Rather than study a relevant process, such as the score process, and show it to look like Brownian motion, we construct such a process so that, by construction alone, we can almost immediately claim to have a key convergence in distribution result. A wide array of tests are possible. The well-known partial likelihood score test obtains as a special case and can even find its justification in this approach. Non-proportional hazards models, partially proportional hazards models, proportional hazards models with intercept and simple proportional hazards models all come under the same heading. Graphical representation of the tests provides a useful additional intuitive tool to inference. Operating characteristics indicating those situations in which certain tests may outperform others are discussed.

8.2   Motivation

For linear regression and other classic models it is relatively easy to break the discussion into one component dealing with the models themselves and another dealing with inference. This is less natural for proportional hazards models although, in as far as it is possible, we try to do this. The reason that this is less natural is that the central interest of the Cox model, and what followed, had to do with inference, in particular estimation techniques that enable us to leave a part of the model unspecified.

Cox (1972) developed an approach to inference that, although correct, was not entirely classical. It remained nonetheless very specific to the model in that attempts to view the inferential approach as a general one (Cox 1975), applicable outside of the model introduced by Cox (1972), were not really successful. To date, whenever the concept of partial likelihood is mentioned, it is all but exclusively tied to the proportional hazards model. Andersen and Gill (1982), following Aalen (1978), took an approach to inference based on counting processes and stochastic integrals. Their approach required no new theoretical results. Even so, the application of this theory to the proportional hazards model, was a quite colossal effort. A generation of students has grappled with this theory and it has to be said that it is not very easy. Our intuition has improved but the whole area of survival analysis became, as a result of the very widespread adoption of the counting process approach, relatively inaccessible.

Our motivation in this chapter is to show how, leaning upon the results of the previous chapter, we can develop a relatively simple approach to inference. It may have a less broad sweep than that of the counting process approach, although it is not easy to come up with problems that it cannot handle. It is very general. In addition it is easier to get a sense of regression effects, both numerically and visually via graphics, using the approach outlined in this chapter. The whole development presented here is relatively recent and there is scope for much further research work; comparing tests in different settings, using the processes to make inferences about the regression function $\beta(t)$ and studying more closely the connections between the approaches.

8.3   Brownian motion approximations

Our first application of the main theorem is in the construction of a simple test of the null hypothesis, $H_0 : \beta = 0$. Considering the partial scores

$$U(\beta, t) = \int_0^t \{\mathcal{Z}(s) - \mathcal{E}_\beta(Z|s)\} d\bar{N}(s) \qquad (8.1)$$

we can see that, under the hypothesis, this will be a sum of zero mean random variables. Under the alternative of some, let's say positive value for β, we will be summing random variables with negative means and these will accumulate leading to useful test statistics. We can work with any function of the random variable Z, the expectation of the function being readily estimated by an application of an immediate generalization of Corollary 7.3. Imagine, for instance, that Z is continuous with some distribution. We can obtain procedures more robust to the impact of this distribution by, say, transforming Z to a binary variable, zero when less than some value, one otherwise.

Note that this would be different to simply recoding from the outset and fitting a model to binary Z. This is because the probabilities $\pi_i(\beta(t), t)$ would not be the same. Generalizations are fairly obvious and, for simplicity of exposition, we keep mostly to the case of a single variable Z. All of our calculations lean on the main theorem and its corollaries. The increments of the process $\int_0^t \mathcal{Z}(s) d\bar{N}(s)$ at $t = X_i$ have mean $E(Z|X_i)$ and variance $V(Z|X_i)$. We can view these increments as being independent (Cox 1975, Andersen and Gill 1982). Thus, only the existence of the variance is necessary to be able to appeal to the functional central limit theorem. We can then treat our observed process as though arising from a Brownian motion process. Simple calculations allow us to also work with the Brownian bridge, integrated Brownian motion and reflected Brownian motion, processes which will be useful under particular alternatives to the model specified under the null hypothesis. Consider the process $U^*(\alpha, \beta, u), (0 < u < 1)$, in which

$$U^*\left(\alpha, \beta, \frac{j}{k}\right) = \frac{1}{\sqrt{k}} \int_0^{t_j} \mathcal{V}_\alpha(Z|s)^{-1/2} dU(\beta, s), \quad j = 1, ..., k.$$

where $t_j = \bar{N}^{-1}(j)$. This process is only defined on k equispaced points of the interval (0,1] but we extend our definition to the whole interval by simply joining the points together by straight lines. By construction

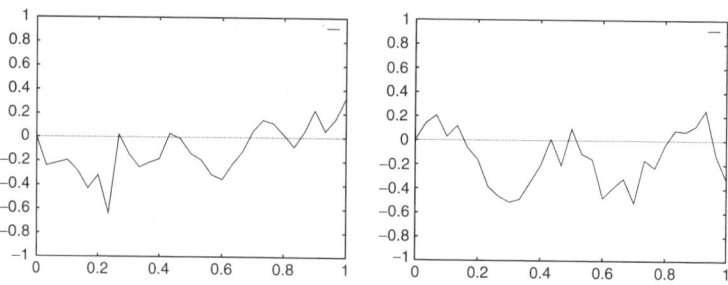

Figure 8.1: Two simulated plots of a process with 30 failures under the hypothesis.

this enables us to claim continuity for the process. More precisely, via linear interpolation for u in the interval j/k to $(j+1)/k$, we write

$$U^*(\alpha, \beta, u) = U^*\left(\alpha, \beta, \frac{j}{k}\right) + \{uk - j\}\left\{U^*\left(\alpha, \beta, \frac{j+1}{k}\right)\right.$$
$$\left. -U^*\left(\alpha, \beta, \frac{j}{k}\right)\right\}.$$

Under the assumption of no effect, a simulation of two such processes is shown in Figure 8.1. As the number of failures increases the process will look more like Brownian motion with zero drift. Should an effect be present then the process would look different showing evidence of a drift. The number of contributions to $U^*(\alpha, \beta, j/k)$ is k and, as mentioned below, k can be less that the total number of failures when these failures provide no further information, i.e., when all the values of Z in the risk set are the same.

Practical calculation of the process

Note that the integral expression of Equation 8.1 is the usual sum with which we are familiar, although only summed up as far as time t rather than over the whole time frame. A more usual form would be;

$$U(\beta, t) = \sum_i \{Z_i(X_i) - \mathcal{E}_\beta(Z|X_i)\} I(X_i \leq t). \qquad (8.2)$$

Standard software provides all of the above ingredients or, otherwise, it is straightforward for the user to derive his or her own routines. Traditionally one calculates the sum, then its variance and, finally, some kind of standardized statistic (i.e., we subtract off the mean and divide by the square root of the variance). Here we proceed slightly differently to obtain not only the same result but also many new results. Rather than standardize at the end of the sum we standardize as we go along. One consequence of this is that we need to be watchful on the contents of the risk sets so that the time point at which the risk set becomes composed of individuals all having the same covariate value (and the estimated variance is zero) we define this point as $t = 1$, the end of our interval. In the multivariate case this point would correspond to the variable for which we are testing while controlling for others via use of the estimates of the previous chapter. No information can be obtained once all the risk set individuals have the same covariate value and the increments in the process become exactly zero. We could more formally say that the ratio of an increment, when zero, to the standard deviation, when zero, is itself zero. Keeping in mind this minor technical consideration we should also take k, often denoting the total number of actual failures (assuming that the ties have been split), to be the total number of failures less any failures taking place beyond the point at which the risk set only contains individuals having the same covariate value. Beyond this point the conditional variances are zero and, although the numerator also becomes zero, we still need to be clear about how we define $0/0$ and about how many actual contributions there are in our final sum. The process is then standardized. Standarizing at the end, rather than as we go along, leads to the partial likelihood score statistic and we consider this in a sub-section of Section 8.5. In the same way that the integral Equation 8.1 could be written in a more familiar way as in Equation 8.2, we can also write;

$$U^*\left(\alpha, \beta, \frac{j}{k}\right) = \sum_i k^{-1/2}\{Z_i(X_i) - \mathcal{E}_\beta(Z|X_i)\}\mathcal{V}_\alpha(Z|X_i)^{-1/2}I(X_i \le t).$$
$$(8.3)$$

The fact that no further information can be obtained from the increments once the risk sets only contain individuals with the same covariate value is true of course even without sequential standardization. Each increment minus the risk set mean necessarily has value zero. Since we do not, at each time point, divide by the square root of the variance, we typically pay no attention to this fact. Here, using sequential standardization, we need take care to ensure that zero divided by zero is assigned a suitable value. The only coherent value is of course zero itself.

Large sample theory

As n goes to infinity, under the usual Breslow and Crowley conditions in which j and k increase at the same rate, i.e., $j \approx kt$ where $0 < t < 1$ and by $\approx$ we mean that we round up to the nearest integer. Although only a minor technical detail, we can be even less restrictive, only requiring that the ratio j/k approaches some constant t strictly between 0 and 1. We then have that, for each j ($j = 1, \ldots, k-1$), $U^*(\alpha, \beta, j/k)$ converges in distribution to a Gaussian process with mean zero and variance equal to t. This follows directly from Donsker's theorem. Replacing β by a consistent estimate leaves asymptotic properties unaltered for all finite dimensional distributions. This follows as an immediate application of Slutsky's theorem. The only part that can be involved mathematically is when we wish to extend the result to the whole continuous process. For this we need the tightness of the process (Billingsley 1968) and this follows here when the variance of the process is estimated consistently and independently of the process itself. This result we can claim to hold as a direct consequence of Tsiatis (1978). There are cases, the goodness-of-fit question being an example, where the numerator and the denominator both involve the same sample-based estimate $\hat{\beta}$ implying a certain degree of dependence. The same arguments nonetheless hold although, in order to keep track of the details, a little more effort is needed (O'Quigley 2003).

Some remarks on the notation $U^*(\alpha, \beta, u)$

Various aspects of the statistic $U^*(\alpha, \beta, u)$ will be used to construct different tests. We choose the * symbol to indicate some kind of standardization as opposed to the non-standardized U. The variance and the number of distinct failure time contributions to the sum are used to carry out the standardization. Added flexibility in test construction is achieved by using the two parameters, α and β, rather than a single parameter β. In practice these are replaced by quantities which are either fixed or estimated under some hypothesis. For goodness-of-fit procedures, which we consider later, we will only use a single parameter, typically $\hat{\beta}$. Goodness-of-fit tests are most usefully viewed as tests of data driven hypotheses of the form $H_0 : \beta = \hat{\beta}$. A test then of a hypothesis $H_0 : \beta = 0$ may not seem very different. This is true in principle. However, for a test of $H_0 : \beta = 0$, we need keep in mind not only behavior under the null but also under the alternative. Because of this it is often advantageous, under a null hypothesis of $\beta = 0$, to work with $\alpha = \hat{\beta}$ and $\beta = 0$ in the expression $U^*(\alpha, \beta, u)$. Under the null, $\hat{\beta}$ remains consistent for the value 0 and, in the light of Slutsky's theorem, the large sample distribution of the test statistics will not be affected. Under the alternative, however, things look different. The increments of the process $\int_0^t \mathcal{Z}(s) d\bar{N}(s)$ at $t = X_i$ no longer have mean $E(Z|X_i)$ and adding them up will indicate departures from the null. But the denominator is also affected and, in order to keep the variance estimate not only correct but also as small as we can, it is preferable to use the value $\hat{\beta}$ rather than zero. Nonetheless, even if less powerful, to use $\alpha = 0$ is not incorrect and corresponds to what is commonly done in current practice.

Some properties of $U^*(\alpha, \beta, u)$

A very wide range of possible tests can be based upon the statistic $U^*(\alpha, \beta, u)$ and we consider a number of these below. Well-known tests such as the so-called partial likelihood score test obtains as a special case. First, we need to make some observations on the properties of $U^*(\alpha, \beta, u)$ under different values of α, β and u.

Lemma 8.1 *The process $U^*(\alpha, \beta, u)$, for all finite α and β is continuous on [0,1]. Also $E\,U^*(\beta, \beta, 0) = 0$.*

Lemma 8.2 *Under model 6.3 $E\,U^*(\hat{\beta}, \beta, u)$ converges in probability to zero.*

Lemma 8.3 *Suppose that $v < u$, then $\mathrm{Cov}\,\{U^*(\hat{\beta}, \beta, v)\,; U^*(\hat{\beta}, \beta, u)\}$ converges in probability to v.*

Since the increments of the process are asymptotically independent we can treat $U^*(\hat{\beta}, \beta, u)$ (as well as $U^*(\beta, \beta, u)$ under some hypothesized β) as though it were Brownian motion.

Brownian motion in reversed time

The process $U^*_-(\alpha, \beta, u)\,, (0 < u < 1)$, derived from $U^*(\alpha, \beta, u)$ in which

$$U^*_-\left(\alpha, \beta, \frac{j}{k}\right) = \frac{1}{\sqrt{k-j}} \int_1^{1-t_j} \mathcal{V}_\alpha(Z|s)^{-1/2} dU(\beta, s)\,, \quad j = 1, \ldots, k, \tag{8.4}$$

and where, as at the beginning of this section, $t_j = \bar{N}^{-1}(j)$ and the linear interpolation is the same, will also approximate Brownian motion under the model. It helps to imagine the process rotated through 180 degrees, the axes held fixed, and then the origin shifted to the last point $U^*(\alpha, \beta, 1)$. We can use such a process to construct a simple test in which effects may be delayed during an initial period.

8.4   Non and partially proportional hazards models

The Brownian motion approximations of the above section extend immediately to the case of non proportional hazards and partially proportional hazards models. The generalization of Equation 8.1 is natural and leads to an unstandardized score;

$$U(\beta(t), t) = \int_0^t \{\mathcal{Z}(s) - \mathcal{E}_{\beta(t)}(Z|s)\} d\bar{N}(s), \qquad (8.5)$$

and, as before, under the null hypothesis that $\beta(t)$ is correctly specified the function $U(\beta(t), t)$ will be a sum of zero mean random variables. The range of possible alternative hypotheses is large and, mostly, we will not wish to consider anything too complex. Often the alternative hypothesis will specify an ordering, or a nonzero value, for just one of the components of a vector $\beta(t)$. In the exact same way as in the previous section, all of the calculations lean on the main theorem and its corollaries. The increments of the process $\int_0^t \mathcal{Z}(s) d\bar{N}(s)$ at $t = X_i$ have mean $E(Z|X_i)$ and variance $V(Z|X_i)$. A little bit of extra care is needed, in practice, in order to maintain the view of the independence of these increments. When $\beta(t)$ is known there is no problem but if, as usually happens, we wish to use estimates, then, for asymptotic theory to still hold, we require the sample size (number of failures) to become infinite relative to the dimension of $\beta(t)$. Thus, if we wish to estimate the whole function $\beta(t)$, then some restrictions will be needed because, full generality implies an infinite dimensional parameter $\beta(t)$. For the stratified model and, generally, partially proportional hazards models, the problem does not arise because we do not estimate $\beta(t)$.

The sequentially standardized process will now be written $U^*(\alpha(t), \beta(t), u)$, $(0 < u < 1)$, in which

$$U^*\left(\alpha(t), \beta(t), \frac{j}{k}\right) = \frac{1}{\sqrt{k}} \int_0^{t_j} \mathcal{V}_{\alpha(t)}(Z|s)^{-1/2} dU(\beta(t), s), \quad j = 1, \ldots, k.$$

where $t_j = \bar{N}^{-1}(j)$. This process can be made to cover the whole interval (0,1] continuously by interpolating in the same way as in the above section. For this process we reach the same conclusion, i.e., that as n goes to infinity, under the usual Breslow and Crowley conditions, we have that, for each j $(j = 1, \ldots, k-1)$, $U^*(\alpha(t), \beta(t), j/k)$ converges in distribution to a Gaussian process with mean zero and variance equal to t where $t \approx j/k$. The only potential difficulty is making use of Slutsky's theorem whereby, if we replace $\alpha(t)$ and $\beta(t)$ by consistent estimates the result still holds. The issue is that of having consistent estimates, which for an infinite dimensional unrestricted parameter we cannot achieve. The solution is simply to either restrict these functions or to work with the stratified models in which we do not need to estimate them. The above subsection headed "Some remarks on the notation" applies equally well here if we replace α and β by $\alpha(t)$ and $\beta(t)$. The lemmas of the above section describing the properties of $U^*(\alpha, \beta, t)$ apply, once again, when working with $U^*(\alpha(t), \beta(t), t)$. Specifically, the process $U^*(\alpha(t), \beta(t), t)$, for all finite $\alpha(t)$ and $\beta(t)$ is continuous on [0,1] and $E\, U^*(\beta(t), \beta(t), 0) = 0$, under model 6.3 $E\, U^*(\hat{\beta}(t), \beta, u)$ converges in probability to zero and for $v < u$, $\text{Cov}\,\{U^*(\hat{\beta}(t), \beta(t), v)\,; U^*(\hat{\beta}(t), \beta(t), u)\}$ converges in probability to v. Since the increments of the process are asymptotically independent we will treat $U^*(\hat{\beta}(t), \beta(t), t)$ (as well as $U^*(\beta(t), \beta(t), t)$ under some hypothesized $\beta(t)$) as though it were Brownian motion.

8.5 Tests based on functions of Brownian motion

Several tests of point hypotheses can be constructed based on the theory of the previous section. These tests can also be used to construct test-based confidence intervals of parameter estimates, obtained as solutions to an estimating equation. Among these tests are the following;

Distance from origin at time t

At time t, under the null hypothesis that $\beta = \beta_0$, often a hypothesis of absence of effect in which case $\beta_0 = 0$, we have that $U^*(\hat{\beta}, \beta_0, t)$ can be approximated by a normal distribution with mean zero and variance t. A p-value corresponding to the null hypothesis is then obtained from

$$\Pr\left\{U^*\left(\hat{\beta}, \beta_0, t\right)/\sqrt{t} > z\right\} = 1 - \Phi(z).$$

This p-value is for a one-sided test in the direction of the alternative $\beta < \beta_0$. For a one-sided alternative in the opposite direction we would use

$$\Pr\left\{U^*\left(\hat{\beta},\beta_0,t\right)/\sqrt{t} < z\right\} = \Phi(z)$$

and, for a two-sided alternative, we would, as usual, consider the absolute value of the test statistic and multiply $1 - \Phi(z)$ by two. Under the alternative, say $\beta > \beta_0$, if we take the first two terms of a Taylor series expansion of $U^*(\hat{\beta},\beta_0,t)$ about β, we can deduce that a good approximation for this would be Brownian motion with drift. At time t this is then a good test for absence of effect (Brownian motion) against a proportional hazards alternative (Brownian motion with drift); good in the sense that type I error is controlled for and, under these alternatives, the test has good power properties.

Under a proportional hazards alternative to the null hypothesis, $U^*(\hat{\beta},\beta_0,t)$ increases in expectation in a way that is very close to linear so that the further out in time t the more powerful our test will be. Specifically the ratio $U^*\left(\hat{\beta},\beta_0,t\right)/\sqrt{t}$ will be approximately increasing linearly in t in the numerator and as $\sqrt{t}$ in the denominator. Power, therefore, will be maximized by using the whole time interval, i.e., taking $t = 1$. Nonetheless, there may be situations in which we may opt to take a value of t less than one. If we know, for instance, that under both the null and the alternative we can exclude the possibility of effects being persistent beyond some time τ say, i.e., the hazard ratios beyond that point should be one or very close to that, then we will achieve greater power by taking t to be less than one, specifically some value around τ. A confidence interval for β_0 can be obtained using normal approximations or by constructing the interval $\{\hat{\beta}^-, \hat{\beta}^+\}$ such that for any point b contained in the interval a test of the null hypothesis is not rejected.

Figure 8.2 illustrates behavior under the alternative for a relatively large sample. In the first case the situation is that of proportional hazards, whereas in the second case the regression effect disappears at some point beyond which the process is approximately parallel to the time axis. After τ there are no expected gains in the numerator and so the test based on the distance from origin at time τ will be more powerful than that at $t = 1$ since, in this case, $t > \tau$. Usually, τ would not be known in advance in which case the following test might be a candidate.

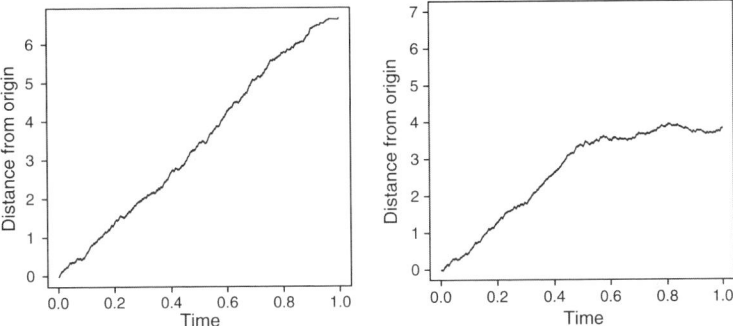

Figure 8.2: Example of processes under proportional and non-proportional hazards and very large samples indicating anticipated large sample behavior.

Maximum distance from origin at time t

In cases where we wish to consider values of t less than one, we may have knowledge of some τ of interest. Otherwise, we may want to consider several possible values of τ. Control on type I error will be lost unless specific account is made of the multiplicity of tests. One simple way to address this issue is to consider the maximum value achieved by the process during the interval $(0, \tau)$. Again, we can appeal to known results for some well known functions of Brownian motion. In particular, approximating $U^*(\hat{\beta}, \beta_0, t)$ by Brownian motion, we have

$$\Pr \left\{ \max_{t \in (0, \tau)} U^* \left(\hat{\beta}, \beta_0, t \right) / \sqrt{t} > z \right\} = 2 \Pr \left\{ U^* \left(\hat{\beta}, \beta_0, \tau \right) / \sqrt{\tau} > z \right\}.$$

Under the null and proportional hazards alternatives this test, as opposed to the usual score test, would lose power comparable to carrying out a two-sided rather than a one-sided test. Under non-proportional hazards alternatives this test can be of use, an extreme example being crossing hazards where the usual score test may have power close to zero. As the absolute value of the hazard ratio increases so would the maximum distance from the origin. This test would be worth considering if we suspect that any potential effects are unlikely to be persistent although we may have little idea as to just how long the effect may last.

Distance from origin with delayed effect

If there is an initial period during which we anticipate, both under the null and the alternative, there to be no regression effect, then following the idea above where we take the upper limit of t to be less than one, we can construct a useful test. We use a Brownian motion in reversed time, starting at $t = 1$ and, rather than go all the way backwards to $t = 0$ we stop short of that at time $1 - \tau$, having chosen τ such that on the interval $(0, \tau)$ we believe that effects have yet to manifest themselves. In practice, rather than actually reverse time, we would simply move the origin to τ and take our process to be $U^*(\hat{\beta}, \beta_0, t) - U^*(\hat{\beta}, \beta_0, \tau)$. Again we would anticipate a gain in power since, effectively, in the same way as above, the expected size of the numerator of the test statistic is not increasing beyond some point. Rather than divide by a larger value of t, using the value $1 - \tau$ results in a larger test statistic.

Arcsine test

Another possibility is to consider the amount of time the whole process is positive (or negative). Under the null hypothesis of no effect we expect, on average, that the process $U^*(\hat{\beta}, \beta_0, t)$ will be positive for half of the interval $(0, t)$. Calling $\Delta(t)$ the percentage of time the process is positive on the interval $(0, t)$ we have (see Section 2.11 on Brownian motion):

$$\Pr\left\{\left[\Delta(u) : U^*\left(\hat{\beta}, \beta_0, u\right) > 0\right] > w\right\} = 2\pi^{-1}\sin^{-1}\sqrt{w}.$$

This test is particularly easy to carry out but does not appear to be very powerful. We have not really studied it in any depth and it may be worthy of further investigation. One reason that may explain a lack of power is that no distinction is made between situations where the process is close to the origin or far from the origin. This extra information, contained in the distance and that is clearly informative concerning effects, is then ignored.

Area under the curve at time t

The area under the curve of the process $U^*(\hat{\beta}, \beta_0, t)$ is given by

$$J(\hat{\beta}, \beta_0, t) = \int_0^t U^*(\hat{\beta}, \beta_0, u)du$$

and is itself a stochastic process. Referring back to the chapter describing the properties of Brownian motion we can conclude immediately that $J(\hat{\beta}, \beta_0, t)$ converges in distribution to integrated Brownian motion, i.e., a Gaussian process with mean zero and covariance process

$$\text{Cov}\left\{J(\hat{\beta}, \beta_0, s), J(\hat{\beta}, \beta_0, t)\right\} = s^2\left(\frac{t}{2} - \frac{s}{6}\right) \quad (s < t). \tag{8.6}$$

At time $t = 1$, rather than base a test on the distance covered by the process $U^*(\hat{\beta}, \beta_0, t)$ we can base a test on the area under the curve. The two are correlated and, if a large distance is covered, we also know that there will be a large area under the curve. A one sided p-value corresponding to the null hypothesis is then obtained from

$$\Pr\left\{\sqrt{3}J\left(\hat{\beta}, \beta_0, t\right)/\sqrt{t} > z\right\} = 1 - \Phi(z).$$

For the other one-sided alternative and for two sided alternatives we proceed as described just above under the heading "Distance from origin at time t."

Under departures from the null which can be exactly described as proportional hazards alternatives the distance from origin test has very slightly greater power than the area under the curve test. This is no longer true when effects decline with time and in many, possibly the majority of practical situations, the area under the curve test has greater power than the distance from the origin test. This can be understood intuitively by considering an exaggerated case of declining effects, a situation in which the effect drops to zero at some time point, t_0. Until that point the process $U^*(\hat{\beta}, 0, u)$ will approximately increase as a straight line with fluctuations (Brownian motion with drift) but, beyond t_0, the line will proceed, on average, as though parallel to the time axis. A test based on the distance from the origin will be impacted rather more strongly than a test based on the area under the curve and will therefore show less power. This follows since, for a distance from the origin test, the numerator, on average, is not increasing, whereas the denominator is.

For an area under the curve test both the numerator and denominator continue to increase, although, of course, the rate of increase of the numerator is slower than it would otherwise be if effects were to be maintained. In both cases the effect, as mirrored by the test, is being diluted and the gain for the area under the curve test stems from the fact that this diluting phenomenon is weaker for the area under the curve test than for a distance from origin test using $t = 1$. An extreme case of non-proportional hazards is one where the effects change direction at some point such that $U^*(\hat{\beta}, 0, 1)$ is very close to zero. A test based on distance from the origin will consequently have no power against such an alternative, whereas an area under the curve test can still detect that the null is contradicted by the observations. This said, for such situations, there are other tests that are mostly to be preferred to the area under the curve test.

Linear combinations of distance from origin and area under curve

It is also useful to note:

Lemma 8.4 *The covariance function of $J(\hat{\beta}, \beta_0, t)$ and $U^*(\hat{\beta}, \beta_0, t)$, converges in probability to $t^2/2$.*

This simple result (see also Section 2.11) enables the construction of a linear combination test. Consider then a class of statistics $D(\theta, t)$ where

$$D(\theta, t) = \theta U^*(\hat{\beta}, \beta_0, t) + (1 - \theta) J(\hat{\beta}, \beta_0, t), \quad 0 \le \theta \le 1,$$

and the following simple lemma which applied under $H_0 : \beta = \beta_0$:

Lemma 8.5 *Under H_0, $D(\theta, t)$ converges in distribution to a normal law with*

$$E\{D(\theta, t) = 0\}; \quad \text{Var } D(\theta, t) = t\theta^2 + \frac{t^3}{3}(1 - \theta)^2 + \frac{t^2}{2}\theta(1 - \theta).$$

The linear combination test will have power against both types of alternative, proportional hazards and non-proportional hazards in which effects decline in time. It can be considered a compromise test between these other two, each one arising as special cases when $\theta = 0$ or $\theta = 1$. The cost of this greater generality is a reduction in power when the true alternative is a clear one of proportional hazards or one of strong non-proportional hazards such as an inversion of the regression effect. The parameter θ would have to be chosen in advance to reflect the kind of alternative we have in mind. It is also possible to allow ourselves to be guided by the observations themselves, in which case we could consider $\sup_\theta D(\theta, t)$, most commonly $\sup_\theta D(\theta, 1)$. For this, unfortunately, the above lemma is no longer applicable and we need to appeal to other notions to obtain the correct null distribution.

Brownian bridge test

Since we are viewing the process $U^*(\hat{\beta}, \beta_0, t)$ as though it were a realization of a Brownian motion, we can consider some other well-known functions of Brownian motion. Consider then the bridged process $U_0^*(\hat{\beta}, \beta_0, t)$;

Definition 8.1 *The bridged process is defined by the transformation*

$$U_0^*(\hat{\beta}, \beta_0, t) = U^*(\hat{\beta}, \beta_0, t) - tU^*(\hat{\beta}, \beta_0, 1)$$

Lemma 8.6 *The process $U_0^*(\hat{\beta}, \beta_0, t)$ converges in distribution to the Brownian bridge, in particular, for large samples, $E\, U_0^*(\hat{\beta}, \beta_0, t) = 0$ and* $\mathrm{Cov}\,\{U_0^*(\hat{\beta}, \beta_0, s),\, U_0^*(\hat{\beta}, \beta_0, t)\} = s(1 - t)$.

The Brownian bridge is also called tied-down Brownian motion for the obvious reason that at $t = 0$ and $t = 1$ the process takes the value 0. Carrying out a test at $t = 1$ will not then be particularly useful and it is more useful to consider, as a test statistic, the greatest distance of the bridged process from the time axis. We can then appeal to:

Lemma 8.7

$$\mathrm{Pr}\left\{\sup_u |U_0^*(\hat{\beta}, \beta_0, u)| \geq a\right\} \approx 2\exp(-2a^2), \tag{8.7}$$

which follows as a large sample result since

$$\mathrm{Pr}\left\{\sup_u |U_0^*(\hat{\beta}, \beta_0, u)| \leq a\right\} \rightarrow 1 - 2\sum_{k=1}^{\infty}(-1)^{k+1}\exp(-2k^2a^2),\ a \geq 0.$$

This is an alternating sign series and therefore, if we stop the series at $k = 2$ then the error is bounded by $2\exp(-8a^2)$ which for most values of a that we will be interested in will be small enough to ignore. For alternatives to the null hypothesis ($\beta = 0$) belonging to the proportional hazards class, the Brownian bridge test will be less powerful than the distance from origin test. It is more useful under alternatives of a non-proportional hazards nature, in particular an alternative in which $U^*(\hat{\beta}, \beta_0, 1)$ is close to zero, a situation we might anticipate when the hazard functions cross over. Its main use, in our view, is in testing goodness-of-fit, i.e., a hypothesis test of the form $H_0 : \beta = \hat{\beta}$.

Reflected Brownian motion

An interesting property of Brownian motion described in Chapter 2 was the following. Let $W(t)$ be Brownian motion, choose some positive value r and define the process $W_r(t)$ in the following way: If $W(t) < r$ then $W_r(t) = W(t)$. If $W(t) \geq r$ then $W_r(t) = 2r - W(t)$. It was shown in Chapter 2 that the reflected process $W_r(t)$ is also Brownian motion. Choosing r to be negative and defining $W_r(t)$ accordingly we have the same result. The process $W_r(t)$ coincides exactly with $W(t)$ until such a time as a barrier is reached. We can imagine this barrier as a mirror and beyond the barrier the process $W_r(t)$ is a simple reflection of $W(t)$. So, consider the process $U^r(\hat{\beta}, \beta_0, t)$ defined to be $U^*(\hat{\beta}, \beta_0, t)$ if $|U^*(\hat{\beta}, \beta_0, t)| < r$ and to be equal to $2r - U^*(\hat{\beta}, \beta_0, t)$ if $|U^*(\hat{\beta}, \beta_0, t)| \geq r$.

Lemma 8.8 *The process $U^r(\hat{\beta}, \beta_0, t)$ converges in distribution to Brownian motion, in particular, for large samples, $E\, U^r(\hat{\beta}, \beta_0, t) = 0$ and* $\mathrm{Cov}\,\{U^r(\hat{\beta}, \beta_0, s),\, U^r(\hat{\beta}, \beta_0, t)\} = s$.

Under proportional hazards there is no obvious role to be played by U^r. However, imagine a non-proportional hazards alternative where the direction of the effect reverses at some point, the so-called crossing hazards problem. The statistic $U^*(\hat{\beta}, 0, t)$ would increase up to some point and then decrease back to a value close to zero. If we knew this point, or had some reasons for guessing it in advance, then we could work with $U^r(\hat{\beta}, \beta_0, t)$ instead of $U^*(\hat{\beta}, \beta_0, t)$. A judicious choice of the point of reflection would result in a test statistic that continues to increase under such an alternative so that a distance from the origin test might have reasonable power. In practice we may not have any ideas on a potential point of reflection. We could then consider trying a whole class of points of reflection and choosing that point which results in the greatest test statistic. We require different inferential procedures for this.

A bound for a supremum type test can be derived by applying the results of Davies (1977, 1987). Under the alternative hypothesis we could imagine increments of the same sign being added together until the value r is reached, at which point the sign of the increments changes. Under the alternative hypothesis the absolute value of the increments is strictly greater than zero. Under the null, r is not defined and, following the usual standardization, this set-up fits in with that of Davies (1977, 1987). We can define γ_r to be the time point satisfying

$U^*(\hat{\beta}, \beta_0, \gamma_r) = r$. A two-sided test can then be based on the statistic $M = \sup_r\{|U^r(\hat{\beta}, \beta_0, 1)| : 0 \le \gamma_r \le 1\}$. Inference can then be based on

$$\Pr\{\sup |U^r(\hat{\beta}, \beta_0, 1)| > c : 0 \le \gamma_r \le 1\} \le \Phi(-c)$$
$$+ \frac{\exp(-c^2/2)}{2\pi} \int_0^1 \{-\rho_{11}(\gamma)\}^{\frac{1}{2}} d\gamma$$

where Φ denotes the cumulative normal distribution function,

$$\rho_{11}(\gamma) = \{\partial^2 \rho(\phi, \gamma)/\partial\phi^2\}_{\phi=\gamma}$$

and where $\rho(\gamma_r, \gamma_s)$ is the autocorrelation function between the processes $U^r(\hat{\beta}, \beta_0, 1)$ and $U^s(\hat{\beta}, \beta_0, 1)$. In general, the autocorrelation function $\rho(\phi, \gamma)$, needed to evaluate the test statistic is unknown. However, it can be consistently estimated using bootstrap resampling methods (O'Quigley and Pessione 1991). For γ_r and γ_s taken as fixed, we can take bootstrap samples from which several pairs of $U^r(\hat{\beta}, \beta_0, 1)$ and $U^s(\hat{\beta}, \beta_0, 1)$ can be obtained. Using these pairs, an empirical, i.e., product moment, correlation coefficient can be calculated. Under the usual conditions (Efron 1981a,b), the empirical estimate provides a consistent estimate of the true value. This sampling strategy is further investigated in related work by O'Quigley and Natarajan (2004). A simpler approximation is available by using the results of Davies (1987) and this has the advantage that the autocorrelation is not needed. This may be written down as

$$\Pr\{\sup |U^r(\hat{\beta}, \beta_0, 1)| > M\} \approx \Phi(-M) + \frac{V_\rho \exp(-M^2/2)}{\sqrt{8\pi}}, \quad (8.8)$$

where $V_\rho = \sum_i |U^r(\hat{\beta}, \beta_0, 1) - U^s(\hat{\beta}, \beta_0, 1)|$, the γ_i, ranging over $(\mathcal{L}, \mathcal{U})$, are the turning points of $T(0, \hat{\beta}; \cdot)$ and M is the observed maximum of $T(0, \hat{\beta}; \cdot)$. Turning points only occur at the k distinct failure times and, to keep the notation consistent with that of the next section, it suffices to take γ_i $(i = 2, \ldots, k)$ as being located half way between adjacent failures, $\gamma_1 = 0$ and γ_{k+1} any value greater than the largest failure time.

Partial likelihood score test

Some discussion of the concept of partial likelihood is given in the following chapter. It turns out that the score test using that concept coincides with one of those given here if we make the following approximation. Suppose that we wish to test $H_0 : \beta = 0$ and instead of $U^*(\hat{\beta}, 0, t)$ we choose to work with $U^*(0, 0, t)$. In the light of Slutsky's theorem it is readily seen that, for all finite divisions of the interval (0,1), the large sample null distributions of the two test statistics are the same. In order to maintain our results for the whole process (filling out the interval) we need tightness (see Section 8.3) which follows here when the denominator is estimated consistently and independently of the numerator. For this we appeal to the main theorem of Tsiatis (1978). Next, instead of standardizing by $\mathcal{V}_0(Z|X_i)$ at each X_i we take a simple average of such quantities, over the observed failures. To see this, first recall Corollary 7.4 indicating that $\int \mathcal{V}_{\hat{\beta}}(Z|t)d\hat{F}(t)$ is consistent for $E \operatorname{Var}(Z|t)$. Rather than integrate with respect to $\hat{F}(t)$ it is more commonly accepted to integrate with respect to $\bar{N}(t)$. It is also more common to fix $\hat{\beta}$ in $\mathcal{V}_{\hat{\beta}}(Z|t)$ at its null value zero. This gives us;

Definition 8.2 *The empirical average conditional variance, $\bar{\mathcal{V}}_0$ is defined as*

$$\bar{\mathcal{V}}_0 = \int_0^1 \mathcal{V}_0(Z|t)d\bar{N}(t)$$

We can choose to include in our test statistics the quantity $\bar{\mathcal{V}}_0$ in place of $\mathcal{V}_0(Z|X_i)$. In this case the distance from origin test described above, at $t = 1$, coincides exactly with the partial likelihood score test. It may even be used to justify the partial likelihood score test since the inferential basis for it stands on an arguably more transparent footing than partial likelihood itself which, to this day, has never really been given a very clear theoretical setting.

Linearly weighted test statistics

A broad class of test statistics based on the first derivative of the logarithm of the partial likelihood consists in linearly weighting the contributions at different time points. Such tests coincide with weighted log-rank statistics. Before considering those tests it is worth pointing out that similar classes, and even the exact same tests can be obtained under the heading of Brownian motion tests. The details follow through in the same way as above and we only consider the basic ideas. Suppose that instead of $U(\beta, t)$ given in equation (8.1) we introduce some time-dependent positive weights, $K(t)$ into the definition and write

$$U(\beta, t) = \int_0^t K(s)\{\mathcal{Z}(s) - \mathcal{E}_\beta(Z|s)\}d\bar{N}(s). \tag{8.9}$$

As before we can see that, under the hypothesis, this will be a sum of zero mean random variables. Under the alternative of some, let's say positive value for β, we will be summing random variables with negative means, the means being now weighted by K. Different choices of K will produce different operating characteristics under the alternative. It is still the case that the increments of the process $\int_0^t \mathcal{Z}(s)d\bar{N}(s)$ at $t = X_i$ have mean $E(Z|X_i)$ and variance $V(Z|X_i)$ and so, as before, we will be able to treat the observed process, after transformation, as though arising from a Brownian motion process. All of the tests based on simple Brownian motion, the Brownian bridge, integrated Brownian motion and reflected Brownian motion will still apply. For this to work we now base calculations on the process $U^*(\alpha, \beta, u)$, $(0 < u < 1)$, in which

$$U^*\left(\alpha, \beta, \frac{j}{k}\right) = \frac{1}{\sqrt{k}} \int_0^{t_j} K^{-1/2}(s)\mathcal{V}_\alpha(Z|s)^{-1/2}dU(\beta, s), \ j = 1, \ldots, k,$$

where $t_j = \bar{N}^{-1}(j)$. We extend this definition on the k equispaced points of the interval $(0,1]$ to the whole interval as before, via linear interpolation. The same formula still applies. The function $K(s)$ can be either deterministic or itself a stochastic process. As long as its variance at time s exists and can be evaluated, or consistently estimated, then this modification introduces no particular additional complexity. Choosing $K(t)$ to be the number of subjects at risk at time t produces a test analogous to the test described by Gehan (1965), whereas the choice of $K(t) = \hat{S}(t)$ produces weights analogous to those of Prentice (1978).

8.6 Multivariate model

In practice it is the multivariate setting that we are most interested in; testing for the existence of effects in the presence of related covariates, or possibly testing the combined effects of several covariates. In this work we give very little specific attention to the multivariate setting, not because we do not feel it to be important but because the univariate extensions are almost always rather obvious and the main concepts come through more clearly in the relatively notationally uncluttered univariate case. Nonetheless, some thought is on occasion required.

The main theorem giving a consistent estimate of the distribution of
the covariate at each time point t applies equally well when the covariate $Z(t)$ is multi-dimensional. Everything follows through in the
same way and there is no need for additional theorems. Let us recall
definition 7.2

Definition 8.3 *The discrete probabilities* $\pi_i(\beta(t), t)$ *are given by*

$$\pi_i(\beta(t), t) = \frac{Y_i(t) \exp\{\beta(t)Z_i(t)\}}{\sum_{j=1}^n Y_j(t) \exp\{\beta(t)Z_j(t)\}}.$$

In the multivariate case the product $\beta(t)Z_i(t)$ becomes a vector or
inner product, a simple linear sum of the components of $Z_i(t)$ and
the corresponding components of $\beta(t)$. Suppose, for simplicity, that
$Z_i(t)$ is two dimensional so that $Z_i(t)^T = \{Z_{1i}(t), Z_{2i}(t)\}$. Then the
$\pi_i(\beta(t), t)$ give our estimate for the joint distribution of $\{Z_{1i}(t), Z_{2i}(t)\}$
at time t. As for any multi-dimensional distribution if we wish to consider only the marginal distribution of, say, $Z_1(t)$ then we simply sum
the $\pi_i(\beta(t), t)$ over the variable $Z_2(t)$. In practice we work with the
$\pi_i(\beta(t), t)$, defined to be of the highest dimension that we are interested in, for the problem in hand, and simply sum over the subsets of
vector Z needed. To be completely concrete let us return to the partial
scores,

$$U(\beta, t) = \int_0^t \{\mathcal{Z}(s) - \mathcal{E}_\beta(Z|s)\}d\bar{N}(s), \tag{8.10}$$

defined previously for the univariate case. Both $\mathcal{Z}(s)$ and $\mathcal{E}_\beta(Z|s)$ are
vectors of the same dimension. So also is $U(\beta, t)$. The vector $U(\beta, t)$
is made up of the component marginal processes any of which we may
be interested in. For each marginal covariate, let's say Z_1, for instance,
we also calculate $\mathcal{E}_\beta(Z_1|s)$ and we can do this either by first working
out the marginal distribution of Z_1 or just by summing over the joint
probabilities. The result is the same and it is no doubt easier to work
out all expectations with the respect to the joint distribution. Let us
then write

$$U(\beta, Z_1, t) = \int_0^t \{\mathcal{Z}_1(s) - \mathcal{E}_\beta(Z_1|s)\}d\bar{N}(s) \tag{8.11}$$

where the subscript "1" denotes the first component of the vector.
The interesting thing is that $\mathcal{E}_\beta$ does not require any such additional

notation, depending only on the joint $\pi_i(\beta(t), t)$. As for the univariate case we can work with any function of the random vector Z, the expectation of the function being readily estimated by an application of an immediate generalization of Corollary 7.3. Note that the process we are constructing is not the same one that we would obtain were we to simply work with only Z_1. This is because the $\pi_i(\beta(t), t)$ involve a univariate Z in the former case and a multivariate Z in the latter. The increments of the process $\int_0^t Z_1(s) d\bar{N}(s)$ at $t = X_i$ have mean $E(Z_1|X_i)$ and variance $V(Z_1|X_i)$. As before, these increments can be taken to be independent (Cox 1975, Andersen and Gill 1982) so that only the existence of the variance is necessary to be able to appeal to the functional central limit theorem.

This observed process will also be treated as though arising from a Brownian motion process. The same calculations as above allow us to also work with the Brownian bridge, integrated Brownian motion and reflected Brownian motion. Our development is entirely analogous to that for the univariate case and we consider now the process $U^*(\alpha, \beta, Z_1, u)$, $(0 < u < 1)$, in which

$$U^*\left(\alpha, \beta, Z_1, \frac{j}{k}\right) = \frac{1}{\sqrt{k}} \int_0^{t_j} \mathcal{V}_\alpha(Z|s)^{-1/2} dU(\beta, Z_1, s), \quad j = 1, \ldots, k,$$

where $t_j = \bar{N}^{-1}(j)$. This process is only defined on k equispaced points of the interval $(0,1]$ and, again, we extend our definition to the whole interval so that, for $u \in [j/k, (j+1)/k]$ we can write $U^*(\alpha, \beta, Z_1, u)$ as

$$U^*\left(\alpha, \beta, Z_1, \frac{j}{k}\right) + \{uk - j\}\left\{U^*\left(\alpha, \beta, Z_1, \frac{j+1}{k}\right)\right.$$
$$\left. -U^*\left(\alpha, \beta, Z_1, \frac{j}{k}\right)\right\}.$$

As n goes to infinity, under the usual Breslow and Crowley conditions, in which j and k increase at the same rate, i.e., $j \approx kt$ where $0 < t < 1$ and by the symbol " $\approx$ " we mean that we round up to the closest integer. In the same way as for the univariate case we can be even less restrictive only requiring that the ratio j/k approaches some constant t strictly between 0 and 1. We then have that, for each j $(j = 1, \ldots, k - 1)$, $U^*(\alpha, \beta, Z_1, j/k)$ converges in distribution to a Gaussian process with mean zero and variance equal to t.

Some further remarks on the notation

The notation $U^*(\alpha, \beta, Z_1, u)$ is a little heavy but becomes even heavier if we wish to treat the situation in great generality. The first component of Z is Z_1 but of course this can be any component. Indeed Z_1 can itself be a vector, some collection of components of Z and, once we see the basic idea, it is clear what to do even though the notation starts to become slightly cumbersome. We prefer to leave those difficulties to the programming stage. Here we are essentially interested in concepts. As for the notation, $U^*(\alpha, \beta, u)$, in which there is only one Z and no need to specify it, the * symbol continues to indicate standardization by the variance and number of failure points. For the multivariate situation the two parameters, α and β, are themselves, both vectors. The parameter α which indexes the variance will be, in practice, the estimated full vector β, i.e., $\hat{\beta}$. Note that, as for the process $U^*(\hat{\beta}, \beta, u)$ we use, for the first argument to this function, the unrestricted estimate. Exactly the same applies here. In the numerator however, under some hypothesis for β_1, say $\beta_1 = \beta_{10}$ then, for the increments $dU(\beta, Z_1, s)$, we would have β_1 fixed at β_{10} and the other components of the vector β replaced by their restricted estimates, i.e., zeros of the estimating equations in which $\beta_1 = \beta_{10}$.

Some properties of $U^*(\alpha, \beta, Z_1, u)$

The same range of possible tests as before can be based on $U^*(\alpha, \beta, Z_1, u)$. To support this it is worth noting three lemmas analogous to those of our initial development on the univariate case.

Lemma 8.9 *The process $U^*(\alpha, \beta, Z_1, u)$, for all finite α and β is continuous on [0,1]. Also $E\, U^*(\beta, \beta, Z_1, 0) = 0$.*

Lemma 8.10 *Under model 6.3 $E\, U^*(\hat{\beta}, \beta, Z_1, u)$ converges in probability to zero.*

Lemma 8.11 *Suppose that $v < u$, then $\mathrm{Cov}\,\{U^*(\hat{\beta}, \beta, Z_1, v)\,;U^*(\hat{\beta}, \beta, Z_1, u)\}$ converges in probability to v.*

Since the increments of the process are asymptotically independent we can treat $U^*(\hat{\beta}, \beta, Z_1, u)$ (as well as $U^*(\beta, \beta, Z_1, u)$ under some hypothesized β) as though it were Brownian motion.

Tests in the multivariate setting

When we carry out a test of $H_0 : \beta_1 = \beta_{10}$ it is important to keep in mind the alternative hypothesis which is usually $H_1 : \beta_1 \neq \beta_{10}$ together with β_j, $j \neq 1$ unspecified. Such a test can be carried out using $U^*(\hat{\beta}, \beta, Z_1, u)$ where, for the second argument β, the component β_1 is replaced by β_{10} and the other components by estimates with the constraint that β_1 is fixed at β_{10}. Assuming our model is correct, or a good enough approximation, then we are testing for the effects of Z_1 having "accounted for" the effects of the other covariates. The somewhat imprecise notion "having accounted for" is made precise in the context of a model. It is not, of course, the same test as that based on a model with only Z_1 included as a covariate.

Another situation of interest in the multivariate setting is one where we wish to test simultaneously for more than one effect. This situation can come under one of two headings. The first, analogous to an analysis of variance, is where we wish to see if there exists any effect without being particularly concerned about which component or components of the vector Z may be causing the effect. As for an analysis of variance if we reject the global null we would probably wish to investigate further to determine which of the components appears to be the cause. The second is where we use, for the sake of argument, two covariates to represent a single entity, for instance three levels of treatment. Testing for whether or not treatment has an impact would require us to simultaneously consider the two covariates defining the groups. We would then consider, for a two-variable model, $\mathcal{Z}(t)$ as a vector with components $\mathcal{Z}_1(t)$ and $\mathcal{Z}_2(t)$, step functions with discontinuities at the points X_i, $i = 1, \ldots, n$, where they take the values $Z_{1i}(X_i)$ and $Z_{2i}(X_i)$ respectively. For this two dimensional case we consider the increments in the one-dimensional process $U(\beta, s)$ where $\beta' = (\beta_1', \beta_2')$ and

$$U(\beta, t) = \int_0^t \{\beta_1 \mathcal{Z}_1(s) + \beta_2 \mathcal{Z}_2(s)\} \, d\bar{N}(s)$$

at $t = X_i$, having mean $\beta_1 E(Z_1|X_i) + \beta_2 E(Z_2|X_i)$ and variance, $\beta_1^2 V(Z_1|X_i) + \beta_2^2 V(Z_2|X_i) + 2\beta_1\beta_2 \text{Cov}(Z_1, Z_2|X_i)$. The remaining steps now follow through just as in the single variable case, β_1 and β_2 being replaced by $\hat{\beta}_1$ and $\hat{\beta}_2$, respectively, and the conditional expectations, variances and covariances being replaced using analogous results to those that we obtained for the single variable case. We can then consider a standardized process, analogous to the single variable case and where $U^*(\alpha, \beta, u), (0 < u < 1)$ is defined as in 8.2 but, instead of 2.13 we have;

$$U^*\left(\alpha, \beta, \frac{j}{k}\right) = \frac{1}{\sqrt{k}} \int_0^{t_j} \mathcal{V}_\alpha(\beta_1 Z_1 + \beta_2 Z_2 | s)^{-1/2} dU(\beta, s), \quad j = 1, \ldots, k.$$
(8.12)

In practice, $\beta = (\beta_1, \beta_2)$ is replaced by consistent estimates or values under some hypothesis. It should now be clear how to deal with yet higher dimensions. It would also be possible to consider functions other than simple ones of the covariate Z. The linear combination $\beta_1 Z_1 + \beta_2 Z_2$, corresponding to the prognostic index (in practice the βs are replaced by estimates), appears to be the most appropriate for testing the combined prognostic effect. However, any other combination would also be valid, implying, according to the combination that is chosen, a slightly different direction in the two dimensional space (Z_1, Z_2) in which we look. If we are more interested in partial effects, testing for instance the effect of Z_1, after having accounted for the effects of Z_2, then we should consider the standardized process of $\mathcal{Z}_1(t)$ and any of the various tests (distance from origin test, area under the curve test and so on) that we can associate with it.

8.7   Graphical representation of regression effects

In linear regression, before we carry out any formal analyses, we will routinely obtain a rough visual idea of the plausibility of there being nonzero effects, as well as to the possible nature of any nonzero effects. This is via the usual scattergram. Such graphical information provides a useful support to our intuition. Although the case of linear regression is possibly the most transparent, graphical support can also be valuable for more complex models, notably the proportional and non-proportional hazards models. We can use the standardized score process to provide some pointers as to the nature of regression effects. The impression gained from the "eyeball factor," although not enough on its own, can provide a lot of insight. In the most straightforward case, we would have a binary Z representing two groups and the Kaplan-Meier plot, perhaps also including rough standard errors, is as good a graphical guide as we have. It can nonetheless be given

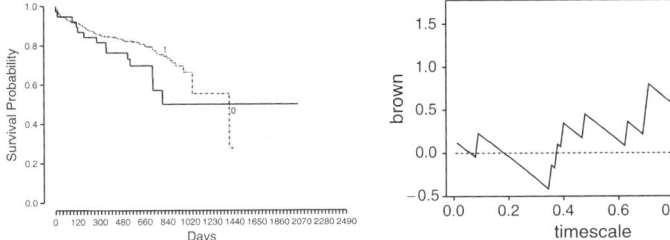

Figure 8.3: Kaplan-Meier plot of a 2-group comparison alongside standardized cumulative score plot suggesting a possibility of a delayed effect.

a supplement. Also, for more general situations, Kaplan-Meier plots may not be available or sufficiently reliable, either because of too few events in some of the subgroups or because of continuous covariates that we would need make discrete, or some combination of the two. We examine survival plots on the basis of covariate information in a later chapter and, for now, we limit attention to plots which tell us something about the test statistics.

Figure 8.3 illustrates part of an analysis stemming from a study in leukemia in which two groups were created on the basis of a tumor marker. There is a suggestion from the Kaplan-Meier plot alone that the survival probabilities may differ between the groups. The difference does not achieve statistical significance on the basis of a partial likelihood score test (which all but coincides with a distance from origin Brownian motion test) and so we would most likely conclude that the observed differences are explicable in terms of random variation. This may be the case. Even so, it is worth plotting the standardized score process shown in the same figure. A glance at this indicates at least one other potential explanation.

Until just beyond the halfway point in the study there appears to be no effect at all. Beyond this the effect may be quite strong, the process climbing close to linearly (looking like Brownian motion with drift) for the remainder of the study. Now, a model with a delayed effect, stipulating that for the first half of the study there is no effect, does in fact achieve statistical significance at the 5% level. The corresponding Kaplan-Meier estimates tell the same story although closer scrutiny of the figure is called for. The observations appear to be compatible with a population model for the survival curves where, for up until close to the marginal median, the curves coincide. Beyond this point the curves may diverge.

As for any regression situation, or indeed any statistical analysis, great caution is called for when we retrospectively attempt to explain the observations. This is the reason we would usually prefer our group of hypotheses to be well identified in advance. The explanation of the above data is plausible but can also easily represent some particular outcome from the set of outcomes that arise when effects are entirely absent. Usually, any such data driven hypothesis lacks enough weight on its own to be convincing. Its weight increases whenever backed up by other external arguments. If, as an example, it is very likely, for clinical reasons, that the two groups identified by the tumor marker, would either show no difference at all or, if there should be any differences, then these would not manifest themselves for some time after the beginning of observation, then this, in conjunction with the statistical results, would be more compelling. Whatever our final conclusion it is clear that the graphical representation adds a helpful angle to the analysis.

In a similar study in leukemia, using a different tumor marker, survival experience is shown in Figure 8.4. As for the above example, the usual partial likelihood score test, and the distance from origin test failed to achieve statistical significance. The two tests gave near identical results. In contrast to the above example there is some evidence that there may be early effects, the Kaplan-Meier plots diverging very noticeably initially, but later coming together and, finally, crossing. Although there is much less data later on, and more noise in the observations, it does not seem unreasonable to suspect that there may be differences between the groups but that these differences are of a non-proportional hazards nature. If that were so then we anticipate

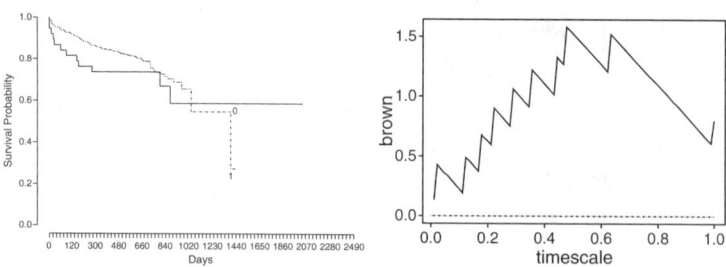

Figure 8.4: Kaplan-Meier plot of a two-group comparison together with standardized cumulative score plot suggesting a possibility of an effect which fails to maintain its presence throughout the study.

that an area under the curve test would be more sensitive than a distance from origin test. However, it turns out that this test also fails to achieve statistical significance. Other tests could of course be attempted but the cautionary note to the first example here becomes even more relevant if we continue to try different tests. In practice, in as much as is possible, we should try to determine in advance the most appropriate test to use.

Estimating the regression function $\beta(t)$

The plots of the standardized score process with time can provide insight into the nature of the regression function $\beta(t)$. If $\beta(t) = 0$ at all time points t then, as argued in the previous sections, the standardized score process will look like a sample of Brownian motion. If we are under the proportional hazards model and $\beta(t) = \beta_0$ then the process will tend to look like Brownian motion with drift, i.e., Brownian motion but, instead of being centered around zero, it is centered around αt where α would relate linearly to β_0. The constant relating α to β_0 would depend mostly on sample size. In any event, for given data, we can approximate this dependence parameter by a constant for that data set. The process, under the proportional hazards alternative, would look approximately linear with some slope.

Next, imagine the following situation. For $t < \tau$, the model coefficient is given by β_0^- and, for $t \geq \tau$, the model coefficient becomes β_0^+. The model is a particularly simple non-proportional hazards model. Suppose that $\beta_0^+ = \beta_0^-/2$. Then, the slope of the standardized score process for $t \geq \tau$ will be approximately one half that of the slope for the process when $t < \tau$. The observed process will manifest this and, having observed two distinct slopes for the standardized score process we could reparameterize the model whereby we write $\beta_0^- = 2\beta_0^+ = \beta_0$. This is even readily achieved using standard software provided the inclusion of time dependent covariates is allowed. In this case we would write a model with time-dependent effects in an equivalent form of a model with time-dependent covariates so that

$$\lambda(t|Z) = \lambda_0(t) \exp[\beta(t)Z] = \lambda_0(t) \exp[\beta_0^- I(t < \tau)Z + \beta_0^+ I(t \geq \tau)Z]$$
$$\equiv \lambda_0(t) \exp[\beta_0 Z(t)] = \lambda_0(t) \exp[\beta_0 \{I(t < \tau)Z + I(t \geq \tau)Z/2\}].$$

In practical examples the ratio of β_0^+ to β_0^- can of course take on any other value. Using a plot as a visual guide we can map out, at least as a first approximation, the function $\beta(t)$ in the simple case just described. However, it is clear that the argument extends immediately to any number of slopes, and not only two, so that an analysis of the standardized score process can give us a lot of information on the unknown function $\beta(t)$. Once we have this function we can consider the standardized score function for the model $\beta(t) = \hat{\beta}(t)$ where $\hat{\beta}(t)$ has been obtained in the above way. A visual inspection of a Brownian bridge process of the transformed standardized score function under $\beta(t) = \hat{\beta}(t)$ can indicate if the fit is adequate.

8.8   Operating characteristics of tests

A thorough study of the size of the tests and how well the large sample approximations work, at the finite samples we commonly work with, remains to be carried out. For large samples we know that we will control correctly for size since the tests all arise as functions of the same process. This process converges in distribution to Brownian motion and the probability model for this is known.

Detailed power studies also remain to be carried out. The number of potential situations is infinitely large and, unlike say the comparison of proportions where we can do our calculations at some values and interpolate for those remaining, only broadly categorized cases can be considered. Nonetheless, it is possible to draw up some general guidelines indicating which tests make the best choice given the kind of situation we believe we may be dealing with.

Under a proportional hazards departure to the null hypothesis of absence of effect the process $U^*(\hat{\beta}, 0, t)$ will look like Brownian motion with drift. A distance from origin test, at time $t = 1$, will have good power. Any other value of t would make a potential test but, since the distance from the origin is increasing at a rate of order t and the standard deviation of $U^*(\hat{\beta}, 0, t)$ is of order $\sqrt{t}$, we maximize power by taking the largest possible value of t. Other contending tests are the partial likelihood score test and $U^*(0, 0, t)$ both of which, on average, will be very slightly less powerful.

There are situations in which we might anticipate an advantage for one group in the early part of the study and a long-term advantage for the other group later in the study. This corresponds to the so-called crossing hazards problem. Two examples come to mind. The first is where surgery may be curative but will have an initial greatly increased hazard associated with the risks of the procedure. Some selection mechanism takes place and those patients who overcome this increased early hazard can then go on to benefit from the curative properties of the procedure. Standard tests such as the partial likelihood score test or the distance from origin test may fail to detect the presence of effects. It can happen that the process $U^*(\hat{\beta}, 0, t)$ increases up to a maximum and then, as the regression effect changes direction, the process declines and may end up very close to zero. A second, more complex, example occurs in marrow transplantation studies. The prognostic factor "graft versus host disease" indicates additional early risk for those patients suffering from this condition. Later on, however, conditionally upon having survived that added risk, the graft versus host effect can be indicating a graft versus leukemia effect, so that those patients have a better chance of aggressively opposing any residual disease.

For the crossing hazards problem the distance test at time $t = 1$ will typically have little or no power. A test based on the maximum of the process $U^*(\hat{\beta}, 0, 1)$, i.e., $\sup_{u \in (0,1)} U^*(\hat{\beta}, 0, u)$ will have good power against this alternative. Also the maximum of the Brownian bridge, which will look much the same if $U^*(\hat{\beta}, 0, 1)$ is close to zero, can be used to test for this specific alternative. Reflected Brownian motion, reflected at a point somewhere in the neighborhood of the maximum will maintain good power against a crossing hazard alternative. Here, the reversal of the sign, at that point, means that the process continues to grow. Under the null hypothesis of no effect, reflected Brownian motion will be the same as Brownian motion and, under the alternative, if the point of reflection corresponds to a change point in the direction of regression effect, then it will look like Brownian motion with drift. In this case then, a test based on distance from the origin at time $t = 1$ will have good power.

A common, possibly the most commonly encountered, alternative to the null hypothesis of no effect is that of an effect which declines with time. If, at some time τ we have reason to believe that the effect has mostly faded, then it might pay to use a distance from origin test at time $t = \tau$ rather than at $t = 1$. If we anticipate that effects, whenever present, are likely to be of the form where they decline through time, then an area under the curve test at time $t = 1$ is likely to be more powerful than a distance from origin test at $t = 1$. For alternatives that may be of a proportional hazards nature or may be of a declining regression effect nature the area under the curve test is still likely to be the most powerful. Indeed, when the alternative is of a proportional hazards nature, the area under the curve test loses relatively little power when compared with the distance from origin test. The area under the curve test is the recommended test in general situations, i.e., those where the alternative can be one with a constant regression effect, a declining regression effect or even an inversion of the regression effect. It is likely to not perform well when regression effect is delayed or increases with time. Such cases are quite rare.

8.9   Goodness-of-fit tests

In this book we take the view that there is nothing very special about goodness-of-fit tests. It is for this reason that the topic is relegated to the status of a section, following the presentation of tests, rather than being given the status of a chapter. We place these kind of tests under the same heading as all of the tests described above. The only real difference is that, instead of testing a hypothesis of the type H_0 : $\beta = \beta_0$, for some β_0 given in advance, we test a data driven hypothesis of the type $H_0 : \beta = \hat{\beta}$. We could also view this as studying some random aspect of the data while conditioning upon the observation $\hat{\beta}$. In this case $\hat{\beta}$ would not be known in advance and so, logically, the goodness-of-fit test follows the usual estimation and testing. This is what is done routinely.

Recall from the above discussion on operating characteristics that we do not expect all tests to perform comparably. Performance will depend on the reality of the situation being faced and, in some cases, certain tests will have high power while others may have power close to zero. The same, of course, then holds for a goodness-of-fit test. Since our estimating techniques will either obtain $\hat{\beta}$ as the solution to $U^*(\hat{\beta}, \beta_0, 1) = 0$, or, depending on how we standardize, to something very close to zero, then a goodness of fit test of the hypothesis $H_0 : \beta = \hat{\beta}$ could be expected to have power zero or power close to zero if based on the distance from the origin test at time $t = 1$. We must look for something else. A good candidate is the Brownian bridge test. A maximum distance from the origin test would also be worthy of study. This is because of the type of behavior we might anticipate under departures from the proportional hazards assumption. The most common departure we would expect to see would be one where effects decline through time, in an extreme case perhaps even changing direction (see example of Stablein et al. below). In such cases we would still expect, at least for an initial part of the study, the standardized cumulative score to behave as under a simple departure from the null of no effects, increasing (or decreasing) steadily. At some point this increase (or decrease) would either halt or, if effects change direction) make its way back toward the origin. A test based on the maximum would have the ability to pick up this kind of behavior. Also, of course, since it is helpful to make use of the cumulative standardized score process, as a visual aid, such behavior would be suggested by the graphs.

A first example is given by the well-known Freireich data, presented in Cox's famous 1972 paper. These data have been examined in detail by many authors and the evidence suggests the proportional hazards approximation to be a very good one. The bridged process is shown in Figure 8.5 and it can be seen to remain well within the 95% bounds shown in the same figure. Given this, alongside an absence of behavior suggesting the presence of any long-term effects which change direction, i.e., any gradual increase or decrease followed by a plateau or a reversal in direction (which under the bridge transformation would look like an increase or decrease followed by a return to the origin), we can conclude that our chosen model is not unreasonable. A larger study in breast cancer, carried out at the Curie Institute, indicated, among other effects, the strong prognostic impact of the variables stage and grade. For a model with both variables included, a bridged process, shown in Figure 8.6, tells us that a two-dimensional proportional hazards model for these data does not provide a good fit. However, if we relax assumptions a little and build a two dimensional model whereby only the stage is included as a proportional hazards effect, the grade

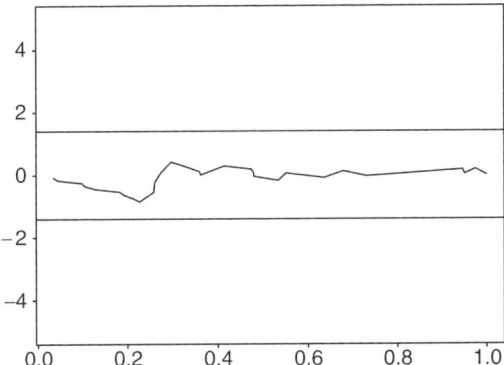

Figure 8.5: A bridged standardized cumulative processes for the Freireich data.

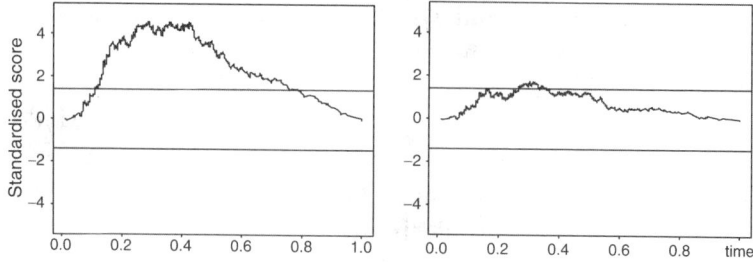

Figure 8.6: A bridged standardized cumulative processes for the Curie breast cancer data. Full PH model (left) and partial PH model (right).

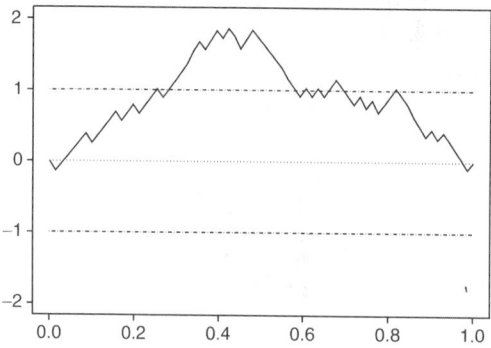

Figure 8.7: A bridged standardized cumulative processes for the Stablein data.

being allowed to exhibit non-proportional hazards behavior, then the fit is very much improved. In both cases the departure can be judged significantly different from that we anticipate for data generated under the fitted model. Even so, given the large sample size, and that the test statistic is only marginally significant, we may judge the fit good enough in practice.

We can still work further to improving fit if we wish. Other quantities, described later, will help us here. These are the goodness-of-fit coefficient and the predictability of the model. A further example is presented in Figure 8.7. The bridged cumulative process corresponds to the data of Stablein et al. (1981). The example is interesting since, in advance of analysis, we might suspect a proportional hazards assumption to fail. There are two groups, surgery and chemotherapy. Whereas the first presents a greater possibility of cure, it is associated with an earlier, higher, hazard rate due to the procedure itself. This example has been studied by a number of authors and comes under the classic heading of the crossing hazards problem. The evidence of a poor fit of a proportional hazards model can be seen almost right away by an inspection of the figure. A test of the null hypothesis of absence of effect fails to achieve significance when we assume a proportional hazards model. Yet there is clear evidence of effects, although the effects are not of a proportional hazards nature.

Finally, the rather limited nature of a goodness-of-fit test should be underlined. For modest and small samples it is often the case that even quite severe departures from model assumptions are not easily detected. For large samples the opposite can occur whereby we are able to detect departures of an arguably inconsequential nature. This alone can argue in favor of a graphical type test where we may pay less attention to, for example, the greatest distance achieved by a process approximating a Brownian bridge than to the overall impression given by the curve. Aside from the greatest distance reached, a curve which crosses several times the time axis indicates a more satisfactory overall fit than one which drifts away from the origin to only return at the end of the observation period. In later chapters (those dealing with explained variation and explained randomness) we consider an approach to getting around the sample size dependence of goodness-of-fit tests. The idea is to deal with population quantities that can be estimated. These provide an index lying between zero and one, a large value indicating a satisfactory fit, a small value an inadequate fit. For small observed values we may judge it worthwhile to investigate some relaxation of the proportional hazards assumption in such a way that a re-calculated index would be much larger.

8.10 Exercises and class projects

1. Carry out integration in Equation 8.1 and express $U(\beta, t)$ as a discrete sum. Also write down $U^*(\alpha, \beta, j/k)$ as a discrete sum.

2. Show that $U^*(\alpha, \beta, u)$, viewed as a function of u, is continuous in u.

3. Describe the reasons for the construction $U^*(\alpha, \beta, u)$ having two parameters, α and β, rather than a construction in which the parametric dependence is expressed through a single parameter, i.e., $\alpha = \beta$.

4. Using a data set of your own, obtain a 95% confidence interval for β, $(\hat{\beta}^-, \hat{\beta}^+)$ on the basis of distance from origin tests at time $t = 1$.

5. For the same data set as in the previous question, calculate 95% confidence intervals using (i) the arcsine test and (ii) the area under the curve test.

6. Show how a linearly weighted test can be used to categorize an area under the curve test. Determine the correct weights for the categorization to be valid.

7. Plot Var $D(\theta, t)$ as a function of θ and show that at $\theta = 0$ and $\theta = 1$ the variance expression is in agreement with those for Brownian motion and integrated Brownian motion.

8. For the Freireich data calculate the p-values for all of the different tests. For the linear combination test use the values; $\theta = 0.3$, $\theta = 0.5$, and $\theta = 0.7$. For reflected Brownian motion use the values $r = 0.2$, $r = 0.8$, and $r = 1.4$. Comment on the results.

9. Using several data sets compare the results for the distance from origin test at $t = 1$ and the partial likelihood score test. Compare the findings. Construct independent or conditionally independent censoring mechanisms that you expect should impact the two tests differently. On the basis of these findings suggest situations in which it is preferable to use one test over the other.

10. Under which circumstances would you anticipate that there may be non-negligible observed differences between the distance from origin test at time $t = 1$ and the partial likelihood score test. Under which circumstances would you anticipate that the two tests would coincide?

11. As a corollary to the previous question, can you construct a situation in which you would expect the distance from origin at time $t = 1$ to have greater power than the partial likelihood score test?

12. In the case of two explanatory variables, Z_1 and Z_2, describe how you would set up a test to test for (i) an overall effect, (ii) the effect of Z_1 given some specific value of Z_2, and (iii) the effect of Z_1 given Z_2 but without specifying the value of Z_2.

13. Describe two group situations in which you would anticipate certain tests performing well and others performing poorly.

14. For the linearly weighted test statistic the choice $K(t)$ being the number at risk at time t (analogous to Gehan 1965) and the choice $K(t) = \hat{S}(t)$, the marginal survival at time t, lead to tests with different operating characteristics. How do you anticipate different censoring mechanism affecting these two tests? If one of the goals in selecting a suitable test is that any dependence on censoring should diminish with increasing sample size, then which test would you choose?

15. Simulate survival data for the two-group problem. For one group the hazard rate is constant and equal to 1. For the other group the hazard rate is equal to 1.7 and changes to the value 0.3 at time τ in the study. Take the total sample size to be 100, and for the censoring to be governed by the same mechanism as survival, so that approximately there will be 50% censoring. Estimate the power of the various tests when $\tau = 0.3$, $\tau = 1.0$, and $\tau = 1.4$.

16. Repeat the calculations of the above question with same sizes equal to 50, 200, and 300. Study also the case when the hazard rate changes from 1.4 to 0.

17. Simulate survival data for the two-group problem. Again, for one group, the hazard rate is constant and equal to 1. For the other group, consider the following situations: (i) the hazard rate is equal to 2.0, (ii) the hazard rate is equal to 2.0 for $t < \tau$, after which time the hazard rate drops in value to 1.0. Take τ to be 0.25, 0.75 and 1.0, and sample size to be 1000 with no censoring. Use graphics to provide a graphical representation of regression effects.

18. For the simulation of the previous question, estimate β and use a Brownian bridge process to carry out a test of the null hypothesis $H_0 \beta = \hat{\beta}$, i.e., a goodness-of-fit test. Comment on your findings under the different situations.

19. For the case of two groups with covariates Z_1 and Z_2 we could study a process of Z_1 arising from a model in which Z_1 occurs on its own or from a process in which the two variables are included and we study the marginal process based on Z_1. Consider the various tests

outlined in this chapter and, when applied to the two situations just described, explain how the interpretation differs.

20. Consider once more the case of two covariates Z_1 and Z_2 included in a two-dimensional model. Find an example (either practical or theoretical) in which the process for the prognostic score $\hat{\beta}_1 Z_1 + \hat{\beta}_2 Z_2$ increases linearly with time indicative of a good model fit but that both of the marginal processes for Z_1 and Z_2 fail tests of fit. How do you explain this phenomenon?

21. As above, simulate survival data for the two-group problem. For the first group, the hazard rate is constant and equal to 1. For the other group, the hazard rate is equal to 2.0 for $t < \tau$, after which time the hazard rate drops in value to 1.0. Take τ to be 0.25, 0.75 and 1.0, and sample size to be 1000 with no censoring. Fit a proportional hazards model and use graphics to provide a graphical representation of regression effects. Show a bridge plot indicating a poor fit. Next, using the observed process, and the observed relationships between the slopes describing the effect, redefine a time-dependent $\beta(t)$. Rework a bridged process and note the improvement in fit.

Chapter 9

Inference: Likelihood

9.1 Summary

Likelihood solutions for parametric survival models are described. These are relatively straightforward. The usual likelihood procedures for inference based on large samples can be applied. It is also possible to construct inference based on the derivative of the log-likelihood together with functions of Brownian motion. This approach is then close to that of the previous chapter. For the fully parametric approach the exponential model is particularly simple. The partial likelihood can be viewed in various ways: (1) as one term of a product, the other term containing little information on the unknown parameter, (2) as an approximation arising from use of the main theorem (Section 7.4), (3) as a marginal likelihood of the ranks, and (4) as a profile likelihood. Other techniques leaning on likelihood, such as conditioning on ancillary statistics as well as Bayesian inference, are highlighted.

9.2 Motivation

For almost any statistical model, use of the likelihood is usually the chosen method for dealing with inference on unknown parameters. Bayesian inference, in which prior information is available, can be viewed as a broadening of the approach and, aside from the prior, it is again the likelihood function that will be used to estimate parameters and carry out tests. Maximum likelihood estimates, broadly speaking, have good properties and, for exponential families, a class to which our models either belong or are close, we can even claim

267

some optimality. A useful property of maximum likelihood estimators of some parameter is that the maximum likelihood estimator of some monotonic function of the parameter is the same monotonic function of the maximum likelihood estimator of the parameter itself. Survival functions themselves come under this heading and so, once we have estimated parameters that provide the hazard rate, then we immediately have estimates of survival. Variance expressions are also obtained quite easily, either directly or by the approximation techniques commonly applied. Keeping in mind that our purpose is to make inference on the unknown regression coefficients, invariant to monotonic increasing transformations on T, we might also consider lesser used likelihood approaches such as marginal likelihood and conditional likelihood. It can be seen that these kind of approaches lead to the so-called partial likelihood. In practice we will treat the partial likelihood as though it were any regular likelihood, the justification for this being possible through several different arguments.

9.3   Likelihood solution for parametric models

For fixed covariates and, in the presence of parametric assumptions concerning $\lambda_0(t)$, inference can be carried out on the basis of the following theorem that simply extends that of Theorem 5.1. We suppose that the survival distribution is completely specified via some parametric model, the parameter vector being say θ. A subset of θ is a vector of regression coefficients, β, to the covariates in the model. The usual working assumption is that of a conditionally independent censoring mechanism, i.e., the pair (T, C) are independent given Z. This would mean, for instance, that within any covariate grouping, T and C are independent but that C itself can depend on the covariate. Such dependence would generally induce a marginal dependency between C and T.

Theorem 9.1 *Under a conditionally independent censoring mechanism the log-likelihood* $\log L(\theta)$ *can be written* $\log L(\theta) = \sum_{i=1}^{n} \log L_i(\theta)$ *where*

$$\log L_i(\theta) = I(\delta_i = 1) \log f(x_i|z_i; \theta) + I(\delta_i = 0) \log S(x_i|z_i; \theta). \quad (9.1)$$

The maximum likelihood estimates obtain as the values of θ, denoted $\hat{\theta}$, that maximize $\log L(\theta)$ over the parameter space. For $\log L(\theta)$ a differentiable function of θ, this value is then the solution to the estimating equation $U(\theta) = 0$ where; $U(\theta) = \sum_i \partial \log L_i(\theta)/\partial\theta$. Next notice that, at the true value of θ, i.e., the value which supposedly generates the observations, denoted θ_0, we have $\mathrm{Var}(U(\theta_0)) = E\,U^2(\theta_0) = E\,I(\theta_0)$ where

$$I(\theta) = \sum_{i=1}^{n} I_i(\theta) = -\partial^2 \log L(\theta)/\partial\theta^2 = -\sum_{i=1}^{n} \partial^2 \log L_i(\theta)/\partial\theta^2.$$

As for likelihood in general, some care is needed in thinking about the meaning of these expressions and the fact that the operators $E(\cdot)$ and $\mathrm{Var}(\cdot)$ are taken with respect to the distribution of the pairs (x_i, δ_i) but with θ_0 fixed. The score equation is $U(\hat{\theta}) = 0$ and the large sample variance is approximated by $\mathrm{Var}(\hat{\theta}) \approx 1/I(\hat{\theta})$. Newton-Raphson iteration is set up by a simple application of the mean value theorem so that

$$\theta_{j+1} = \theta_j + I(\theta_j)^{-1}U(\theta_j), \quad j \geq 1, \tag{9.2}$$

where θ_1 is some starting value, often zero, to the iterative cycle. The iteration is brought to a halt once we achieve some desired level of precision. Note that likelihood theory would imply that we work with the expected information (called Fisher information) $E\{I(\theta)\}$ but in view of Efron and Hinkley (1978) and the practical difficulty of specifying the censoring we usually prefer to work with a quantity allowing us to consistently estimate the expected information, in particular the observed information.

Large sample inference can be based on any one of the three tests described in Chapter 2. For the score test there is no need to carry out parameter estimation or to maximize some function. Many well established tests can be derived in this way. In exponential families, also the so-called curved exponential families (Efron 1975), such tests reduce to contrasting some observed value to its expected value under the model. Good confidence intervals (see Cox and Hinkley 1974, page 212) can be constructed from "good" tests. For the exponential family class of distributions the likelihood ratio forms a uniformly most powerful test and, as such, qualifies as a "good" test in the sense of Cox and Hinkley. The other tests are asymptotically equivalent so that confidence intervals based on the above test procedures will agree as sample size increases. Also, we can use such intervals for other quantities of interest such as the survivorship function since this function depends on these unknown parameters.

Estimating the survival function for parametric models

We can estimate the survival function as $S(t; \hat{\theta})$. If Θ_α provides a $100(1 - \alpha)\%$ confidence region for the vector θ, then we can obtain a $100(1 - \alpha)\%$ confidence region for $S(t; \theta)$ in the following way. For each t let

$$S_\alpha^+(t; \hat{\theta}) = \sup_{\theta \in \Theta_\alpha} S(t; \theta), \quad S_\alpha^-(t; \hat{\theta}) = \inf_{\theta \in \Theta_\alpha} S(t; \theta), \qquad (9.3)$$

then $S_\alpha^+(t; \hat{\theta})$ and $S_\alpha^-(t; \hat{\theta})$ form the endpoints of the $100(1 - \alpha)\%$ confidence interval for $S(t; \theta)$. Such a quantity may not be so easy to calculate in general, simulating from Θ_α or subdividing the space being an effective way to approximate the interval. Some situations nonetheless simplify. The most straightforward is where the survival function is a monotonic function of the one dimensional parameter θ. As an illustration, the scalar location parameter, θ, for the exponential model corresponds to the mean. We have that $S(t; \theta)$ is monotonic in θ. For such cases it is only necessary to invert any interval for θ to obtain an interval with the same coverage properties for $S(t; \theta)$. Denoting the upper limit of the $100(1 - \alpha)\%$ confidence interval for θ as θ_α^+ and the lower limit of the $100(1 - \alpha)\%$ confidence interval for θ as θ_α^-, we can then write; $S_\alpha^+(t; \hat{\theta}) = S(t; \theta_\alpha^-)$ and $S_\alpha^-(t; \hat{\theta}) = S(t; \theta_\alpha^+)$. Note that these intervals are calculated under the assumption that t is fixed. For the exponential model, since the whole distribution is defined by θ, the confidence intervals calculated pointwise at each t also provide confidence bands for the whole distribution. If we wish to obtain confidence bands, valid for a range of values of t, then more work is needed.

9.4   Likelihood solution for exponential models

As for the case of a single group, an analysis based on the exponential model is particularly simple. For this reason alone it is of interest but also (see section below on the non-parametric exponential model) the results are much more general than is often supposed. We restrict attention to the two-group case in order to enhance readability. The two groups are defined by the binary covariate Z taking the value either zero or one. The extension to higher dimensions is all but immediate. For the two-group case we will only need to concern ourselves with two parameters, λ_1 and λ_2, which, once we have them or consistent estimates of them, we have the whole survival experience (or estimates of this) for both groups.

Expressing the model in proportional hazards form we can write; $\lambda_1 = \lambda$ and $\lambda_2 = \lambda \exp(\beta)$. Referring to Equation 9.1 then, if individual i corresponds to group 1, his or her contribution to the likelihood is $f(x_i; \lambda) = \lambda \exp(-\lambda x_i)$ when $\delta_i = 1$, whereas for $\delta_i = 0$, the contribution is $S(x_i; \lambda) = \exp(-\lambda x_i)$. If the individual belongs to group 2 the likelihood contribution would be either $\lambda \exp(\beta) \exp(-e^\beta \lambda x_i)$ or $\exp(-e^\beta \lambda x_i)$ according to whether δ_i is equal to one or zero. We use the variable $w_i = I(z_i = 1)$ to indicate which group the subject is from. From this we have:

Lemma 9.1 *For the 2-sample exponential model, the likelihood satisfies;*

$$\log L(\lambda, \beta) = k \log \lambda + \beta k_2 - \lambda \left\{ \sum_{j=1}^n x_j (1 - w_j) + e^\beta \sum_{j=1}^n x_j w_j \right\},$$

where there are k_1 distinct failures in group 1, k_2 in group 2, and $k = k_1 + k_2$. Differentiating the log-likelihood with respect to both λ and β and equating both partial derivatives to zero we readily obtain an analytic solution to the pair of equations. From this we have:

Lemma 9.2 *The maximum likelihood estimates $\hat{\beta}$ and $\hat{\lambda}$ for the two-group exponential model are written*

$$\hat{\beta} = \log \frac{k_2}{\sum_{j=1}^n x_j w_j} - \log \frac{k_1}{\sum_{j=1}^n x_j (1 - w_j)} ; \qquad \hat{\lambda} = \frac{k_1}{\sum_{j=1}^n x_j (1 - w_j)}.$$

It follows immediately that; $\hat{\lambda}_1 = \hat{\lambda}$ and that $\hat{\lambda}_2 = \hat{\lambda} \exp(\hat{\beta}) = k_2 / \sum_{j=1}^n x_j w_j$. In order to carry out tests and construct confidence intervals we construct the matrix of second derivatives of the log-likelihood, $I(\lambda, \beta)$, obtaining

$$\begin{cases} -\partial^2 \log L(\lambda, \beta)/\partial \lambda^2 & -\partial^2 \log L(\lambda, \beta)/\partial \lambda \partial \beta \\ -\partial^2 \log L(\lambda, \beta)/\partial \lambda \partial \beta & -\partial^2 \log L(\lambda, \beta)/\partial \beta^2 \end{cases} = \begin{cases} k/\lambda^2 & e^\beta \sum_j x_j w_j \\ e^\beta \sum_j x_j w_j & \lambda e^\beta \sum_j x_j w_j \end{cases}.$$

The advantage of the two parameter case is that the matrix can be explicitly inverted. We then have:

Lemma 9.3 *Let $D = \lambda^{-1} e^\beta \sum_j x_j w_j \{k - \lambda e^\beta \sum_j x_j w_j\}$. Then, for the two-group exponential model the inverse of the information matrix is given by*

$$I^{-1}(\lambda, \beta) = D^{-1} \begin{cases} \lambda e^\beta \sum_j x_j w_j & -e^\beta \sum_j x_j w_j \\ -e^\beta \sum_j x_j w_j & k/\lambda^2 \end{cases}.$$

The score test is given by $X_S^2 = U'(\hat{\lambda}, 0)I^{-1}(\hat{\lambda}, 0)U(\hat{\lambda})$. Following some simple calculations and recalling that $\exp(-\hat{\beta}) = \sum_{j=1}^{n} x_j w_j / k_2$, we have:

Lemma 9.4 *For the two-group exponential model the score test is given by*

$$X_S^2 = k^{-1} k_1 k_2 \exp(-\hat{\beta})\{1 - \exp(\hat{\beta})\}^2.$$

At first glance the above expression, involving as it does $\hat{\beta}$, might appear to contradict our contention that the score statistic does not require estimation of the parameter. There is no contradiction although we consider in the above expression λ to be a nuisance parameter not specified under the null hypothesis. This parameter value *does* require estimation, although still under the null. The regression parameter itself turns out to have a simple explicit form, and it is this same term that appears in the score statistic. Typically, for other models, the maximum likelihood estimate would not have an explicit analytic form. We do not need estimate it in order to evaluate the score statistic. On the other hand, both the Wald test and the likelihood ratio test do require estimation under the alternative. The calculations in this specific case can be carried out straightforwardly and we also have a relatively simple, and again explicit solution (i.e., not requiring the finding of an iterative solution to the likelihood equation) for the likelihood ratio test. We have then the following lemma:

Lemma 9.5 *For the two-group exponential model the likelihood ratio test X_L^2 is given by;*

$$X_L^2 = 2 \left(k_2 \log \frac{k_2}{\sum_j x_j w_j} + k_1 \log \frac{k_1}{\sum_j x_j (1 - w_j)} - k \log \frac{k}{\sum_j x_j} \right).$$

The third of the tests based on the likelihood, the Wald test, is also straightforward to calculate and we have the corresponding lemma:

Lemma 9.6 *For the two-group exponential model, the Wald test is given by;*

$$X_W^2 = k^{-1} k_1 k_2 \, \hat{\beta}^2.$$

For large samples we anticipate the three different tests to give very similar results. For smaller samples the Wald test, although the most commonly used, is generally considered to be the least robust. In particular, a monotonic transformation of the parameter will, typically, lead to a different value of the test statistic.

Exponential analysis of Freireich data

The maximum likelihood estimates of the hazard rates in each group are

$$\hat{\lambda}_1 = 9/359 = 0.025 , \quad \text{Var}(\hat{\lambda}_1) = 9/(359)^2 = 0.000070,$$
$$\hat{\lambda}_2 = 21/182 = 0.115 , \quad \text{Var}(\hat{\lambda}_2) = 21/(182)^2 = 0.00063.$$

We might note that the above results are those that we would have obtained had we used the exponential model separately in each of the groups. In this particular case then the model structure has not added anything or allowed us to achieve any greater precision in our analysis. The reason is simple. The exponential model only requires a single parameter. In the above model we have two groups and, allowing these to be parameterized by two parameters, the rate λ and the multiplicative factor $\exp(\beta)$, we have a saturated model. The saturated model is entirely equivalent to using two parameters, λ_1 and λ_2, in each of the two groups separately. More generally, for exponential models with many groups or with continuous covariates, or for other models, we will not usually obtain the same results from separate analyzes than those we obtain via the model structure. The model structure will, as long as it is not seriously misspecified, usually lead to inferential gains in terms of precision of parameter estimates.

Since $\exp(\hat{\beta}) = 21/182 \times 359/9 = 4.60$ we have that the estimate of the log relative risk parameter, $\hat{\beta}$ is 1.53. We also have that the score test, $X_S^2 = 17.8$, the Wald test $X_W^2 = 14.7$ and the likelihood ratio test $X_L^2 = 16.5$. The agreement between the test statistics is good and, in all cases, the significance level is sufficiently strong to enable us to conclude in favor of clear evidence of a difference between the groups.

Had there been no censoring then $k = n$, the sample size, and $\sum_{j=1}^{n} t_j$ corresponds to a sum of n independent random variables each exponential with parameter λ. We could therefore treat $n/\hat{\lambda}$ as a gamma variate with parameters (λ, n). In view of the consistency of $\hat{\lambda}$, when there is censoring, we can take $k/\hat{\lambda}$ as a gamma variate with parameters (λ, k), when $k < n$. This is not an exact result, since it hinges on a large sample approximation, but it may provide greater accuracy than the large sample normal approximation.

As described earlier, we can make use of standard tables by multiplying each term of the sum by 2λ. The result of this product is a sum of n exponential variates in which each component of the sum has variance equal to 2. This corresponds to a gamma $(2, n)$ distribution which is also equivalent to a chi-square distribution with $2n$ degrees of freedom. Taking the range of values of $2k\lambda/\hat{\lambda}$ to be between $\chi_{\alpha/2}$ and $\chi_{1-\alpha/2}$ gives a $100(1-\alpha)\%$ confidence interval for λ. For the Freireich data we obtained as a 95% $CI = (0.0115, 0.0439)$. On the basis of intervals for λ, we can obtain intervals for the survivorship function which is, in this particular case, a monotonic function of λ. The upper and lower limits of the $100(1-\alpha)\%$ confidence interval are denoted by $S_\alpha^+(t;\hat{\lambda})$ and $S_\alpha^-(t;\hat{\lambda})$ respectively. We write,

$$[S_\alpha^+(t;\hat{\lambda}), S_\alpha^-(t;\hat{\lambda})] = \left[\exp\left\{-\left(\frac{\hat{\lambda}\chi_{\alpha/2}}{2k}\right)t\right\}, \exp\left\{-\left(\frac{\hat{\lambda}\chi_{1-\alpha/2}}{2k}\right)t\right\}\right].$$
(9.4)

A different approximation was described in Chapter 5 in which

$$S_\alpha^+(t;\hat{\lambda}) \approx \exp\left\{-\left(\frac{\hat{\lambda}}{\sqrt{k}}\right)\left(\sqrt{k} - z_{1-\alpha/2}\right)t\right\},$$
(9.5)

where the corresponding expression for $S_\alpha^-(t;\hat{\lambda})$ obtains from using the percentiles, $z_{1-\alpha/2}$ by $z_{\alpha/2}$ of the standard normal distribution. Agreement between these two approximations appears to be very close. It would be of interest to have a more detailed comparisons between the approaches.

Non-parametric exponential analysis

For the one sample case we have already seen how, referring to Section 2.5 and Theorem 2.8, we are able to make use of the result that, for any continuous positive random variable T, with distribution function $F(t)$, the variate $\Lambda(T) = \int_0^T f(u)/[1 - F(u)]du$ has a standard exponential distribution. For the one sample case we can work with the empirical survival function, $\hat{S}(t)$ appealing to the result that the observations $-\log \hat{S}(X_i)$ can be taken to have been sampled from a standard exponential distribution.

   In the two-group and, by extension, many group case it will be necessary to transform observations from a group other than that giving rise to the group estimate, say $\hat{S}_G(t)$. To facilitate this, as described in Chapter 5, we work with the continuous version of the Kaplan-Meier estimate, $\bar{S}(t)$. Note that a two-group proportional hazards model can be expressed as; $S_2(t) = S_1^\alpha(t)$ where $\alpha = \exp(\beta)$. Taking logarithms then enables an analytic expression for β as;

$$\beta = \log\{-\log S_2(t)\} - \log\{-\log S_1(t)\}.$$
(9.6)

For the case of three groups, defined by the pair of binary indicator variables, Z_1 and Z_2, the model states that $S(t|Z_1, Z_2) = S_1^\alpha(t)$ where, in this more complex set-up, $\log \alpha = \beta_1 Z_1 + \beta_2 Z_2$. Here, in exactly the same way, we obtain analytic expressions for β_1 and β_2 as the arithmetic difference between the log-log transformations of the respective marginal survival curves.

For two independent groups G_1 and G_2 we can consider two separate estimators, $\hat{S}_1(t)$ and $\hat{S}_2(t)$. Since we are assuming a proportional hazards model, we will carry over this restriction to the sample based estimates whereby $\hat{S}_2(t) = \hat{S}_1^\alpha(t)$ and where, as before, $\alpha = \exp(\beta)$. In view of the above result we have:

Lemma 9.7 *A consistent estimator of β obtains from*

$$\hat{\beta} = \frac{1}{n_m} \sum_{i=1}^{n_m} \left[\log\{- \log \bar{S}_2(X_i)\} - \log\{- \log \bar{S}_1(X_i)\} \right], \qquad (9.7)$$

where $n_m = \sum_i I(X_i \leq m)$, $m_\ell = \max\{X_i | X_i \in G_\ell\}$, and $m = \min(m_1, m_2)$.

All of the simple results that are available to us when data are generated by an exponential distribution can be used. In particular, if we wish to compare the means of two distributions, both subject to censoring, then we can transform one of them to standard exponential via its empirical survival function, then use this same transformation on the other group. The simple results for contrasting two censored exponential samples can then be applied even though, at least initially, the data arose from samples generated by some other mechanism.

9.5 Semi-parametric likelihood solution

Recalling that the observed data are the triplets $\{Z_i(t), Y_i(t), X_i ; i = 1, \ldots, n\}$, we might view any problem of interest in survival analysis to be summarized by the joint density of (T, Z), let's say $a_\theta(t, z)$. The marginal covariate distribution for Z is typically fixed by design and all of our models express some relation for the conditional distribution of T given Z. We are then able to write that $a_\theta(t, z) = f_\theta(t|z)g(z)$. This is straightforward in the parametric case and corresponds to what is described above. We limit our attention to $f_\theta(t|z)$ alone because we assume that $g(z)$ does not depend on any of the unknown parameters of interest to us.

In the log-likelihood expression, $g(z)$ would then appear as a constant and disappear upon differentiation with respect to the model parameters. Instead of considering $f_\theta(t|z)$ though, we could also write $a_\theta(t, z) = h_\theta(z|t)f(t)$. The marginal quantity $f(t)$ will typically include some information on the unknown parameters. This follows as a consequence of our models postulating conditional dependence between the covariate and survival time, this dependence expressed via $f_\theta(t|z)$. Thus, applying the law of total probability, we will, in general, find some complicated form of dependence between $f(t) = \int f_\theta(t|z)g(z)dz$ and the unknown parameters. However, we might imagine, or even postulate, any such dependence to be either very weak or simply non existent. In place of a full likelihood based upon $a_\theta(t, z)$ we can work with just the conditional distributions based only on $h_\theta(z|t)$.

Constructing likelihoods based on $f_\theta(t|z)$ or $h_\theta(z|t)$, rather than the full density $a_\theta(t, z)$, can be described as leading to a *semi-parametric* solution. This is because any parametric dependence via $f(t)$ or $g(z)$ is either ignored or deemed not to exist. Furthermore, inference based on $f_\theta(t|z)$ is unaffected by monotonic increasing transformations of Z, and inference based on $h_\theta(z|t)$ is, in turn, unaffected by monotonic increasing transformations of T. The first is then considered to be *rank invariant* with respect to Z, wheras the second is *rank invariant* with respect to T. As long as the ranks remain the same, which is the case for arbitrary monotonic increasing transformations, then inference is unaffected. This is often felt to be something desirable in survival modeling where we have poor indicators concerning an appropriate choice for $f_\theta(t|z)$ and we believe that the essential information is contained in the observed ranks. Although, for the observations, we are not able to write down $h_\theta(Z_i|X_i)^{\delta_i}$ (because we do not know h) we can apply the main theorem (Section 7.4) leading to our best approximation of this likelihood.

Definition 9.1 *The product, $L\{\beta(t)\}$, of the discrete probabilities $\pi_i(\beta(t), X_i)$,*

$$\pi_i(\beta(t), t) = \frac{Y_i(t)\exp\{\beta(t)Z_i(t)\}}{\sum_{j=1}^n Y_j(t)\exp\{\beta(t)Z_j(t)\}},$$

over the observed failures is called the partial likelihood.

Conjecture 9.1 *The product, $L\{\beta(t)\}$, of the discrete probabilities $\pi_i(\beta(t), X_i)$ over the observed failures has the same properties as usual likelihood.*

The conjecture has been shown to hold in many particular cases (Cox 1975, Hill et al. 1990, Andersen et al 1993). Under (6.3), i.e. the constraint $\beta(t) = \beta$, the product of the π's over the observed failure times gives the likelihood described by Cox (1972, 1975). We return to the concept "partial likelihood" itself below, but note that in view of the main theorem (Section 7.4) and our estimating equations it is not in any sense crucial to the basic construction. The concept is difficult and, since it is not in any way essential to the development of estimating equations, we do not dwell upon it. For the purposes of this book, whenever we write the term "partial likelihood" we mean the above product without implying any deeper ideas relating to how the product arises. Strictly speaking the partial likelihood has only been motivated for time-independent $\beta(t)$, i.e., a constant, but the form is the same and, in the absence of a better name, it seems appropriate to refer to all such expressions as those of a partial likelihood.

The result is not of primary interest and we do not need it to justify our estimating equations. These arise naturally as moment estimators as a consequence of the main theorem (Section 7.4). The solutions to these estimating equations have good properties. It is of interest in as much as it is common to talk about partial likelihood and that we can take this to be defined as in the conjecture. The actual concept *partial likelihood* itself is discussed in the following paragraph. Taking the logarithm and the derivative with respect to β, we obtain the estimating function,

$$U(\beta) = \sum_{i=1}^{n} \delta_i \left\{ Z_i - \frac{\sum_{j=1}^{n} Y_j(X_i) Z_j \exp(\beta Z_j)}{\sum_{j=1}^{n} Y_j(X_i) \exp(\beta Z_j)} \right\} \tag{9.8}$$

that we refer to as a score function in view of its connection to a likelihood. The equation, upon setting equal to zero, can generally be solved without difficulty using the Newton-Raphson method, to obtain the maximum partial likelihood estimate (MPLE) $\hat{\beta}$ of β. To make inference about β, the simplest and most common approach is to treat $\hat{\beta}$ as asymptotically normally distributed with mean β and variance $I(\hat{\beta})^{-1}$, where $I(\beta)$, called the information in view of the analogy with classical likelihood, is minus the second derivative of $L(\beta)$ with respect to β, i.e., letting

$$I_i(\beta) = \frac{\sum_{j=1}^n Y_j(X_i) Z_j^2 \exp(\beta Z_j)}{\sum_{j=1}^n Y_j(X_i) \exp(\beta Z_j)} - \left\{ \frac{\sum_{j=1}^n Y_j(X_i) Z_j \exp(\beta Z_j)}{\sum_{j=1}^n Y_j(X_i) \exp(\beta Z_j)} \right\}^2 \quad (9.9)$$

then $I(\beta) = \sum_{i=1}^n \delta_i I_i(\beta)$. Inferences can also be based on likelihood ratio methods. A third possibility that is sometimes convenient, is to base tests on the score $U(\beta)$, which in large samples can be considered to be normally distributed with mean zero and variance $I(\beta)$. In the above expressions for $U(\beta)$ and $I(\beta)$ we have taken the covariable Z to not depend upon time. This is simply in order to keep the presentation uncluttered and, if Z depends on t, then, in all of the above expressions, we would have, for instance, $Z_j(X_i)$. Multivariate extensions are completely natural, with the score being a vector and I an information matrix. In this setting, i.e., parametric or semi-parametric likelihood, less emphasis is given to the role played by time and more to the the individuals themselves, indexed by i. The connections are nonetheless strong and note for instance that $I_i(\beta) = \mathcal{V}_\beta(Z|X_i)$.

Log-rank and associated test procedures

The usual univariate and multivariate likelihood testing procedures can be applied on the basis of $L(\beta)$, $U(\beta)$ and $I(\beta)$, enabling the testing of complex hypotheses as well as the construction of interval estimates. For non-proportional hazards models we simply replace β by $\beta(t)$. For stratified models, we calculate $L(\beta)$, $U(\beta)$ and $I(\beta)$ within each stratum, calling these for the sake of argument; $L_s(\beta)$, $U_s(\beta)$ and $I_s(\beta)$ for stratum s. We then sum across the strata.

One encouraging side to inference based on $U(\beta)$ is that a broad class of procedures, known as log-rank type tests, are recovered as special cases. These procedures were developed without any specific model in mind as a series of two by two tables, drawn up at each of the failure points. We can do this at each of the observations X_i, $i = 1, \ldots, n$ since for X_i where $C_i < T_i$, the contribution from the table will be zero. Table 9.1 summarizes the required information. In group 1 we observe a total of $m_1(i)$ failures. In the simplest cases, and

	Dead at time X_i	Alive after X_i	Totals
Group 1	$m_1(i)$	$n_1(i) - m_1(i)$	$n_1(i)$
Group 2	$m_2(i)$	$n_2(i) - m_2(i)$	$n_2(i)$
Totals	$m(i)$	$n(i) - m(i)$	$n(i)$

Table 9.1: Basic 2×2 survival table at time X_i.

where we split ties for the sake of clarity, $m_1(i)$ is a binary random variable $(0,1)$. The total number of subjects at risk at time X_i is $n(i)$, this being divided into $n_1(i)$ for group 1 and $n_2(i)$ for group 2.

Lemma 9.8 *Under the null hypothesis of independence between group indicator and survival status, and disallowing the possibility of ties, the random variable $m_1(i)$ is Bernoulli with mean $e_1(i)$ and variance $v(i)$, where*

$$e_1(i) = \frac{m(i)n_1(i)}{n(i)} \; ; \quad v(i) = m(i)\frac{n_1(i)n_2(i)}{n^2(i)} \frac{\{n(i) - m(i)\}}{\{n(i) - 1\}}.$$

In the presence of ties, i.e., allowing for $m(i)$ to be greater than one, the result for the mean and variance still holds since $m_1(i)$ is then a hypergeometric variable.

Definition 9.2 *The weighted log-rank test is based on approximating*

$$\left\{ \sum_i K_i^2 v(i) \right\}^{-1} \left\{ \sum_i K_i \{m_1(i) - e_1(i)\} \right\}^2$$

for some positive weights, K_i, by a chi-squared variate on one degree of freedom.

When there are no ties, the result is of interest in view of:

Lemma 9.9 *When $K_i = 1$ for all i, the log-rank test coincides with the score test based on $U(0)$ and $I(0)$.*

The first obvious generalization of the score test is then weighted score tests involving the weights, K_i. The choice $K_i = 1$ corresponds to the classic log-rank test. The choice $K_i = n(i)$ corresponds to the test proposed by Gehan (1965) and the choice $K_i = \hat{S}(X_i)$ corresponds to the choice proposed by Prentice (1978). Under the null hypothesis, regardless of choice of K, for large samples the test statistic will be closely approximated by a chi-square. Under an alternative closely approximated by proportional hazards, i.e., effects remain constant in time, then the choice $K_i = 1$ will maximize power against local alternatives. If early effects are greater than later effects, then either of the choices of Prentice (1978) or Gehan (1965) will, generally, lead to increased power. The optimal test to choose regarding power depends on the nature of the particular alternative. Choosing the weights of Prentice will maximize power when the decline in effect with t mirrors the decline in $S(t)$ with t. Choosing the Gehan weights maximizes power for a decline in effect that depends on the censoring and so this is more difficult to make general recommendations for. Stratified log-rank tests are obtained by summing the $m_1(i)$, the $e_1(i)$ and $v(i)$ across strata. Before such summing all of the calculations are carried out within any given stratum.

More elaborate log-rank type tests can be written down in much the same way, enabling tests of homogeneity for several groups, tests of trends, as well as adjusted tests. Since these are all obtainable as special cases of a score test based on the model, little additional insight is to be found in detailing these various calculations. They are straightforward and can be found for example in Hill et al. (1990).

9.6   Other likelihood expressions

The Bayesian view of a multi-parameter likelihood is simply that of a joint density for the parameters given the observations (having standardized the function so that the area under the curve is equal to one). Unless a prior distribution is degenerate, conclusions based on the Bayesian view and those based on the classic view coincide for large samples and so, for our discussion here, we take the Bayesian view since it is more transparent. A full likelihood includes all the parameters and we can take the maximum likelihood estimates to be the modes of the multi-parameter density. Other estimators would be available, the means, medians, for instance, all coinciding for large samples. That, however, is not so much our concern as that of making inferences for some of the parameters while, in some sense, accounting for others.

Conditional likelihood

A conditional likelihood, that we may like to view as a conditional density, simply fixes some parameters at certain values. An obvious example is that of a parameter that is a function of some variance, a quantity which indicates the precision in an estimate of other parameters but tells us nothing about where such parameters may be located. It has been often argued that it is appropriate to condition on such quantities. The variance parameter is then taken as fixed and known and the remaining likelihood is a conditional likelihood.

Cox (1962) gives an example of two instruments measuring the same quantity but having very different precision. His argument, which is entirely persuasive, says that we should use the conditional likelihood and not the full likelihood that involves the probabilities of having chosen one or the other instrument. The fact that we may have chosen the less precise instrument a given rate of the time is neither here nor there. All that matters are the observations and the particular instruments from which they arise. Another compelling example arises in binomial sampling. It is not at all uncommon that we would not know in advance the exact number n of repetitions of the experiment. However, we rarely would think of working with a full likelihood in which the distribution of n is explicitly expressed. Typically, we would simply use the actual value of n that we observed. This amounts to working with a conditional likelihood.

Fisher (1934) derived an exact expression for the distribution of the maximum likelihood estimate for a location or scale parameter conditional on observed spread in the data. Fisher's expression is particularly simple and corresponds to the likelihood function itself standardized so that the integral with respect to the parameter over the whole of the parameter space is equal to one. It is quite an extraordinary result in its generality and the fact that it is exact. Mostly we are happy to use large sample approximations based on central limit theorems for the score statistic and Taylor series approximations applied to a development around the true value of an unknown parameter. Here we have an exact result regardless of how small our sample is as long as we are prepared to accept the argument that it is appropriate to condition on the observed spread, the so-called "configuration" of the sample. For a model in which the parameter of interest is a location parameter the configuration of the sample is simple the set of distances between the observations and the empirical mean. For a location-scale family this set consists of these same quantities standardized by the square root of the variance.

Fisher's results were extended by Hinkley (1988) to the very broad class of models coming under the heading of curved exponential family models (Efron 1984). This extension was carried out for more general situations by conditioning on the observed information in the sample (a quantity analogous to the information contained in the set of standardized residuals). Although no longer an exact result the result turns out to be very accurate and has been extensively studied by Bandorff-Nielsen (1988). For the proportional hazards model we can use these

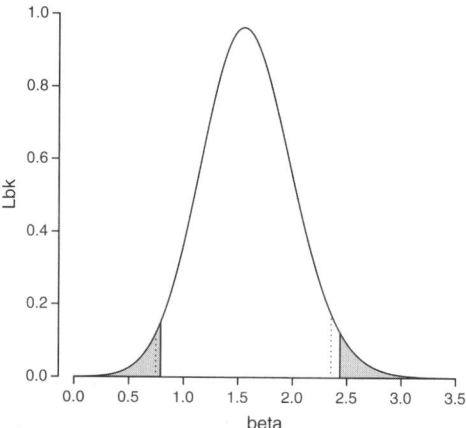

Figure 9.1: The standardized (i.e., area integrates to one) partial likelihood function for the proportional hazards model based on the Freireich data.

results to obtain $g(\beta)$ as the conditional distribution of the maximum likelihood estimator via the expression

$$g(u) = \prod_i \pi(u, X_i) \Big/ \int_u \Big\{ \prod_i \pi(u, X_i) \Big\} du.$$

In Figure 9.1 we illustrate the function $g(u)$, i.e., the estimated conditional distribution of the maximum likelihood estimator for the proportional hazards model given the two sample data from Freireich et al. (1963). It is interesting to compare the findings with those availability via more standard procedures. Rather than base inference on the mode of this distribution (i.e., the maximum partial likelihood estimate) and the second derivative of the log-likelihood we can consider

$$\tilde{\beta} = \int_u u g(u) du\,, \quad v(\beta) = \int_u u^2 g(u) du - \Big(\int_u u g(u) du \Big)^2,$$

and base tests, point, and interval estimation on these. Estimators of this kind have been studied extensively by Pitman (1948) and, generally, have smaller mean squared error than estimators based on the mode. Note also, in a way analogous to the calculation of bootstrap percentile intervals, that we can simply consider the curve $g(u)$ and tick off areas of size $\alpha/2$ on the upper and lower halves of the function to obtain a $100(1 - \alpha)\%$ confidence interval for β. Again, analogous to bootstrap percentile intervals, these intervals can pick up asymmetries and provide more accurate confidence intervals for small samples than those based on the symmetric large sample normal approximation. For the Freireich data, agreement between the approaches is good and that is to be expected since, in this case, the normal curve is clearly able to give a good approximation to the likelihood function.

For higher dimensions the approach can be extended almost immediately, at least conceptually. A potential difficulty arises in the following way: suppose we are interested in β_1 and we have a second parameter, β_2, in the model. Logically we would just integrate the two-dimensional density with respect to β_2, leaving us with a single marginal density for β_1. This is straightforward since we have all that we need to do this, although of course there will usually be no analytical solution and it will be necessary to appeal to numerical techniques. The difficulty occurs since we could often parameterize a model differently, say incorporating β_2^2 instead of β_2 alone. The marginal distribution for β_1, after having integrated out β_2, will not generally be the same as before. Nonetheless, for the proportional hazards model at least, the parameterization is quite natural and it would suffice to work with the models as they are usually expressed.

Partial likelihood

Again, starting from a full likelihood, or full density, we can focus interest on some subset, "integrating out" those parameters of indirect interest. Integrating the full density (likelihood) with respect to those parameters of secondary interest produces a marginal likelihood. Cox (1975) develops a particular expression for the full likelihood in which an important component term is called the partial likelihood. Careful choices lead to us back to conditional likelihood and marginal likelihood and Cox then describes these as special cases. For a given problem, Cox provides some guidelines for finding partial likelihoods: (1) the omitted factors should have distributions depending in an essential way on nuisance parameters and should contain no information about the parameters of interest and (2) incidental parameters, in particular the nuisance parameters, should not occur in the partial likelihood.

Lemma 9.10 *For the proportional hazards model with constant effects,*

$$L(\beta) = \prod_{i=1}^{n} \left\{ \frac{\exp(\beta Z_i)}{\sum_{j=1}^{n} Y_j(X_i) \exp(\beta Z_j)} \right\}^{\delta_i} \tag{9.10}$$

satisfies the two guidelines provided by Cox (1975).

For this particular case then the term "partial likelihood" applies. We use the term more generally for likelihoods of this form even though we may not be able to verify the extent to which the guidelines apply. Unfortunately, at least when compared to usual likelihood, marginal and conditional likelihood, or profile likelihood (maximizing out rather than integrating out nuisance parameters), partial likelihood is a very difficult concept. For general situations, it is not at all clear how to proceed. Furthermore, however obtained, such partial likelihoods are unlikely to be unique. Unlike the more commonly used likelihoods, great mathematical skill is required and even well steeled statistical modelers, able to obtain likelihood estimates in complex applied settings, will not generally be able to do the same should they wish to proceed on the basis of partial likelihood. For counting processes (Andersen et al. 1993), it requires some six pages (pages 103-109), in order to formulate an appropriate partial likelihood.

However, none of this impedes our development since the usual reason for seeking an expression for the likelihood is to be able to take its logarithm, differentiate it with respect to the unknown parameters and, equating this to zero, to enable the construction of an estimating equation. Here we already have, via the main theorem (Section 7.4), or stochastic integral considerations, appropriate estimating equations. Thus, and in light of the above mentioned difficulties, we do not put emphasis on partial likelihood as a concept or as a general statistical technique. Nonetheless, for the main models we consider here, the partial likelihood, when calculated, can be seen to coincide with other kinds of likelihood, derived in different ways, and that the estimating equation, arising from use of the partial likelihood, can be seen to be a reasonable estimating equation in its own right, however obtained. The above expression was first presented in Cox's (1972) original paper where it was described as a conditional likelihood.

Cox's conditional likelihood

At any time point t there is a maximum of n subjects under study. We can index the jth subject as having hazard rate $\lambda_j(X_i)$ at time point X_i. In other words we let $Y_j(t)$ take the value 1 if an individual is available to make a transition, zero otherwise. Then, for the j th subject we have an intensity function $\alpha_j(t) = Y_j(t)\lambda_j(t)$. The n patients can be viewed as a system. At time point t the system either remains the same (no failure), changes in a total of n possible ways, in which a single patient fails or the system may change in a more complicated way in which there can be more than a single failure.

If we are prepared to assume that at any given point there can be no more than a single failure, then the system can change in a maximum of n ways. This defines a simple stochastic process. Notice that, conditional upon there being a transition, then a straightforward application of Bayes formula (Chapter 15.6 Johnson and Johnson 1980) enables us to deduce the probability that the transition is of type j. This is simply

$$\frac{\alpha_j(X_i)}{\sum_{\ell=1}^{n} \alpha_\ell(X_i)} = \frac{Y_j(X_i)\lambda_j(X_i)}{\sum_{\ell=1}^{n} Y_\ell(X_i)\lambda_\ell(X_i)} = \frac{\exp(\beta Z_j)}{\sum_{\ell=1}^{n} Y_\ell(X_i)\exp(\beta Z_\ell)}.$$

For subjects having either failed or being lost to follow-up before time t we can still carry out the sum over all n subjects in our evaluation at time t. This is because of the indicator variable $Y_j(t)$ that takes the value zero for all such subjects, so that their transition probabilities become zero. The same idea can be expressed via the concept of risk sets, i.e., those subjects alive and available to make the transition under study. However, whenever possible, it is preferable to make use of the indicator variables $Y_j(t)$, thereby keeping the sums over n.

Multiplying all the above terms over the observed failure times produces $L(\beta)$. In his 1972 paper Cox described this as a conditional likelihood and suggested it be treated as a regular likelihood for the purposes of inference. In their contribution to the discussion of Cox's paper, Kalbfleisch and Prentice point out that the above likelihood does not have the usual probabilistic interpretation. If we take times to be fixed then $\exp(\beta Z_j)/\sum_\ell Y_\ell(X_i)\exp(\beta Z_\ell)$ is the probability of the subject indexed by j failing at X_i and that all other subjects, regardless of order, occur after X_i. Cox's deeper study (Cox 1975) into $L(\beta)$ led to recognition of $L(\beta)$ as a partial likelihood and not a conditional likelihood in the usual sense.

The flavor of $L(\beta)$ is nonetheless that of a conditional quantity, even if the conditioning is done sequentially and not all at once. Cox's discovery of $L(\beta)$, leading to a host of subsequent applications (time-dependent effects, time-dependent covariates, random effects), represents one of the most important statistical advances of the twentieth century. Although it took years of subsequent research in order to identify the quantity introduced by Cox as the relevant quantity with which to carry out inference and, although it was argued that Cox's likelihood was not a conditional likelihood in the usual sense (where all of the conditioning is done at once), his likelihood was all the same the right quantity to work with.

Marginal likelihood of ranks

Kalbfleisch and Prentice (1973) pointed out, that for fixed covariates Z the partial likelihood coincides with the likelihood, or probability, of the rank vector occurring as observed, under the model. We then have an important result;

Theorem 9.2 *Under an independent censoring mechanism the probability of observing the particular order of the failures is given by $L(\beta)$.*

Alternatively we can express the likelihood as a function of the regression vector β and the underlying failure rate $\lambda_0(t)$. Writing this down we have:

$$L(\beta, \lambda_0(t)) = \prod_{i=1}^{n} \left[\lambda_0(t) e^{\beta Z_i} S_0(X_i)^{\exp(\beta Z_i)} \right]^{\delta_i} \left[S_0(X_i)^{\exp(\beta Z_i)} \right]^{1-\delta_i}.$$

From this we can break the likelihood into two components. We then have:

Theorem 9.3 *$L(\lambda_0, \beta)$ can be written as the product $L_\lambda(\lambda_0, \beta)L(\beta)$.*

This argument is also sometimes given to motivate the idea of partial likelihood, stating that the full likelihood can be decomposed into a product of two terms, one of which contains the nuisance parameters $\lambda_0(t)$ inextricably mixed in with the parameter of interest β and a term that only depends upon β. This second term is then called the partial likelihood. Once again, however, any such decomposition is unlikely to be unique and it is not clear how to express in precise mathematical terms just what we mean by "inextricably mixed in" since this is more of an intuitive notion suggesting that we do not know how to separate the parameters. Not knowing how to separate the parameters does not mean that no procedure exists that might be able to separate them. And, if we were to sharpen the definition by stating, for example, that within some large class there exists no transformation or re-parameterization that would separate out the parameters, then we would be left with the difficulty of verifying this in practice, a task that would not be feasible.

9.7 Goodness-of-fit of likelihood estimates

The question of goodness-of-fit addresses the appropriateness of the class of chosen models rather than the properties of estimates based on any particular technique. In organizing the material in this text an initial plan was to group together all material relating to goodness-of-fit in a single chapter. There is an extensive literature on goodness-of-fit for survival models (see O'Quigley and Xu 1998 for a non-exhaustive review) and, in this work, rather than bring together the various ideas in a single chapter, it is preferred to describe some techniques that sit well with the chosen inferential procedure. This is very clear for inference based on functions of Brownian motion processes where graphical illustrations tell us pretty much all we need to know, supplemented if desired by more formal tests, themselves based on properties of particular functions of Brownian motion. For inference based more directly on likelihood it is less clear to identify, in some sense, related goodness-of-fit techniques. That said, the techniques described here work well with parametric and semi-parametric models in which the estimating technique appeals to likelihood.

As already mentioned in the previous chapter we often take the view that a goodness-of-fit test is simply a test of a particular kind of hypothesis, notably a data driven hypothesis where certain parameters, under the original specification of the model, have been replaced by estimates. We then place these kind of tests under the same likelihood heading as the other tests appealing to likelihood. More formally, instead of testing a hypothesis of the type $H_0 : \beta = \beta_0$, for some β_0 or set of β_0 given in advance, we test a hypothesis of the type $H_0 : \beta = \hat{\beta}$. Referring to the discussion of the previous chapter on operating characteristics, note that we do not expect all tests to perform comparably and a good choice was that based on a process approximating a

Brownian bridge. The construction was based upon being able to estimate the first two moments of the covariate at each failure point X_i. For parametric models using likelihood the dependence is typically expressed via the conditional distributions of time given the covariate, i.e., $f(t|z)$, that is unlike the case based on the main theorem and the processes described in the previous chapter. However, by applying Bayes theorem successively we can obtain necessary expectations in terms of the conditional probabilities of T given Z, i.e., via the use of $f(t|z)$. Specifically we have:

$$E(Z|T = t) = \int_{\mathcal{Z}} \left\{ z - \frac{\int_{\mathcal{Z}} z f(t|z) dG(z)}{\int_{\mathcal{Z}} f(t|z) dG(z)} \right\} \left\{ \frac{f(t|z)}{\int_{\mathcal{Z}} f(t|z) dG(z)} \right\} dG(z).$$
(9.11)

Figure 9.2 shows a bridged process for the Freireich data where the fitted model is a parametric Weibull model. The very good fit is confirmed by inspection of the curve. The only difference between the likelihood approach in this chapter and the approach of the previous chapter is in the calculation of the expectations. In one case we lean upon the main theorem (Section 7.4) and the moments that follow from it, in the other we work with any parametric model and use it to obtain the necessary expectations, these expectations now coming via an application of Bayes theorem. In the two-group case with constant hazard rates and where there is no censoring the above expression can be simplified. Letting the proportion of the first group be given by π_1, the second by π_2 such that $(\pi_1 + \pi_2 = 1)$ and writing $\psi(v) = a(v)/\{\pi_1 e^{-v} + a(v)\}$ where $a(v) = \pi_2 e^{\beta} \exp(-v e^{\beta})$ then

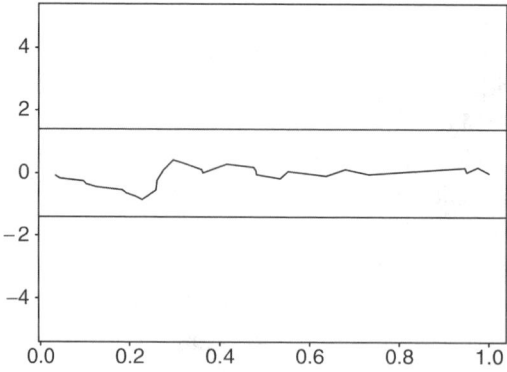

Figure 9.2: A bridged standardized cumulative processes for the Freireich data using expectations based on a Weibull model.

$$E\left\{[Z - \mathcal{E}_\beta(Z|t)]^p\right\} = \pi_1 \int_0^t \{-\psi(v)\}^p \, e^{-v} dv$$
$$+\pi_2 \int_0^t \{1 - \psi(v)\} \, e^\beta \exp(-ve^\beta) dv.$$

The most immediate departure from proportional hazards would be one where effects decline through time and, as mentioned before, perhaps even changing direction (Stablein et al. 1981). The standardized cumulative score, however the moments are calculated, would, under such departures, increase (or decrease) steadily until the increase (or decrease) dies away and the process would proceed on average horizontally or, if effects change direction) make its way back toward the origin. A test based on the maximum would have the ability to pick up this kind of behavior. Again, the visual impression given by the cumulative standardized score process can, of itself, suggest the nature of the departure from proportional hazards. This kind of test is not focused on parameters in the model other than the regression effect given by β or possibly $\beta(t)$. The goal is to consider the proportionality or lack of such proportionality and, otherwise, how well the overall model may fit is secondary.

Non-proportional hazards model with intercept

In the review of goodness-of-fit tests for survival models, O'Quigley and Xu (1998) gave particular consideration to the non proportional hazards model with intercept described in Chapter 6. Using this model and the usual likelihood procedures we can test for several specific departures from proportional hazards. For the sake of simplicity of exposition we continue to limit attention to the single variable case. Extension to higher dimensions is immediate. Also we will sometimes write Z, instead of $Z(t)$, in order for the notation to not become over cluttered. It can always be replaced by $Z(t)$ without additional concerns. We write

$$\lambda(t|Z) = \lambda_0(t) \exp\{[\beta + \alpha Q(t)]Z(t)\}, \tag{9.12}$$

where $Q(t)$ is a function of time that does not depend on the parameters β and α. Under a null hypothesis that a proportional hazards model is adequate, i.e., $H_0 : \alpha = 0$ we recover a proportional hazards model. In the context of sequential group comparisons of survival data, the above model has been considered by Tsiatis (1982) and Harrington et al. (1982). In keeping with the usual notation we denote $\mathcal{E}_{\beta,\alpha}(Z|t)$ to be the expectation taken with respect to the probability distribution $\pi_i(\beta, \alpha, t)$, where

$$\pi_i(\beta, \alpha, t) = \frac{Y_i(t) \exp\{[\beta + \alpha Q(t)]Z_i(t)\}}{\sum_{j=1}^n Y_j(t) \exp\{[\beta + \alpha Q(t)]Z_j(t)\}}. \tag{9.13}$$

Lemma 9.11 *The components of the score vector $U(\beta, \alpha)$ can be expressed as*

$$U_\beta(\beta, \alpha) = \sum_{i=1}^n \delta_i\{Z_i(X_i) - \mathcal{E}_{\beta,\alpha}(Z|X_i)\}, \tag{9.14}$$

$$U_\alpha(\beta, \alpha) = \sum_{i=1}^n \delta_i Q(X_i)\{Z_i(X_i) - \mathcal{E}_{\beta,\alpha}(Z|X_i)\}. \tag{9.15}$$

A test would be carried out using any one of the large sample tests arising from considerations of the likelihood. If we let $\hat{\beta}$ be the maximum partial likelihood estimate of β under the null hypothesis of proportional hazards, i.e., $H_0 : \alpha = 0$, then $U_\beta(\hat{\beta}, 0) = 0$. The score test statistic arising under H_0 is $B = U_\alpha(\hat{\beta}, 0)G^{-1}U_\alpha(\hat{\beta}, 0)$, where $G = I_{22} - I_{21}I_{11}^{-1}I_{12}$ and G^{-1} is the lower right corner element of I^{-1}. The elements of the information matrix required to carry out the calculation are given below in Lemma 9.12. Under H_0, the hypothesis of proportional hazards, B has asymptotically a χ^2 distribution with one degree of freedom.

Lemma 9.12 *Taking $k = 1, 2$ and $\ell = 1, 2$, the components of I are*

$$I(\beta, \alpha) = -\begin{pmatrix} U_{\beta\beta} & U_{\beta\alpha} \\ U_{\alpha\beta} & U_{\alpha\alpha} \end{pmatrix} = \begin{pmatrix} I_{11} & I_{12} \\ I_{21} & I_{22} \end{pmatrix} \quad \text{where}$$

$$I_{k\ell}(\beta, \alpha) = \sum_{i=1}^n \delta_i Q(X_i)^{k+\ell-2}\{\mathcal{E}_{\beta,\alpha}(Z^2|X_i) - \mathcal{E}_{\beta,\alpha}^2(Z|X_i)\}.$$

The literature on the goodness-of-fit problem for the Cox model has considered many formulations that correspond to particular choices for $Q(t)$. The first of these was given in the founding paper of Cox (1972). Cox's suggestion was equivalent to taking $Q(t) = t$. Defining $Q(t)$ as a two dimensional vector, Stablein et al. (1981) considered $Q(t) = (t, t^2)'$, and Brown (1975), Anderson and Senthiselvan (1982), O'Quigley and Moreau (1984), Moreau et al. (1985), and O'Quigley and Pessione (1989) assumed $Q(t)$ to be constant on predetermined intervals of the time axis, i.e. $Q(t)$ is a step function.

Although in the latter cases there is more than one parameter associated with $Q(t)$, the computation of the test statistic is similar. Murphy (1993) studied the size and the power of the test of Moreau et al. (1985) and found that, although it is consistent against a wide class of alternatives to proportional hazards, the Moreau test is nonetheless an omnibus test that is used to greatest advantage when there is no specific alternative in mind.

We can choose $Q(t)$ to be an unknown function and use the available data to provide an estimate of this function. For instance, Breslow et al. (1984) chose $Q(t) = \Lambda(t)$ and replaced this unknown function by the Nelson estimator. Because the estimates at time t depend only on the history of events up to that time, the development of Cox (1975) and Andersen and Gill (1982) and thus, the usual asymptotic theory, still applies. Breslow et al. (1984) showed that their choice $Q(t) = \Lambda(t)$ has good power against the alternative of crossing hazards. Tsiatis (1982), Harrington et al. (1982), and Harrington and Fleming (1982) used score processes based on the non-proportional hazards model with intercept to derive sequential tests. They showed that after $Q(t)$ has been replaced by its estimate, the score process at different time points converges in distribution to a multivariate normal. Harrington and Fleming focus particular interest on the G^ρ family, where $Q(t) = S(t)^\rho$.

Another special case is described in O'Quigley and Pessione (1991), O'Quigley and Natarajan (2004) and O'Quigley (1994), where $Q(t) = I(t \leq \gamma) - I(t > \gamma)$ with γ an unknown change point. When γ is known the test statistic can be evaluated with no particular difficulty. For γ unknown, we would maximize over all possible values of γ and special care is required for the resulting inference to remain valid. We consider this issue on its own in Chapter 12. O'Quigley and Pessione (1991) showed that tests using a changepoint model can be powerful for testing the equality of two survival distributions against the specific alternative of crossing hazards. Also, such tests suffer only moderate losses in power, when compared with their optimal counterparts, if the alternative is one of proportional hazards.

Lin (1991) and Gill and Schumacher (1987) have taken a slightly different approach to working with the function $Q(t)$ and introduce it directly into a weighted score. This can be written as

$$U_Q(\beta) = \sum_{i=1}^{n} \delta_i Q(X_i)\{Z_i(X_i) - E(Z|X_i; \beta)\}, \qquad (9.16)$$

where the only key requirement is that $Q(\cdot)$ be a predictable process (see Section 3.6) that converges in probability to a non-negative bounded function uniformly in t. Let $\hat{\beta}_Q$ be the zero of (9.16) and $\hat{\beta}$ be the partial likelihood estimate. Under the assumption that the proportional hazards model holds and that $(X_i, \delta_i, Z_i)\,(i = 1, \ldots, n)$ are i.i.d. replicates of (X, δ, Z), $n^{1/2}(\hat{\beta}_Q - \hat{\beta})$ is asymptotically normal with zero mean and covariance matrix that can be consistently estimated. It then follows that a simple test can be based on the standardized difference between the two estimates. Lin (1991) showed such a test to be consistent against any model misspecification under which $\beta_Q \neq \beta$, where β_Q is the probability limit of $\hat{\beta}_Q$. In particular, it can be shown that choosing a monotone weight function for $Q(t)$ such as $\hat{F}(t)$, where $\hat{F}(\cdot)$ is the Kaplan-Meier estimate, is consistent against monotone departures (e.g., decreasing regression effect) from the proportional hazards assumption.

9.8   Exercises and class projects

1. Describe why the assumption of marginal independence is a stronger assumption than that of conditional independence. Describe situations in which each of these assumptions appears to be reasonable. How is the likelihood function impacted by the assumptions?

2. Suppose that the censoring mechanism is not independent of the survival mechanism, in particular suppose that

$$\log \Pr\left(T > x + u | C = u, T > u\right) = 2 \log \Pr\left(T > x + u | T > u\right).$$

Write down the likelihood for a parametric model for which the censoring mechanism is governed by this equation. Next, suppose that we can take the above equation to represent the general form for the censoring model but that, instead of the constant value 2, it depends on an unknown parameter, i.e., the number 2 is replace by α. What kind of data would enable us to estimate the parameter α?

3. As a class project, simulate data with a dependent censoring mechanism as above with an unknown parameter α. Investigate the distribution of $\hat{\alpha}$ via 1000 simulations.

4. Fit a Weibull proportional hazards model to data including at least two binary regressors; Z_1 and Z_2. Calculate a 90% confidence intervals for the probability that the most unfavorable prognosis among the 4 groups has a survival greater than the marginal median. Calculate a 90% confidence interval that a subject chosen randomly from either the most unfavorable, or the second most unfavorable, group has a survival greater than the marginal median.

5. As a generalization of the previous exercise, consider a parametric model with covariate vector Z of dimension p. The p dimensional model is considered to provide a good approximation to the underlying mechanism generating the observations. On the basis of the fitted p dimensional model, write down an expression for a confidence interval for the probability of surviving longer than t for a subject in which Z_1, the first component of Z, is equal to z_1.

6. For the two-sample exponential model, write down the likelihood and confirm the maximum likelihood estimates given in Section 9.4. Calculate the score test, the likelihood ratio test and Wald's test, and compare these with the expressions given in Section 9.4. For a data set with a single binary covariate calculate and compare the three test statistics.

7. On the basis of data, estimate the unknown regression coefficient, β, as the expected value of the conditional likelihood (see Section 9.6). Do this for both an exponential based likelihood and the partial likelihood. Next, consider the distribution of $\log \beta$ in this context and take an estimate as $\exp E(\log \beta)$. Do you anticipate these two estimators to agree? Note that the corresponding maximum likelihood estimators do agree exactly. Comment.

8. Try different weights in the weighted log-rank test and apply these to a data set. Suppose we decide to use the weight that leads to the most significant result. Would such an approach maintain control over Type I error under the null hypothesis of no association? Suggest at least two ways in which we might improve control over the Type I error rate.

Chapter 10

Inference: Stochastic integrals

10.1 Summary

Further to the background theory on counting processes, martingales and stochastic integrals, outlined in Sections 2.12 and 3.6, we note that the score statistic arising from the log partial likelihood can be seen to come under the heading of a stochastic integral. The terms of this integral are then equated with those elements composing a stochastic integral and for which we can appeal to known large sample results. In the case of the multiplicative model, replacing the unknown background cumulative hazard by the Nelson-Aalen estimate produces the same result as that obtained from use of the observed information matrix. This is not the case for the additive model. In this case the variance estimator using martingale theory is generally to be preferred over the information based estimator. Considerable flexibility results from the martingale concept of conditioning on the accumulated history. One example is that of multistate processes which are easily dealt with. A second is that of obtaining procedures which are non-parametric with respect to both time and the covariate (O'Brien 1978, O'Quigley and Prentice 1991).

10.2 Motivation

Aalen (1978), followed by Andersen and Gill (1982), pioneered the counting process approach to inference in the context of proportional hazard models. Writing the score statistic as a stochastic integral, they

295

showed how a framework for inference can be structured. This framework uses the results from martingale theory described in Sections 2.12 and 3.6. Asymptotic normality and an expression for the large sample variance are readily obtained. For the multiplicative model this variance expression is the same as that obtained by working with the second derivative of the partial likelihood (Cox, 1972). In this chapter we will see that this is not the case for the additive model and that, in this case, the variance expression derived from martingale theoretic considerations is, in general, more stable than that derived on the basis of the second derivative. Indeed, for a model with an additive risk, the estimated variance of the score statistic, when viewed as a stochastic integral, differs from that obtained from the inverse of the information matrix and has the advantage of always being positive, a property not shared by the information matrix estimate. They can be seen to be equivalent asymptotically.

The stochastic integral representation enables us to recover the great majority of those results we already have from exploiting likelihood theory, but it can take us further in different ways. First, the different kinds of weightings proposed by Prentice (1978), some fixed ahead in time, some evaluated dynamically, are readily accommodated within the martingale theory. Also, higher-order moments are easily derived and allow us to obtain corrected percentiles of the score distribution through the use of a Cornish-Fisher expansion. This simple procedure can yield more accurate inference for the score statistic in small samples. We return to this in the following chapter dealing with inference in small samples.

10.3   Counting process framework to the model

The reader may first wish to recall the background of counting processes, martingales and stochastic integrals from Sections 2.12 and 3.6. The greatest generality can be obtained by working with a multivariate counting process $N(t)$, the n components of which are, typically, the independent individual counting processes. This is not always the case, though, and, in particular the individual processes need not be independent. Large sample theory follows by working with the multivariate martingale central limit theorem (Rebolledo 1980). This approach has now been widely adopted in making inferences for survival models. It is described in detail in Fleming and

Harrington (1991) and Andersen et al. (1993). Here, we take a less general approach working instead with collections of univariate counting processes and appealing to more classic central limit theorems for sums of independent but not identically distributed random variables. This leads to a very considerable gain in clarity and the cost of this gain is very small. For instance all of the main developments of the proportional hazards model (stratification, inclusion of random effects, multistate models, repeated events, cure models) can be handled by working with the usual central limit theorem for independent variables and verifying the Lindeburg condition rather than working with the multivariate martingale central limit theorem.

We take $N_i(t), i = 1, \ldots, n$ to be n univariate counting processes, starting off at zero, i.e., $N(0) = 0$, and making a unit jump at the distinct failure time point X_i, if $I(C_i > X_i) = 1$, so that $N(t) = 1$ for $t \geq X_i$. The intensity process $\alpha_i(t)$ can be expressed in terms of the counting process and $\mathcal{F}_{t-}$, the total accumulated information on failures, censorings and covariate processes available an instant before time t, as

$$\alpha_i(t) = \lim_{\Delta t \to 0} (\Delta t)^{-1} \Pr\{N_i(t + \Delta t) - N_i(t) = 1 | \mathcal{F}_{t-}\}. \qquad (10.1)$$

Note that the set $\mathcal{F}_{t-}$ in the simplest case corresponds to the information that $T > t$. The form $\mathcal{F}_{t-}$ allows for great generality and enables us to easily deal with complex situations such as multistate models, repeated events when subjects move in and out of risk sets, and particular kinds of conditioning such as the use of covariate ranks. A proportional hazards model with more general risk for this intensity function can be written as

$$\alpha_i(t) = Y_i(t)\lambda_0(t)R\{\beta Z_i(t)\}, \qquad (10.2)$$

where, as before, $Y_i(t)$ is an indicator function taking the value 1 if the ith subject is at risk at time t and 0 otherwise, $\lambda_0(t)$ is the fixed "baseline" hazard function and β, the parameter, or vector of parameters, to be estimated. $Z_i(t)$ is, as usual, the covariate for the ith subject at time t and $R(r)$ the relative risk function; equal to $\exp(r)$ for the multiplicative model and $1 + r$ for the additive model.

10.4   Some nonparametric statistics

The inferential background described in Section 3.6 paid particular attention to Aalen's (1978) multiplicative intensity model. Recall that for the i.i.d. case, the compensator for $N_i(t)$ is $\alpha_i(t) = Y_i(t)\lambda(t)$, where $\lambda(t)$ is the hazard rate. The compensator, $\bar{A}(t)$, for $\bar{N}(t) = \sum_{i=1}^{n} N_i(t)$ is:

$$\bar{A}(t) = \int_0^t \{\textstyle\sum_{i=1}^{n} Y_i(s)\}\lambda(s)ds = \int_0^t \bar{Y}(s)\lambda(s)ds,$$

from which we see the intensity process for $\bar{N}(t)$ to be $\bar{Y}(t)\lambda(t)$. This simple structure can be used to underpin several nonparametric tests. Using a martingale as an estimating equation (see Sections 3.6 and 3.7) we can take $\bar{N}(t)$ as an estimator of its compensator $\bar{A}(t)$. Dividing by the risk set (assumed to be always greater than zero), we have that,

$$\int_0^t \frac{d\bar{N}(t)}{\bar{Y}(t)} - \int_0^t \lambda(s)ds$$

is a martingale and has, in consequence, zero expectation. An estimate of the cumulative risk $\hat{\Lambda}(t)$ is given by $\hat{\Lambda}(t) = \int_0^t \bar{Y}(t)^{-1}d\bar{N}(t)$ which is the estimator of Nelson-Aalen. Simple nonparametric statistics for the comparison of two groups can be obtained immediately from this. If we consider some predictable weighting process (for an explanation of what we mean by "predictable" see Sections 2.12 and 3.6), $W(s)$, then define

$$K(t) = \int_0^t W(s)\left\{\frac{d\bar{N}_1(s)}{\bar{Y}_1(s)} - \frac{d\bar{N}_2(s)}{\bar{Y}_2(s)}\right\} \tag{10.3}$$

where a subscript 1 denotes subjects from group 1 and a 2 from group 2. The choice of the weighting function $W(s)$ can be made by the user and might be chosen when some particular alternative is in mind. The properties of different weights were investigated by Prentice (1978). Referring back to Section 3.6, we can readily claim that $K(\infty)$ converges in probability to $\mathcal{N}(0, \sigma^2)$. We estimate σ^2 by $\langle K \rangle(\infty)$ where

$$\langle K \rangle(t) = \int_0^t \left\{\frac{W(s)}{\bar{Y}_1(s)}\right\}^2 d\bar{N}_1(s) + \int_0^t \left\{\frac{W(s)}{\bar{Y}_2(s)}\right\}^2 d\bar{N}_2(s). \tag{10.4}$$

The choice $W(s) = \bar{Y}_1(s)\bar{Y}_2(s)/[\bar{Y}_1(s) + \bar{Y}_1(s)]$ leads to the log-rank test statistic and would maintain good power under a proportional hazards alternative of a constant group difference as opposed to the null hypothesis of no group differences. The choice $W(s) = \bar{Y}_1(s)\bar{Y}_2(s)$ corresponds to the weighting suggested by Gehan (1965) in his generalization of the Wilcoxon statistic. This test may offer improved power over the log-rank test in situations where the group difference declines with time. These weights and therefore the properties of the test are impacted by the censoring and, in order to obtain a test free of the impact of censoring, Prentice (1978) suggested the weights, $W(s) = \hat{S}_1(s)\hat{S}_2(s)$. These weights would also offer the potential for improved power when the regression effect declines with time.

10.5 Stochastic integral representation of score statistic

The score statistic, based on Cox's (1975) partial likelihood, can be written as

$$U(\beta) = \sum_{i=1}^{n} N_i(X_i)\{Z_i(X_i) - \mathcal{E}_\beta(Z|X_i)\}, \qquad (10.5)$$

where the definition of $\mathcal{E}_\beta(Z|X_i)$ is the same as in earlier chapters. Defining $H_i(t) = Z_i(t) - \mathcal{E}_\beta(Z|t)$, we see that Equation (10.5) can be rewritten as $U(\beta) = U(\beta, \infty)$ where

$$U(\beta, t) = \sum_{i=1}^{n} \int_0^t H_i(s)dN_i(s). \qquad (10.6)$$

The martingale approach to inference can be very briefly summarized as follows: The process $N_i(t)$ counts events on the ith subject, in our case 0 or 1 and very clearly, for $u > 0$, $E\{N_i(s+u)|\mathcal{F}_s\} \geq E\{N_i(s)|\mathcal{F}_s\}$. Recall that $\mathcal{F}_{s-}$ denotes the set containing the total information we have concerning $N_i(s)$ from all time points strictly less than s, so that $E\{dN_i(s)|\mathcal{F}_{s-}\} = Y_i(s)\lambda_i(s)ds$, the infinitesimal probability of the occurrence of an event for a Bernoulli variable at time point s. The Doob-Meyer decomposition theorem whereby we can express the submartingale $N_i(s)$ in terms of a compensator, $A_i(s)$ and a martingale, $M_i(s)$, enables us to write

$$N_i(t) = M_i(t) + \int_0^t Y_i(u)\lambda_i(u)du. \qquad (10.7)$$

Also recall that, if the function $H_i(s)$, at each time point s, contains no information that is not contained in $\mathcal{F}_{s-}$ and is left continuous (not a practical constraint), then it is predictable, in other words, at time points s and beyond, a fixed constant. A stochastic integral of the above form can then be always made into a martingale via this decomposition since, if $M_i(t)$ is a martingale and $H_i(t)$ a predictable process, then the stochastic integral $\int_0^t H_i(s)dM_i(s)$ will also be a martingale. Using the known results concerning inference for martingales and submartingales (which can be turned into martingales via the Doob-Meyer theorem), for large samples, we have in particular that:

Theorem 10.1 *The large sample distribution of $U(\beta, \infty)$ is normal, having first two moments given by*

$$E\{U(\beta, \infty)\} = \int_0^\infty \sum_{i=1}^n E\{Y_i(s)H_i(s)\}\lambda_i(s)ds \qquad (10.8)$$

$$E\{U^2(\beta, t)\} = \int_0^t \sum_{i=1}^n E\{Y_i(s)H_i^2(s)\}\lambda_i(s)ds. \qquad (10.9)$$

The variance is then $\mathrm{Var}\{U(\beta, \infty)\} = E\{U^2(\beta, \infty)\} - E^2\{U(\beta, \infty)\}$. These results are very general and enable us to consider a wide range of situations. Below we examine the flexibility granted to us by different possible choices for H. The results can be used to test hypotheses, and thereby obtain interval estimates for parameters via inversion of tests, or for point estimation. In order to carry out these prescriptions we will need replace $\lambda_i(t)$ by estimates and we look at this below.

Large sample approximations

Note that at the true value of β, $E\{Y_i(s)H_i(s)\} = 0$. In order to estimate the variance, we replace $\lambda_i(s)ds$ by $d\hat{\Lambda}_i(s)$, for any consistent estimate $\hat{\Lambda}_i(t)$ of the cumulative hazard, in particular the one suggested by Breslow (1972), Aalen (1978), and Andersen and Gill (1982), expressed via the lemma:

Lemma 10.1 $\Lambda_i(t)$ *is consistently estimated by* $\hat{\Lambda}_i(t) = \hat{\Lambda}_0(t)R\{\hat{\beta}z_i(t)\}$ *where*

$$\hat{\Lambda}_0(t) = \int_0^t \frac{d\bar{N}(s)}{\sum_{i=1}^n Y_i(s)R\{\hat{\beta}Z_i(s)\}} = \sum_{i=1}^n \frac{I(C_i > X_i)}{\sum_{j=1}^n Y_j(X_i)R\{\hat{\beta}Z_j(X_i)\}},$$

the integral simplifying to the sum given on the right-hand side since $d\bar{N}(s)$ is always zero apart from at values where s is equal to one of the failure points X_i $(C_i > X_i)$. Using this and defining $W_i(t) = Z_i(t)$ for the multiplicative model, $W_i(t) = Z_i(t)/(1 + \beta Z_i(t))$ for the additive model, we can write:

Lemma 10.2 *A consistent estimate of* $E\{U^2(\beta, \infty)\}$ *is given by*

$$\hat{E}\{U^2(\beta, \infty)\} = \sum_{i=1}^{n} \sum_{j=1}^{n} N_i(X_i) Y_j(X_i) \{W_j(X_i) - \mathcal{E}_\beta(W|X_i)\}^2 \pi_j(\beta, X_i),$$

where the weights obtain from $\pi_j(\beta, t) = Y_j(t) R\{\beta Z_j(t)\}/\sum_{\ell=1}^{n} Y_\ell(t) R\{\beta Z_\ell(t)\}$.

Noting that $\sum_{j=1}^{n} Y_j(X_i) \pi_j(\beta, X_i) = 1 \; \forall i$, developing the bracket and collecting terms, we obtain the more familiar term for this second moment:

Corollary 10.1

$$\hat{E}\{U^2(\beta, \infty)\} = \sum_{i=1}^{n} \sum_{j=1}^{n} N_i(X_i) Y_j(X_i) \{W_j^2(X_i) - \mathcal{E}_\beta^2(W|X_i)\} \pi_j(\beta, X_i). \tag{10.10}$$

This estimate coincides with that based on the partial likelihood information matrix for the multiplicative model (Cox 1972) when we replace β by $\hat{\beta}$. This is not so however for the additive model where minus the second derivative of the partial log-likelihood with respect to β equals

$$\sum_{i=1}^{n} N_i(X_i) \left\{ \frac{Z_i^2(X_i)}{(1 + \beta Z_i(X_i))^2} - \frac{(\sum_{j=1}^{n} Y_j(X_i) Z_j(X_i))^2}{\sum_{j=1}^{n} Y_j(X_i)(1 + \beta Z_j(X_i))^2} \right\}.$$

Some elementary algebra produces the following lemma;

Lemma 10.3 *For the additive model, the information based estimate gives*

$$\hat{E}\{U^2(\beta, \infty)\} = \sum_{i=1}^{n} N_i(X_i) \{W_i^2(X_i) - \mathcal{E}_\beta^2(W|X_i)\}. \tag{10.11}$$

It is instructive to compare the estimates based on equations (10.10) and (10.11). Assuming the model generating the data is correctly specified then, under the usual conditions on the censoring, the law of large numbers indicates that the difference between them will tend to zero. While the estimate based on Equation 10.10 is always positive in view of the Cauchy-Schwartz inequality, it should be noted that an estimate based on (10.11) can be negative in finite samples.

Logit-rank and associated test procedures

The log-rank and associated test procedures described under the heading of likelihood methods can all be also be obtained under the counting process heading as weighted stochastic integrals. The only restriction on the weightings is that they be predictable. The resulting weighted log-rank type tests are all linear combinations of the components to the score statistic. It is, however, possible to consider much more general classes of tests, as long as predictability is maintained. This idea motivated the class of tests described by O'Quigley and Prentice (1991). One of the main purposes was to derive tests both non-parametric, i.e., rank invariant, to the covariate scale as well as the time scale.

The tests of O'Quigley and Prentice (1991) derive from the transformed covariate models considered previously. In particular we consider functions $\psi_j(\mathcal{F}_t)$ for the jth subject at time t, i.e., functions that depend upon all available information at time t and assign thereby the value $\psi_j(\mathcal{F}_t)$ to the subject j at time t. The score statistic at $\beta = 0$ based on the partial likelihood can be written as

$$U(0) = \sum_{i=1}^{n} \delta_i \{\psi_i(\mathcal{F}_{X_i}) - \mathcal{E}_0(\psi|X_i)\}, \qquad (10.12)$$

where

$$\mathcal{E}_0(\psi|X_i) = \sum_{j=1}^{n} Y_j(X_i)\psi_j(\mathcal{F}_{X_i}) / \sum_{j=1}^{n} Y_j(X_i).$$

The corresponding observed information matrix at $\beta = 0$ is

$$I(0) = \sum_{i=1}^{n} \delta_i \left(\sum_{j=1}^{n} Y_j(X_i)\psi_j^2(\mathcal{F}_{X_i}) / \sum_{j=1}^{n} Y_j(X_i) \right) - \mathcal{E}_0^2(\psi|X_i). \quad (10.13)$$

A broad class of tests can now be constructed on the basis of choices for $\psi_j(\mathcal{F}_t)$. O'Brien's logit-rank test (O'Brien 1978) arises from his definitions for p_{ii} and y_{ii} which we can re-express in terms of $\psi_j(\mathcal{F}_t)$.

Definition 10.1 *Let $p_{i\ell}$ be the ranking of the ℓth subject at time X_i if at risk, and is zero otherwise. Then*

$$p_{i\ell} = \psi_\ell(\mathcal{F}_{X_i}) = \sum_{j=1}^{n} I(Z_j \le Z_\ell) Y_j(X_i).$$

The specification $Z_\ell(t) = \psi_\ell(t, \mathcal{F}_t) = y_{i\ell}$ when $t_{(i-1)} < t \le t_{(i)}$ for $i = 1$ to k and is equal to $y_{k\ell}$ when $t > t_{(k)}$ where $t_{(0)} = 0$ and $y_{i\ell} = 0$ at times at which the ℓth subject is not at risk, leads exactly to O'Brien's logit-rank test. In particular, since $E(t_i) = 0$, $i = 1, \ldots, k$, it follows that this specification gives $U(0) = \sum_{i=1}^{k} y_{ii}$. Furthermore we have:

Lemma 10.4 *Under the null hypothesis the information can be written as*

$$I(0) = \sum_{i=1}^{k} \sum_{j=1}^{n_i} n_i^{-1} \left\{ \text{logit} \left(j n_i^{-1} - (2n_i)^{-1} \right) \right\}^2 = \sum_{i=1}^{k} v_i.$$

Thus, the standardized logit-rank test is a partial likelihood score test. An asymptotic standard normal distribution can then be asserted, under a broad range of censoring distributions, on the basis of martingale convergence results (e.g. Rebolledo 1980, Andersen and Gill 1982) upon noting that the processes defined in the above are predictable, as follows since the sample paths for each $\psi_\ell(t, \mathcal{F}_t)$ are left continuous with right-hand limits. Notice that O'Brien's initial suggestion of basing a test upon $\sum_{i=1}^{k} p_{ii}$ also obtains as a score test under the proportional hazards model upon defining $\psi_\ell(t, \mathcal{F}_t) = p_{i\ell}$ when $t_{(i-1)} < t \le t_{(i)}$ for $i = 1, \ldots, k$ and is given the value $p_{k\ell}$ when $t > t_{(k)}$ and where $p_{i\ell} = 0$ at times at which the ℓth subject is not at risk. In this case we can use the fact that $\sum_{j=1}^{m} j = m(m+1)/2$ and that $\sum_{j=1}^{m} j^2 = m(m+1)(2m+1)/6$ to obtain a simple expression for the variance of $\sum_{i=1}^{n} p_{ii}$ as a corollary to the above.

Corollary 10.2

$$\text{Var} \left(\sum_{i=1}^{k} p_{ii} \right) = \sum_{i=1}^{k} (4n_i + 1)(n_i - 1)/12n_i.$$

Note that the mean of $\sum_{i=1}^{k} p_{ii}$ under H_0 is $k/2$ so that the following lemma provides an easily evaluable approximate test.

Lemma 10.5 *Under the null hypothesis of absence of association,*

$$\frac{36(2\sum_{i=1}^{k} p_{ii} - k)^2}{\sum_{i=1}^{k}(4n_i + 1)(n_i - 1)/n_i}$$

has a distribution converging to that of a chi-square on one degree of freedom.

The two definitions for $Z_\ell(t)$ given above correspond to O'Brien's original suggestions. Many other candidates or classes of candidates are possible. For instance, we may choose $\psi_\ell(t, \mathcal{F}_t)$ not to depend on information on subjects $i \neq \ell$, in which case scores chosen at the start of the study remain unchanged throughout the study. Tests analogous to the logit-rank tests would then obtain by simply transforming the marginal empirical distribution of the Z_i to be uniform over $(0, 1)$ or to have standardized logistic distributions. The use of normal order statistics in place of the original measurements amounts to a transformation to normality conditional on the risk sets or unconditionally if applied only once at the start of the study. Given this interpretation of the effect of different kinds of rankings, another approach, although not rank invariant, would be to replace the original x_i measurements by $G^{-1}\{\tilde{F}(Z_i)\}$ where $\tilde{F}(\cdot)$ is some consistent estimate of the cumulative distribution function for Z and $G(\cdot)$ is the cumulative distribution function we are transforming to, for example, the standard normal. In the case of stratification of certain covariates Z^c and having defined m classes the extension is straightforward. We would assume that, within a class, little further information on prognosis is obtainable from the actual values Z_i^c i.e., each class is, in some sense, homogeneous with respect to survival. Then base tests on the model

$$\lambda_{ij}(t) = Y_i(t)\lambda_{0j}(t) \exp\{\beta\psi_i(t, \mathcal{F}_t)\} \ ; \ j = 1, \ldots, m$$

using the usual methods applied to the sum of the log-likelihoods over strata. In an obvious notation the two versions of the logit-rank test for stratified data w_s and w_s' say become

$$w_s = \sum_{j=1}^{m}\sum_{i=1}^{k_j} y_{iij} / \left(\sum_{j=1}^{m}\sum_{i=1}^{k_j} v_{ij}\right)^{\frac{1}{2}} , \quad w_s' = \frac{36(2\sum_{j=1}^{m}\sum_{i=1}^{k_j} p_{iij} - \sum_{j=1}^{m} k_j)^2}{\sum_{j=1}^{m}\sum_{i=1}^{k_j}(4n_{ij} + 1)(n_{ij} - 1)/n_{ij}}.$$

It is of interest to quantify the size of losses in efficiency when using a scoring system other than that generating the data. This also enables us to assess the utility of re-ranking the covariate values at each failure time, in analogy with O'Brien's original scheme, as opposed to the simpler procedure of ranking, or scoring, at the outset and then treating the covariates as one that is constant with time. Suppose that the model generating the observations is as above and we derive a score test for β on the basis of the partial likelihood and the function $\psi_i^*(t, \mathcal{F}_t)$ instead of $\psi_i(t, \mathcal{F}_t)$. We can then use a result from O'Quigley and Prentice (1991) which extends a basic result of Lagakos and Schoenfeld (1984).

Theorem 10.2 *An expression for asymptotic relative efficiency is given by*

$$e(\psi, \psi^*) = \frac{(\int \phi[g\{Z(t)\}, g^*\{Z(t)\}]\lambda_0(t)dt)^2}{\int \phi[g\{Z(t)\}, g\{Z(t)\}]\lambda_0(t)dt \int \phi[g^*\{Z(t)\}, g^*\{Z(t)\}]\lambda_0(t)dt}.$$

In the above expression we assume that, $\lim_{n\to\infty} \psi(t, \mathcal{F}_t) = g\{Z(t)\}$ and that $\lim_{n\to\infty} \psi^*(t, \mathcal{F}_t) = g^*\{Z(t)\}$, where $g(\cdot)$ and $g^*(\cdot)$ are continuous monotonic functions. All of the integrals are over $(0, \infty)$ and

$$\phi(a(t), b(t)) = E\{Y(t)a(t)b(t)\} - E\{Y(t)a(t)\}E\{Y(t)b(t)\}/E\{Y(t)\},$$

the expectations being taken with respect to the distribution of $\mathcal{F}_t$ and $e(\psi, \psi^*)$ denotes asymptotic relative efficiency. Applying the above expression for $e(\psi, \psi^*)$, to time-independent covariates, we find that the asymptotic relative efficiency of the logit-rank procedure to the simplified logit-rank procedure equals one. The same applies to other pairs of procedures, the normal-rank and simplified normal rank for instance. To see this, denote by $\xi_{\ell m}(Z)$ the score, under some chosen system and at some time point, given to the ℓth ranked subject out of a total of m subjects. At a later time point during which time there have been a further s combined failures and censorings, then, under re-ranking, the subject will be given a score between $\xi_{\ell, m-s}(Z)$ and $\xi_{\ell-s, m-s}(Z)$ (inclusive). However, providing $\ell/m \to \theta(Z)$ ($0 < \theta(Z) < 1$) then $\xi_{\ell m}(Z) \to \tilde{g}(Z)$ say, the same limit as for $\xi_{\ell, m-s}(Z)$, $\xi_{\ell-s, m-s}(Z)$ and all intermediary values, so that this same limit is also the expected value of the score asymptotically (O'Quigley and Prentice 1991). Using normal scores instead of logit scores and vice versa leads to an asymptotic relative efficiency of 0.97 in the absence of censoring.

O'Quigley and Prentice (1991) used three different models in order to gain insight into the relative small sample power properties of the simplified versions of the logit-rank and normal rank procedures. Three models were used for values of β not too far removed from the null, concretely for β between 0.0 and 0.8. The first model, described as model (a), stipulated that $\Pr(T > t) = \exp(-\lambda(Z)t)$, $\Pr(C > c) = \exp(-\gamma\lambda(Z)t)$, $\lambda(Z) = \exp(\beta Z)$ where $\log Z \sim N(0,1)$ and $\gamma = 0.3, 1.2, 2.0$. Model (b) was the same except that $\lambda(Z) = \exp(\beta)$ when $Z < 1$ and was equal to $\exp(\beta Z)$ otherwise. Model (c) was the same as model (a) except that $\lambda(Z) = \exp(\beta Z)$ when $Z < 1$ and was equal to $\exp(\beta)$ otherwise. The motivation for considering model (a) was to see what happens when the explanatory variable is skewly distributed having relatively few high values but ones that strongly affect survival. Models (b) and (c) are plateau models. For model (b) the lower values of the covariable are not associated with survival. Some threshold value needs to be reached beyond which survival is more or less strongly associated with the explanatory variable according to the value of β. The threshold was taken to occur at the median of the explanatory variable. Model (c) is, in some sense, a mirror image of model (b) such that maximum effect occurs at the median, beyond which further increases in the explanatory variable have no additional effect. The probability of censoring, $\Pr(T > C)$, can be calculated as $\gamma/(1 + \gamma)$. The total sample size was taken to be 50, the same as in O'Brien's power comparisons. The maximum amount of censoring allowable ($\gamma = 2.0$) was 67% so that "effective sample size" varied between just over 16 and just under 39 ($\gamma = 0.3$).

O'Quigley and Prentice (1991) contrasted the logit-rank with the simplified logit-rank procedure for models (a) and (b). The respective power curves all but coincide. The same was done for model (c). Once again the curves are almost indistinguishable. The evidence then indicates that nothing is lost in using the simplified logit-rank procedure in preference to the one originally suggested by O'Brien. The same conclusion holds when comparing the normal-rank and the simplified normal-rank procedures.

Using the data of Example 1. from O'Brien's paper we obtain a value of 2.66 as the square root of the score statistic based on the logit-rank procedure. Apart from the sign, O'Brien obtained a value of 2.68, the difference being attributable to rounding errors. Our suggested simplified logit-rank procedure resulted in the value 2.69, confirming, in this example at least, our conclusions from the previous section.

Tests based on the normal-rank and simplified normal-rank procedures produced the values 2.66 and 2.69 respectively, identical in fact, up to the fourth decimal place, to those obtained above. Given the sample size of 7, this is not surprising as, for small samples, normal order statistics and uniform order statistics look very similar.

The second example concerns 335 patients, diagnosed as suffering from aplastic anemia and treated by allogeneic bone marrow transplantation in Seattle, Washington. At diagnosis the majority of patients were between 10 and 30 years of age; only six patients were aged over 49. One question concerned the possibility of a trend association between age at diagnosis and prognosis, but since, under the alternative hypothesis, it is not obvious what form any trend should take, it seemed appropriate to apply some of the approaches outlined here. Two strata were established: less than or equal to twenty years of age and greater than twenty years of age. Ignoring the strata, the overall logit-rank score test was equal to 12.48 (to be compared with a chi-squared variate on one degree of freedom), the simplified logit-rank was equal to 10.15. The corresponding normal-rank score tests gave 10.58 and 9.25 respectively. The stratified logit-rank and simplified version produced the values 15.49 and 14.10, whilst for the normal-rank analogues these figures became 13.46 and 12.96 respectively (all values again to be compared with a chi-squared variate on one degree of freedom).

The above examples and the results of the previous section would suggest little is, in general, to be gained (or lost) by re-ranking at each failure, whatever scores are to be given to the resulting ranks. To construct a counter example we would need a failure mechanism whose features were mirrored in the re-ranking process. This could arise for a certain class of time-dependent relative risk functions characterized by severe departures from the proportional hazards assumption. To see this, consider a finite sample where uniform scores are allocated at the outset and the data generated under a distant alternative with β positive. The higher scores will tend to fail first leading, under proportional hazards and a non re-ranking scheme, to a progressive decrease in the mean score of the remaining survivors, a reduction in variance and an induced skewness. For non-proportional hazards, whereby log relative-risk functions increase or decrease throughout the study, providing there is no change in sign, the same situation prevails. Re-ranking the covariates then, such that the variance remains the same and skewness is redressed might give an advantage in some situations although, apart from some small gains in controlling for size, it is not very clear which ones.

10.6   Exercises and class projects

1. Make plots of the intensity function $\alpha_i(t)$ for (i) $Z_i(t)$ a binary indicator and (ii) $Z_i(t)$ sampled from a uniform on (0,6). Do these for both the multiplicative and additive models. What is the rôle played by $Y_i(t)$? Take β to be one.

2. Fill in the steps in deriving Equation 10.6 from Equation 10.5. In Equation 10.5 replace $N_i(X_i)$ by the jump in the Kaplan-Meier curve at time X_i. Now derive an analogous expression to Equation 10.6 involving the estimated marginal survival $\hat{F}(t)$.

3. Following from the previous question, suppose that $Z_i(t)$ is an indicator variable for groups i; $i = 1, 2$. For each group we calculate a Kaplan-Meier curve $\hat{F}_i(t)$. In Equation 10.5 we now replace $N_i(X_i)$ by the jump in the Kaplan-Meier curve of the group from which the failure occurred. Derive an expression analogous to Equation 10.6.

4. Show that, if model assumptions are correct, then all of the estimators based on solving the above corresponding estimating equations converge to the same population quantity β. When β is in truth a function of time, $\beta(t)$, and the model is only an approximation, discuss the relative merits of the different estimating equations and the populations quantities to which their solutions converge.

5. Suppose that $N_i(t)$ counts events on subject i and take values greater than one but never greater than some upper limit, n say. Show that the Doob-Meyer decomposition must always apply in this case. Next, let n be allowed to increase without bound. Under which circumstances might the Doob-Meyer decomposition break down?

6. Using an actual data set involving two groups (for example the Freireich data) carry out an analysis finding estimates based on the three different estimating equations described in questions 2 to 4. Simulate non-proportional hazards data using a piecewise exponential model increasing the amount of non-proportionality by increasing the change in the absolute value of the regression coefficients. Discuss the results.

7. Within the counting process framework it is common to allow t in $U(\beta, t)$ to be unbounded. It is this feature, alongside a potentially unbounded covariate, which requires the imposition of conditions on $\mathcal{F}_t$ in order for the relevant probability spaces to be correctly specified, in particular to not fail Kolmogorov's axiom of additivity. In our other approaches to inference we have taken t to have some upper bound, written, without loss of generality as 1. Discuss the disadvantages, or advantages, of imposing similar restrictions in this context, i.e., working with $U(\beta, 1)$.

8. Using the technique of double expectation show that, when the model is correctly specified, $E\{Y_i(s)H_i(s)\} = 0$.

9. Simulate censored and uncensored observations from the exponential model with a single continuous covariate. Obtain the covariate values from three simulated distributions: a uniform, an exponential, and a log-normal. For each simulation, carry out O'Brien's test for the presence of effect. Next, replace the continuous covariate values by their ranks at the outset. Comment on the similarity between the test statistics. How do these compare to a test based on the original covariate left untransformed? Finally, replace the ranks by the normal order statistics and, again, compare the test statistics. Do this for the three suggested covariate distributions.

10. Equation 10.10 would seem to be using more information than Equation 10.11. Is this so? Which of the two equations would seem to be the more dependable? If the model were seriously misspecified which of the two equations would be the more useful? Give reasons.

11. Rather than replace a continuous covariate by its ranks we could simply derive a new covariate being binary and created from the original covariate by means of a cut-off. How would you find any such cut-off? Under which circumstances would an approach based on a cut-off lead to a more powerful test than an approach based on ranking. Under which situations would the converse be true?

Chapter 11

Inference: Small samples

11.1 Summary

The finite sample distribution of the score statistic is considered more closely. Since the other test statistics are derived from this, and the regression coefficient itself a monotonic function of the score, it is enough to restrict attention to the score statistic alone. One direct approach leads to a simple convolution expression which can be evaluated by interated integrals. It is also possible to make improvements to the large sample normal approximation via the use of saddlepoint approximations or Cornish-Fisher expansions. For these we can use the results of Corollaries 7.3 and 7.4. Corrections to the distribution of the score statistic can be particularly useful when the distribution of the explanatory variable is asymmetric. Corrections to the distribution of the score equation have a rather small impact in the case of the fourth moment but can be of significance in the case of the third moment. The calculations themselves are uncomplicated and simplify further in the case of an exponential distribution. Since we can transform an arbitrary marginal distribution to one of the exponential form, while preserving the ranks, we can then consider the results for the exponential case to be of broader generality.

11.2 Motivation

The focus of our inferential efforts, regardless of the particular technique we choose, is mostly the score statistic. For this statistic, based on the properties of the estimating equation, we can claim large sample normality. Recall that our underlying probability model is focused on the distribution of the covariate, or covariate vector, at each time t given that the subject is still at risk at this time. From this the mean and the variance of the conditional distribution can be consistently estimated and this is typically the cornerstone for the basis of any tests or confidence interval construction. Implicitly we are summarizing these key distributions by their means and variances or, at least, our best estimates of these means and variances. The fact that it is the distributions themselves that are of key interest, and not just their first two moments, suggests that we may be able to improve the accuracy of any inference if we were to take into account higher order moments. As a consequence of the main theorem it turns out that this is particularly straightforward, at least for the third moment and relatively uncomplicated for the fourth moment. In fact, corrections based on the fourth moment seem to have little impact and so only the third moment might be considered in practice.

We can incorporate information on these higher moments via a Cornish-Fisher expansion or via the use of a saddlepoint approximation. Potential improvements over large sample results would need to be assessed on a case-by-case basis, often via the use of simulation. Some limited simulations are given in O'Quigley and Benichou (2007) and suggest that these small sample corrections can lead to more accurate inference, in particular for situations where there is strong group imbalance. A practical illustration is provided in the context of an actual study concerning the prognostic impact of certain tumor markers in gastric cancer. The corrections to the usual large sample theory appear to provide more accurate inference.

11.3   Additive and multiplicative models

The likelihood for the multiplicative model falls in naturally with those for exponential families and for this reason we might anticipate large sample approximations to work well. This is indeed the case although for small samples we may still be able to obtain improvements. The additive model does not fit naturally into the exponential family class (Moolgavkar and Venzon 1987) and large sample behavior approaching normality is attained much more slowly. Here then there is room for even greater improvements over large sample approximations when applied to small samples. Distinguishing between the additive and the multiplicative models we write the intensity process, $\alpha_i(t)$, as $\alpha_i(t) = \lim_{\Delta t \to 0} (\Delta t)^{-1} \Pr\{N_i(t + \Delta t) - N_i(t) = 1 | \mathcal{F}_{t-}\}$, and our models as, $\alpha_i(t) = Y_i(t)\lambda_0(t)R\{\beta Z_i(t)\}$, where $R(r)$, the relative risk function, is equal to $\exp(r)$ for the multiplicative model and $1 + r$ for the additive model. Recall the expression for the score statistics as, $U(\beta) = U(\beta, \infty)$ where $\mathcal{Z}(t) = Z_i(t)I(t = X_i)$ and where

$$U(\beta, t) = \int_0^t \{\mathcal{Z}(s) - \mathcal{E}_\beta(Z|s)\}d\bar{N}(s) = \sum_{i=1}^n N_i(X_i)\{Z_i(X_i) - \mathcal{E}_\beta(Z|X_i)\}.$$

The stochastic integral approach to inference comes from defining $H_i(t) = Z_i(t) - \mathcal{E}_\beta(Z|t)$, noting that the score equation can be rewritten as the stochastic integral $U(\beta) = U(\beta, \infty)$ where $U(\beta, t) = \sum_{i=1}^n \int_0^t H_i(s)dN_i(s)$. Having claimed normality for the large sample distribution of $U(\beta, \infty)$ we will consequently limit attention to the first two moments, $E\{U(\beta)\}$ and $E\{U^2(\beta)\}$, given, respectively, by

$$\int_0^\infty \sum_{i=1}^n E\{Y_i(s)H_i(s)\}\lambda_i(s)ds \; ; \; \int_0^\infty \sum_{i=1}^n E\{Y_i(s)H_i^2(s)\}\lambda_i(s)ds.$$

From this we obtain the variance as, $\text{Var}\{U(\beta)\} = E\{U^2(\beta)\} - E^2\{U(\beta)\}$. These results are very general and enable us to consider a wide range of situations. The results can be used for point estimation or to test hypotheses, and thereby obtain interval estimates for parameters via inversion of tests. In order to carry out these prescriptions we will need replace $\lambda_i(t)$ by estimates. We discuss these below. Before that, in the following section, we consider again the quantities $E\{Y_i(s)H_i(s)\}$ and $E\{Y_i(s)H_i^2(s)\}$. The above moments of U require us to consider, for example, $E\{Y_i(s)H_i(s)\}$. By the use of double expectation we can see that this is equal to $E\{H_i(s)|Y_i(s) = 1\}\Pr\{Y_i(s) = 1\}$ and so, fixing s according to the outer integral, our task becomes that of studying the conditional distribution of $Z(t)$ given $(T = t, C > t)$. This has already been studied in Chapter 7 in the case of the multiplicative model. For both models we need to study more closely the distribution $P(Z \le z | T = t)$. The conditional distribution of $Z(t)$ given $T = t$ obtains in the general risk function case, in the same way as for the multiplicative model, using the definition:

Definition 11.1 *The discrete probabilities $\pi_i(\beta, t)$ are given by*

$$\pi_i(\beta, t) = \frac{Y_i(t)R\{\beta Z_i(t)\}}{\sum_{j=1}^n Y_j(t)R\{\beta Z_j(t)\}}. \tag{11.1}$$

As before, based upon the $\pi_i(\beta, t)$ we have:

Definition 11.2 *Moments of Z with respect to $\pi_i(\beta, t)$ are given by*

$$\mathcal{E}_\beta(Z^k|t) = \sum_{i=1}^{n} Z_i^k(t)\pi_i(\beta, t), \quad k = 1, 2, \ldots, \tag{11.2}$$

We also need the same definition that we gave in Equation 7.5 in order to distinguish conditionally independent censoring from independent censoring, and this was that: $\phi(z, t) = [\Pr\ (C \geq t|z)]^{-1} \int \Pr\ (C \geq t|z)$ $g(z)dz$ although, in none of the applications in this work, do we need either use or try to estimate $\phi(z, t)$. We manage to work our way around requiring full generality. The first way in which this can be accomplished, as already mentioned in Section 7.4, is under the provision that the censoring does not depend upon z. In this case, $\phi(z, t)$ will depend on neither z nor t and is, in fact, equal to one. Otherwise, under a conditionally independent censoring assumption, we can consistently estimate $\phi(z, t)$ and we call this $\hat{\phi}(z, t)$. The main theorem (Section 7.4), already applied in the case of the multiplicative model, carries over readily to the additive model.

Theorem 11.1 *Under the model and assuming that β is known, the conditional distribution function of $Z(t)$ given $T = t$ is consistently estimated by*

$$\hat{P}\{Z(t) \leq z|T = t\} = \frac{\sum_{z_i \leq z} Y_i(t)R\{\beta z_i(t)\}\hat{\phi}(z_i, t)}{\sum_{j=1}^{n} Y_j(t)R\{\beta z_j(t)\}\hat{\phi}(z_j, t)}. \tag{11.3}$$

Corollary 11.1 *For a conditionally independent censoring mechanism we have*

$$\hat{P}(Z(t) \leq z|T = t, C > t) = \sum_{j=1}^{n} \pi_j(\beta, t)I(Z_j(t) \leq z). \tag{11.4}$$

Slutsky's theorem enables us to deduce that the result still holds for β replaced by any consistent estimate. When the hypothesis of proportionality of risks is correct then the result holds for the estimate $\hat{\beta}$. We also have a further important corollary to Theorem 11.1 which holds when the model is correctly specified:

Corollary 11.2 *For $k = 1, 2, \ldots$, $\mathcal{E}_{\hat{\beta}}(Z^k|t)$ consistently estimates $E_\beta(Z^k(t)|t)$. In particular $\mathcal{E}_{\hat{\beta}}(Z|t)$ provides consistent estimates of $E_\beta(Z(t)|T = t)$ and, letting $\mathcal{V}_\beta(Z|t) = \mathcal{E}_\beta(Z^2|t) - \mathcal{E}_\beta^2(Z|t)$, then $\text{Var}(Z|t)$ is consistently estimated by $\mathcal{V}_{\hat{\beta}}(Z|t)$.*

As a consequence of Theorem 11.1 we can replace the distribution function $G(z|t)$ and the density $g(z|t)$ by consistent estimates, say $\hat{G}(z|t)$ and $\hat{g}(z|t)$ under an assumption of independent censoring. For a conditionally independent censoring mechanism these conditional distributions would also be conditioned by $C > t$ as well as $T = t$. Finally, it is convenient to subtract off the means, the zero mean distribution and density estimates denoted by $\hat{G}_0(z|t)$ and $\hat{g}_0(z|t)$.

11.4 Estimation: First two moments

Inference is based on $U(\hat{\beta}, t)$ which has been standardized to have mean zero. Our focus is then on:

$$U(\beta, t) = \int_0^t \{\mathcal{Z}(s) - \mathcal{E}_\beta(Z|s)\} d\bar{N}(s) = \sum_{i=1}^n N_i(X_i) \{Z_i(X_i) - \mathcal{E}_\beta(Z|X_i)\}$$

and the approximating formulae we work with, either Cornish-Fisher expansions or saddlepoint approximations, will require estimation of the conditional moments of Z. In the case of approximations based on a constant baseline hazard, i.e., the exponential distribution, we calculate the third and fourth moments after having subtracted off the mean and the square root of the variance. In consequence, in order to use these theoretical results in practice, we multiply by this square root and add on the mean. Note that at the true value of β, $E\{Y_i(s)H_i(s)\} = 0$. In order to estimate the variance, we replace $\lambda_i(s)ds$ by $d\hat{\Lambda}_i(s)$, for any consistent estimate $\hat{\Lambda}_i(t)$ of the cumulative hazard, in particular the one suggested by Breslow (1972), Aalen (1978) and Andersen and Gill (1982). For this we refer to Lemma 10.1 of the previous chapter. Referring again to the previous chapter we were able to use Lemma 10.1 is order to obtain a consistent estimate of $E\{U^2(\beta, \infty)\}$. Recall that this estimate coincides with that based on the partial likelihood information matrix for the multiplicative model (Cox 1972) when we replace β by $\hat{\beta}$. For the additive model this was no longer the case where minus the second derivative of the partial log-likelihood with respect to β led to Equation 10.11. The estimates based on equations (10.10) and (10.11) converge to the same population counterpart when the model generating the data is correctly specified. As pointed out in the previous chapter though, while the estimate based on Equation 10.10 is always positive in view of the Cauchy-Schwartz inequality, in finite samples, negative estimate based on (10.11) can occur. When the sample sizes are small we might anticipate this behaviour to be all the more acute and, in practice, it is often associated with difficulties in actually obtaining our estimates since the numerical algorithms themselves can begin to diverge once the routine hits any running estimate which is an inadmissible value according to the model. Transformations and approximations can be a way to alleviate these problems.

11.5 Edgeworth and saddlepoint approximations

For the proportional hazards model it is relatively easy to obtain Edgeworth corrections to the score statistic. For the particular case of the multiplicative link function the saddlepoint approximation is also straightforward. In either case we make progress by evaluating the cumulant generating function $K(\theta)$ which is defined as the expectation,

$$K(\theta) = \log E(e^{\theta Z}) = \log EE(e^{\theta Z}|t) = \log \int E(e^{\theta Z}|t)dF(t),$$

the trick of double expectation leading to quantities we can readily estimate in view of the results of Section 7.4. Furthermore, applying Corollary 7.3, we have that the difference between $E(e^{\theta Z}|t)$ and $\mathcal{E}_{\hat{\beta}}(e^{\theta Z}|t)$ or $\mathcal{E}_{\beta}(e^{\theta Z}|t)$ converges in probability to zero. For the multiplicative model we can then use the estimate $\hat{K}(\theta)$ where

$$\exp\{\hat{K}(\theta)\} = \int \left\{ \frac{\sum_i \sum_j Y_j(X_i)\exp[(\theta + \beta)Z_j]}{\sum_i \sum_j Y_j(X_i)\exp\{\beta Z_j\}} \right\} d\hat{F}(t),$$

leading to the results, after some elementary manipulation, that

$$\hat{K}'(0) = \int \mathcal{E}_{\beta}(Z|t)d\hat{F}(t); \quad \hat{K}''(0) = \int \mathcal{V}_{\beta}(Z|t)d\hat{F}(t).$$

The following results enable us to obtain the needed terms.

Lemma 11.1 *Letting* $A(\theta) = \exp\{K(\theta)\}$ *then:*

$$\left\{ \frac{\partial^p A(\theta)}{\partial \theta^p} \right\}_{\theta=0} = \int \mathcal{E}_{\beta}(Z^p(t)|t)d\hat{F}(t) \qquad (11.5)$$

The first two derivatives of $A(\theta)$ are well known and widely available from any software which fits the proportional hazards model. The third and fourth derivatives are a little fastidious although, nonetheless straightforward to obtain. Pulling all of these together we have:

Lemma 11.2 *The first four derivatives of $K(\theta)$ are obtained from:*

$$A'(\theta)/A(\theta) = K'(\theta),$$
$$A''(\theta)/A(\theta) = [K'(\theta)]^2 + K''(\theta),$$
$$A'''(\theta)/A(\theta) = [K'(\theta)]^3 + 3K'(\theta)K''(\theta) + K'''(\theta),$$
$$A^{(4)}/A(\theta) = [K'(\theta)]^4 + 6[K'(\theta)]^2 K''(\theta) + 4K'(\theta)K'''(\theta)$$
$$+ 3[K''(\theta)]^2 + K^{(4)}(\theta).$$

We can use these results in either a saddlepoint approximation or an Edgeworth approximation allowing us to gain greater accuracy than that provided by the usual assumption of large sample normality. In the light of several comparative studies of the relative merits of the two kinds of approximation we cannot really anticipate obtaining any clear answer as to which is the best to use. It will depend on the particular case under study and, typically, there is very little to choose between the two. The above results can be used in both cases. Once the particular parameters of a study (total sample size, number of groups, group imbalance, approximate distribution of the covariates) are known, then simulations can help answer this question. Simulations show that real advantages, in particular for the additive model, can be obtained by making these adjustments. Some preliminary work can lead to further simplification and, subtracting off the mean allows us to ignore $K'(\theta)$, which is equal then to zero at $\theta = 0$. Instead of referring the test statistic to the percentage points Z_α of the normal distribution we use

$$L_\alpha = Z_\alpha + (Z_\alpha^2 - 1)E\{U^3(\hat{\beta}, \infty)\}/6 + (Z_\alpha^3 - 3Z_\alpha)\left[E\{U^4(\hat{\beta}, \infty)\}\right.$$
$$\left. -3E^2\{U^2(\hat{\beta}, \infty)\}\right]/24.$$

$$(11.6)$$

The adjustment can be used either when carrying out a test of a point hypothesis or when constructing test-based confidence intervals. In the simulations we can see that the correction, relatively straightforward to implement, leads to improved control on type I error in a number of situations.

11.6　Distribution of estimating equation

The parameter β is estimated by equating to zero the score function, $U(\beta)$. Thus, $U(\hat{\beta}) = 0$ and the distribution of $\hat{\beta}$ can be investigated via the estimating equation, since

$$\Pr\left(\hat{\beta} < b\right) = \Pr\left\{U(b) < 0\right\}$$

We can use what we know about U to make statements about $\hat{\beta}$. This only works because $U(\beta)$ is monotonic in β and the same will continue to hold in the multivariate setting when considering components of the vector U. Formulating the probability statement about $\hat{\beta}$ in terms of U is particularly convenient. Two simple illustrations of this are: (1) Bayesian inference and (2) the exact distribution of a sum of independent, not necessarily identically distributed, random variables. If we have prior information on β, in the form of the density $q(\beta)$, then we can write

$$\Pr\left(\hat{\beta} < b\right) = \int \Pr\left\{U(b) < 0\right\} q(b) db,$$

being an expression in terms of total probability rather than the usual Bayes formula since, instead of a data statistic depending on the model parameter, we have a direct expression for the parameter estimate. For the small sample exact distribution of the sum we use the following lemma:

Lemma 11.3 *Let $U_1, \ldots, U_n$ be independent, not necessarily identically distributed, continuous random variables with densities $p_1(x)$ to $p_n(s)$ respectively. Let $S_n = \sum_{j=1}^{n} U_j$. Then the density, $q_n(s) = dQ_n(s)$, of S_n is given by*

$$dQ_n(s) = \int_{-\infty}^{\infty} dQ_{n-1}(s - u) dP_n(u) du.$$

We use the above form $dQ_n(s)$ in order to accommodate the discrete and the continuous cases in a single expression. The lemma is proved by recurrence of an elementary convolution result (see for example Kendall and Stewart, 1977). Following Cox (1975) and Andersen and Gill (1982) we will take the contributions to the score statistic, $U(X_i)$ to be independent with different distributions given by Theorem 7.1. We can then apply the result by letting $U_i = H_i(X_i)$ where the i indices now run over the k, rather than n failure times. The distribution of U_1 is given by $G_0(X_1)$, of U_2 by $G_0(X_2)$ and we can then construct a sequence of equations based on the above expression to finally obtain the distribution $Q_k(s)$ of the sum. Any prior information can be incorporated in this expression in the same way as before.

Higher-order moments

It is possible, in a way similar to that leading to the variance of the score statistic, to obtain third and fourth moments. We can then use the estimated cumulative hazard (equation to derive data-based estimates. Since, under the model, we have that $E\{U(\beta, t)\} = 0$ then

$$E\{U^3(\beta, \infty)\} = \int_0^\infty \sum_{i=1}^n E\{Y_i(s)H_i^3(s)\}\lambda_i(s)ds. \qquad (11.7)$$

We can estimate $E\{U^3(\beta, \infty)\}$ consistently by replacing $\lambda_i(s)ds$ and $\lambda_i(s_1)\,\lambda_j(s_2)\,ds_1ds_2$ by $R\{\hat{\beta}Z_i(s)\}d\hat{\Lambda}_0(s)$ and $R\{\hat{\beta}Z_i(s_1)\}R\{\hat{\beta}Z_j(s_2)\}$ $d\hat{\Lambda}_0(s_1)d\hat{\Lambda}_0(s_2)$ respectively.

$$E\{U^4(\beta, \infty)\} = \int_0^\infty \sum_{i=1}^n E\{Y_i(s)H_i^4(s)\}\lambda_i(s)ds$$

$$+ 6\int_0^\infty \int_0^\infty \sum_{i=1}^n \sum_{j>i} E\{Y_i(s_1)H_i^2(s_1)\}E\{Y_j(s_2)H_j^2(s_2)\}$$

$$\times \lambda_i(s_1)\lambda_j(s_2)ds_1ds_2. \qquad (11.8)$$

Again, we can estimate $E\{U^4(\beta, \infty)\}$ consistently by replacing $\lambda_i(s)ds$ and $\lambda_i(s_1)\,\lambda_j(s_2)ds_1ds_2$ by $R\{\hat{\beta}Z_i(s)\}d\hat{\Lambda}_0(s)$ and $R\{\hat{\beta}Z_i(s_1)\}$ $R\{\hat{\beta}Z_j(s_2)\}d\hat{\Lambda}_0(s_1)d\hat{\Lambda}_0(s_2)$ respectively. Note that we can replace $E\{Y_i(s)H_i^2(s)\}$, $E\{Y_i(s)H_i^3(s)\}$, $E\{Y_i(s)H_i^4(s)\}$ by observed values at time of failure or by an average taken over the risk set.

Integral transform of the baseline hazard

Note that for an arbitrary distribution function, $F(t)$, the probability integral transform tells us that the variable $Y = -\log\{1 - F(T)\}$ has a standard exponential distribution. In the case of two groups, and using $\hat{F}$ in place of F, we can transform a Kaplan-Meier curve into one approaching a standard exponential for one group and, using the same transformation, into one approaching an exponential distribution with parameter $\exp(\beta)$ in the other. For this reason a study of the moment adjustments for the special case of an exponential distribution can be of value since our interest in the baseline hazard itself is only accessory. The necessary calculations simplify.

Being a special case of the proportional hazards model, an analysis based on the case of a constant hazard enables us to compare our estimates with fixed population values. Furthermore, the estimates can still be useful when the true model is different from the exponential one. Suppose that we have a binary covariate $(0,1)$ denoting group membership of which there are π_1 in the first group and π_2 in the second $(\pi_1 + \pi_2 = 1)$, survival time is distributed exponentially with underlying hazard equal to 1 in the first group and e^β in the second and there is no censoring. Then:

$$E\left\{\left[Z - \mathcal{E}_\beta^p(Z|t)\right]^p\right\} = \pi_1 \int_0^t \{-\psi(v)\}^p\, e^{-v} dv$$

$$+ \pi_2 \int_0^t \{1 - \psi(v)\}\, e^\beta \exp(-ve^\beta) dv,$$

where

$$\psi(v) = \frac{\pi_2 e^\beta \exp(-ve^\beta)}{\pi_1 e^{-v} + \pi_2 e^\beta \exp(-ve^\beta)}.$$

Evaluating the above formula under $\beta = 0$, we have:

Corollary 11.3 *The second, third, and fourth moments of U are given by the following where U_s is U standardized to have unit variance:*

$$E\{U^2(0,\infty)\} = n\pi_1\pi_2 ; \qquad E\{U_s^3(0,\infty)\} = n^{-1/2}(\pi_1 - \pi_2)(\pi_1\pi_2)^{-1/2}$$
$$E\{U_s^4(0,\infty)\} = n^{-1}(\pi_1^3 + \pi_2^3)(\pi_1\pi_2)^{-1} + 3n^{-1}(n-1).$$

More generally, consider the case of a continuous variable Z with support I and density f. Furthermore, suppose that survival time is distributed exponentially with underlying hazard equal to $\lambda_0 R(\beta z)$ with $R(\beta z) = 1$ for $\beta = 0$, and that there is no censoring. Then, for $\beta = 0$ and $p \geq 2$, we have:

$$\int_0^\infty \sum_{i=1}^n E\{Y_i(s)H_i^p(s)\}\lambda_i(s)ds = \int_I \left[\int_0^\infty \{z - E(Z)\}^p \lambda_0 e^{-\lambda_0 t} dt\right] f(z)dz$$

and this integral can be readily evaluated so that the right-hand term becomes:

$$\int_I \{z - E(Z)\}^p \left(\int_0^\infty \lambda_0 e^{-\lambda_0 t} dt\right) f(z)dz = \int_I \{z - E(Z)\}^p f(z)dz$$

$$= E\{Z - E(Z)\}^p.$$

Therefore, the required terms can easily be evaluated from the central moments of Z. Specifically, taking the subscript s to refer to the standardized variable, we obtain:

Corollary 11.4 *The second, third, and fourth moments of U are given, respectively, by:*

$$E(U^2(0, \infty)) = nE\{Z - E(Z)\}^2,$$

$$E(U_s^3(0, \infty)) = n^{-1/2} E\{Z - E(Z)\}^3 \left[E\{Z - E(Z)\}^2\right]^{-3/2},$$

$$E(U_s^4(0, \infty)) = n^{-1} E\{Z - E(Z)\}^4 \left[E\{Z - E(Z)\}^2\right]^{-2} + 3n^{-1}(n-1),$$

where the subscript s refers to the standardized variable. Note that these results also hold for a discrete variable Z.

Integral transform of the conditional covariate distribution

Consider the case of a continuous covariate Z. A more sure way, albeit a more onerous one, to correct for asymmetries in the conditional covariate distribution, is to, once again, lean on the probability integral transform. The need to evaluate higher order moments then disappears since, by construction, the odd order moments will be zero and the fourth very close to its normal counterpart. Denote by $\hat{G}(z|t)$ the estimated conditional distribution of Z given that $T = t, C > t$, i.e.,

$$\hat{G}(z|t) = \hat{P}(Z(t) \leq z | T = t, C > t) = \sum_{j=1}^{n} \pi_j(\beta, t) I(Z_j(t) \leq z). \quad (11.9)$$

Note that the definition of $\pi_j(\beta, t)$ restricts the subjects under consideration to those in the risk set at time t. The cumulative distribution $\hat{G}(z|t)$ is restricted by both z and t. We will need to invert this function, at each point X_i corresponding to a failure. Assuming no ties in the observations (we will randomly break them if there are any) then, at each time point X_i, we order the observations Z in the risk set. We express the order statistics as $Z_{(1)} < Z_{(2)} < \ldots < Z_{(n_i)}$ where there are n_i subjects in the risk set at time X_i. We define the estimator $\tilde{G}(z|X_i)$ at time $t = X_i$ and for $z \in (Z_{(m)}, Z_{(m+1)})$ by

$$\tilde{G}(z|X_i) = \hat{G}(Z_{(m)}|X_i) + \frac{z - Z_{(m)}}{Z_{(m+1)} - Z_{(m)}} \left\{\hat{G}(Z_{(m+1)}|X_i) - \hat{G}(Z_{(m)}|X_i)\right\},$$

noting that, at the observed values $Z_{(m)}, m = 1, \ldots, n_i$, the two estimators coincide so that $\tilde{G}(z|X_i) = \hat{G}(z|X_i)$ for all values of z taken in the risk set at time X_i. Otherwise, $\tilde{G}(z|X_i)$ linearly interpolates between adjacent values of the observed order statistics $Z_{(m)}$, $m = 1, \ldots, n_i$. Also, we are assuming no ties, in which case, the function $\tilde{G}(z|X_i)$, between the values $Z_{(1)}$ and $Z_{(n_i)}$, is a strictly increasing function and can thereby be inverted. We denote the inverse function by $\tilde{G}^{-1}(\alpha)$, $0 < \alpha < 1$.

Our purpose is achieved by using, instead of $\tilde{G}^{-1}(\alpha)$ which would take us back to where we began, the inverse of the cumulative normal distribution $\Phi^{-1}(\alpha)$. We define the transform

$$Z^*_{(m)} = \Phi^{-1}\tilde{G}(Z_{(m)}|X_i), \tag{11.10}$$

noting that the transform is strictly increasing so that the order of the covariate observations in the risk set is respected. We are essentially transforming to normality via the observed empirical distribution of the covariate in the risk set. Under the null hypothesis that $\beta = 0$ the cumulative distribution $\tilde{G}(Z_{(m)}|X_i)$ is discrete uniform where each atom of probability has mass $1/n_i$. Thus, the $Z^*_{(m)}$, $m = 1, \ldots, n_i$ will be close (the degree of closeness increasing with n_i) to the expectation of the mth smallest order statistic from a normal sample of size n_i. The statistic $U(\beta)$ is then a linear sum of zero mean and symmetric variables that will be closer to normal than that for the untransformed sequence. At the same time any information in the covariate is captured via the ranks of the covariate values among those subjects at risk and so local power to departures from the null would be model dependent. Under the null the suggested transformation achieves our purpose, the mean of $U(0)$ is zero and the distribution of $U(0)$ is symmetric. Under the alternative, however, we would effectively have changed our model by the transformation and a choice of model which coincides with the mechanism generating the selection from the risk set would maximize power. The above choice would not necessarily be the most efficient. An expression for the statistical efficiency of using some particular covariate transformation model when another one generates the observations is given in O'Quigley and Prentice (1991).

The simplest way to maintain exact control over type I error using $Z^*_{(m)} = \Phi^{-1}\tilde{G}(Z_{(m)}|X_i)$ is to consider, at each observed failure time, alongside $Z^*_{(m)}$, its reflection in the origin $-Z^*_{(m)}$, such values, and any more extreme in absolute value, arising with the same probability under the null hypothesis of no effect. A non-parametric test considers the distribution of the test statistic under all possible configurations of the vector of dimension equal to the number of the observed failures having entries $Z^*_{(m)}$ or $-Z^*_{(m)}$.

The number of possibilities grows exponentially so that it is possible, with even quite small samples, to achieve almost exact control over type I error. The significance level is simply the number of tests with more extreme values than those obtained by the configuration that corresponds to the observed data themselves. This approach would be very attractive apart from the drawback of the intensity of calculation. With as few as 10 observations per group, in a two-group case, the number of cases to evaluate is over one million. Finding, say, the most extreme five percent of these requires comparisons taking us into the thousands of billions. Approximations are therefore unavoidable.

Robinson (1982) developed a simple saddlepoint approximation to the densities corresponding to paired data. Relabelling the elements of this sum as y_i, $i = 1, \ldots, 2^k$, and the null density as $\psi(u)$ then, following Robinson's development and regrouping terms, we obtain

$$\psi(u) = \left\{ \frac{k^2}{2\pi \sum y_j^2 \operatorname{sech}^2(\lambda_u y_j)} \right\}^{\frac{1}{2}} \exp\left(\sum \log \cosh(\lambda_u y_j) - \lambda_u u \right),$$

$$(11.11)$$

where $\sum y_j \tanh(\lambda_u y_j) = uk$. For any values of u, λ_u is obtained as a solution to this second equation. In the above expressions all sums range from 1 to 2^k. We can obtain the significance level by numerical integration, in particular via the use of orthogonal polynomials (Abramowitz and Stegun 1970, Ch. 25). Alternatively we can work directly with an approximation to the cumulative distribution itself (Daniels 1987). Both require numerical approximation and the results, and effort involved, are, for practical purposes, the same.

11.7 Simulation studies

O'Quigley and Benichou (2007) carried out extensive simulations on the various possible corrections to the score statistic. The conclusion is that the most effective way of correcting the score statistic is via the use of an adjustment based on the third conditional moment together with either a Cornish-Fisher expansion or a saddlepoint approximation. The impact of the fourth moment appears weak, at least in the cases we have investigated. An adjustment based on the integral transform of the conditional covariate distribution produced essentially the same effect but involved slightly more work. We considered the distribution based on the large number of simulations to be the exact distribution. This was then contrasted with the uncorrected distribution based on large sample theory and, also, the distribution corrected in particular by the use of third and fourth moments although, in view of its weak impact, corrections involving the fourth moment are not reported. Two situations defined by the nature of the covariate Z (binary or continuous) were investigated. Throughout this section, the focus is on the score statistic $U(\infty; 0)$, often expressed simply as U.

Binary Covariate

For a binary covariate $Z(0, 1)$ defining group membership, data were generated from an exponential survival model with hazard given by $\exp\beta Z$, and there was no censoring. For each case, 10,000 replications were performed. Table 1 in O'Quigley and Benichou (2007) gives theoretical values and estimates of the second, third and fourth moments of the score statistic under H_0 (defined by $\beta = 0$) and for several values of n_1 and n_2, the sample sizes in groups 1 ($Z = 0$) and 2 ($Z = 1$). From the tables we could conclude that the second moment tends to be underestimated for small sample sizes. For instance, the ratio of the mean variance estimate to the theoretical variance was 0.79 for $n_1 = n_2 = 5$ and increased to 0.86 for $n_1 = n_2 = 10$, 0.92 for $n_1 = n_2 = 20$ and 0.96 for $n_1 = n_2 = 50$. This underestimation of the variance of the score statistic has also been observed by Latta (1981). As for the third and fourth moments however, results show very good agreement between theoretical values and estimates, even for very small sample sizes (e.g. $n_1 = 5$ and $n_2 = 10$). Moreover, values of the third moment show evidence of skewness for imbalanced groups and small sample sizes (e.g., $E(U_s^3) = -1.83$ for $n_1 = 5$ and $n_2 = 10$) and some evidence of positive or negative kurtosis for small sample sizes (e.g., $E(U_s^4) = 2.90$ for $n_1 = 5$ and $n_2 = 10$). These findings suggest the possibility of making improvements over the normal approximation to the distribution of the score statistic for small sample sizes.

Table 2 from O'Quigley and Benichou (2007) studies the small-sample properties of the score test (or logrank test) under $H_0(\beta = 0)$ and for several values of n_1 and n_2. For each case, the table gives the number of replications out of 10,000 for which the standardized score statistic $U/(\text{Var}U)^{1/2}$ was greater than critical values, corresponding to the right half of the distribution, or lower than critical values, corresponding to its left half. The first figure was obtained by making no correction to the critical value, that is, by using critical values for the normal distribution that U is assumed to follow, as for the usual logrank test. The second figure was obtained by correcting the critical value with the third moment estimate via a Cornish-Fisher expansion. The table does not show the correction which takes simultaneously account of the third and fourth moment since this additional correction turned out to be small.

For balanced groups, i.e., $(n_1 = n_2)$, the distribution of the standardized statistic is symmetric but the test with no correction (usual logrank test) has levels of significance that are too high. For instance, for $n_1 = n_2 = 10$, we obtained 0.002, 0.015, 0.035, 0.630, 0.116, 0.113, 0.061, 0.033, 0.014 and 0.002 instead of 0.001, 0.010, 0.025, 0.050, 0.100, 0.100, 0.050, 0.025, 0.010 and 0.001 respectively. This feature of the logrank test was also observed by Latta (1981) who obtained very similar results with a uniform rather than exponential survival distribution. It is due to the underestimation of the variance of the score statistic. The correction based on the third moment improved matters to some extent, but not completely. For instance, for $n_1 = n_2 = 10$, the significance levels were 0.030 and 0.030 as compared to 0.035 and 0.033 with no correction, for a theoretical value of 0.025. Additionally, taking the fourth moment into account yielded almost no change at all.

For imbalanced groups, $(n_2 > n_1)$, the distribution of the logrank statistic is skewed and biased toward the larger sample size. For instance for $n_1 = 10$ and $n_2 = 50$, the significance levels were 0.044 and 0.019 instead of 0.025, and 0.073 and 0.039 instead of 0.050. Again, similar results were observed by Latta (1981) for uniform survival distributions. As could be expected, given the skewness of the distribution, the correction based on the third moment substantially corrected the asymmetry. For instance, for $n_1 = 10$ and $n_2 = 50$, the significance levels were 0.033, and 0.023 and 0.063 and 0.043 for respective theoretical values of 0.25 (former two) and 0.050 (latter two).

Table 3 from O'Quigley and Benichou (2007) showed results for data generated with $\beta = \log 1.5$. The power of the test of $\beta = 0$ can be increased or decreased by the third moment correction depending on the direction of the imbalance in the sample sizes. Here U was defined as the difference between observed and expected events in group 2 and the power was increased by the correction when $n_2 > n_1$. As before, the additional fourth moment correction had very little impact on the results.

Continuous Covariate

O'Quigley and Benichou (2007) considered a covariate Z with a symmetric distribution, namely a uniform (0,1) distribution. Results were very similar to those observed with a binary covariate and balanced groups. We calculated the theoretical values and estimates of the second, third, and fourth moments of the score statistic under $H_0(\beta = 0)$ for several values of the sample size n. As shown above, the expectation of the kth moment of the score statistic $E(U^k)$ is equal to $E(Z-EZ)^k$, the central moment of Z. The second moment was underestimated for small sample sizes. The ratios of the variance estimate to the theoretical variance were 0.71, 0.82, 0.91 and 0.95 for $n = 10, 20, 50,$ and 100, respectively. However, the third moment (equal to 0) and the fourth moment were estimated quite precisely.

Table 4 of O'Quigley and Benichou (2007) studies the small-sample properties of the score statistic under $H_0 : \beta = 0$, and $H_1 : \exp(\beta) = 1.5$. Results are analogous to those for a binary covariate with balanced groups. The distribution of the standardized score statistic is symmetric, but because of the underestimation of the variance of U, the test with no correction has significance levels that are too high. For instance, for $n = 10$, we obtained 0.002, 0.014, 0.030, 0.062, 0.114, 0.120, 0.064, 0.033, 0.015 and 0.002 instead of 0.001, 0.010, 0.025, 0.050, 0.100, 0.100, 0.05, 0.025, 0.010 and 0.001, respectively. The problem was partially corrected by taking into account the third moment. For instance, the significance levels were 0.027 and 0.030 as compared to 0.030 and 0.033 with no correction for a theoretical value of 0.025. The additional fourth-moment correction had virtually no effect. Results under $H_1(\exp \beta = 1.5)$ were parallel to results under H_0. There was a minor loss of power due to the third-moment correction and the fourth-moment correction had again virtually no effect at all.

For a covariate Z with a skewed distribution, namely a standard exponential distribution, the main results were similar. Data were generated as before. Very similar results were obtained to the case of a binary covariate with imbalanced groups. Table 5 of O'Quigley and Benichou (2007) shows an underestimation of the variance of the score statistic, with ratios of the variance estimate to the theoretical variance respectively equal to 0.71, 0.83, 0.92 and 0.95 for $n = 10, 20, 50$ and 100 respectively. However, in contrast with the case of a binary covariate, the third and fourth moments were also underestimated, severely so for a small sample size. As for the case of an imbalanced covariate, there was a marked asymmetry in the distribution of the score statistic under H_0 (Table 6 from O'Quigley and Benichou (2007)). For instance, for $n = 20$, the significance levels were 0.019 and 0.050 instead of 0.025, and 0.043 and 0.079 instead of 0.050. The third moment correction made the distribution more symmetric and partially got rid of this problem despite the underestimation of the third moment. For instance, the significance levels for $n = 20$ were 0.024 and 0.036 instead of 0.025 and 0.048 and 0.069 instead of 0.050. Again, the fourth moment correction had virtually no impact which was partly due to the underestimation in the fourth moment.

In Table 6 of O'Quigley and Benichou it was possible to observe some loss of power for $\exp \beta = 1.5$, particularly for a small sample size, when the third moments correction is applied. These studies indicate that when the covariate Z has a symmetric distribution (whether discrete or continuous), the third moment correction, while being on average 0, tends to bring the significance levels of the score test down, hence closer to the nominal value under H_0. When the covariate has a skewed distribution, the third moment correction reduces the asymmetry in the distribution of the score statistic under H_0. In both cases, the fourth moment correction has a much weaker impact and, for practical purposes, can be ignored.

11.8 Example

Rashid et al. (1982) investigated the prognostic influence of the levels of five different acute phase reactant proteins on survival in gastric cancer. The currently recognized most important prognostic factor in the disease is stage, determined at the time of operation. A staging classification simpler than the usual TNM one was developed by Rashid et al. and proved itself to be very predictive of survival. On the basis of this system, it was possible to strongly discriminate among different prognostic groups. However, since the staging information was only available at the time of surgical intervention, it could not be used to identify the patients with a poorer prognosis for whom intervention should possibly be avoided. The rationale for the use of the acute phase reactant protein is that any such information can be used preoperatively. A statistical difficulty, however, arises due to the highly non-normal, in particular highly skewed, distributional behavior exhibited by these protein measurements. It has been noted (Kalbfleisch and Prentice, 1980) that outliers can have an unbounded effect on estimates and test statistics.

A first approach might be to dichotomize, whereby all the high values are grouped together, and indeed this is the usual way in which such data are handled, the notion of "normal" and "raised" levels being common in the medical literature. Such an approach, however, sacrifices information and, given that power may be lacking due to modest sample sizes, this may not be the best approach. A second approach (O'Brien 1978, O'Quigley and Prentice 1991) transforms the explanatory variables to some familiar scale (uniform or normal order statistics for example). A third approach, that suggested in this paper, leaves the covariate scale as observed and makes higher-order corrections to the score statistic to compensate for the induced lack of normality. This third approach has an advantage over the second in that the (arbitrary) choice of scaling, necessarily impacting the result, is avoided. The second approach has an advantage over the third in that inference is rank invariant, not only with respect to the time variable, but also with respect to the explanatory variable. The data are taken from Rashid et al. (1982) where the focus was on the continuously measured variable C-reactive protein and its impact on prognosis. In the original study of Rashid et al. (1982), in addition to the well-known prognostic indicators such as stage and tumor histology, there was interest in the degree to which the pre-operative biological measurements might on their own indicate prognostic effects. In such studies it is common to define some kind of cut-off for such measures below which the patients are considered to be within the normal range, and beyond which the tumor is suspected of being particularly aggressive. The reason to consider the original measurements rather than a new variable defined on the basis of a cut-off is that there may be a gradual worsening of prognosis rather than any sudden phenomenon in effect.

A two-sided test based on the score statistic produces the value $p = 0.034$, in reasonably close agreement with the Wald test ($p = 0.042$), although sufficiently removed from the likelihood ratio test ($p = 0.066$) to suggest that the large sample approximations may be slightly suspect. Carrying out a third moment correction to the score statistic, the value $p = 0.060$ was obtained and increases to $p = 0.063$ when we apply a fourth moment correction. This is in closer agreement with the likelihood ratio test and indicates that the uncorrected score statistic may be slightly underconservative. The values $p = 0.06$ or 0.07 appear then to more accurately reflect the percentile under the null. Whether making a test based decision or using a test as a means to construct confidence intervals, inference will be more accurate when the p-value is more accurately obtained. It could be argued of course that, in this particular case, it would have been more straightforward to just calculate the likelihood ratio test which seems to be more accurate. However, in other cases, in which there is lack of agreement between the likelihood ratio test and the score test, we have no way of knowing which is the more reliable. Indeed all of the corrections outlined here could be equally well applied to the likelihood ratio test (via a Taylor expansion) instead of the score statistic.

11.9 Further points

The most useful tool in assessing which of the several approaches is likely to deliver the best rewards is that of simulation. It is difficult otherwise because, even when we can show that taking into account higher moments will reduce the order of error in an estimate, the exact value of these moments is not typically known. The further error involved in replacing them by estimates involving error can often lead us back to an overall order of error no less than we had in the first place. In some cases we can carry out exact calculation. Even here though caution is needed since if we need to evaluate integrals numerically, although there is no statistical error involved, there is a risk of approximation error. Among the three available tests based on the likelihood; the score test, likelihood ratio, and the Wald test, the score test is arguably the most satisfactory. Although all three are asymptotically equivalent, the Wald test's sensitivity to parameterization has raised questions as to its value in general situations. For the remaining two, the score test (log-rank test) has the advantage of not requiring estimation under the alternative hypothesis and has nice interpretability in terms of simple comparisons between observed and expected quantities. Indeed it is this test, the log-rank test in the case of a discrete covariate, that is by far the most used. The higher moments are also evaluated very easily, again not requiring estimation under the alternative hypothesis, and therefore it is possible to improve the accuracy of inference based on the score test at little cost. Only tests of the hypothesis $H_0 : \beta = 0$ have been discusssed. More generally, we may wish to consider testing $H_0 : \beta = \beta_0; \beta_0 \neq 0$, such a formulation enabling us to construct confidence intervals about non-null values of β. The same arguments apply

to this case also and, by extension, will lead to intervals with more accurate coverage properties. This contention has yet to be investigated
in simulations which, necessarily, would be rather more involved than
those discussed here.

11.10   Exercises and class projects

1. Use a two-sample data set such as the Freireich data and carry out
a one-sided test at the 5% level. How does the p-value change if we
make the Edgeworth correction given in Equation 11.6.

2. Repeat the above question but this time using a saddlepoint approximation.

3. Using bootstrap resampling, calculate a 95% confidence interval for
the estimated regression coefficient for the above data by the percentile
method on $\hat{\beta}$. Compare this to a 95% confidence interval obtained by
inverting the monotone function $U(b)$ and by determining values of
b for which the estimated $\Pr\{U(b) < 0\} \leq 0.025$ and $\Pr\{U(b) >
0\} \leq 0.025$.

4. Consider the following two priors on β : (1) $\Pr(\beta < -1) = 0.1$;
$\Pr(\beta > 2) = 0.1$; $\Pr(-1 < \beta < 2) = 0.8$, (2) $\Pr(\beta < 0) = 0$;
$\Pr(\beta > 1) = 0.2$; $\Pr(0 < \beta < 1) = 0.8$. Using these priors repeat the
above confidence interval calculations and comment on the impact of
the priors.

5. In clinical trials and many epidemiological investigations we are often in a position to know ranges of implausible values of the regression
coefficient. Should we incorporate this knowledge into our inferential
calculations, and if so, how?

6. Either use an existing data set or generate censored data with a
single continuous covariate. Evaluate the empirical distribution of the
covariate at each failure time in the risk set. Use several transformations of this distribution, e.g., to approximately normal, exponential
or uniform, and take as a test statistic the maximum across all considered transformations. How would you ensure correct control of type
I error for this test? What are the advantages and drawbacks to this
test?

Chapter 12

Inference: Changepoint models

12.1 Summary

Changepoint models are considered for three situations in which the models allow departures from proportional hazards to be approximated in a simple way. The first situation is a so-called time-covariate qualitative interaction (O'Quigley and Pessione 1991) or crossing hazards problem. The second situation considers a decline in regression effect that is modeled by a sudden change in effect at some unknown time point (O'Quigley and Natarajan 2004). The third situation, common in prognostic modeling, deals with inference when we wish to simplify a continuous covariate into two classes, above and below some threshold. The results of Davies (1977, 1987) are particularly useful and the conditions under which these results can be applied are described.

12.2 Motivation

We can lean on the non-proportional hazards model with intercept (see Section 6.8) in order to express the simplest departure from proportional hazards in a direction of a monotone effect. We choose, for the function $Q(t)$ in

$$\lambda(t|Z) = \lambda_0(t) \exp\{[\beta + \alpha Q(t)]Z(t)\}, \qquad (12.1)$$

the particularly simple form, $Q(t) = I(t \leq \gamma) - I(t > \gamma)$ for some value of time γ lying strictly on the interior of an interval $(\mathcal{L}, \mathcal{U})$. Such a simple structure can be of interest in a number of clinical applications. It can help us to investigate whether or not the effect of some treatment is only transitory or to what extent an initially measured prognostic variable maintains its effect throughout the study period. In studies in childhood acute lymphoblastic leukemia, Sather et al. (1981) showed the importance of re-evaluation. They demonstrated the very strong prognostic effect of leukocyte count at diagnosis and how, re-evaluating 12, 18 and 24 months after achievement of remission, this effect gradually disappeared. Indeed the Kaplan-Meier survival curves, conditional upon having survived at least 24 months, were almost indistinguishable. These same curves, initially, indicated differences so strong as to give the lower count group an estimated 70 th percentile survival beyond 5 years, to be contrasted with the higher count group of about 18 months. Gore, Pocock and Kerr (1984), in a detailed study on survival in breast cancer, give a convincing demonstration of the need, when modeling or establishing prognostic indices, to give consideration as to how strength of effect might depend on time. The studies of Gore et al. and of Sather et al. have in common substantial data bases (936 children in the leukemia study and 3922 patients in the breast cancer study) and very thorough follow-up. In such studies, complex modeling is not only warranted and useful, it is also feasible. Many prognostic studies will be much smaller, usually with samples of 100 or fewer patients, and the power needed to fit complex non-proportional hazards models may be lacking. In these situations, all that may reasonably be asked of the data are simpler questions such as are effects persistent and, if not, then is it possible to have some idea as to how long effects are maintained.

The simple changepoint model, although only ever a first approximation to an inevitably more complex reality, can then be used effectively to obtain broad and useful inferences. Effects can even change direction in certain situations and a change point model allows us to say for how long, on average, the effects are in one direction, how long in the opposite direction and at which point in time does the change in direction take place. Other non-proportional hazards models could address such questions but would be more involved from the construction and interpretation point of view. From these angles the changepoint model is much simpler. The price, however, for such simplicity is somewhat added complexity concerning the question of inference.

12.3 Some changepoint models

Anderson and Senthilselvan (1982) studied a non-proportional hazards model in which, $\lambda(t|Z(t)) = \lambda_0(t) \exp\left(\beta(t) Z(t)\right)$ and where $\beta(t) = \beta_{(1)} I(t \leq \gamma) + \beta_{(2)} I(t > \gamma)$. Thus $\beta_{(1)}$ and $\beta_{(2)}$, the regression coefficients, quantify the risk before and after the changepoint γ. In their work they estimated γ from the data and then proceeded as though γ were fixed and known. This is then relatively straightforward, not requiring any special inferential considerations, but leads clearly to procedures that do not adequately control for type 1 error as well as estimation procedures that underestimate the true error of estimates. This is because the estimates for $\beta_{(1)}$ and $\beta_{(2)}$ depend upon the estimate for the changepoint γ, and no allowance for this variability in estimation is being catered for. O'Quigley and Pessione (1991), O'Quigley (1994), and O'Quigley and Natarajan (2004) took γ as being unknown and requiring estimation. We can see that under the definition $Q(t) = I(t \leq \gamma) - I(t > \gamma)$, we can re-express the model of Anderson and Senthilselven so that $\beta_{(1)} = \beta + \alpha$ and $\beta_{(2)} = \beta - \alpha$. The model then comes under the heading of a non-proportional hazards model with intercept.

The particular problem of crossing hazards can be investigated using this model. In this case however, it can be advantageous to impose further restrictions on the parameterization (O'Quigley and Pessione 1991) and fix, in advance, $\beta = 0$. Such a parameterization, imposing an effect, before and after the changepoint, to be of the same magnitude appears very inflexible. Such inflexibility, though, can lead to a test with good power properties when specifically targeted to detect departures from the null of no effects in the direction of an alternative where effects change direction at some point but may be comparable in size before and after the changepoint.

The case of linear regression in which a continuous independent (explanatory) variable is dichotomized into two groups has been well studied. For this situation the null hypothesis is written: $H_0 : \Pr(Y \leq y|Z \leq \gamma) = \Pr(Y \leq y|Z > \gamma)$ for all y and γ in the domains of Y and Z respectively. An alternative hypothesis is structured around a shift effect so that we write $\Pr(Y \leq y|Z \leq \gamma) = \Pr(Y - \mu \leq y|Z > \gamma)$ for all y and γ in the domains of Y and Z respectively. Inferential techniques for changepoint models can be used to carry out tests or to calculate confidence intervals for μ and γ. For proportional hazards models the closest analogue to a shift model arises from the $\log - \log$

transformation. Consider then some continuous covariate z and the model,

$$\log \lambda(t|Z > \mu) = \log \lambda(t|Z \le \mu) + \beta, \tag{12.2}$$

that essentially regroups all the survival probabilities for those subjects having a value of z below some threshold γ and relates it to the marginal survival probability of all other subjects. It is straightforward to add confounding covariates to the above threshold model.

12.4   Inference when γ is known

For γ fixed and known, the model of Equation 12.1 can be fit in with the framework of Harrington et al. (1982). We fix β at the value $\hat{\beta}$, the maximum likelihood estimate of β under a proportional hazards assumption and $Q(t) = I(t \le \gamma) - I(t > \gamma) = Q_\gamma$. As before, we have the score $U_\alpha(0, \hat{\beta} : Q_\gamma)$ and the information, $\mathrm{Var}\{U_\alpha(0, \hat{\beta} : Q_\gamma)\}$, which in practice is replaced by the estimate $\widehat{\mathrm{Var}}\{U_\alpha(0, \hat{\beta} : Q_\gamma)\}$, obtained by using observed values in the place of expected ones and where $\lambda_0(s)\, ds$ is replaced by the Nelson-Aalen estimate of $d\hat{\Lambda}_0(t)$. Tests can then be carried out by referring the statistic

$$T(0, \hat{\beta} : Q_\gamma) = U_\alpha(0, \hat{\beta} : Q_\gamma) / \sqrt{\widehat{\mathrm{Var}}\{U_\alpha(0, \hat{\beta} : Q_\gamma)\}} \tag{12.3}$$

to standard normal tables. In the case of a single binary variable taking the values 0 or 1 to indicate group membership, the calculations take on a particularly simple form. At the distinct failure times indexed by i, define n_{1i}, n_{2i} to be the risk set sizes for groups 1 and 2 respectively. Letting $\psi_i = n_{2i} \exp(\hat{\beta}) / \{n_{1i} + n_{2i} \exp(\hat{\beta})\}$ and $v_i = \delta_i n_{1i} \psi_i^2 \exp(-\hat{\beta}) / n_{2i}$ we have:

Lemma 12.1 *The variance of $U_\alpha(0, \hat{\beta} : Q_\gamma)$ is $\sum_i v_i$ and we can write,*

$$U_\alpha(0, \hat{\beta} : Q_\gamma) = \sum_{X_i \le \gamma} \delta_i \{I(Z_i = 1) - \psi_i\} - \sum_{X_i > \gamma} \delta_i \{I(Z_i = 1) - \psi_i\}. \tag{12.4}$$

The test is then quite straightforward and does not bring into play any non standard techniques. We will in fact rely on the result for when γ is known to construct a rather more involved, but correct, expression for the probabilities for the distribution of the test statistic when we do not specify in advance any particular value for γ. In some sense we integrate over, or sum out, the possible values that γ can take.

12.5 Inference when γ is unknown

In real applications the time point, γ, at which the risk coefficients change in sign and/or magnitude, will not be known. Procedures that take into account all the possible values that γ might take need to be considered. A Bayesian approach could put different weights on these values. The unusual feature of the problem is that, for the specific value of α studied under the null hypothesis, i.e. $\alpha = 0$, the parameter γ is undefined. Davies (1977, 1987) considers hypothesis testing in just such situations, (namely, where there is a parameter, γ, that enters the model only under the alternative hypothesis) and studies tests of the other parameters with γ fixed. A class of tests then exists, each member of which corresponds to some particular fixed value of γ.

Under the null hypothesis, Davies (1977) shows how the overall probability we can associate with given outcomes can be evaluated by considering the limit of Bonferroni type combinations of tests. The distribution of the maximum of these can then be obtained and this was the focus of Davies' attention. He obtained a precise expression for the asymptotic formula for this distribution under the following three conditions: (1) for fixed γ, increasing values of the test statistic, $T(\gamma)$ imply increasing evidence against the null and in favor of the alternative, (2) for fixed γ and for large samples the distribution of the test statistic tends to normality, and (3) the test statistic is a continuous function of γ with a continuous derivative almost everywhere.

For the problem of concern here, we view γ as being neither known nor estimable under the null, nor can an appeal be made to large sample theory. The simplest way to proceed is to make an appeal to the results of Davies, verifying the three conditions needed for his results to apply. Firstly, note that for known γ it is easy to formulate the appropriate test based on standard theory using the above results. For a fixed value of γ lying in the interval between two successive failures, we are able to calculate this statistic $T(\gamma)$. The partial likelihood remains unchanged for values of γ lying between two successive distinct failures and hence it is sufficient to consider any fixed value of γ in this interval and in practice we will take the midpoint between adjacent failures. We will then base our test on the distribution of the statistic M where $M = \sup\{T(\gamma) : \mathcal{L} \le \gamma \le \mathcal{U}\}$ and where $T(\gamma)$ is the appropriate test statistic when γ is known and $[\mathcal{L}, \mathcal{U}]$ is the range of plausible values for γ over which we wish to focus our attention.

Lemma 12.2 *The test statistic $T(0, \hat{\beta} : Q_\gamma)$, defined in Equation 12.3, can be taken to satisfy the three conditions of Davies.*

As stated the lemma cannot be exactly correct since the third condition of Davies requires continuity of $T(0, \hat{\beta} : Q_\gamma)$, as a function of γ, with a continuous derivative everywhere. This is not immediate. However, note the following. The function $T(0, \hat{\beta} : Q_\gamma)$ is constant between failures and so we can fix a finite set of γ_i of interest: these are such that γ_i is located exactly mid-way between the $(i-1)$th and the ith distinct failure time, $i = 1, \ldots, k-1$. We may view the function $T(0, \hat{\beta} : Q_\gamma)$ as a stochastic process, indexed by γ. The first two conditions require no additional work. As for the third condition, we will first approximate the process $T(0, \hat{\beta} : Q_\gamma)$ that is a step-function in γ by considering $T_1(0, \hat{\beta} : Q_\gamma)$ for $\gamma \in [\gamma_i, \gamma_{i+1}]$ defined by

$$T_1(0, \hat{\beta} : Q_\gamma) = T(0, \hat{\beta} : Q_{\gamma_i}) + \left\{ \frac{\gamma - \gamma_i}{\gamma_{i+1} - \gamma_i} \right\} \left\{ T(0, \hat{\beta} : Q_{\gamma_{i+1}}) - T(0, \hat{\beta} : Q_{\gamma_i}) \right\}$$

It is readily verified that $T_1(0, \hat{\beta} : Q_\gamma)$ is continuous in γ and is equal to $T(0, \hat{\beta} : Q_{\gamma_\ell})$ when $\ell = i$ or $i+1$. By an application of the Weierstrauss approximation theorem (see for example Bartle 1976, Ch. 4), there exists a function $T_c(0, \hat{\beta} : Q_\gamma)$ that can be made arbitrarily close to $T_1(0, \hat{\beta} : Q_\gamma)$ and which is twice differentiable in γ. By arbitrarily close we mean, for any given $\epsilon > 0$, $\sup_\gamma |T_c(0, \hat{\beta} : Q_\gamma) - T(0, \hat{\beta} : Q_\gamma)| < \epsilon$. Clearly, $T_c(0, \hat{\beta} : Q_\gamma)$ meets the third condition of Davies. Finally, we can choose ϵ to depend upon k such that it converges to zero at a rate no slower than $\sqrt{k}$. It then follows that $T_c(0, \hat{\beta} : Q_\gamma) - T(0, \hat{\beta} : Q_\gamma)$ converges in probability to zero, at each γ_i, $0 < i < k+1$, from which (see for example Serfling 1980, Chapter 1) we can conclude that the large sample distribution of T_c is the same as that for T. The required derivatives are approximated numerically using only $T(0, \hat{\beta} : Q_{\gamma_i}) i = 1, \ldots, k+1$ so that the function T_c is only required in the conceptual construction and need not be obtained explicitly. Thus, as prescribed by Davies (1977, 1987), for γ unknown, an appropriate two-sided test should be based on the statistic $M = \sup\{T(0, \hat{\beta} : Q_\gamma) : \mathcal{L} \leq \gamma \leq \mathcal{U}\}$ where $(\mathcal{L}, \mathcal{U})$ is the range of possible values of γ. In order to calculate the relevant probabilities we use:

Lemma 12.3 *Taking $\rho(\phi, \gamma)$ to be the auto-correlation function between $T(0, \hat{\beta} : Q_\gamma)$ and $T(0, \hat{\beta} : I_\phi)$, the distribution of $M = \sup T(0, \hat{\beta} : Q_\gamma)$ is approximated by,*

$$\Pr\{M > c : \mathcal{L} \leq \gamma \leq \mathcal{U}\} \leq \Phi(-c)$$
$$+\frac{\exp(-c^2/2)}{2\pi} \int_{\mathcal{L}}^{\mathcal{U}} \{-\rho_{11}(\gamma)\}^{-1/2} d\gamma, \qquad (12.5)$$

where Φ denotes the cumulative normal distribution function and where

$$\rho_{11}(\gamma) = \{\partial^2 \rho(\phi, \gamma)/\partial \phi^2\}_{\phi=\gamma}.$$

For a two-sided test we would consider the distribution of $|T(0, \hat{\beta} : Q_\gamma)|$ rather than that of $T(0, \hat{\beta} : Q_\gamma)$. We can do this, once again, for each fixed value of γ and note that, both $T(0, \hat{\beta} : Q_\gamma)$ and $-T(0, \hat{\beta} : Q_\gamma)$, under a two-sided alternative, satisfy the first condition of Davies, the other two conditions being immediate. In practice we would work with $|T(0, \hat{\beta} : Q_\gamma)|$, positive values of c in Equation 12.5 and simply multiply the one-sided p-value by 2. This would be good enough although the issue of a limiting chi-square distribution and two-sided tests was explicitly addressed by Davies. We return to this question below. Davies suggests an approximation in which the auto-correlation function is not required. This approximation arises from:

Lemma 12.4 *The distribution of the statistic $T(0, \hat{\beta} : Q_\gamma)$ can be approximated by:*

$$\Pr\{\sup |T(0, \hat{\beta} : Q_\gamma)| > m\} \approx \Phi(-m) + (8\pi)^{-1/2} V_\rho \exp(-m^2/2),$$
$$(12.6)$$

where $V_\rho = \sum_i |T(0, \hat{\beta} : Q_{\gamma_i}) - T(0, \hat{\beta} : Q_{\gamma_{i-1}})|$, the γ_i, ranging over $(\mathcal{L}, \mathcal{U})$, are the turning points of $T(0, \hat{\beta} : \cdot)$ and m is the observed maximum of $T(0, \hat{\beta} : \cdot)$.

Turning points only occur at the k distinct failure times and, to keep the notation consistent, it suffices to take γ_i ($i = 1, \ldots, k+1$) as being located halfway between adjacent failures, $\gamma_1 = 0$ and, to take care of the edge effect, we define $\gamma_{k+1} = 2t_k - \gamma_k$.

In general, the function $\rho(\phi, \gamma)$, needed to evaluate the test statistic is unknown and the auto-correlation, ρ_{11}, is not known. It can nonetheless be consistently estimated using bootstrap resampling methods. For

ϕ and γ, taken as fixed, we can take bootstrap samples from which several pairs of $T(0, \hat{\beta} : Q_\gamma)$ and $T(0, \hat{\beta} : Q_\phi)$ can be obtained. Using these pairs, an empirical, i.e. product moment, correlation coefficient can be calculated. The bootstrap estimator of $\rho(\phi, \gamma)$ is denoted by $R(\phi, \gamma)$. Under the usual conditions for convergence of bootstrap estimates (Efron 1981), our empirical estimate provides a consistent estimate of the true value.

12.6   Maximum of log-rank type tests

As discussed in Section 9.5 one of the most commonly used and popular tests based on $U(\beta)$ is the log-rank test, a test that is in fact one of a broader family of tests that we call log-rank type tests. These procedures were developed without any specific model in mind as a series of two by two tables, drawn up at each of the failure points. For a changepoint problem we can write down a series of tables for all cutpoints located between adjacent values of the ordered covariate Z as in Table 12.1. For every cutpoint γ we can set up the series of tables, and, for simplicity of notation, we do this at each of the observations X_i, $i = 1, \ldots, n$. The only non-zero contributions arise at the actual failures, i.e., values of X_i for which $C_i > T_i$. We observe a total of $m_\gamma^-(i)$ failures for subjects having $Z \leq \gamma$ and $m_\gamma^+(i)$ failures for subjects having $Z > \gamma$. Following the reasoning of Section 9.5 we can consider $m_\gamma^-(i)$ to be a binary random variable (0,1). The total number of subjects at risk at time X_i is $n(i)$, this being divided into $n_\gamma^-(i)$ for those subjects with a covariate value less than γ and $n_\gamma^+(i)$ for those with a covariate value greater than γ.

Lemma 12.5 *Under the null hypothesis of independence between the covariate Z and survival, and disallowing the possibility of ties, for every γ, the random variable $m_\gamma^-(i)$ is Bernoulli with mean $e_\gamma^-(i)$ and variance $v_\gamma(i)$ where:*

	Dead at time X_i	Alive after X_i	Totals
$Z \leq \gamma$	$m_\gamma^-(i)$	$n_\gamma^-(i) - m_\gamma^-(i)$	$n_\gamma^-(i)$
$Z > \gamma$	$m_\gamma^+(i)$	$n_\gamma^+(i) - m_\gamma^+(i)$	$n_\gamma^+(i)$
Totals	$m(i)$	$n(i) - m(i)$	$n(i)$

Table 12.1: One member of a series of 2×2 tables arising from dichotomizing the explanatory variable Z into two groups: $Z \leq \gamma$ and $Z > \gamma$.

$$e_\gamma^-(i) = \frac{m(i)n_\gamma^-(i)}{n(i)} \quad : \quad v_\gamma(i) = m(i)\frac{n_\gamma^-(i)n_\gamma^+(i)}{n_\gamma^2(i)}\frac{\{n(i)-m(i)\}}{\{n(i)-1\}}.$$

In the presence of ties, i.e., allowing for $m(i)$ to be greater than one, the result for the mean and variance still holds since $m_\gamma^-(i)$ is then a hypergeometric variable (see for example Hill et al. 1990). For some given set of positive weights, K_i, in particular the usual log-rank weights, $K_i = 1$, the test at any given value of γ (see Section 9.5) is:

$$T(\gamma) = \left\{\sum_i K_i^2 v_\gamma(i)\right\}^{-1}\left\{\sum_i K_i\{m_\gamma^-(i) - e_\gamma^-(i)\}\right\}^2, \quad (12.7)$$

which, under the null hypothesis, would have a chi-square distribution on one degree of freedom. Not knowing which value of γ to use we maximize over all possible values taking as a test statistic $M = \sup_\gamma T(\gamma)$. Note that in much informal statistical analysis it is very common practice to use some statistic that is of the form of a maximum, or a collection of maxima. Often, the investigator will consider several, possibly a large number of potential relationships, use some simple test such as the t-test and then eliminate from consideration all those relationships that fail to achieve some significance level.

More formally this amounts to working with the order statistics and, as we have already seen, the distribution of the order statistics can be very different from the parent distribution from which they come. In micro-array analyses and proteomic studies these questions have become more sharply focused since the number of tests we may wish to carry out can be large and the difference between the parent distribution and the distribution of the larger order statistics can be great. For maximum type tests then we should explicitly take account of the maximum order statistic and use the expression given in Lemma 12.3.

12.7 Computational aspects

Several computational issues are discussed in O'Quigley and Pessione (1991). In order to evaluate the expression given in Lemma 12.3 we need calculate the auto-correlation function. This can be estimated via the use of bootstrap techniques and will require the generation of N bootstrap samples. For each such sample, the Cox parameter $\hat\beta$ is estimated and a family of test statistics, $T(0, \hat\beta : Q_\gamma)$ (indexed by γ) using Equation 12.3 is computed. The empirical correlation coefficient between $T(0, \hat{\beta} : Q_\gamma)$ and $T(0, \hat{\beta} : Q_\phi)$ can then be computed using the N pairs generated by the bootstrap procedure. The $k + 1$ ordered changepoints γ_i $(i = 1, \dots, k + 1)$ are determined before carrying out bootstrap sampling and remain fixed and independent of any chosen bootstrap sample.

For a particular bootstrap sample, some of the distinct failures will most likely not be represented. Nevertheless, in estimating $\rho(\gamma_i, \gamma_j)$ by $\hat{\rho}(\gamma_i, \gamma_j)$, the $k + 1$ positions are maintained, whether or not the distinct failure times with which they are associated are represented in a particular subsample. The term, $\rho_{11}(\cdot)$, the second derivative of the auto-correlation function needs to be evaluated and a convergent slope estimate of this derivative is given by $r_{11}(\gamma_i) = k^2 \{ \hat{\rho}(\gamma_{i+1}, \gamma_i) + \hat{\rho}(\gamma_{i-1}, \gamma_i) - 2 \}$. Note that $\rho(\phi, \gamma)$ remains constant for different values of ϕ and γ separated by the same failure points which means that we need only calculate $\hat{\rho}(\gamma_i, \gamma_j)$ for γ_i, arbitrarily located between two failures. We then define $\hat{\rho}(\tilde{\gamma}, \gamma_j) = \hat{\rho}(\gamma_i, \gamma_j)$ for points $\tilde{\gamma}$ belonging to a real line segment containing γ_i and such that the intervals $(\gamma_i, \tilde{\gamma})$ $(\gamma_i < \tilde{\gamma})$ or $(\tilde{\gamma}, \gamma_i)$ $(\gamma_i > \tilde{\gamma})$ contain no failures. The expression for the test then becomes

$$\Pr\{ M > m : \gamma \in (0, t^*) \} \leq \Phi(-m) + (2\pi k)^{-1} \exp(-m^2/2) \sum \{ -r_{11}(\gamma_i) \}^{\frac{1}{2}},$$

where $M = \sup |T(0, \hat{\beta} : I_\gamma)|$. In carrying out the bootstrap, we consider γ_{i-1}, γ_i and γ_{i+1}, generate a new sample for which we can calculate the pairs $\{ T(0, \hat{\beta} : Q_{\gamma_{i+1}}), T(0, \hat{\beta} : Q_{\gamma_i}) \}$ and $\{ T(0, \hat{\beta} : Q_{\gamma_{i-1}}), T(0, \hat{\beta} : Q_{\gamma_i}) \}$, repeat the process for the chosen number of bootstrap samples and, on the basis of this, calculate the relevant empirical correlation coefficients.

To make the derivation clearer, k has been left in the above formulae. It does, in fact, cancel and does not enter into the calculation in practice. For the two sample problem, a formula similar to that of Equation 12.4 enables us to successively calculate the scores. Denoting the jth bootstrapped score at point γ_i by $T_j^*(0, \hat{\beta} : Q_{\gamma_i})$, this can be written $T_j^*(0, \hat{\beta} : Q_{\gamma_i}) = T_j^*(0, \hat{\beta} : Q_{\gamma_{i+1}}) - (\sum_k v_k)^{-1} 2\delta_{i+1}^* \{ I(Z_{i+1} = 1) - \psi_{i+1}^* \}$, where δ_i^* and ψ_i^* are the bootstrapped realizations of δ_i and ψ_i. If, in a particular bootstrap sample, the distinct failure point between γ_i and γ_{i+1} is not represented, then clearly $\delta_{i+1}^* = 0$ and $T_j^*(0, \hat{\beta} : Q_{\gamma_i}) = T_j^*(0, \hat{\beta} : Q_{\gamma_{i+1}})$.

For the approximation in which we require the turning points of $T(0, \hat{\beta} : Q_\gamma)$, taking again γ_i as being located half way between adjacent failures and that $T(0, \hat{\beta} : Q_{\gamma_1})$ gives the standard score test for a proportional hazards model, we can successively calculate $T(0, \hat{\beta} : Q_{\gamma_i})$ via

$$T(0, \hat{\beta} : Q_{\gamma_{i+1}}) = T(0, \hat{\beta} : Q_{\gamma_i}) + 2 \sum \delta_i (1 - \psi_i), \qquad (12.8)$$

the sum being over the possibly tied failures at the distinct failure time indexed by i. If we call P_1 the upper bound for the estimated probability of the one sided test described just above then, once again following Davies (1987), we can write down the p-value of a two-sided test from:

$$P_2 = \Pr\left\{\sup_\gamma T(\gamma) > c^2 : \mathcal{L} \le \gamma \le \mathcal{U}\right\} \approx \Pr\left(\chi_p^2 > c^2\right) + \int_{\mathcal{L}}^{\mathcal{U}} \psi(\gamma) d\gamma$$

where $T(\gamma) = \sum_{i=1}^p Z_i^2(\gamma)$, the $Z_i(\gamma)$ being standardized normal variates for each value of γ and where $\psi(\gamma) = E(|\eta(\gamma)|)c^{(p-1)}\pi^{-\frac{1}{2}}2^{-\frac{1}{2}p}\exp(-c^2/2)/\Gamma(p/2+1/2)$, in which the parameter vector is of dimension p. The $\eta(\gamma)$ are independent normal variates with mean zero and variances given by the eigenvalues of $H(\gamma) - A(\gamma)'A(\gamma)$ where

$$\mathrm{Var}\begin{pmatrix} Z(\gamma) \\ \partial Z(\gamma)/\partial\gamma \end{pmatrix} = \begin{pmatrix} I & A(\gamma) \\ A'(\gamma) & H(\gamma) \end{pmatrix}.$$

In the particular case of the crossing hazards problem there is only one covariate and the calculations simplify. In this case $p = 1$ and the eigenvalues, $\lambda(\gamma)$, are therefore all equal (indeed there is only one of them) and we can write, $E(|\eta(\gamma)|) = (2\lambda(\gamma))^{\frac{1}{2}}\Gamma(p/2+1/2)/\Gamma(p/2) = (2\lambda(\gamma)/\pi)^{\frac{1}{2}}$. Using this simplification in Davies formula we have that $\psi(\gamma) = \exp(-c^2/2)\pi^{-1}\lambda^{\frac{1}{2}}(\gamma)$ so that

$$P_2 = \Pr\left\{\sup_\gamma T(\gamma) > c^2 : \mathcal{L} \le \gamma \le \mathcal{U}\right\} \approx \Pr\left(\chi_1^2 > c^2\right) + \Omega^* \exp(-c^2/2),$$

where $\lambda(\gamma_i) = k^2 R\{S(\gamma_{i+1}) - S(\gamma_i), S(\gamma_{i+1}) - S(\gamma_i)\} - R^2\{S(\gamma_{i+1}), k(S(\gamma_{i+1}) - S(\gamma_i))\}$ and $\Omega^* = (\pi)^{-1}k^{-1}\sum_i \lambda^{\frac{1}{2}}(\gamma_i)$. O'Quigley (1994) studied the approximation that consists in multiplying the one sided p-value by two with the p-value from the two sided test and concluded that the two results are so close as to be almost indistinguishable in practice. The direct two-sided test presents a gain in power but it can be considered to be negligible. Twice the one-sided p-value is easier to calculate.

12.8   Two groups with crossing hazards

For the two-group case we can let n_{1i} and n_{2i} indicate the numbers of subjects at risk, in groups 1 and 2 respectively, at the time point X_i. The variable δ_i is the usual indicator variable taking the value one should the ith subject leave the study as a result of a failure, and is zero otherwise. We use the notation $r^-(\gamma)$ to denote the number of distinct failure times in group 2 taking place at times less than γ, $r^+(\gamma)$ the number of distinct failure times in this same group taking place at times greater than γ. Note that the partial log-likelihood, L, written as $L_\gamma(\beta)$, can be decomposed so that we have:

Lemma 12.6 *We can express $L_\gamma(\beta)$ as the sum of $L_\gamma^-(\beta)$ and $L_\gamma^+(\beta)$ where*

$$
L_\gamma^-(\beta) = r^-(\gamma)\beta - \sum_\gamma^- \delta_i \log(n_{1i} + n_{2i}\, e^\beta),
$$
$$
L_\gamma^+(\beta) = \sum_\gamma^+ \delta_i \log(n_{1i} + n_{2i}\, e^{-\beta}) - r^+(\gamma)\beta,
$$

and where $\sum_\gamma^+$ denotes summation over subjects having a distinct failure time greater than γ, $\sum_\gamma^-$ summation over subjects having a distinct failure time less than or equal to γ.

Furthermore if we define $\psi_i = n_{2i}/(n_{1i}+n_{2i})$ and $v = \sum_{i=1}^n \delta_i n_{1i}\psi_i^2/n_{2i}$ we have:

Lemma 12.7 *The test statistic, $T(\gamma)$, can be written explicitly as*

$$
T(\gamma) = v^{-\frac{1}{2}}\left(r^-(\gamma) - r^+(\gamma) - \sum_\gamma^- \delta_i\psi_i + \sum_\gamma^+ \delta_i\psi_i\right).
$$

For γ unknown, we calculate, $M = \sup\{T(\gamma) : \mathcal{L} \le \gamma \le \mathcal{U}\}$ where $(\mathcal{L},\mathcal{U})$ is the range of possible values of γ. A one sided test can be based on $\Pr\{\sup T(\gamma) > c : \mathcal{L} \le \gamma \le \mathcal{U}\}$ using Equation 12.6.

Resampling strategies to evaluating p-values

For the two-sample problem it may seem natural to resample from the two estimated marginal distributions, $\hat{F}_1(t)$ and $\hat{F}_2(t)$, separately. Otherwise, we could work with the combined conditional distributions, $\hat{F}(t, z = 0)$ and $\hat{F}(t, z = 1)$, making up the overall marginal distribution. Either approach is valid from the viewpoint of large sample bootstrap theory. Following the investigation of Akritas (1986) the most satisfactory approach to resampling from censored data is that outlined by Efron (1981). Here we consider the atoms of the parent bootstrap distribution to be the pairs of time and censoring indicator. Even in this case, though, we still need to decide whether to sample from the groups 1 and 2 separately or whether to use the combined sample together with a binary variable indicating group membership. For small sample sizes the higher-order inaccuracies consequent upon use of the bootstrap would best be handled by working with a single sample. For larger sample sizes power considerations would tend to suggest preserving the initial constitution of the groups, resampling separately. Either approach is valid.

For group 1 the data consists of the n_1 pairs of observations (t_{1i}, δ_{1i}) where t_{1i} is the observed, censored or uncensored, failure time for the ith patient in this group, δ_{1i} takes the value one for an uncensored failure time and is zero otherwise. An analogous definition holds for the n_2 pairs of observations (t_{2i}, δ_{2i}) for group 2.

For the mth bootstrap take a sample of size n_j $(j = 1, 2)$ from group j with replacement $m = 1, \cdots, N$. Denote the ith observation by $(t^*_{jim}, \delta^*_{jim}), j = 1, 2 : i = 1, \cdots, n_j$. For the mth bootstrap combine the two samples to obtain a single ranked sample, thereby respecting the initial proportions as observed. This corresponds to considering the sample sizes n_1 and n_2 as being ancillary, and the conditionality principle would then give us added grounds for the separate sampling approach. Rather than lean on the asymptotic normality of $\hat{\beta}$, an intermediary solution would consist in technically carrying out the joint estimating procedure of Anderson and Senthilselvan on bootstrap samples, replacing the true sampling distribution of the pair $(\hat{\beta}, \hat{\gamma})$ by its bootstrap distribution. This would be the most non parametric approach although, for smaller sample sizes, as for instance with non parametric estimates of cumulative distributions such as the Kaplan-Meier estimate, we may anticipate poor efficiency. The procedure is appealing though because of its simplicity. For the mth bootstrap sample denote the estimates of β and γ by $\hat{\beta}^*_m$ and $\hat{\gamma}^*_m$. Over the N bootstrap samples summary estimates could be obtained although, for our parameterisation, we can exploit an, at first sight, curious phenomenon (O'Quigley and Pessione 1991) whereby, under $H_0 : \beta = 0$, the limiting bootstrap distribution turns out to be bimodal and symmetrically distributed. This is not a phenomenon attenuated by large sample sizes and is reflective of the constraints of the particular parameterization chosen along with the estimation process.

An idealized representation of this phenomenon is shown in O'Quigley and Pessione (1991) in which γ is fixed at the median survival time of the reference group, having baseline hazard from an exponential distribution. Moving away from the null hypothesis bimodality is maintained but not symmetry, mass being progressively transferred in the direction of the alternative hypothesis until such a point that all the mass is around the alternative, unimodality being thereby recovered. Since the total mass under the bimodal curve equals one, a simple test amounts to equating the p-value with the area under the smallest component of the bimodal pair.

Specifically denote the empirical bootstrap density of $\hat{\beta}$ by $b^*(\hat{\beta})$ and let $B^*(u) = \int_{-\infty}^{u} b^*(w)dw$. Then a test of H_0 can be based on $\alpha = \min\{B^*(0), 1 - B^*(0)\}$. In words, of our n_B bootstrap repetitions, there will be n_B^- giving rise to values of $\hat{\beta}$ less than zero and n_B^+ to values of $\hat{\beta}$ greater than zero. Assuming none actually give rise to the value zero itself, an event of zero probability unconditionally regardless of H_0 and asymptotically of zero probability conditional on the observed data, then:

$$\alpha = \min(n_B^+, n_B^-)/n_B : \quad n_B = n_B^+ + n_B^-. \tag{12.9}$$

For small values of α the null hypothesis is rejected and we see that for a two-sided test the significance level is simply 2α. Simulations (O'Quigley and Pessione 1991) based on an underlying two-stage exponential model for total sample size of 25 and 50 subjects in each group, and for varying degrees of censoring, indicate the nominal significance level to be accurately approximated by 2α. Furthermore, under H_0 the null distribution of α, based on 200 simulations, looks to be very close to the uniform.

The direct bootstrap approach gains in simplicity, although following Efron's recommendations when estimating percentiles, the cost of this simplicity is to be paid in greatly increased bootstrap simulation. The same broad phenomenon is observed for changepoints other than the median, albeit with bimodality becoming less and less marked as we move away from the median.

Comparison of different procedures

For the purposes of comparison, we call T_1 the test based on Lemma 12.3, T_2 the test based on Equation 12.6 and T_3 the test based on Equation 12.9. Additional potential competitors are to be found in the tests of Fleming et al. (1980), referred to here as T_4 and the procedure of Breslow et al. (1984), referred here to as T_5. All tests are two-sided, i.e., we do not specify which of the two treatment groups will fare better initially. For T_1 and T_2 the two sided approximation obtains upon multiplying the one-sided value by two. As constructed, T_3 is naturally two-sided.

O'Quigley and Pessione (1991) carried out several simulations in order to compare the different approaches. Since the bootstrap procedure is consistent, then $r_{11}(\gamma)$ will converge to $\rho_{11}(\gamma)$. For large samples this enables us to appeal to the findings of Davies whereby T_2 will be strictly conservative compared with T_1. Davies found this loss of power to be rather small. As shown in O'Quigley and Pessione (1991), the distribution of the p-value for T_3, under the null, is well approximated by the uniform.

To obtain more idea as to the null behavior of the three procedures, T_1, T_2, and T_3, 400 simulations were carried out. Data were generated under a standard exponential distribution and, for the censored case, the same distribution was used for the censoring mechanism. This second set of values was generated independently of the first so that the probability of censoring was equal to 0.5. The results are illustrated in Table 1 in O'Quigley and Pessione (1991). The main points to emerge are the following. All three tests appear to adequately control for type 1 error, T_3 being consistently closer to the nominal level than either T_1 or T_2. At times T_3 is slightly over conservative and, at small effective sample sizes, tends to be slightly lower than the nominal level. We might expect this to reflect itself in power and this is discussed in the next paragraph. The censored and uncensored cases appear to behave similarly when making a sample size adjustment for the degree of censoring.

Moving away from the null hypothesis, the question becomes one of power. Some alternative hypothesis needs to be borne in mind and again some general remarks are in order. The alternatives examined by Breslow et al. (1984), by Fleming et al. (1980) and by Stablein and Koutrouvelis (1985) lead to strong differences at different time points, so that even crossing hazards look like alternatives of the Lehmann type over large portions of the data. These tests can then be valuable since they can detect different kinds of departures from the null of no effects. T_1, T_2 and T_3 are well structured to detect possibly small departures from the null of a crossing hazards nature that may easily be confused with no effect. Note that the simulations of Breslow et al.

(1984) can only address the issue of early crossing hazards since the standard exponential model and Weibull models they use can only cross at values of t less than 1. There is a slight error in their statement at the bottom of page 1055 where, instead of $t = 1$, should be written $t = 0.25$. At this point the reference group only has a slightly greater than 20% chance of failing before this time.

Three broad situations were considered by O'Quigley and Pessione (1991): early crossing hazards (25th percentile for the reference group), median crossing hazards (50th percentile for the reference group) and late crossing hazards (75th percentile for the reference group). Data were then generated according to a piecewise exponential model having log-relative risks, before and after the change, equal to $(-0.2, 0.4)$, $(0.3, -0.5)$ and $(0.6, -0.6)$, referred to as models 1, 2, and 3 respectively. The reason for choosing these three models is to examine the relative merits of the tests in the presence of relatively weak to moderate effects. Model 3 corresponds exactly to the chosen parameterization, whereas models 1 and 2 deal with not particularly strong interactions for which the chosen parameterization only represents a first approximation. The results are summarized in Table 2 of O'Quigley and Pessione (1991).

The main points that emerged were the following. Tests T_1 and T_2 are more powerful than T_4 and T_5. The tests T_1, T_2, and T_3 are particularly impressive for changes occurring around the median and here their power advantage is far from negligible. In most cases in which the changepoint is around the median, the simple direct bootstrap test T_3 is more powerful than all the other tests. For model 3, though, its power does not exceed that of T_2 and is slightly less than that of T_1. In other situations T_3 performs less well, its power diminishing as the changepoint moves away from the median. In contrast, T_1 and T_2 retain comparable power regardless of the location of the changepoint. As predicted, T_2 is systematically, although not by very much, less powerful than T_1. The trade off will, nonetheless, often be worthwhile given the greater ease with which T_2 is calculated.

The sample sizes used and the alternatives chosen are such that none of the tests are very powerful. The numerical intensity of these procedures makes it not easy to carry out a thorough study at a large range of potential sample sizes. It nonetheless is reasonable to assume that the power advantage of T_1, T_2, and T_3 over T_4 and T_5 will be maintained. The context of this power advantage needs to be kept in mind, that of a weak to moderate effect that changes direction during

the course of the study. A very early or very late change in direction
of effect may be missed without the help of large samples. Even so,
an inference indicating a consistently positive (or negative) regression
effect, failing to detect a change in direction of this effect, would remain
broadly correct.

Proportional hazards alternatives

For proportional hazards alternatives, the proposed procedures, T_1 and
T_2, again perform very well, perhaps surprisingly at first glance in view
of the specific parameterization. Clearly some power is lost, this loss
stemming from the second term in Equation 12.5. For such alterna-
tives, the estimated changepoint will be given by γ_1 or γ_{k+1}, or values
close to these and the absolute value of the relative risk is estimated
consistently.

   In contrast the log-rank test, while being optimal in some sense for
proportional hazards, has greatly reduced, if not close to zero, power
for crossing hazard alternatives. In an analogous way, the power of T_3
is severely diminished in the presence of proportional hazards. This
was highlighted in simulations of O'Quigley and Pessione (1991) on
the basis of the exponential model described above, the "changepoint"
being fixed at the origin and a log-relative risk after this changepoint
of 0.3, 0.5 and 0.6, respectively. The probability of censoring was 50%
for the reference group leading to average overall censoring varying between 42% and 46%. As before there were 40 subjects in each group.
For the three cases the power of the log-rank test was 0.15, 0.37 and
0.53 respectively. For tests (T_1, T_2, and T_3), the corresponding powers
were (0.12, 0.10, 0.06), (0.31, 0.28, 0.05) and (0.47, 0.44, 0.05), respectively, based on 400 simulations. Whereas T_1 and T_2 perform well,
albeit less well than the log-rank test, T_3 behaves as it would under
the null hypothesis.

12.9   Illustrations

Some examples of changepoint inference are illustrated for three different situations: (1) a null hypothesis of absence of effect against a
specific alternative of crossing hazards, (2) a null hypothesis of no
effect against a specific alternative of declining regression effect and
(3) a null hypothesis of no effect against a cut-point alternative for the
covariate.

Examples involving alternatives of crossing hazards

Stablein et al. (1985) studied 90 patients in a trial comparing chemotherapy with combined chemotherapy and radiation therapy in the treatment of locally unresectable gastric cancer. Their test was significant at the 0.01 level. Here, we find that T_1, on the basis of 200 bootstrap repetitions, gave a one-sided significance level of 1.28 $\times$ 10^{-3}. For T_2 the one-sided significance level was 1.68 $\times$ 10^{-3}. A two-sided test, obtained by multiplying these values by two, is thus highly significant in agreement with the findings of Stablein et al. (1985). A significance level of 2×10^{-3} was obtained for T_3, but this required 2000 bootstraps. With only 200 bootstraps the value 0 was obtained indicating a one-sided significance less than 0.005.

The second example, for which the two methods of Breslow et al. (1984) resulted in p-values 0.018 and 0.035, concerned 35 patients with ovarian cancer, having either low or high grade tumors. For these data, the test of Fleming et al. led to a *p*-value equal to 0.002. On the basis of T_1 and 200 bootstraps we found $p = 7.7 \times 10^{-3}$, T_2 resulted in $p = 9.5 \times 10^{-3}$ and for T_3 we found $p = 1.45 \times 10^{-1}$. The third example, concerning bile-duct cancer is described by both Breslow et al. (1984) and by Fleming et al. (1980). There is a minor error in the data presented by Breslow et al., for the second group the observation 257 is missing. We found $p = 0.031, 0.038$ and 0.035 for T_1, T_2 and T_3, respectively. These values are very similar to the values obtained using their suggested approach.

Sanders et al. (1985) investigated factors associated with survival in 114 children with acute lymphoblastic leukemia. All patients were treated by allogeneic bone marrow transplant from HLA identical siblings. Associated with this therapy is the risk of developing acute graft versus host disease (AGVHD), a serious immunological complication initiated by T lymphocytes in the donor marrow. All patients routinely receive immunosuppresive therapy, in particular methotrexate, in an attempt to reduce the incidence of AGVHD. Nonetheless, some 30% will present moderate to severe symptoms of AGVHD after receiving marrow from an HLA identical sibling. A Kaplan-Meier plot of two groups, those developing AGVHD and those not (O'Quigley and Pessione 1989) shows no apparant survival differences, either in the short (1 year), medium (5 years), or long (beyond 8 years) terms.

Note that although AGVHD can be treated as a time-dependent covariate, since onset, if at all, is soon after transplant, a more straightforward analysis is appropriate. Even so, the visual impression from the Kaplan-Meier plot is backed up by a test from the proportional hazards model ($\chi^2 = 0.28$). It may well be, though, that things are somewhat more complicated than they first appear. Patients who develop AGVHD have an increased incidence of fatal cytomegalovirus pneumonia; some 36% of the children in the Sander's study suffered from this complication of which 80% subsequently died. It is hypothesized though that, conditional on surviving this early period, AGVHD may turn out to be a long-term negative risk factor, associated with increased survival probability, since we would anticipate it either being associated with or indicative of a graft-versus-leukemia effect. The null hypothesis of no effect is then worth testing against the very specific alternative of "inversion of the regression effect." In the context of testing against specific departures from proportional hazards, O'Quigley and Pessione (1989) have already considered this question for this same data set. It was necessary in that work to incorporate external information about the onset distribution of cytomegalovirus pneumonia. Specifically, the authors took the changepoint as being fixed and known at 80 days. In general, such information will not be available. In any event it is worth re-analyzing the data in the absence of any such assumption.

We find that $\sup T(\gamma) = 3.7$ and that $\exp\{-(3.7)^2/2\}\sum_i\{-r_{11}(\gamma_i)\}^{\frac{1}{2}} = 0.018$ after having estimated each $r_{11}(\gamma_i)$ on the basis of 200 bootstrap samples. The value of 3.7 for $T(\gamma)$ is attained when $\gamma = 96$ days. The null hypothesis is thus rejected at $p < 0.05$. The approximation suggested by Davies yields a p-value of 0.064. The simpler approach using Equation 12.9 led to $\alpha \leq 0.005$, i.e., $p < 0.01$, on the basis of the same number of bootstrap repetitions, of which all 200 produced estimates of $\hat{\beta}$ the same side of zero. We limited ourselves here to 200 repetitions. In practice, without the support of the other methods, and following Efron's (1987) guidelines when estimating a percentile, it is recommended to use a minimum of 1000 repetitions.

An example involving an alternative of diminishing effect

Risk factors affecting survival in breast cancer were studied in a large clinical trial at the Institut Curie in Paris, France. Histology grade is known to be one of the major factors to affect survival rates and in the Curie study there were data on the histology grade of 3908 breast cancer patients. This information took the form of five categories of grade 0, 1, 2, 3, 4 giving information on the cell differentiation of the tumor.

For an initial data analysis, the five categories were collapsed into two groups, defining a binary covariate Z such that, $Z = 0$ corresponds to histology grades 0, 1 and 2 and $Z = 1$ to the others. There were 2336 and 1572 patients in groups $Z = 0$ and $Z = 1$, respectively. As a first step, a proportional hazards model was fit to the data and the coefficient for Z was significantly different from 0, having a point estimate $\widehat{\beta} = 0.763$. Such a value indicates an estimated relative risk for the group $Z = 1$ more than double that for the group $Z = 0$. The Kaplan-Meier curves and the survival curves from the proportional hazards model are superimposed in Figure 12.1. The goodness-of-fit using the bridged score process described in Chapter 8 is shown in Figure 12.2 and Figure 12.3 and we would conclude that a changepoint model provides a significant improvement on the fit over the simple proportional hazards model.

A formal test based on the Brownian bridge is highly statistically significant although this is of less interest than the visual impression of the curve and of the improvement provided by the more elaborate model. In terms of a test based on the expression given in Lemma 12.3 (leaning on the model $Q(t) = I(t \leq \gamma) - I(t > \gamma)$) it was possible to definitely reject the null hypothesis of proportional hazards ($p < 0.001$) in favor of the alternative of a changepoint model. The

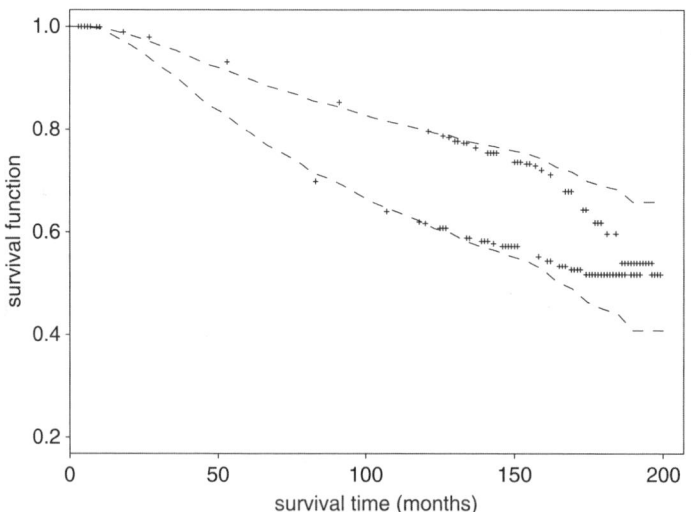

Figure 12.1: Kaplan-Meier and changpoint model estimates of the marginal survival functions for breast cancer patients based on the risk factor grade.

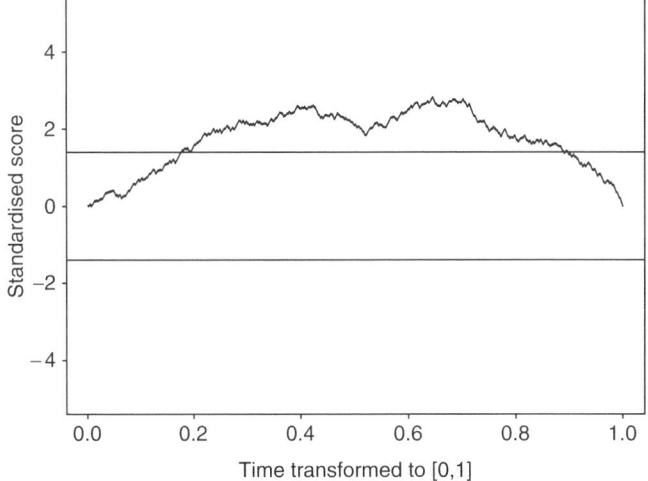

Figure 12.2: Breast cancer data: Fit of the model without changepoint. Parallel lines indicate 95% confidence intervals for the corresponding Brownian bridge process.

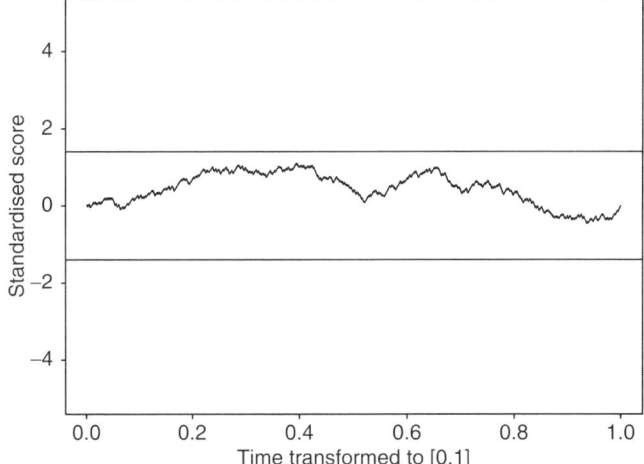

Figure 12.3: Institut Curie breast cancer data: Fit of a PH model with changepoint. The improvement in fit when compared with Figures 12.2 is noticeable.

changepoint, γ, itself was estimated at around 70.5 months, whereas the total study time was over 180 months. Further, the maximum likelihood estimate of the coefficient α was $\hat{\alpha} = 0.376$ indicating that the regression coefficient is 1.14 before the changepoint and 0.39 after. This implies that the relative risk is much higher earlier on than would be concluded from a simple proportional hazards analysis and diminishes with time such that, after about 70 months, the relative risk is a little less than half of its value during the early months of the study.

Examples involving a cut-point in a continuous covariate

We return to the data from the Curie Institute. Among the many questions concerning these data is one that has given rise to much discussion: how is prognosis impacted by the age of the patient at diagnosis? While some have argued that younger women tend to show a more aggressive form of the disease, others have claimed that differential effects are simply reflecting different patterns of incidence coupled with time-dependent regression effects. These issues are addressed to some extent by the previous example. In this example we address the simpler issue of whether or not we can consider the continuous variable age, once dichotomized into two groups, as a factor having prognostic importance. The cut-point of the continuous variable age that maximized the log-rank test was obtained at age 41 and corresponded to a p-value less than 10^{-6}. The two groups defined by this cutpoint, i.e., those aged less than 41 versus those having an age at diagnosis greater than 41 years, are significantly different. The Kaplan-Meier estimates of the survival functions of the groups defined by the estimated cut-point are represented in Figure 12.5. The estimated coefficient for age is $\hat{\beta} = -0.39$. Using the methods of explained variation (see Chapter 13) we can construct a plot of R^2 as a function of the cut-point. Note that were we to do this for a linear model the maximum of R^2 and the estimated cut-point based on maximizing the log-likelihood, would coincide. Here the residual sum of squares is not constant (as is the case in ordinary linear regression) and, as a result, the maximum of R^2 and the estimated cut-point do not coincide exactly. Nonetheless they are very close as can be seen from a visual inspection of Figure 12.4. Lausen and Schumacher (1996) study the prognostic value of the S-phase fraction for a sub-population of breast cancer patients treated at the Departement of Gynecology of the University of Freiburg. The study group is described in Pfisterer et al. (1995). The end point of interest is time from the operation to the first of the following events: recurrence, metastasis, second malignancy or death, i.e. observed recurrence-free survival. The variable S-phase fraction had an observed range of (13, 230), an empirical mean of 64, and a standard deviation of 44.4.

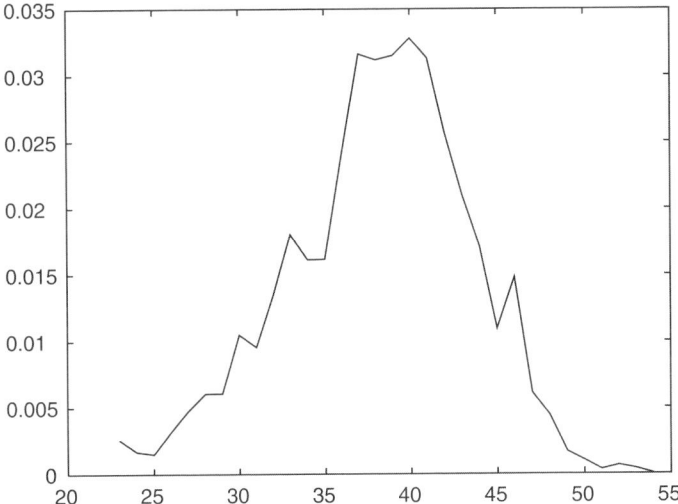

Figure 12.4: R^2 plot as a function of changepoint in age for Institut Curie breast cancer data. Likelihood and max R^2 estimates are very similar in this example.

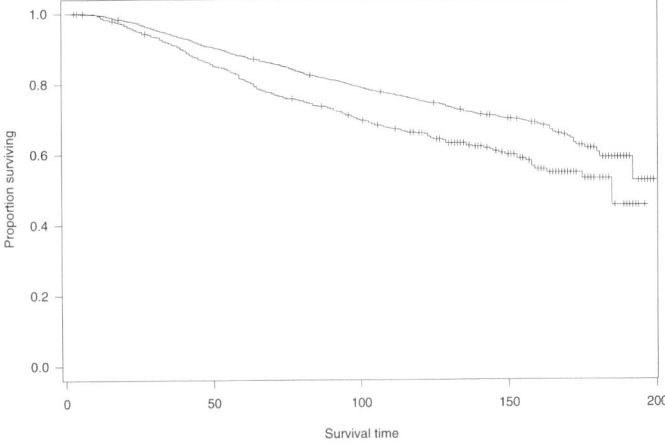

Figure 12.5: Survival for breast cancer patients according to whether age at diagnosis is greater or less than age 41.

Using the standardized log-rank estimator, Lausen and Schumacher (1996) find a cutpoint in the S-phase fraction at 107. Their approach required that they ignore a small percentage of the observations, in this case 10% of possible cut-points corresponding to the highest or lowest values. They obtained a significance level of $p = 0.123$. Contal and O'Quigley (1999), using the methods described in this chapter, obtained a slightly different cut-point of 55, although this result was also not significant ($p = 0.192$) and so the results from the two approaches are not inconsistent.

Contal and O'Quigley (1999) show that the approach of Lausen and Schumacher, when only 1% of the data are excluded, produces a result identical to theirs, also thereby underlining an observation from Lausen and Schumacher that the choice of the admissible interval can have an influence on the result. If no correction is made for the fact that the cut-point was not known and obtained via a maximization procedure we obtain a p-value of 0.007 which highlights the findings of Altman et al. (1994) that failure to statistically take account of the way in which a cut-point is identified can be very misleading.

12.10   Some guidelines in model and test selection

As a general recommendation, for simple changepoint models, the test based on the Davies approximation (referred to as T_2 and given by Equation 12.6), could be used. This test has the advantage of being easy to calculate and of only making light demands on computational power. Further power can be obtained if required, at the expense of heavier calculations, via the test T_2 obtained by using Lemma 12.3.

The direct bootstrap method, based on Equation 12.9, has the advantage of great simplicity as well as apparent gains in power for situations in the changepoint is around the median. The direct bootstrap method also appears to give more accurate control over type 1 error. Unlike T_1 and T_2 the test T_3 is lacking in power when testing a null hypothesis of no effect against an alternative in which the effect is of a proportional hazards nature. O'Quigley and Pessione (1991) pointed out that T_3 makes no use of the observed magnitude of $\hat{\beta}$ and that, in a bid to gain further power, it may be possible to construct a modified version that would incorporate such information. To the author's knowledge no study of this has been carried out.

For the simple qualitative interaction, i.e., crossing hazards, and the parameterization that fixes $\beta = 0$ and $Q(t) = I(t \leq \gamma) - I(t > \gamma)$, i.e.,

$$\lambda(t|Z) = \lambda_0(t) \exp\{\alpha Z(t)[I(t \leq \gamma) - I(t > \gamma)]\}, \qquad (12.10)$$

constraining the log-relative risk to be of the same magnitude but different sign, is a strong one. Some thought needs to be given as to the implications of this. We might consider to what extent the proposed procedures work when models other than that of Equation 12.10 hold. Suppose that the more general model of Anderson and Senthilselvan (1982) holds, i.e., $\beta(t) = \beta_1 I(t \leq \gamma) + \beta_2 I(t > \gamma)$. For local departures, still describing covariate-time qualitative interactions, although of different magnitude before and after the changepoint, the simulations of O'Quigley and Pessione (1991) indicate that the test procedure will continue to work well.

The issues involved in using one parameter to represent a model in which two parameters correctly describe the mechanism governing survival was the subject of an investigation by Gail et al. (1984), further developed by Struthers and Kalbfleisch (1986) and Xu and O'Quigley (2000). Within a wide class of possibilities, if β_1 and β_2 have the same sign, then the interaction is of a quantitative nature and despite some loss in power by using a proportional hazards model, testing procedures will be broadly correct, the single β coefficient being expressible, to a first approximation, as a weighted sum of β_1 and β_2.

Xu and O'Quigley (2000) point out that this weighting will be influenced by the censoring even when the censoring mechanism is an independent one. Even so, the weights are positive and so it remains true that we have some kind of average, even if dependent on the censoring. For β_1 and β_2 having opposite signs we have a qualitative interaction and the same idea prevails when using the model suggested above in place of a proportional hazards one. Thus, we can give a sensible interpretation to a single β, at the very least for local alternatives.

For non-local alternatives the estimate of γ, under the constraint of a single β will be inconsistent (Xu and O'Quigley 2000). The idea here, though, is not so much one of modeling, in which consistency may be required, but more to provide a powerful test against a broad range of local alternative. Once the null hypothesis is rejected we may very well wish to explore further, and the number of parameters we would want to employ in this exploration, will depend largely on the available data.

Type 1 error and power

O'Quigley and Pessione (1991) and O'Quigley and Natarajan (2004) carried out extensive simulations to study type 1 error and power.

From 1000 repetitions, each with 100 bootstraps and for a standard exponential model having 10 subjects in each group, the null hypothesis was rejected 53 times, i.e., $p = 0.053$, using the direct two sided test. Multiplying the one sided p-value by one led to $p = 0.052$. In order to get a handle on power we firstly define

$$\mathcal{E}\{S(\gamma_i), S(\gamma_j)\} = ER\{S(\gamma_i), S(\gamma_j\}$$

and noting the following results:

$$\mathcal{E}\{S(\gamma_i) - S(\gamma_j), S(\gamma_i) - S(\gamma_j)\} = \mathcal{E}\{S(\gamma_i), S(\gamma_i)\} + \mathcal{E}\{S(\gamma_j), S(\gamma_j)\} \\ -2\mathcal{E}\{S(\gamma_i), S(\gamma_j)\}$$

$$\mathcal{E}\{S(\gamma_i), k(S(\gamma_i) - S(\gamma_j))\} = k(\mathcal{E}\{S(\gamma_i), S(\gamma_i)\} - \mathcal{E}\{S(\gamma_i), S(\gamma_j)\})$$

and that for local alternatives $\mathcal{E}\{S(\gamma_j), S(\gamma_j)\} = 1$, $j = 1, \ldots, k+1$.

We then have that $\lambda(\gamma_i) = k^2(1 - \mathcal{E}^2\{S(\gamma_{i+1}), S(\gamma_i)\})$. Noting that the difference between $-r_{11}(\gamma_i)$ and $k^2(2 - \mathcal{E}\{S(\gamma_{i+1}), S(\gamma_i)\} - \mathcal{E}\{S(\gamma_{i-1}), S(\gamma_i)\})$ converges in probability to zero and letting $A(c) = \exp(-c^2/2)(\pi k)^{-1}$ we have (O'Quigley 1994) that the difference between the p-values, $2P_1 - P_2$ equals the product $kA(c)$ and

$$\sum_i (2 - \mathcal{E}\{S(\gamma_{i+1}), S(\gamma_i)\} - \mathcal{E}\{S(\gamma_{i-1}), S(\gamma_i)\})^{\frac{1}{2}} \\ -(1 - \mathcal{E}^2\{S(\gamma_{i+1}), S(\gamma_i)\})^{\frac{1}{2}},$$

a term that is strictly positive, plus a negligible term of order smaller than $1/k^2$ whenever the following condition:

$$\mathcal{E}\{S(\gamma_{i+1}), S(\gamma_i)\}(1 - \mathcal{E}\{S(\gamma_{i+1}), S(\gamma_i)\}) < 1 - \mathcal{E}\{S(\gamma_{i-1}), S(\gamma_i)\} \tag{12.11}$$

holds for all values of i. The expression is a quadratic form in $\mathcal{E}\{S(\gamma_{i+1}), S(\gamma_i)\}$ and can always seen to be verified whenever $\mathcal{E}\{S(\gamma_{i-1}), S(\gamma_i)\} < 0.75$. It is very difficult to construct a situation in which this condition is not met and it could only possibly occur for very small values of k, certainly no more than ten. Otherwise, with increasing sample size, $\mathcal{E}\{S(\gamma_{i-1}), S(\gamma_i)\}$ will take increasingly larger values, quickly tending to one. Under the classic and simple heading, where we put an upper bound on observable survival time, we have that, for $i = kq$ and $0 < q < 1$, that $S(\gamma_{i+1})$ and $S(\gamma_i)$ approach the same value. The same applies to $S(\gamma_{i-1})$ and $S(\gamma_i)$. Moreover since $\mathcal{E}\{S(\gamma_{i+1}), S(\gamma_i)\}$ and $\mathcal{E}\{S(\gamma_{i-1}), S(\gamma_i)\}$ are close to one, then the equation approaches zero and in consequence $2P_1 - P_2 \to 0$. The preceding results require that, as sample size n increases, k/n approaches some constant between 0 and 1 (i.e., the usual requirement needed to obtain asymptotic results with censored data: see for example Breslow and Crowley (1974)). However, even though in the limit $2P_1 = P_2$, we would still anticipate that for large samples $2P_1 - P_2 > 0$, since $\mathcal{E}\{S(\gamma_{i+1}), S(\gamma_i)\}$ goes to $\mathcal{E}\{S(\gamma_{i-1}), S(\gamma_i)\}$ and since on the left-hand side of Equation 12.11 we have that $\mathcal{E}\{S(\gamma_{i+1}), S(\gamma_i)\} < 1$, then the inequality will be maintained.

Table 12.2 is taken from O'Quigley and Natarajan (2004) and shows results from simulations studying type 1 error and power according to different types of decline of effect and sample size. The tests were based on 500 simulations where group 1 is standard exponential and various scenarios allowed for group 2.

It would be worth making a power comparison of the above results, obtained via the use of a simple changepoint model and those we would obtain had the exact relative risk form of model been used. Although this is pure speculation, not backed up by empirical or theoretical arguments, the author believes that the use of a "correct" model rather than an under-parameterized, over simplified model, would produce little in the way of power gains.

Table 12.2: Power estimates for different rates of decline of effect using simple changepoint model. First row uses test based on lemma 12.3, the second row the test based on Equation 12.6: 50% independent censoring: RR = relative risk.

Distribution in second group	Number of subjects		
	$n = 20$	$n = 40$	$n = 100$
H_0: Exponential with log RR = 0.7	0.014	0.022	0.046
	0.018	0.028	0.066
H_0: Exponential with log RR = 1.4	0.016	0.014	0.008
	0.024	0.016	0.012
H_1: Piecewise exponential:	0.060	0.238	0.608
piecewise log RR = (1.1, 0.9, 0.7, 0.4, 0.0)	0.086	0.252	0.614
H_1: Piecewise exponential:	0.112	0.340	0.710
piecewise log RR = (1.6, 1.4, 1.1, 0.7, 0.0)	0.128	0.370	0.722
H_1: Piecewise exponential:	0.172	0.506	0.802
piecewise log RR = (2.2, 1.9, 1.6, 1.1, 0.0)	0.204	0.512	0.810

12.11   Exercises and class projects

1. In the equation for the non-proportional hazards model with intercept fix $\beta = \lambda_0(t) = \alpha = 1$ and take $Z(t)$ to be a binary indicator variable for two groups. For each group evaluate the survival function when: (i) $Q(t) = 0$, (ii) $Q(t) = -t$, (iii) $Q(t) = I(t \leq \gamma) - I(t > \gamma)$ and where γ is the marginal median. Discuss the similarity between (ii) and (iii).

2. Describe real situations in which you would anticipate the regression effect to: (i) remain constant in time, (ii) to decrease in time and (iii) to increase in time. How would you employ a changepoint model to address this third question? Using actual data in which regression effects decline in time, fit a changepoint model. Compare the fit of the changepoint model to a model imposing constant effects.

3. Explicitly derive the results given by Equation 12.4.

4. Referring to the data of Stablein et al. (1981) carry out a proportional hazards analysis for the two groups followed by a goodness of fit test. Next, carry out an analysis that allows for an unknown changepoint. Carry out the goodness-of-fit test for this situation. Discuss the implications of the results in particular taking account of the clinical context.

5. Explicitly derive the two lemmas given in Section 12.8 for the crossing hazards problem.

6. Simulate data with a continuous covariate and with $\beta = 1.0$. Take the covariate distribution at $T > 0$ to be: (i) exponential, (ii) lognormal and (iii) extreme value. Fit a proportional hazards model based on the assumption that the covariate distribution is uniform. Describe how the biases in the estimate of β depend on the nature of the underlying distribution of the covariate. Next, use a simple changepoint model to recode the covariate into two groups. Estimate the corresponding β in this case and investigate how the estimate is influenced by the underlying covariate distribution. Comment on the simulation results.

Chapter 13

Explained variation

13.1 Summary

Some suggestions on possible measures of explained variation which have appeared in the literature are considered. Following this an outline of the recommended approach is given. Leaning upon the theory of explained variation detailed in Chapter 2 and in particular 3.9 we show how a solid theory of explained variation for proportional and non-proportional hazards regression can be established. This contrasts with a substantial body of literature on this topic, almost entirely constructed around intuitive improvisations and ad-hoc modifications to sample based quantities gleaned from classical linear regression. The main reference here is the paper by O'Quigley and Flandre (1994) which showed how the Schoenfeld residuals provide the required ingredients for the task in hand. The properties of population quantities and sample based estimates have been studied thoroughly (O'Quigley and Xu 2001) and these provide the user with the necessary confidence for their practical use.

13.2 Motivation

Referring back to Chapter 2 and Section 3.9 it is clear that the concept explained variation is a fundamental one, directly quantifying the notion of predictive ability. This quantification is a consequence of the Chebyshev inequality. As an example of a practical setting in which we are motivated to look at this, consider a study of 2174 breast cancer patients, followed over a period of 15 years at the Institut Curie

359

in Paris, France. A large number of potential and known prognostic factors were recorded. Detailed analyses of these data have been the subject of a number of communications and we focus here on a limited analysis on a subset of prognostic factors, identified as having some prognostic importance. These factors were: (1) age at diagnosis, (2) histology grade, (3) stage, (4) progesterone receptor status, and (5) tumor size. In addition to the usual model fitting and diagnostic tools, it seems desirable to be able to present summary measures estimating the percentage of explained variation and the relative importance of the different prognostic factors. We would like to be able to say, for example, that stage explains some 20% of survival but that, once we have taken account of progesterone status, age, and grade, then this figures drops to 5%. Or that adding tumor size to a model in which the main prognostic factors are already included then the explained variation increases, say, a negligible amount, specifically from 32% to 33%. Or, given that a suitable variable indicates predictability, then to what extent do we lose (or gain), in terms of these percentages, by recoding the continuous prognostic variable, age at diagnosis, into discrete classes on the basis of cutpoints.

For our situation, in which inference is rank invariant with respect to monotonic transformations on time, then from Section 3.9, we can see that this implies evaluation of the explained variation in the covariate given time rather than, the apparently more natural, explained variation of time given the covariate. For normal models the two are the same anyway and, here, we would anticipate them as being very close. In addition, we have all that is needed if we prefer to consider the explained variation of time given the covariates.

It helps to keep in mind the implication of working with the conditional distribution of the covariate given time rather than the other way around. It means that explained variation, translated as predictability as a consequence of Chebyshev's inequality, refers to the predictability of the failure ranks. Absence of effect should then translate as 0% predictability; perfect prediction of the correct ordering of the survival ranks should translate as 100%; and intermediate values are to be interpretable as providing an ordered scale, any point of which indicates precisely the amount of predictive strength in the model. These concepts are outlined below.

13.3   Finding a suitable measure of R^2

Some suggestions in the literature

The R^2 measure of explained variability, or predictive capability, is well known under a normal linear model. As pointed out by Korn and Simon (1990), and in contrast to what is oftentimes taught and written, such measures are only indirectly concerned with fit. They are directly concerned with predictability. For the proportional hazards model some correlation measures were first suggested by Harrell (1986) although it turned out that his measures depend heavily on independent censoring and can not be easily interpreted. Kent and O'Quigley (1988) developed a measure based on the Kullback-Leibler information gain and this could be interpreted as the proportion of randomness in the observed survival times explained by the covariates.

The principal difficulty in Kent and O'Quigley's measure was its complexity of calculation although a very simple approximation was suggested and appeared to work well. The Kent and O'Quigley measure was not able to accommodate time-dependent covariates. Xu and O'Quigley (1999) developed a similar measure based on information gain, using the conditional distribution of the covariates given the failure times. The measure accommodates time-dependent covariates, and is computable using standard softwares for fitting the Cox model. We consider this measure in the following chapter.

Korn and Simon (1990) suggested a class of potential functionals of interest, such as the conditional median, and evaluated the explained variation via an appropriate distance measuring the ratio of average dispersions with the model to those without a model. Their measures are not invariant to time transformation, nor could they accommodate time-dependent covariates. In this context these disadvantages are quite severe. Schemper (1990, 1994) introduced the concept of individual survival curves for each subject, with the model and without the model. Interpretation is very difficult. As with the Harrell measure, the Schemper measures depend on censoring, even when the censoring mechanism is completely independent of the failure mechanism. Schemper and Kaider (1997) proposed to estimate the correlation coefficient between failure rankings and the covariates via multiply imputing the censored failure times. Although numerically complex, and, again, not readily affording any clear interpretation, this latter coefficient of Schemper and Kaider shows promise and may be worthy of

further study. It is possible to remove the dependence on the censoring
and this has been considered by O'Quigley, Flandre and Reiner (1999)
and Schemper and Henderson (2000).

Distance measures

Explained variation is clearly based on a measure of distance. Some
authors have preferred to directly address the question of predictive
ability of any model via classes of distance measures. This is the case
for Harrell (1986), Korn and Simon (1990), Schemper (1990, 1992)
and Graf and Schumacher (1995). Apart from the measure of Harrell,
which relates to measures of information gain described in the following
chapter, all of these measures relate to those described by Schemper.

In this description of the Schemper measures we keep to his notation (Schemper 1990) in order to facilitate any comparative study the
reader may be interested in carrying out. Schemper defined S_{ij}, interpretable as an "empirical survivorship function" per individual, for
subject i at observed failure time point t_j $(j = 1, \ldots, k_i)$. The quantity
k_i will be the total number of failures should individual i correspond
to a failure; otherwise k_i is the number of failures occurring prior to
the censoring time of the individual i. $S_{ij} = 1$ for individual i at all
time points t_j for which the individual is still alive, drops to 0.5 at the
point at which the individual fails, and thereafter $S_{ij} = 0$. Note that
changing the definition of S_{ij} so that it drops to zero rather than 0.5
at the observed failure time will have a negligible impact in practice
and an impact approaching zero as sample size (number of failures)
increases.

Denote further $\bar{S}_j$ to be the Kaplan-Meier estimate of survival at
time t_j and $\bar{S}_{ij}$ the estimate of survival for individual i at time point t_j
derived from the proportional hazards model. Two different measures
of the *proportion of variability explained* were suggested, V_1 and V_2
where, for $\ell = 1, 2$:

Definition 13.1 *Schemper's proportion of variability explained is*

$$V_\ell = 1 - \frac{\sum k_i^{-1} \sum |S_{ij} - \bar{S}_{ij}|^\ell}{\sum k_i^{-1} \sum |S_{ij} - \bar{S}_j|^\ell}; \quad \ell = 1, 2. \tag{13.1}$$

For an exponential model and different relative risks, values of V_1 and
V_2 were tabulated on the basis of a single large simulation (Schemper
1990). The entries for V_2 turned out not to be based on a sum of

squares, as the above expression and Schemper's original paper indicate, but in fact on a rather less classical squared sum (Schemper 1994). Thus, the original definition for V_2 was considered to be in error by Schemper (1994) and replaced by an alternative one, say V_2^*, for when $\ell = 2$, replacing $\sum |S_{ij} - \bar{S}_{ij}|^\ell$ by $\sum (k_i^{-1} \sum |S_{ij} - \bar{S}_{ij}|)^2$ in the numerator and $\sum k_i^{-1} \sum |S_{ij} - \bar{S}_j|^\ell$ by $\sum (k_i^{-1} \sum |S_{ij} - \bar{S}_j|)^2$ in the denominator. There is something unusual, requiring further justification it would seem, in working with distances defined in terms of squared sums rather than sums of squares. The merits of such a definition were not detailed by Schemper (1994) although subsequent work (O'Quigley, Flandre and Reiner 1999; Schemper and Henderson 2000) suggest the original definition should be retained as the correct one. In support of this is the interesting observation that, for an exponential model and no censoring, the population equivalents of V_1 and V_2 converge to the same quantity.

Schemper's coefficients can be seen to depend on the unknown independent censoring mechanism (O'Quigley, Flandre and Reiner 1999, Schemper and Henderson 2000). This can however be remedied and we look at this in a later section. The Schemper coefficients are generally bounded by a number strictly less than one. This is also true in the uncensored case and, for the cases studied by Schemper (1990), the population values of V_1 and V_2 are bounded by 0.5.

Relationship between distance measures

Discussion of the relationships between different coefficients based on some measure of distance is given in Graf and Schumacher (1995). A study of the Schemper proposal and its large sample properties is enough to deduce the properties we would anticipate from closely associated measures. We return to this in Section 13.10 and point out here the way in which these coefficients are connected. It is useful to consider the population equivalents of V_1 and V_2 and we do this by considering the probability limits of the numerator and denominator in definition 13.1. If, for $\ell = 1, 2$, the numerator converges in probability to N_ℓ and the denominator to D_ℓ then we can study the population parameter θ_ℓ where $\theta_\ell = 1 - D_\ell^{-1} N_\ell$. We look at this in more detail in Section 13.10. For now we simply consider the form of N_ℓ as this brings out the relationship between the distance measures.

Korn and Simon (1990) considered squared error to be a particular kind of loss function and therefore other kinds of loss function, such as absolute error, might also be considered. The main development is around integrated squared error loss. For the numerator in their expression, let's call it N_{KS} here, we have

$$\tilde{N}_{KS} = \int \int \tilde{S}(u|z)\{1 - \tilde{S}(u|z)\}du dH_n(z). \qquad (13.2)$$

In the absence of censoring, for the population equivalent of V_2, we can construct a theoretical numerator, $\tilde{N}_2$ given by

$$\tilde{N}_2 = \int \int \int \{Y_t(u) - \tilde{S}(u|z)\}^2 d\tilde{F}(u) d\tilde{F}(t|z) dH_n(z).$$

In the uncensored case then the distance measures are closely related. The differences arise as a result of the weightings. For the Schemper coefficients these are given in terms of increments in $\tilde{F}(t)$ rather than increments in t itself. This we deduce from taking the above integral one step further where we see that:

$$\tilde{N}_2 = \int \int \tilde{S}(u|z)\{1 - \tilde{S}(u|z)\}d\tilde{F}(u) dH_n(z), \qquad (13.3)$$

which we can then compare with Equation 13.2. The same conclusion has also been obtained by Graf and Schumacher (1995). Note that monotonic transformations of t would typically impact the Korn and Simon measures, whereas the increments in $\tilde{F}(t)$, and thereby V_ℓ itself, remain unaffected. Given that inference under the proportional hazards model has this invariance property, it may be considered a desirable property of V_ℓ. Furthermore, for the broad class proposed by Korn and Simon (1990), it would be straightforward to extend their measures by adopting such a modification, in order to accommodate such a property if deemed necessary.

Recommended approach

The most transparent approach, interpretable in terms of explained variation, is that described by O'Quigley and Flandre (1994). This approach, in tune with the general theory of Section 3.9, studies the explained variation in T given the covariate vector Z, or, in order to maintain rank invariance, the explained variation of the prognostic index (Z alone in the univariate case) given T. If we stray from this we lose interpretability and, although many of the other suggestions have merit, they can run into all sorts of problems such as unknown bounds on the index, negative values, strong dependence on the censoring, even when independent of the failure mechanism and, simply, no way to interpret them. Thus, a value of 0.03, under one set of circumstances, may indicate a stronger effect than a value of 0.5, obtained under a different set of circumstances. A more solid approach can be constructed by keeping the basic theory in mind from Section 3.9. Leaning on that basic theory we can anticipate obtaining indices with meaningful properties. Even so, it is still important to investigate any properties deemed desirable, and not automatically inherited by virtue of Section 3.9.

Our recommended approach is essentially that outlined in O'Quigley and Flandre (1994). Their motivation came from linear regression where we denote $r_i(\hat{\beta})$ to be the fitted residual, i.e., the difference between the observation and its model based expectation evaluated under $\beta = \hat{\beta}$. The null residual $r_i(0)$ obtains by putting instead $\beta = 0$ and this corresponds to replacing all expectations by the overall mean. Next we calculate the average squared deviation of the observations from their model based predictions, $\sum r_i^2(\beta)/n$, leading to the well known expression for R^2, written as $R^2(\beta)$ in order to make explicit the dependence on $\hat{\beta}$, from

$$R^2(\beta) = 1 - \frac{\sum r_i^2(\beta)}{\sum r_i^2(0)}. \qquad (13.4)$$

Some additional work was needed in order for the R^2 measure of O'Quigley and Flandre to be consistent in general situations. This is achieved by weighting things correctly and this is described below. We discuss all the needed statistical properties for the measure including obtaining confidence intervals with coverage properties asymptotically the same as those for the regression coefficient estimate itself. A sum of squares decomposition, an expression for explained variation and the relationship between increasing values of the measure and predictability of the survival ranks all help form the basis for a more solid interpretation. Via simulations we compare this measure with some of the measures mentioned above. Those aspects particular to the multicovariate case are examined more closely and some general recommendations are given. The measure can also be easily extended to other relative risk models.

13.4   An R^2 measure based on Schoenfeld residuals

Recall the Schoenfeld residuals as the discrepancy between the observed value of the covariate, viewed of as having been sampled at time point X_i and its expected value,

$$r_i(\beta) = Z_i(X_i) - \mathcal{E}_\beta(Z|X_i), \tag{13.5}$$

for $\delta_i = 1$ at each observed failure time X_i. The expectation $\mathcal{E}_\beta(Z|X_i)$ is worked out with respect to an exponentially tilted distribution. The stronger the regression effects the greater the tilting, and the smaller we might expect, on average, the values $r_i^2(\beta)$ to be when compared with the residuals under the null model $\beta = 0$. Based on these residuals, a measure of explained variation, analogous to the coefficient of determination for the linear model, can be defined (O'Quigley and Flandre 1994).

Since the semiparametric model leaves inference invariant under monotonic increasing transformations of the time axis, and being able to predict at each failure time which subject is to fail is equivalent to being able to predict failure rankings of all the failed subjects, it is sensible to measure the discrepancy between the observed covariate at a given failure time and its expected value under the model. In the absence of censoring the quantity $\sum_{i=1}^n r_i^2(\hat{\beta})/n$ can be viewed as the average discrepancy between the observed covariate and its expected value under the model, whereas $\sum_{i=1}^n r_i^2(0)/n$ can be viewed as the average discrepancy without a model. This consideration led O'Quigley and Flandre (1994) to define

$$R^2(\beta) = 1 - \frac{\sum r_i^2(\beta)}{\sum r_i^2(0)} \tag{13.6}$$

This is then a clear analogue to that of R^2 for linear regression. That of itself would not be enough since there may be other possible generalizations. We need study its properties and show that an interpretation for the population equivalent in terms of explained variation holds.

Investigating the impact of censoring

The effect of censoring for large samples on $R^2(\beta)$ was studied by O'Quigley and Flandre (1994) and is so small that it can be ignored in practice, even for rates of censoring between ninety to ninety nine percent. However, if we are to obtain exact asymptotic results, in which our estimator converges to a quantity unaffected by an independent censoring mechanism, then we need to do a little extra work. This work amounts to weighting the squared Schoenfeld residuals by the increments of any consistent estimate of the marginal failure time distribution function F. Therefore, let $\hat{F}$ be the left-continuous Kaplan-Meier estimate of F, and define $W(t) = \hat{S}(t)/\sum_1^n Y_i(t)$ where $\hat{S} = 1 - \hat{F}$. Then $W(t)$ is a non-negative predictable stochastic process and, assuming there are no ties, it is straightforward to verify that $W(X_i) = \hat{F}(X_i+) - \hat{F}(X_i)$ at each observed failure time X_i, i.e., the jump of the Kaplan-Meier curve. In practice, ties, if they exist, are split randomly. We then define the quantity $\mathcal{I}(b)$ for $b = 0, \beta$ by

$$\mathcal{I}(b) = \sum_{i=1}^n \int_0^\infty \{Z_i(t) - \mathcal{E}_b(Z|t)\}^2 d\hat{F}(t)$$

or, in the more familiar counting process notation by,

$$\mathcal{I}(b) = \sum_{i=1}^n \int_0^\infty W(t)\{Z_i(t) - \mathcal{E}_b(Z|t)\}^2 dN_i(t) = \sum_{i=1}^n \delta_i W(X_i) r_i^2(b).$$
$$(13.7)$$

These quantities are, as before, averages of squared residuals, under the null model and under the best fitting model, the only difference being that the average here is weighted with respect to the increments $d\hat{F}(t)$. For large samples we will be able to assert that $\hat{F}(t)$ will be close to $F(t)$ and so our average is taken over time. With this in mind we then appeal to a broadened definition for R^2 in which:

$$R^2(\beta) = 1 - \frac{\sum_{i=1}^n \delta_i W(X_i) r_i^2(\beta)}{\sum_{i=1}^n \delta_i W(X_i) r_i^2(0)} = 1 - \frac{\mathcal{I}(\beta)}{\mathcal{I}(0)}. \qquad (13.8)$$

The definition given by O'Quigley and Flandre (1994) would be the same as above if we defined $W(t)$ to be constant and, of course, the two definitions coincide in the absence of censoring. The motivation for the introduction of the weight $W(t)$ is to obtain large sample properties of R^2 that are unaffected by an independent censoring mechanism. Viewing R^2 as a function of β turns out to be useful. In practice, we are mostly interested in $R^2(\hat{\beta})$ where $\hat{\beta}$ is a consistent estimate of β such as the partial likelihood estimate.

Population parameter Ω^2

The population parameter $\Omega^2(\beta)$ of $R^2(\hat{\beta})$ was given in O'Quigley & Flandre (1994). $R^2(\hat{\beta})$ can be considered a semi-parametric estimate of $\Omega^2(\beta)$ in as much as it is unaffected by monotonic increasing transformations on time (see Section 3.9). We will see that $\Omega^2(\beta)$ is unaffected by an independent censorship mechanism. If in addition Z is time-invariant, we also see that

$$\Omega^2(\beta) = 1 - \frac{E\{E[Z - E(Z|\mathcal{A}(T))]^2\}}{E\{E[Z - E(Z|\mathcal{B}(T)]^2\}}, \qquad (13.9)$$

where $\mathcal{A}(t) = \{t\}$ and $\mathcal{B}(t) = \{u : u \geq t\}$ so that, in view of equation (3.32), $\Omega^2(\beta)$ has the interpretation of the proportion of explained variation. This of itself would not be interesting enough and we also show that this choice of $\mathcal{B}$ is a sensible one. In fact, the results for the above choice, chosen to accommodate sequential conditioning on the risk sets, are very close to those arising under the definition $\mathcal{B}(t) = \mathcal{T}$ (see Table 13.1). Indeed, for practical purposes of interpretability we can take $\Omega^2(\beta)$ to be defined as in the following equation where the approximation symbol is replaced by an equality symbol, i.e.,

$$\Omega^2(\beta) \approx \frac{\text{Var}\{E(Z|T)\}}{\text{Var}(Z)}.$$

O'Quigley and Flandre showed that $\Omega^2(\beta)$ depends only relatively weakly on different covariate distributions, and values of $\Omega^2(\beta)$ give a good reflection of strength of association as measured by β, tending to 1 for high but plausible values of β. The numerical results support the conjecture that Ω^2 increases with the strength of effect, thereby agreeing with the third stipulation of Kendall (1975, p. 4) for a measure of rank correlation. The first two stipulations were that perfect agreement or disagreement should reflect itself in a coefficient of absolute

Table 13.1: Ω^2 as a function of β.

covariate*	c	c	d	c	c	c	d
β	0	0.7	0.7	1.4	2.8	4.2	4.2
$\mathcal{B}(t) = \{u : u \geq t\}$	0.0002	0.0990	0.0979	0.2844	0.5887	0.7577	0.8728
$\mathcal{B}(t) = \mathcal{T}$	0.0018	0.0998	0.0985	0.2848	0.5889	0.7578	0.8728

* Covariate distribution: d – binary, c – uniform. Data are simulated under the same mechanism as that described below.

value 1; the third stipulation that for other cases the coefficient should have absolute value less than 1, and in some acceptable sense increasing values of the coefficient should correspond to increasing agreement between the ranks. Here we have a squared coefficient, and Kendall's stipulations are considered in a broader sense because we are not restricted to the ranks of the covariates in the semiparametric context. In the next section we will show that $\Omega^2(\beta) \to 1$ as $|\beta| \to \infty$ and that it increases with the ability to explain survival rankings by the covariates. Before that, we look at a closely related quantity which turns out to be of use.

Alternative measure $R_{\mathcal{E}}^2$

For mostly theoretical purposes we also consider an alternative definition to R^2, in which we use the expected (with respect to the π's) rather than the observed squared residuals. Consider then

$$\mathcal{J}(\beta, b) = \int_0^\infty W(t) \sum_{j=1}^n \pi_j(\beta, t) \{Z_j(t) - \mathcal{E}_b(Z|t)\}^2 d\bar{N}(t)$$

$$= \sum_{i=1}^n \delta_i W(X_i) \mathcal{E}_\beta \{r_i^2(b)|X_i\}$$

and define

$$R_{\mathcal{E}}^2(\beta) = 1 - \frac{\sum_{i=1}^n \delta_i W(X_i) \mathcal{E}_\beta \{r_i^2(\beta)|X_i\}}{\sum_{i=1}^n \delta_i W(X_i) \mathcal{E}_\beta \{r_i^2(0)|X_i\}} = 1 - \frac{\mathcal{J}(\beta, \beta)}{\mathcal{J}(\beta, 0)}. \quad (13.10)$$

Our experience indicates that when the proportional hazards model correctly generates the data, $R_{\mathcal{E}}^2$ will be very close in value to R^2. Indeed we will show, under the model, that $|R^2(\hat{\beta}) - R_{\mathcal{E}}^2(\hat{\beta})|$ converges to zero in probability. This coefficient is of interest in its own right although our main purpose here is to use it for developing properties of the next section. It can also be used to construct confidence intervals for the population quantity $\Omega^2(\beta)$, intervals which have, for increasing sample size, exactly the same coverage properties of those for $\hat{\beta}$ itself. Another angle to understand $\mathcal{J}(\beta, b)$ follows from taking the expectation of $\mathcal{I}(b)$ under the model, using the results for counting processes (see for example Fleming and Harrington 1991) we have

$$E\{\mathcal{I}(b)\} = \sum_{i=1}^n \int_0^\infty E\{W(t)[Z_i(t) - \mathcal{E}_b(Z|t)]^2 Y_i(t) \exp[\beta Z_i(t)]\} d\Lambda_0(t),$$

where $\Lambda_0(t) = \int_0^t \lambda_0(s)ds$. If we replace the unknown Λ_0 by the Nelson-Aalen estimate (Breslow 1972, 1974) and the expectations under the integral by the observed quantities, then we recover $\mathcal{J}(\beta, b)$ as an estimate of $E\{\mathcal{I}(b)\}$. It is also straightforward to verify that $\mathcal{J}(\beta, \beta)$ is the weighted information of the Cox model.

13.5   Finite sample properties of R^2 and $R_{\mathcal{E}}^2$

We have the following immediate lemmas:

Lemma 13.1 *Viewing R^2 as a function of β then: $R^2(0) = 0$ and $R^2(\beta) \leq 1$.*

Lemma 13.2 *$R^2(\beta)$ is invariant under linear transformations of Z and monotonically increasing transformations of T.*

The following lemma is not a precise result, although we have a precise equivalent for large samples. It provides some insight into R^2, viewed as a function of β. It also indicates why, apart from theoretical interest, only $R^2(\hat{\beta})$ need concern us.

Lemma 13.3 *$R^2(\beta)$ as a function of β, reaches its maximum around $\hat{\beta}$.*

Proofs of the above are similar to those given by O'Quigley and Flandre (1994). More details are provided in the chapter on proofs. Note that R^2, unlike $R_{\mathcal{E}}^2$ and Ω^2, cannot be guaranteed to be non-negative. A negative value for R^2 is nonetheless difficult to obtain in practice, corresponding to the unusual case where the best fitting model, in a least squares sense, provides a poorer fit than the null model. $R^2(\hat{\beta})$ will only be slightly negative in such cases if $\hat{\beta}$ is very close to zero.

Lemma 13.4 *An approximate sums of squares decomposition holds for r_i^2 and holds exactly in the following expression:*

$$\mathcal{E}_\beta\{r_i^2(0)|X_i\} = \mathcal{E}_\beta\{r_i^2(\beta)|X_i\} + \{\mathcal{E}_\beta(Z|X_i) - \mathcal{E}_0(Z|X_i)\}^2. \quad (13.1)$$

Both the approximate and the exact sum of squares decomposition, outlined in more detail below, are valuable in underlining the great similarity between proportional hazards models and linear models. Although we do not pursue the idea it would be quite possible to develop for the proportional hazards model a whole theory for testing and fit based on sums of squares and analysis of variance type decompositions. Even F-tests can be constructed, although, at the present time, there appears to be no obvious advantage to any such alternative approach. One consequence of the above breakdown is:

Lemma 13.5 *The coefficient $R_\mathcal{E}^2(\beta)$ can be reexpressed as:*

$$R_\mathcal{E}^2(\beta) = \frac{\sum_{i=1}^n \delta_i W(X_i)\{\mathcal{E}_\beta(Z|X_i) - \mathcal{E}_0(Z|X_i)\}^2}{\sum_{i=1}^n \delta_i W(X_i)\mathcal{E}_\beta\{r_i^2(0)|X_i\}}.$$

The re-expression of $R_\mathcal{E}^2(\beta)$ in the lemma is helpful in obtaining the further lemmas:

Lemma 13.6 *As a function of β, $0 \leq R_\mathcal{E}^2(\beta) \leq 1$, and $R_\mathcal{E}^2(0) = 0$.*

Whereas $R^2(\beta)$ depends on the observations directly, $R_\mathcal{E}^2(\beta)$ is a function of expectations across the observations and although, at least for correctly specified models, there will be close agreement between the $R^2(\hat{\beta})$ and $R_\mathcal{E}^2(\hat{\beta})$ (a result made more precise below), the two coefficients behave very differently when viewed as functions of β. In particular, in contrast to Lemma 13.3, we have:

Lemma 13.7 *As $|\beta| \to \infty$ then $R_\mathcal{E}^2(\beta) \to 1$.*

We also have:

Lemma 13.8 *$R_\mathcal{E}^2(\beta)$ is invariant under linear transformations of Z and monotonically increasing transformations of T.*

The proof of the linearity property follows in the same way as for R^2 (O'Quigley and Flandre 1994), and an outline of the proof of monotonicity is given in the chapter on proofs. The figure helps illustrate the contrasting behaviors of the two coefficients, seen as functions of β. It is clear that $R^2(\beta)$ as a function of β does not increase to 1 as $|\beta| \to \infty$, but rather reaches its maximum near $\hat{\beta}$. The monotonicity property of $R_\mathcal{E}^2(\beta)$ also has an interesting connection to the literature on the efficiency of the Cox model, which has also noted that the information $\mathcal{J}(\beta, \beta) \to 0$ as $|\beta| \to \infty$ (Efron 1977, Oakes 1977, Kalbfleisch and Prentice 1980 Section 4.7).

13.6   Large sample properties

The most straightforward approach is to define the population parameter $\Omega^2(\beta)$ as the probability limit of $R_\mathcal{E}^2(\beta)$ as $n \to \infty$. We can then investigate separately how meaningful is $\Omega^2(\beta)$, in particular how it can be viewed as an index of explained variation. We then need to show that $R^2(\hat{\beta})$ converges in probability to $\Omega^2(\beta_0)$ where β_0 is the "true" value under which the data are generated. Let

$$S^{(r)}(\beta, t) = n^{-1} \sum_{i=1}^{n} Y_i(t) e^{\beta Z_i(t)} Z_i(t)^r, \quad s^{(r)}(\beta, t) = ES^{(r)}(\beta, t),$$

for $r = 0, 1, 2, 3, 4$. We assume that the Andersen-Gill conditions hold. First it is straightforward to establish that: $\mathcal{E}_\beta(Z|t) = S^{(1)}(\beta, t)/S^{(0)}(\beta, t)$. Next we have:

Lemma 13.9 *The coefficient $\mathcal{J}(\beta, b)$ can be reexpressed as:*

$$\mathcal{J}(\beta, b) = \int W(t) \left\{ \frac{S^{(2)}(\beta, t)}{S^{(0)}(\beta, t)} - 2 \frac{S^{(1)}(\beta, t) S^{(1)}(b, t)}{S^{(0)}(\beta, t) S^{(0)}(b, t)} + \frac{S^{(1)}(b, t)^2}{S^{(0)}(b, t)^2} \right\} d\bar{N}(t).$$

Theorem 13.1 *As $n \to \infty$ $\mathcal{J}(\beta, b)$ converges in probability to $J(\beta, b)$ where*

$$J(\beta, b) = \int w(t) \left\{ \frac{s^{(2)}(\beta, t)}{s^{(0)}(\beta, t)} - 2 \frac{s^{(1)}(\beta, t) s^{(1)}(b, t)}{s^{(0)}(\beta, t) s^{(0)}(b, t)} + \frac{s^{(1)}(b, t)^2}{s^{(0)}(b, t)^2} \right\}$$
$$s^{(0)}(\beta, t) \lambda_0(t) dt$$

and where $w(t) = S(t)/s^{(0)}(0, t)$.

The value to which $R_{\mathcal{E}}^2(\beta)$ converges for large samples, i.e.,

$$R_{\mathcal{E}}^2(\beta) \xrightarrow{\text{P}} 1 - \frac{J(\beta, \beta)}{J(\beta, 0)}, \tag{13.2}$$

leads to a natural definition for the relevant population parameter via:

Definition 13.2 *Let us take*

$$\Omega^2(\beta) = 1 - \frac{J(\beta, \beta)}{J(\beta, 0)}, \tag{13.3}$$

and, from this, we obtain the important convergence in probability result:

Theorem 13.2 $|R_{\mathcal{E}}^2(\beta) - \Omega^2(\beta)| \xrightarrow{P} 0$. *In particular, $\mathcal{J}(\beta, \beta)$ and $\mathcal{J}(\beta, 0)$ converge in probability to $J(\beta, \beta)$ and $J(\beta, 0)$, respectively.*

Corollary 13.1 $0 \leq \Omega^2(\beta) \leq 1$, $\Omega^2(0) = 0$, *and as* $|\beta| \to \infty$, $\Omega^2(\beta) \to 1$. *Additionally* $\Omega^2(\beta)$ *is invariant under linear transformations of* Z *and monotonically increasing transformations of* T.

We now show that $R^2(\hat{\beta})$ and $R^2_{\mathcal{E}}(\hat{\beta})$ are asymptotically equivalent; therefore $R^2(\hat{\beta})$ is consistent for $\Omega^2(\beta_0)$.

Theorem 13.3 *Under the Andersen-Gill conditions,* $|R^2(\hat{\beta}) - R^2_{\mathcal{E}}(\hat{\beta})| \xrightarrow{P} 0$.

In our own practical experience, when the proportional hazards model holds, there is very close agreement between the coefficients $R^2(\hat{\beta})$ and $R^2_{\mathcal{E}}(\hat{\beta})$ (see the examples below). When discrepancies arise, this is indicative of a failure in model assumptions. We also have that:

Corollary 13.2 $R^2(\hat{\beta})$ *consistently estimates* $\Omega^2(\beta_0)$. *In particular,* $\mathcal{I}(\hat{\beta})$ *and* $\mathcal{I}(0)$ *consistently estimate* $J(\beta_0, \beta_0)$ *and* $J(\beta_0, 0)$, *respectively.*

Theorem 13.4 $R^2(\hat{\beta})$ *and* $R^2_{\mathcal{E}}(\hat{\beta})$ *are asymptotically normal.*

Monotonicity of Ω^2

As strength of association increases so should the measure of correlation or explained variation. We know, from the results of Section 3.9 that Ω^2 is quantifying predictability. We can obtain further insights into this by considering additional properties of Ω^2. For instance we have that increasing strength of association manifests itself via an increasing $|\beta_0|$, once the covariate scale has been fixed. We have

Theorem 13.5 $\Omega^2(\beta_0)$ *as a function of* β_0, *increases with* $|\beta_0|$.

In fact, we will show below that Ω^2 increases with the predictability of survival rankings, which corresponds to Kendall's third stipulation (in the context of the semiparametric Cox regression). Let $Z_j > Z_i$ be the covariates for two subjects in the study, and assume $\beta_0 > 0$ without loss of generality. We can transform all the survival times to exponentially distributed via the transformation $\Lambda_0(\cdot)$, where Λ_0 is the baseline cumulative hazard function. Such a transformation preserves the ranking of the failures so that $\Omega^2(\beta_0)$ is unchanged. Then conditional, on the covariates, a simple calculation shows that

$$\Pr(T_i > T_j) = \frac{\exp(\beta_0 Z_j)}{\exp(\beta_0 Z_i) + \exp(\beta_0 Z_j)},$$

which increases strictly with β_0.

From the above we see that, given the covariates, as the predictability of the survival rankings increases, so does Ω^2. Furthermore, as a result of Theorem 13.5, we can obtain confidence intervals of $\Omega^2(\beta_0)$ from those for β_0, since $\Omega^2(\beta_0)$ is an increasing function of $|\beta_0|$. Only the absolute value conveys information concerning strength of effect and we can then simply invert intervals for β_0, obtained by the usual methods, into intervals for $\Omega^2(\beta_0)$. The coverage properties will then be the same as those already established for the log relative-risk estimate. Since $R_{\mathcal{E}}^2(\beta)$ is consistent for $\Omega^2(\beta)$ for any β then, in practice, we only need to "plug" the two endpoints of the β-confidence interval into $R_{\mathcal{E}}^2$. This gives an approximate confidence interval for $\Omega^2(\beta_0)$. We have not carried out detailed investigation of the coverage properties of such intervals, but in the examples below, we see that such "plug-in" method gives a confidence interval that agrees very well with inference based on bootstrap resampling.

Independent censoring

Here, we assume that C is independent of T and Z. An important property is that the population parameter $\Omega^2(\beta)$ be not affected by the censorship. In order to show this, it helps to recall our earlier discussion on the two roles that time plays in the model. First, $Z(\cdot)$ in general is a stochastic process with respect to time, meaning that $Z(t)$ is a random variable at any fixed t and may have different distributions at different times t. Secondly, the failure time variable T is a non-negative random variable denoting time. While it is immediate to understand the distribution of T given the covariates, we have at any fixed time t two different conditional distributions of $Z(t)$ on T that are of interest to us. One is conditioning on $T \geq t$ under the independent censoring assumption this can be interpreted as given all the subjects that have survived at least until time t and can be estimated by the empirical distribution of $Z(t)$ in the risk set at time t.

Another kind of conditional distribution of interest is that of $Z(t)$ given $T = t$. Under the assumption that T has a continuous distribution we usually observe only one failure at a time and it is difficult to estimate this latter conditional distribution based on a single observation, or a few in the case of ties. We can, however, obtain a consistent estimate by leaning on the model and the main theorem of proportional hazards regression of Section 7.4, one of whose corollaries is: under the model and an independent censorship, the conditional distribution function of $Z(t)$ given $T = t$ is consistently estimated by

$$\hat{F}_t(z|t) = \hat{P}(Z(t) \le z|T = t) = \sum_{\{j:Z_j(t) \le z\}} \pi_j(\hat{\beta}, t).$$

Note that the corollary also applies to multiple dimensional covariates. As a consequence, we also have that:

Corollary 13.3

$$\frac{s^{(1)}(\beta, t)}{s^{(0)}(\beta, t)} = E_\beta\{Z(t)|t\}, \quad \frac{s^{(2)}(\beta, t)}{s^{(0)}(\beta, t)} = E_\beta\{Z(t)^2|t\}, \quad \frac{s^{(1)}(0, t)}{s^{(0)}(0, t)} = E_0\{Z(t)|t\}.$$

Corollary 13.4 *The cumulative distribution for T can be expressed as*

$$F(t) = \int_0^t w(t)s^{(0)}(\beta, t)\lambda_0(t)dt.$$

Lemma 13.10 *For b in $J(\beta, b)$ taking the values β, 0:*

$$J(\beta, b) = \int E_\beta\{[Z(t) - E_b(Z(t)|t)]^2|t\}dF(t).$$

Corollary 13.5 *We can now rewrite $\Omega^2(\beta)$ as:*

$$\Omega^2(\beta) = 1 - \frac{\int E_\beta\{[Z(t) - E_\beta(Z(t)|t)]^2|t\}dF(t)}{\int E_\beta\{[Z(t) - E_0(Z(t)|t)]^2|t\}dF(t)}. \tag{13.4}$$

We can deduce from the corollary that $\Omega^2(\beta)$ does not involve the censoring distribution. It is therefore unaffected by changes in any independent censoring mechanism, in particular its removal as a mechanism impacting our ability to make observations on T.

13.7 Interpretation

In order to be completely assured before using R^2 in practice it is important to know that R^2 is consistent for Ω^2, that $\Omega^2(0) = R^2(0) = 0$, $\Omega^2(\infty) = 1$, that Ω^2 increases as strength of effect increases, and that Ω^2 is unaffected by an independent censoring mechanism. This enables us to state that an Ω^2 of 0.4 translates greater predictability than an Ω^2 of 0.3. We do, however, need one more thing. We would like to be able to say precisely just what a value such as 0.4 corresponds to. That is the purpose of this section.

A sum of squares decomposition

In the definition of $R^2(\beta)$, $\sum_{i=1}^{n} \delta_i W(X_i) r_i^2(\beta)$ can be considered as a residual sum of squares analogous to the linear regression case, while $\sum_{i=1}^{n} \delta_i W(X_i) r_i^2(0)$ is the total sum of squares. Notice that

$$\sum_{i=1}^{n} \delta_i W(X_i) r_i^2(0)$$

$$= \sum_{i=1}^{n} \delta_i W(X_i) r_i^2(\beta) + \sum_{i=1}^{n} \delta_i W(X_i) \{\mathcal{E}_\beta(Z|X_i) - \mathcal{E}_0(Z|X_i)\}^2$$

$$+ 2 \sum_{i=1}^{n} \delta_i W(X_i) \{\mathcal{E}_\beta(Z|X_i) - \mathcal{E}_0(Z|X_i)\}\{Z_i(X_i) - \mathcal{E}_\beta(Z|X_i)\}.$$

The last term in the above is a weighted score and therefore converges asymptotically to zero. It is this result which will enable us to break down the total sum of squares into two components: a residual sum of squares and a regression sum of squares. To make this precise we introduce the following definition which, immediately, can be seen to be analogous to those with which we are familiar from ordinary linear regression.

Definition 13.3 *The total, residual, and regression sum of squares are defined by:*

$$SS_{\text{reg}} = \sum_{i=1}^{n} \delta_i W(X_i) \{\mathcal{E}_{\hat{\beta}}(Z|X_i) - \mathcal{E}_0(Z|X_i)\}^2$$

$$SS_{\text{tot}} = \sum_{i=1}^{n} \delta_i W(X_i) r_i^2(0), \qquad SS_{\text{res}} = \sum_{i=1}^{n} \delta_i W(X_i) r_i^2(\hat{\beta}).$$

From this definition we obtain an asymptotic decomposition of the total sum of squares into the residual sum of squares and the regression sum of squares, i.e.

Lemma 13.11 *Asymptotically, the above three quantities are related by:*

$$SS_{\text{tot}} = SS_{\text{res}} + SS_{\text{reg}}. \tag{13.5}$$

We can then conclude that R^2 is asymptotically equivalent to the ratio of the regression sum of squares to the total sum of squares. Notice that for $R_{\mathcal{E}}^2(\beta)$, even with finite samples, we have an exact decomposition of the sum of the squares. Therefore $R_{\mathcal{E}}^2$ can be expressed exactly as the ratio of a regression sum of squares to the total sum of squares.

Explained variation

For time-invariant covariates and independent censoring, the coefficient $\Omega^2(\beta)$ has a simple interpretation in terms of explained variation. In this case, $Z(t) \equiv Z$ and, letting $\mathcal{A}(t) = \{t\}$ and $\mathcal{B}(t) = \{u : u \geq t\}$ then we have that:

$$
\begin{aligned}
J(\beta, \beta) &= E\{E[Z - E(Z|\mathcal{A}(T))]^2\} \\
J(\beta, 0) &= E\{E[Z - E(Z|\mathcal{B}(T))]^2\}
\end{aligned}
$$

The first equation is immediate and the second follows since $E_0(Z|t) = E_\beta(Z|T > t)$. We can then claim that Ω^2 is indeed a measure of explained variation, the above expressions fitting in precisely with equation (3.32). It is then clear, and backed up further by the simulations of Table 13.1, that

$$
\Omega^2(\beta) \approx 1 - \frac{E\{\mathrm{Var}(Z|T)\}}{\mathrm{Var}(Z)} = \frac{\mathrm{Var}\{E(Z|T)\}}{\mathrm{Var}(Z)}. \tag{13.6}
$$

What is more, there is nothing to stop us defining explained variation as in the right-hand side of the equation since the marginal distribution of Z and T can be estimated by the empirical and the Kaplan-Meier estimator, while the conditional distribution of Z given $T = t$ by the $\pi_i(\hat{\beta}, t)$. However, it is not clear that there is any advantage to this and we recommend that all calculations be done via the Schoenfeld residuals, evaluated at $\beta = \hat{\beta}$ and $\beta = 0$.

The agreement shown in the table between the different ways of conditioning is rather remarkable. One almost suspects that there may be an actual equality and that the observed differences are simply due to rounding errors. But we have not been able to show as much. The important thing to conclude is that we have a very clear, and precise, interpretation in terms of explained variation.

Explained variation in T given Z

As just described we can interpret our coefficient as an estimate of the variation in Z explained by T. In the context of proportional hazards regression where inference is not impacted by any arbitrary monotonic increasing transformation on T, then the variances and mean squared errors of Z given T are the correct quantities to use in order to quantify predictive strength. This is not immediately intuitive however and it is frequently argued that what is required is a coefficient built around the variances and mean squared errors of T given Z. In response to that viewpoint, it could be argued that this amounts to wanting to have your cake and eat it, since by making an appeal to the proportional hazards model we are implying that we wish to suppress or ignore the distributional properties of T given Z and that our model (especially in the light of the main theorem of Section 7.4) only describes the conditional distribution of Z given T.

However, at very little cost and effort, we can, if we wish, base our construction on the same quantities we have worked with so far together with an appeal to Bayes rule. This results in a coefficient with an interpretation as the explained variation in T given Z. Recall that for the case of a bivariate normal distribution the two different ways of defining explained variation result in identical population quantities Ω^2. For other distributions (as is the case here) we nonetheless expect that agreement will be strong. This has been the case in our practical experience. We need two quantities: $\mathrm{Var}(T)$ and $E\,\mathrm{Var}(T|Z)$. The first is readily estimated and often we may wish to estimate it by restricting the time interval to have some upper limit. As for $E\,\mathrm{Var}(T|Z)$, note that:

$$E\{\mathrm{Var}(T|Z)\} = \int_T \int_Z \left\{ t - \int_T t\,dF(t|z) \right\}^2 dF(t|z)dG(z). \quad (13.7)$$

If there is no censoring then consistent estimates for $\Omega_T^2(Z)$ are found by replacing $F(t)$, $G(z)$ and $F(t|z)$ by the empirical estimates $F_n(t)$, $G_n(z)$ and $F_n(t|z)$ to obtain an estimate, let's call it R^2. By virtue of the Helly-Bray theorem R^2 will provide a consistent estimate of Ω^2. Two major problems arise. The first is that, if the dimension of z is high or even continuous, then the estimates $F_n(t|z)$ may be too unreliable to be of practical use. If we wish to appeal to the proportional hazards model then any estimate of $F(t|z)$ will necessarily involve the unspecified $\lambda_0(t)$. Censoring simply adds to the difficulties. However all of these hurdles are readily overcome by a simple appeal to Bayes rule whereby we can write:

$$E\{\mathrm{Var}(T|Z)\} = \int_T \int_Z \left\{ t - \frac{\int_T ug(z|u)dF(u)}{\int_T g(z|u)dF(u)} \right\}^2 dG(z|t)dF(t). \quad (13.8)$$

Consistent estimates for $\Omega_T^2(Z)$ follow if we can consistently estimate the conditional distribution $G(z|t)$ and the marginal distribution $F(t)$. For the marginal distribution of $F(t)$ we have of course the Kaplan-Meier estimate. This makes an assumption of independence between the censoring and the failure mechanisms. If we wish to make the weaker assumption of conditional independence then, rather than use the Kaplan-Meier estimate, we appeal to the law of total probability and use a weighted combination of within group Kaplan-Meier estimates. In practice, making this relaxing assumption, has a negligible impact on the estimates of Ω^2. It is simpler then to work with an independent censoring assumption. The main theorem of Section 7.4 enables us to replace $g(z|u)$, at each failure point $u = X_i$, by $\pi_i(\beta, X_i)$ as a result of the expression for $\hat{P}(Z(t) \leq z|T = X_i)$ given by Equation 7.6. All of the calculations involve the very same quantities used to construct the coefficient of explained variation in terms of Z given T. The specificity of the model is made use of via the same appeal to the main theorem of Section 7.4. For estimation purposes, all of the integrals in Equation 13.8 reduce to simple sums beginning with the outer integral which, upon replacing $F(t)$ by the stepwise Kaplan-Meier estimate, means that we sum over the observed failure times. The weights will be the step size of the Kaplan-Meier decrement. The empirical cumulative distribution of the $\pi_i(\beta, X_i)$ is also a step function so that, within the outer sum, we also have an inner sum to approximate the integral. There is quite clearly more work to do in order to obtain the coefficient with a direct interpretation as the explained variation in T given Z and, since the results are anticipated to be very close, it is a matter for the user to decide just how important that precise interpretion is.

13.8 Simulation results

It is helpful to recall some simulations comparing the behavior of R^2 with some of the measures mentioned earlier. We make use of some of the results from Table II of Schemper and Stare (1996). In Table 13.2, data are generated with hazard function $\lambda(t) = \exp(-\beta Z)$, where $\beta = 0, \log 2, \log 4, \log 16, \log 64$, and Z distributed as either uniform $[0, \sqrt{3}]$ ("c") or dichotomous 0,1 with equal probabilities ("d"). These two covariate distributions have identical variances and thus allow comparison of the results for continuous and dichotomous covariates. Censoring mechanisms are uniform $[0, \tau]$, where τ is chosen to achieve a certain percentage of censoring.

As in Schemper and Stare (1996), there were 100 simulation for each entry of the results. In the table, R^2 is the measure proposed here, ρ^2 is the measure of dependence based on information gain (Xu

Table 13.2: A simulated comparison of different measures ($n = 5000$).

$\exp(\beta)$	% censored	Covariate	R^2	ρ^2	ρ^2_W	$\rho^2_{W,A}$	r^2_{pr}	KS
1	0%	c	0.000	0.000	0.000	0.000	0.000	0.000
	50%	c	0.000	0.000	0.000	0.000	0.000	0.000
	90%	c	0.002	0.002	0.000	0.000	0.000	0.000
2	0%	c	0.098	0.102	0.096	0.119	0.092	0.101
	50%	c	0.101	0.108	0.089	0.122	0.093	0.088
	90%	c	0.104	0.105	0.103	0.099	0.074	0.015
	0%	d	0.099	0.102	0.113	0.118	0.096	0.095
	50%	d	0.105	0.110	0.114	0.121	0.096	0.089
	90%	d	0.112	0.106	0.125	0.100	0.076	0.016
4	0%	c	0.281	0.295	0.304	0.338	0.272	0.231
	50%	c	0.303	0.334	0.298	0.344	0.274	0.267
	90%	c	0.325	0.340	0.279	0.342	0.278	0.063
16	0%	c	0.586	0.598	0.623	0.664	0.584	0.354
	50%	c	0.623	0.690	0.622	0.668	0.584	0.564
	90%	c	0.703	0.723	0.605	0.670	0.585	0.188
64	0%	c	0.757	0.758	0.785	0.815	0.754	0.397
	50%	c	0.790	0.848	0.790	0.815	0.730	0.717
	90%	c	0.863	0.876	0.763	0.816	0.694	0.321
	0%	d	0.870	0.681	0.777	0.814	0.707	0.319
	50%	d	0.873	0.860	0.776	0.815	0.718	0.861
	90%	d	0.941	0.756	0.795	0.792	0.701	0.135

and O'Quigley 1999). The last four columns are from Schemper and Stare (1996), where ρ^2_W and $\rho^2_{W,A}$ are from Kent and O'Quigley (1988), the measure r^2_{pr} is from Schemper and Kaider (1997) and 'KS' from Korn and Simon (1990) based on quadratic loss. From Table 13.2 we see that overall there is mostly good agreement among these particular coefficients except for KS.

Unlike all the other measures included, the KS measure does not remain invariant to monotone increasing transformation of time. This measure is most useful when the time variable provides more information than just an ordering. There is noticeably close agreement between ρ^2 and R^2 for the majority of the cases. This may have its root in the fact that both measures are semiparametric and calculated using the conditional probability π's. The numerical results for dichotomous covariates with high hazard ratio 64 reflects the fact that for discrete covariates ρ^2 is bounded away from one as $|\beta|$ increases. However, as discussed in Xu and O'Quigley (1999) as well as Kent (1983), in practice ρ^2 can usually be interpreted without paying special attention to the discreteness of the distribution.

There are most likely theoretical grounds for anticipating some level of agreement among R^2, ρ^2, ρ_W^2 and $\rho_{W,A}^2$. Roughly speaking, R^2 has at it base something like a score statistic while the three versions of ρ^2 a likelihood ratio statistic. Large sample agreement for such statistics has been documented and further exploration may shed light on this. The values of r_{pr}^2 tend to be slightly lower than these four coefficients, although the strength of association reflected is similar. The measure r_{pr}^2 requires more computation than all the other ones in the table because of the multiple imputation technique employed. Some work has been done (Xu and O'Quigley 1999) on establishing the statistical and interpretative properties of ρ^2. Such work remains to be carried out on the other contenders before they could be proposed for routine implementation.

13.9 Extensions

Multiple coefficients

Assume a multicovariate proportional hazards model with β and $Z(t)$ being $p \times 1$ vectors. Under this model, the dependence of the survival time variable on the covariates is via the prognostic index (Andersen et al. 1983, Altman and Andersen 1986)

$$\eta(t) = \beta' Z(t).$$

So we can imagine that each subject in the study is now labelled by η. The value R^2 as a measure of explained variation or, predictive capability, should evaluate how well the model predicts which individual or equivalently, its label, is chosen to fail at each observed failure time. This is equivalent to predicting the failure rankings given the prognostic indices. When $p = 1$, Z is equivalent to η, therefore we can construct the R^2 using residuals of the Z's. But for $p > 1$, the model does not distinguish between different vector Z's as long as the corresponding η's are the same. So instead of residuals of Z, we define the multiple coefficient using residuals of η. Recall that, in the multivariate setting, the main theorem of Section 7.4 provides us with the estimated joint distribution of the covariate vector Z given time. The most useful way of summarizing this vector is via the linear combination corresponding to the prognostic index. We then proceed very much as for the univariate setting, making the more general definitions of the coefficients.

Definition 13.4 *For the multivariate case we define* $\mathcal{I}(b)$ *as*

$$\mathcal{I}(b) = \sum_{i=1}^{n} \int_{0}^{\infty} W(t)\{\eta_i(t) - \beta'\mathcal{E}_b(Z|t)\}^2 dN_i(t). \qquad (13.9)$$

Definition 13.5 *For the multivariate case we define* $\mathcal{J}(\beta, b)$ *as*

$$\mathcal{J}(\beta, b) = \int_{0}^{\infty} W(t) \sum_{j=1}^{n} \pi_j(\beta, t)\{\eta_j(t) - \beta'\mathcal{E}_b(Z|t)\}^2 d\bar{N}(t). \quad (13.10)$$

For the univariate case we recover the previous definitions apart from a constant multiple which will cancel. We then have:

Definition 13.6

$$R^2(\beta) = 1 - \frac{\mathcal{I}(\beta)}{\mathcal{I}(0)}; \qquad R_{\mathcal{E}}^2(\beta) = 1 - \frac{\mathcal{J}(\beta, \beta)}{\mathcal{J}(\beta, 0)}. \qquad (13.11)$$

Definition 13.7 *In order to describe probability limits we define* $J(\beta, b)$ *to equal*

$$\int w(t)\beta' \left\{ \frac{s^{(2)}(\beta, t)}{s^{(0)}(\beta, t)} - 2\frac{s^{(1)}(\beta, t) \otimes s^{(1)}(b, t)}{s^{(0)}(\beta, t)s^{(0)}(b, t)} + \frac{s^{(1)}(b, t)^{\otimes 2}}{s^{(0)}(b, t)^2} \right\}$$
$$\beta s^{(0)}(\beta, t)\lambda_0(t)dt,$$

where $a^{\otimes 2} = aa'$ *and* $a \otimes b = ab'$ *for vectors a and b.*

The definition leads to:

Lemma 13.12 *Under the Andersen-Gill conditions; letting* $n \to \infty$, *we have*

$$\Omega^2(\beta) = 1 - \frac{J(\beta, \beta)}{J(\beta, 0)},. \qquad (13.12)$$

Notice that although $R^2(\beta)$ and $R_{\mathcal{E}}^2(\beta)$ are not defined for $\beta = 0$, the limits exist and are equal to zero as $\beta \to 0$. So we can define $R^2(0) = R_{\mathcal{E}}^2(0) = \Omega^2(0) = 0$. As in the one-dimensional case, we have the following similar properties:

Theorem 13.6 $|R_\mathcal{E}^2(\beta) - \Omega^2(\beta)| \xrightarrow{P} 0$. *In particular, $\mathcal{J}(\beta, \beta)$ and $\mathcal{J}(\beta, 0)$ converges in probability to $J(\beta, \beta)$ and $J(\beta, 0)$, respectively.*

Corollary 13.6 $0 \leq \Omega^2(\beta) \leq 1$, $\Omega^2(0) = 0$, *and as $|\beta| \to \infty$, $\Omega^2(\beta) \to 1$. Additionally $\Omega^2(\beta)$ is invariant under linear transformations of Z and monotonically increasing transformations of T.*

We have that $R^2(\hat\beta)$ and $R_\mathcal{E}^2(\hat\beta)$ are asymptotically equivalent, therefore $R^2(\hat\beta)$ is consistent for $\Omega^2(\beta_0)$.

Theorem 13.7 *Under the Andersen-Gill conditions, $|R^2(\hat\beta) - R_\mathcal{E}^2(\hat\beta)| \xrightarrow{P} 0$.*

In our own practical experience, when the proportional hazards model holds, there is very close agreement between the coefficients $R^2(\hat\beta)$ and $R_\mathcal{E}^2(\hat\beta)$ (see the examples below). When discrepancies arise, this would seem to be indicative of a failure in model assumptions. We can also see that

Corollary 13.7 $R^2(\hat\beta)$ *consistently estimates $\Omega^2(\beta_0)$. In particular, $\mathcal{I}(\hat\beta)$ and $\mathcal{I}(0)$ consistently estimate $J(\beta_0, \beta_0)$ and $J(\beta_0, 0)$, respectively.*

Theorem 13.8 $R^2(\hat\beta)$ *and $R_\mathcal{E}^2(\hat\beta)$ are asymptotically normal.*

Lemma 13.13 *All three quantities; $R^2(\beta)$, $R_\mathcal{E}^2(\beta)$ and $\Omega^2(\beta)$ are invariant under linear transformations of Z and monotonically increasing transformations of T.*

Finally, a sum of squares decomposition can be obtained for both R^2 and $R_\mathcal{E}^2$, in the same way as in the one-dimensional case.

Partial coefficients

The partial coefficient can be defined via a ratio of multiple coefficients of different orders. Specifically, and in an obvious change of notation just for the purposes of this subsection, let $R^2(Z_1, \ldots, Z_p)$ and $R^2(Z_1, \ldots, Z_q)$ $(q < p)$ denote the multiple coefficients with covariates Z_1 to Z_p and covariates Z_1 to Z_q, respectively. Note that $R^2(Z_1, \ldots, Z_p)$ is calculated using $\hat\beta_1, \ldots, \hat\beta_p$ estimated when $Z_1, \ldots, Z_p$ are included in the model, and $R^2(Z_1, \ldots, Z_q)$ using $\hat\beta_{10}, \ldots, \hat\beta_{q0}$ estimated when only $Z_1, \ldots, Z_q$ are included. Define the partial coefficient $R^2(Z_{q+1}, \ldots, Z_p | Z_1, \ldots, Z_q)$, the correlation after having accounted for the effects of Z_1 to Z_q by

$$1 - R^2(Z_1, \ldots, Z_p) = [1 - R^2(Z_1, \ldots, Z_q)][1 - R^2(Z_{q+1}, \ldots, Z_p | Z_1, \ldots, Z_q)].$$

The above coefficient, motivated by an analagous expression for the multivariate normal model, makes intuitive sense in that the value of the partial coefficient increases as the difference between the multiple coefficients increases, and takes the value zero should this difference be zero. Partial $R_{\mathcal{E}}^2$ and partial Ω^2 can be defined in a similar way.

We can also derive the above definition directly. Following the discussion of multiple coefficients, we can use the prognostic indices obtained under the model with $Z_1, \ldots, Z_p$ and that with $Z_1, \ldots, Z_q$. This would be equivalent to defining $1 - R^2(Z_{q+1}, \ldots, Z_p | Z_1, \ldots, Z_q)$ as $\mathcal{I}(Z_1, \ldots, Z_p)/\mathcal{I}(Z_1, \ldots, Z_q)$, the ratio of the numerators of $1 - R^2(Z_1, \ldots, Z_p)$ and $1 - R^2(Z_1, \ldots, Z_q)$. However, since the two numerators are on different scales, being inner products of vectors of different dimensions, their numerical value require standardization. One natural way to standardize is to divide these numerators by the denominators of $1 - R^2(Z_1, \ldots, Z_p)$ and $1 - R^2(Z_1, \ldots, Z_q)$, respectively. This gives the above definition.

Stratified model

The partial coefficients of the previous section enable us to assess the impact of one or more covariates while adjusting for the effects of others. This is carried out in the context of the assumed model. It may sometimes be preferable to make weaker assumptions than the full model and adjust for the effects of other multilevel covariates by stratification. Indeed it can be interesting and informative to compare adjusted R^2 measures, the adjustments having been made either via the model or via stratification. For the stratified model the basic definitions follow through readily. To be precise, we define a stratum specific residual for stratum s ($s = 1, \ldots, S$), where, in the following, a subscript is in place of i means the ith subject in stratum s. Thus we have

$$r_i(b; s) = Z_{is}(X_{is}) - \mathcal{E}_b(Z | X_{is}) \tag{13.13}$$

where $\mathcal{E}_b(Z | X_{is})$ is averaged within stratum s over the risk set at time X_{is}, and we write

$$\mathcal{I}(b) = \sum_i \sum_s \int_0^\infty W(t)\{Z_{is}(t) - \mathcal{E}_b(Z | t)\}^2 dN_{is}(t) = \sum_i \sum_s \delta_{is} W(X_{is}) r_i^2(b, s).$$

From this we can define

$$R^2(\beta) = 1 - \frac{\sum_i \sum_s \delta_{is} W(X_{is}) r_i^2(\beta, s)}{\sum_i \sum_s \delta_{is} W(X_{is}) r_i^2(0, s)} = 1 - \frac{\mathcal{I}(\beta)}{\mathcal{I}(0)}. \qquad (13.14)$$

Note that we do not use a stratum specific $W(t)$ and, as before, we work with an assumption of a common underlying marginal survival distribution. The validity of this hinges on an independent, rather than a conditionally independent, censoring mechanism. Under a conditionally independent censoring mechanism, a weighted Kaplan-Meier estimate (Murray and Tsiatis, 1996) of the marginal survival distribution could be used instead. We would not anticipate this having a great impact on the calculated value of $R^2(\beta)$ but this has yet to be studied.

Other relative risk models

It is straightforward to generalize the R^2 measure to other relative risk models, with the relative risk of forms such as $1 + \beta z$ or $\exp\{\beta(t)z\}$. Denote $r(t; z)$ a general form of the relative risk. Assume that the regression parameters involved have been estimated, and define $\pi_i(t) = Y_i(t)\hat{r}(t; Z_i)/\sum_{j=1}^n Y_j(t)\hat{r}(t; Z_j)$. Then we can similarly define $\mathcal{E}_\beta(Z|t)$ and form the residuals, thereby defining an R^2 measure similar to (13.8). In addition, it can be shown that under an independent censorship, the conditional distribution of $Z(t)$ given $T = t$ is consistently estimated by $\{\pi_i(t)\}_i$, so properties such as being unaffected by an independent censorship are maintained.

It is particularly interesting to study the use of such an R^2 measure under the time-varying regression effects model, where the relative risk is $\exp\{\beta(t)z\}$. Different approaches have been proposed to estimate $\beta(t)$ (Sleeper and Harrington 1990, Zucker and Karr 1990, Murphy and Sen 1991, Gray 1992, Hastie and Tibshirani 1993, Verweij and Van Houwelingen 1995, Sargent 1997 and Gustafson 1998). In this case we can use R^2 to compare the predictability of different covariates as we do under the proportional hazards model; we can also use it to guide the choice of the amount of smoothness, or the "effective degrees of freedom" as it is called by the some of the aforementioned authors, in estimating $\beta(t)$. As a brief illustration, suppose that we use the sieves method which estimates $\beta(t)$ as a step function, and that we are to choose between two different partitions of the time axis, perhaps one finer than the other.

Denote the two estimates obtained under these two partitions by $\hat{\beta}_1(t)$ and $\hat{\beta}_2(t)$, the latter corresponding to the finer partition. We can measure the extra amount of variation explained by fitting $\hat{\beta}_2(t)$ versus fitting $\hat{\beta}_1(t)$, by

$$R_{\text{ex}}^2 = 1 - \frac{\mathcal{I}(\hat{\beta}_2(\cdot))}{\mathcal{I}(\hat{\beta}_1(\cdot))}.$$

This can be thought of as a partial coefficient, if we look at the "dimension" of $\beta(t)$ through time. The use of R_{ex}^2 in estimating $\beta(t)$ has recently been explored in Xu and Adak (2000).

13.10   Theoretical construction for distance measures

The distance measures described in Section 13.3 were defined empirically with no population model in mind. However, it is quite straightforward to set up a theoretical structure enabling ready conclusions concerning large sample behavior (O'Quigley, Flandre and Reiner 1999; Schemper and Henderson 2000). The simulation results of Schemper (1990) are confirmed. Also we will see that the measures can be expected to have an upper bound less than 1 as hinted at by Schemper's empirical investigation and that, for example, in the uncensored case the measures V_1 and V_2 estimate the same population quantity. The theoretical setting makes it clear that, unless further modification is undertaken, the population equivalents of the distance measures are affected by censoring, whether or not independent of the failure mechanism.

Uncensored case

As usual we define the empirical distribution function of survival by $F_n(t)$, the empirical survival distribution conditional on the covariate z by $F_n(t|z)$ and the empirical distribution of the covariate z by $H_n(z)$. Also, we have $S_n(t) = 1 - F_n(t)$ and $S_n(t|z) = 1 - F_n(t|z)$. Finally the individual observations can be re-expressed via the function $Y_t(u)$ where $Y_t(u)$ takes the value 1 when $0 < u < t$, the value 0.5 when $u = t$ and the value 0 otherwise. We keep the definition $Y_t(u) = 0.5$ at the value $u = t$ in order to facilitate comparison with Schemper's work (1990, 1992, 2000). However, as far as large sample theory is concerned, we could define this to be either zero or one at $u = t$ without impacting population quantities.

Referring to Section 13.3, observe that for an observed survival time t, the function $Y_t(u)$ corresponds to the empirical survival function S_{ij} in which the ith subject fails at time t and the argument u is given values corresponding to the observed failure times.

Note that the inner and outer sums contain n elements where n is the number of independently observed survival times. We have $k_i = n$, $\forall i$. Things become more transparent when we multiply the outer sum by n^{-1} in both numerator and denominator. The weak law of large numbers then indicates that these quantities converge in probability to expectations. For the inner sum for example $k_i^{-1} \sum_i |S_{ij} - \bar{S}_j|^\ell$, $\ell = 1, 2$ converges in probability, as $k_i(= n) \to \infty$, to the mean absolute ($\ell = 1$) or quadratic ($\ell = 2$) distance between the marginal survival curve at point u and a randomly chosen subject's empirical curve $Y_t(u)$. This mean is calculated over all possible values of u i.e., with respect to the marginal density of survival. Analogously $k_i^{-1} \sum_i |S_{ij} - \bar{S}_{ij}|^\ell$, $\ell = 1, 2$ converges in probability to a distance between the conditional survival curves (given the covariate) and $Y_t(u)$, once again over all values of u. The outer sums, multiplied by n^{-1} also converge to expectations. In the uncensored case inner and outer expectations are with respect to the same density, that governing survival. It is then natural to have:

Definition 13.8 *The population quantity θ_ℓ is expressed via the ratio of a denominator D_ℓ and a numerator N_ℓ so that $\theta_\ell = 1 - D_\ell^{-1} N_\ell$ where we write:*

$$N_\ell = \int \int \int |Y_t(u) - S(u|z)|^\ell dF(u) dF(t|z) dH(z), \quad (13.15)$$

$$D_\ell = \int \int |Y_t(u) - S(u)|^\ell dF(u) dF(t). \quad (13.16)$$

The simplest situation in which we can readily see how to obtain a consistent estimate of θ_ℓ arises when z takes a small number of finite values. For each value, we can consider the corresponding empirical quantities: $F_n(t|z)$, $S_n(t|z)$, $F_n(t)$ and $H_n(t)$ and then, in the above equation, we can replace the population quantities; $F(t)$, $F(t|z)$ and $H(z)$ by $F_n(t)$, $F_n(t|z)$ and $H_n(z)$ respectively. We can denote such an estimate by $\hat{\theta}_\ell$ and conclude that it is a consistent estimate for θ_ℓ (O'Quigley, Flandre and Reiner 1999). The consistency follows from standard results for weak convergence (see Section 3.3) whereby $F_n(t) \to F(t)$, $F_n(t|z) \to F(t|z)$ and $H_n(t) \to H(t)$ at all continuity points t of $F(t)$, $F(t|z)$ and $H(t)$, all arrows indicating convergence in probability.

The above expressions make no appeal to any model and, as such, can be considered to be completely non parametric. Not forgetting that we are still dealing with the uncensored case, we could nonetheless view $S_n(t|z) = 1 - F_n(t|z)$ as stratified estimates under the Cox model, since, the stratified model has no constraints and the corresponding survival estimates reduce to the usual empirical ones. This is artificial but consider the following; the above arguments only require that our estimates be consistent. If the Cox model is correct then the stratified model (essentially no model for discrete z) and the usual model both produce consistent estimates for $F(t|z)$. Thus, if we were to redefine $\hat{\theta}_\ell$ to be as above but with $\tilde{S}(t|z)$, the estimate based on the Cox model (see Chapter 15), in place of $S_n(t|z)$, then the consistency property is unchanged.

Censored case

For the empirical quantities presented by Schemper (1990, 1992) the sums were taken over both the observed censored and failure times. This appears attractive in that as much as the information as possible is being used. However, as shown by O'Quigley, Flandre and Reiner (1999), and in an analogous demonstration using counting process notation (Schemper and Henderson 2000), the property of consistency is lost. To see this we deal separately with the sums of censored observations and those that are uncensored. The quantities denoted k_i still count the number of terms in the respective sums so that we can again make a simple appeal to the weak law of large numbers. The standardized "censored" sum converges to an expectation taken with respect to the density $f_{U|U<t}(u|U < t)$, the conditional density of failure time U given that it is less than t. The standardized "uncensored" sum converges to an expectation taken with respect to $f(u)$. The outer sums concern all observations so that the expectations to which these standardized sums converge is taken with respect to the distribution of the minimum of observed survival and censoring times. The survival distribution for censoring, denoted $G(u)$, though enters explicitly into the calculations. The denominator converges to the sum of two terms: an "uncensored" term and a "censored" term which we can write (O'Quigley, Flandre and Reiner 1999) as:

$$\int\int |Y_t(u) - S(u)|^\ell f(u)G(t)dudF(t)$$

$$+ \int\int_0^t (1 - F(t))^{-1}|Y_t(u) - S(u)|^\ell f(u)dudG(t).$$

The censoring distribution appears explicitly in this expression and any resulting evaluation would be impacted by this distribution. An expression for the numerator can also be worked out (O'Quigley, Flandre and Reiner 1999) and again it involves the unknown censoring mechanism. It would be nice if the censoring distribution were to factor out leading to the property we are aiming for but this is not the case.

Convergence in the censored case

Let us define $\tilde{S}(t)$ to be the usual Kaplan-Meier estimate and $\tilde{S}(t|z)$ to be the proportional hazards estimate of conditional survival, given z. If the model correctly generates the observations, then both $\tilde{S}(t)$ and $\tilde{S}(t|z)$ converge to their population counterparts, $S(t)$ and $S(t|z)$.

Lemma 13.14 *The parameter θ_ℓ is consistently estimated by $\tilde{\theta}_\ell$ where $\tilde{\theta}_\ell = 1 - \tilde{D}_\ell^{-1}\tilde{N}_\ell$ and where $\tilde{N}_\ell$ and $\tilde{D}_\ell$ are defined by:*

$$\tilde{N}_\ell = \int\int\int |Y_t(u) - \tilde{S}(u|z)|^\ell d\tilde{F}(u)d\tilde{F}(t|z)dH_n(z), \qquad (13.17)$$

$$\tilde{D}_\ell = \int\int |Y_t(u) - \tilde{S}(u)|^\ell d\tilde{F}(u)d\tilde{F}(t). \qquad (13.18)$$

Note that although we have taken $\tilde{F}$ to be the Kaplan-Meier estimator the arguments hold for any other consistent estimator of the true underlying marginal survival curve. Under stronger model assumptions we can work even with a parametric estimator. We might anticipate the Nelson estimator of the survivorship function to produce similar results to those for the Kaplan-Meier estimator. Since $|\hat{\theta}_\ell - \tilde{\theta}_\ell| \to 0$, it follows that $\tilde{\theta}_\ell$ is consistent for θ_ℓ and that, under independent censoring, unlike V_ℓ, it is estimating the same quantity it would have estimated were it possible to remove the censoring. Attempts to extract more information from the censorings, in the absence of further, necessarily strong assumptions, leads to inconsistency if we agree, in this context to take inconsistency to mean that estimators converge to population quantities different to those to which they would have converged were it possible to remove the censoring.

13.11   Isolation method for bias-reduction

In order to motivate this section we first recall the relationship between multiple and partial coefficients which holds in the linear case. When $R^2(Z_{q+1}, \ldots, Z_p | Z_1, \ldots, Z_q)$, is the remaining or partial correlation between the outcome and Z_{q+1} to Z_p after having taken into account the effects of Z_1 to Z_q and $R^2(Z_1, \ldots, Z_p)$ is the multiple correlation with all Z_1 to Z_p in the model, then:

$$1 - R^2(Z_1, \ldots, Z_p) = [1 - R^2(Z_1, \ldots, Z_q)][1 - R^2(Z_{q+1}, \ldots, Z_p | Z_1, \ldots, Z_q)]$$

This expression holds exactly for the linear model and so, whether we build the multiple correlation by constructing increasingly complex partial coefficients or we define the partial coefficient by increasingly simpler multiple coefficients the final answer is the same. Unfortunately this equation does not hold for other situations which is why there is more than one way to define partial and multiple correlation.

Our suggestion for the multiple coefficient is to reduce it formally to a univariate coefficient via use of the prognostic index. We then defined the partial coefficient via the same expression as the above equation. In any event, whether exact or as an approximation, we can use the equation to make the following simple observation. As we add new variables to the expression for multiple correlation, the value of multiple R^2 will almost certainly increase. Only if the partial correlation for the newly included variable is identically equal to zero will the multiple coefficient stay the same. Sampling error will inevitably lead to squared partial correlations more or less removed from zero and, in turn, for an increasingly biased estimate for the multiple correlation itself. This bias pulls the coefficient in the direction of one and so, in practice, estimated coefficients of explained variation can be quite inflated.

The phenomenon of inflation in the multivariate setting is well known and there are several suggestions for tackling the bias. The most well known remedies are the Akaike Information Criterion, the Bayes Information Criterion, the Schwartz Criterion and Cross-Validation. None of these remedies does very well. For smaller sample sizes they will, typically, over adjust and can even lead to negative squared correlations and, for larger samples, they will mostly not make enough of an adjustment. Apart from Cross-Validation, the scope of these corrections is also very limited and, in the main, is concerned only with biases due to the dimension of the explanatory variable in relation to the sample size.

In the practical setting of model building the dimension of the co-variate vector is only the most immediate and often the least important of several factors which result in inflated estimates. There are indeed many other factors among which: (1) the size of the potential covariate pool from which those used in the model form a subset, (2) the data based transformations on continuous or ordered covariates (3) the stepwise algorithms used to make a selection from the covariate pool, (4) the use of cut-offs to define new derived variables and (5) the inclusion of some relaxation of model assumptions in the light of goodness-of-fit procedures. None of these five factors is usually taken into consideration and yet their impact is far greater than that of the dimension of the final model, in particular when the model has been constructed from a very large data base.

A way which addresses all five factors together with the sixth, the dimension of the covariate vector, is the following. However obtained, a final model is viewed as having two quite distinct underlying construction components. The first of these - the most important in any investigator's eye - is the true strength of effect of the multivariate relationship, however formulated, and which finds its expression in the final model. The second component concerns everything involved in the process which led to writing down that final model. All six of the above factors and any others we may have overlooked are deemed to be a part of this second component.

Let's look a little more closely at this second component. Imagine an investigator who decides to fit a model of dimension five from a data set with one hundred individuals and twenty measured covariates. A second investigator is studying a similar problem on one hundred entirely comparable individuals but this time, instead of twenty covariates to choose from, he has two hundred. A third investigator finds him or herself in a situation comparable to that of the second investigator but has results from two separate data sets. It is clear that the bias here is increasing. It is also difficult to have any idea as to what the size of this bias might be. None of the usual techniques address this form of bias. Next, suppose an investigator decides that all the skew distributions should be subject to log-transformations and, if such a transformation leads to a more significant result then the transformation is maintained, otherwise we leave the scaling as it was. He or she then decides that, for the purposes of interpretation, some continuous variables will be broken down into categorical variables. If the effect across the ordered categories is comparable, as judged by the regression coefficients, then the $p - 1$ binary variables describing p groups are replaced by a single ordinal variable. Note also that, if the spacing of the effect is not the same, it can be made so by rescaling.

A model for true and overfitted effects

There are almost endless ways of fine tuning any model and, in the process, as many ways of inflating our idea as to how predictive the model really is. The kind of transformations just suggested will typically indicate a more predictive model than is really warranted. They are also used very frequently by investigators.

We suppose the following model for these two components: the first being the true strength of effect and the second, everything else involved in the construction of the model. The covariate vector of interest is Z and, as usual, we would calculate $R^2(Z)$, a quantity which we observe. However, we would really like to calculate the multiple correlation given the fitting. We write this as $R^2(Z|F)$ where F is not something we measure, or observe, but is a conceptual quantity indicating the sum total of all the actions taken during the fitting process. We might consider these actions taken on their own in which case we would have $R^2(F)$. The observed multiple $R^2(Z)$ simultaneously involves, as well as the real effects, the fitting process, an important fact made explicit by writing, $R^2(Z) = R^2(Z, F)$. Note that:

$$1 - R^2(Z, F) = [1 - R^2(F)][1 - R^2(Z|F)] \qquad (13.19)$$

There are three quantities in the above expression and only one of them, $R^2(Z, F)$, can be observed. If we were able to obtain $R^2(F)$ then the quantity we are really interested in, $R^2(Z|F)$, the true impact of the covariates having removed those effects due to the fitting, becomes immediately available from the above equation.

Estimating the overfitted effects

As a first approximation we can suppose that the fitting effects themselves are orthogonal to the true effects. By this we mean that the amount of inflation, as measured by $R^2(F)$, only depends on the fitting procedures and extraneous factors such as sample size etc. As true effect increases, the population equivalent of $R^2(Z, F)$ will, of course, also increase but it would not be unreasonable to suppose that the pure inflation factor alone, as measured by $R^2(F)$, depends only weakly, if at all, on any true effect. In particular we will calculate $R^2(F)$ in the absence of any effect and use this value when there are non-zero effects. Recalling the main theorem of Section 7.4 we have that, at each ranked failure time t, the probabilities of choosing individual with covariate vector $Z_i(t)$ obtains from:

$$\pi_i(\beta(t), t) = \frac{Y_i(t) \exp\{\beta(t) Z_i(t)\}}{\sum_{j=1}^{n} Y_j(t) \exp\{\beta(t) Z_j(t)\}}. \tag{13.20}$$

This mechanism is assumed to generate the observations. Suppose that there is no effect. Then the coefficient vector, $\beta(t)$ is identically equal to zero. At each failure time t, letting $n(t) = \sum Y_i(t)$ be the number of subjects in the risk set then, from Equation 13.20, the probability that any individual is chosen is simply $1/n(t)$.

We keep the risk sets fixed, i.e., we condition on the observed risk sets and, from these we sample individuals, each with probability $1/n(t)$ at time point t, thereby establishing a simulated data set in which the true effect is zero. On the basis of these data, the investigator can proceed to use the same fitting procedures, and strategies, that he or she has used on the unmodified data. Stepwise searches, transformations, maximizations, eliminations following goodness-of-fit, categorizations and any other used modeling strategy is replicated on this same data set. For the resulting multivariate model, corresponds an R^2 coefficient. We write this as $R^2(F)$. The more involved, elaborate and exhaustive the fitting technique the higher, on average, we anticipate $R^2(F)$ to be. Overfitting the data manifests itself directly in the coefficient $R^2(F)$.

Some further observations on this whole process are worth making. Firstly, we do not have just a single value of $R^2(F)$. Under a further replication we would, typically, obtain a different value of $R^2(F)$. Under a large number of replications we would obtain a whole, simulated, distribution of values of $R^2(F)$. If we denote by u any one of these replicated values and by $H(u)$ the empirical distribution function of the replications, then we can take $R^2(Z|F)$ by using Equation 13.19 to be:

$$R^2(Z|F) = \int \left(\frac{R^2(Z, F) - u}{1 - u} \right) dH(u). \tag{13.21}$$

It can also be interesting to consider the whole distribution of $R^2(Z|F)$, as induced from $H(u)$, rather than just the mean. Another point to note is that, by conditioning on the risk sets, we allow the possibility that in the replications, the same subject could be selected more than once. This may seem odd if we are interpreting being selected as a failure (which indeed it is) but this is only a formal procedure, respecting the probability model which we assume to be generating the observations. That the same subject could not in practice fail more than once is something which does not impact our construction and is in fact required if we do not wish to include complex calculations involving the censoring mechanism. This is not unlike the bootstrap which can also involve repetitions which the design itself could not have produced. Finally, in Equation 13.21 a good approximation would arise from taking the mean $\bar{u}$ across the replications of $R^2(F)$ and writing $R^2(Z|F) \approx [R^2(Z, F) - \bar{u}]/(1 - \bar{u})$.

Bias reduction

We call the above the isolation method by which the effects of interest are isolated from those which are artificially generated through the process of model construction. The basic idea is derived from the chaotization principle developed by Kipnis (1977). Kipnis studied the tails of the distribution of an R^2 type measure and how changes in this distribution, occurring by the inclusion of additional factors, could be anticipated by the fitting process alone. His focus was on the significance level of the multivariate coefficient rather than bias reduction itself but the central idea is the same. It requires replication under a model of no association. Kipnis's idea was to use permutation distributions which could be generated under an assumption of no effect whereby all permutations would be considered equally likely. For our particular case, we do not need to carry out any permutation. It is enough to sample based on the probabilities given by Equation 13.20 in which we fix $\beta(t)$ at the value zero.

The approach can lead to significant reduction in bias caused by overfitting, especially when dealing with a large number of covariates. The method is easy to implement and can be adapted readily to deal with more complex situations. For example, we may wish to focus on some factor after having taken account of several factors already included in the model. Here, in order to generate the relevant distribution for $R^2(F)$, and referring to the multivariate version of Equation

13.20, we would fix at zero the coefficient corresponding to the factor of interest and allow those factors for which we are adjusting to be replaced by estimates. These would be constrained estimates in that the coefficient corresponding to the additional factor of interest is always fixed at zero. We then use Equation 13.20 with these values in order to generate the distribution $H(u)$.

13.12   Illustrations from studies in cancer

Study in leukemia

The first example concerns the Freireich (1963) data, which records the remission times of 42 patients with acute leukemia treated by 6-mercaptopurine (6-MP) or placebo. The estimate of the regression coefficient is $\hat{\beta} = 1.53$, and $R^2(\hat{\beta}) = 0.386$ and $R_{\mathcal{E}}^2(\hat{\beta}) = 0.371$. The 95% confidence interval for $\Omega^2(\beta)$, obtained using the monotonicity of $R_{\mathcal{E}}^2(\beta)$, i.e., inverting the interval for β, is $(0.106, 0.628)$. On the basis of 1000 bootstrap samples we find a simple percentile interval as $(0.154, 0.714)$ using R^2, and $(0.154, 0.715)$ using $R_{\mathcal{E}}^2$. The bootstrap mean is 0.413 for R^2 and 0.405 for $R_{\mathcal{E}}^2$, which gives estimated bias of 0.028 and 0.034, respectively. This suggests that bias correction may be necessary, and employing Efron's bias-corrected accelerated bootstrap (BCa) method we have confidence interval $(0.111, 0.631)$ using R^2, and $(0.103, 0.614)$ using $R_{\mathcal{E}}^2$. We see that these have very good agreement with one another (suggesting that the proportional hazards assumption is a reasonable one) as well as the interval obtained through monotonicity.

In Figure 13.1 we plot the values of $R^2(\beta)$ (dots) and $R_{\mathcal{E}}^2(\beta)$ (circles) for the Freireich data versus different values of β. The figure illustrates well the facts that $R^2(\beta)$ reaches a maximum at around $\beta = 1.5$, which is the value of our estimate $\hat{\beta}$ and that $R_{\mathcal{E}}^2(\beta)$ increases with β, approaching 1 as $\beta \to \infty$. Notice that $R^2(\beta) = R_{\mathcal{E}}^2(\beta)$ occurs somewhere between $\beta = 1.5$ and 1.6, again around our estimate $\hat{\beta}$. This is to be anticipated in view of Theorem 13.7.

The above $R^2(\hat{\beta})$ can be compared with some of the other suggestions mentioned in the introduction. For the same data the measure proposed by Kent and O'Quigley (1986) resulted in the value 0.37, and the measure of explained randomness (Xu and O'Quigley 1999), described in the following chapter, obtains the value of 0.40. The explained variation proposals of Schemper (1990), based on empirical

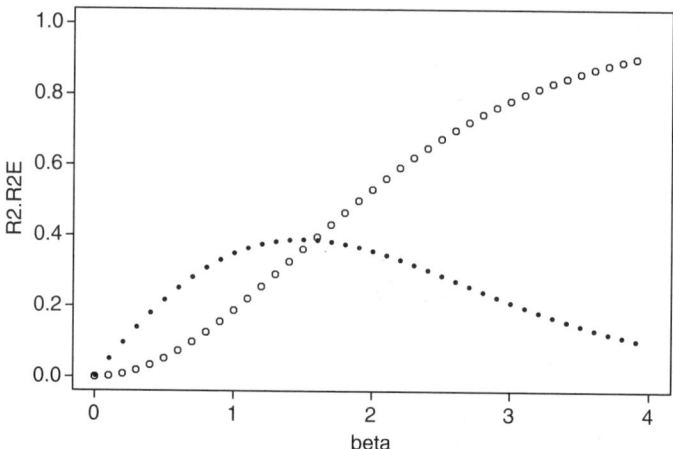

Figure 13.1: A plot of R^2 and $R^2_{\mathcal{E}}$ as functions of β given the observations.

survival functions per subject, resulted in (his notation) $V_1 = 0.20$ and $V_2 = 0.19$ and Schemper's later correction (1994) resulted in $V_2 = 0.29$, although all three of the Schemper measures depend heavily on the censoring (O'Quigley et al. 1999, Schemper and Henderson 2001). The measure of Schemper and Kaider (1997) resulted in $r^2_{pr} = 0.34$. The measure of Korn and Simon (1990), based on quadratic loss, gave the value 0.32. This measure does not remain invariant to monotone increasing transformation of time. For these data the value 0.32 drops to 0.29 if the failure times are replaced by the square roots of the times.

Study in breast cancer

The available data consist of 1504 patients with complete covariate information, among whom there were 357 recorded deaths. The 5 and 10 year survival rates were 0.83 and 0.70, respectively. Of the five covariates, age has a range of 23-55 years, with a median of 45 years. About 6%, 20%, 28%, 27% and 19% of the patients had histology grade 0, 1, 2, 3, and 4, respectively. About 45%, 24%, 23%, 5%, and 2% of the patients had stage 1, 2, 3, 4, and 5 disease, respectively. Out of the 1504 patients, 1075 (71%) had positive progesterone receptor status. The maximum tumor size was 170mm, with a median of 30 mm.

In univariate analysis under the proportional hazard model, all variables are highly significant (Table 13.3). We also calculated the

Table 13.3: R^2 analysis of breast cancer data. Upper part of the table shows results for univariate analyzes. Lower part shows the nested multivariate coefficients.

Single covariate	$\hat{\beta}$	p-value	R^2
Age	-0.24	<0.01	0.005
Histology	0.37	<0.01	0.12
Stage	0.53	<0.01	0.20
Receptor	-0.73	<0.01	0.07
Size	0.02	<0.01	0.18

Covariates in multivariate model	R^2	partial R^2
Age	0.01	
Age and histology	0.12	0.12
Age, histology, and stage	0.26	0.16
Age, histology, stage, and receptor	0.33	0.09
Age, histology, stage, receptor, and size	0.33	0.01

univariate R^2's, and we see that the predictive powers are quite different. Stage and tumor size, as one might expect, have reasonably high predictability. Histology grade also has predictive power, although this covariate has been shown to have a non-proportional regression effect. This might explain the observed discrepancy between R^2 and $R_{\mathcal{E}}^2$. So we fit a simple two-stage model with the regression effect dropping to zero after a certain change point. When the change point is chosen at 24 months, R^2 from the fitted model turns out to be 0.238, and $R_{\mathcal{E}}^2$ 0.332. On the other hand, age has very weak predictive capability, though significant. This estimated weak effect could be due to: (1) a population weak effect, or (2) a suboptimal coding of the covariate. We investigated this second possibility via two recoded models. The first, making a strong trend assumption, coded age as 1 (0-33), 2 (34-40) and 3 (41 and above). The second model, making no assumptions about trend, used two binary variables to code the three groups. All three models gave very similar values of R^2. In consequence only the simplest model is retained for subsequent analysis, i.e., the age groups 1-3. In the lower part of Table 13.3, we calculated the multiple R^2 for a set of nested models. It also contains the values of the partial R^2 when each additional covariate is added to the existing model. The partial coefficient for tumor size having accounted for the other four variables is only 0.006, suggesting that the extra amount of variation

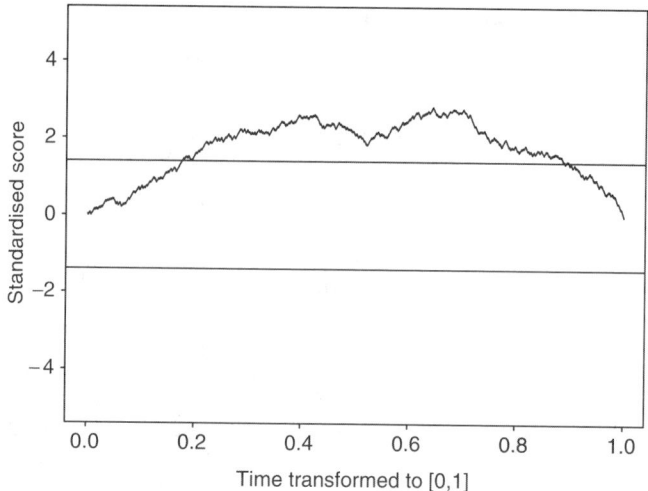

Figure 13.2: *Goodness-of-fit plot for the covariate grade.*

in survival explained by the patient's tumor size is small. Some co-variates in this dataset are known to have non-proportional regression effect. Figure 13.2 shows a goodness-of-fit process (see Chapter 8) for the proportional hazards assumption. Recall that the maximum absolute value exceeding the boundary of 1.36 corresponds to the 0.05 significance level of an underlying Brownian bridge when the model is correctly specified. The variation in survival times of these breast cancer patients are mainly explained by three of the five covariates: histology grade, stage, and progesterone receptor status. Xu and Adak (2001) examined closely the time-varying regression effects of these three variables using a tree-based approach. The tree method gives piecewise constant estimated log relative risks. After obtaining a set of nested trees, as one of the methods for selecting a final tree, they used the coefficient R^2_{ex} defined above to arrive at a final tree with two cutpoints at 27 and 46 months. The estimated piecewise-constant regression effects of the three covariates are reproduced in Table 13.4.

While the R^2 from a proportional hazards model with these three covariates is 0.32, the R^2 from the above fitted three piece $\beta(t)$ model is 0.51. For the latter R^2 the calculation used $\hat{\beta}_1(t) \equiv 0$ and $\hat{\beta}_2(t) = \hat{\beta}(t)$. as given in Table 13.4. The improvement in explained variation here reflects also an improvement in fit, underlining the relationship between predictability and goodness-of-fit.

Table 13.4: $\hat{\beta}(t)$ (standard error) from Xu and Adak (2001).

Variable	$t \in [0, 27]$	$t \in [28, 46]$	$t \in [47, 165]$
Histology	0.653 (0.125)	0.362(0.094)	0.201 (0.069)
Stage	0.607 (0.094)	0.349(0.089)	0.365 (0.071)
Receptor	-0.803 (0.225)	-0.708(0.201)	-0.293 (0.164)

Study in gastric cancer

In a study on prognostic factors in gastric cancer (Rashid et al., 1982) certain acute phase reactant proteins were measured pre-operatively. Five covariates were studied: stage together with the proteins α_1-anti chimotrypsin (ACT), carcino embryonic antigen (CEA), C-reactive protein (CRP) and α_1 glyco-protein (AGP). Surgery is needed in order to determine the stage of the cancer, a clinical factor known to strongly influence survival, and one of the purposes of the study was to find out how well the four protein covariates, available pre-operatively, are able to explain survival in the absence of information on stage. A logarithmic transformation for CEA was found to be necessary. This is also reflected in a R^2 increasing from 0.10 to 0.20 after the transformation.

Table 13.5 shows that each of the five covariates has reasonable predicting power, with R^2 for stage alone to be 0.48. A direct calculation of sample correlation shows that ACT, CRP and AGP are highly correlated, which is supported by biological evidence. In addition, fitting the Cox model with all four protein covariates shows that CRP and AGP are no longer significant in the presence of the other covariates. These two variables can then be dropped from further study. The value of R^2 for a model with ACT and log(CEA) is 0.37; this increases to 0.54 when stage is also included, and the corresponding partial R^2 is equal to 0.27. In conclusion, there is strong prognostic information in the pre-operative measurements ACT and log(CEA), but this only partially captures the information contained in stage.

Study in multiple myeloma

A further example is motivated by the increasing number of studies carried out in cancer research to correlate the outcome with multi-dimensional molecular and genetic markers. As we see the predictability by an individual marker is generally low, with the highest R^2 of 0.08 from plasma cell labelling (PCL) index; in particular, the Durie-Salmon stage has the smalles R^2 of 0.004. When all 13

Table 13.5: Univariate analysis of gastric cancer data (Rashid et al. 1982).

Covariate	$\hat{\beta}$	p-value	R^2
Stage	1.78	<0.01	0.48
ACT	2.26	<0.01	0.29
log(CEA)	0.30	<0.01	0.20
CRP	0.02	<0.01	0.26
AGP	0.70	<0.01	0.14

covariates are included in a multivariate Cox model, only six of them remain significant (p-value < 0.08), with the multivariate $R^2 = 0.202$. In particular, the traditional staging system is no longer significantly predictive of survival given the laboratory measurements. Leaving out the non-significant variables in a Cox model gives $R^2 = 0.18$. As an illustration of variable selection using R^2, we build hierarchical models starting with PCL index which has the highest univariate R^2. We then choose the variable among the remaining five that has the highest partial R^2, and so on. The lower part of Table 13.6 gives the nested models and the corresponding R^2's. The data come from a clinical trial (EST 9486) of multiple myeloma conducted by the Eastern Coorperative Oncology Group (Oken et al. 1999). The trial enrolled 653 patients to three randomized arms; VBMCP alone, VBMCP with added HiCy and rIFNα2. No significant survival difference were found across the three arms. The trial collected laboratory measurements on patients' myeloma cells, including measurements from blood or serum: albumin (1 if $\geq$ 3g/l, 0 otherwise), β_2 microglobin 1 if $\geq$ 2.7mg/dl, 0 otherwise), creatinine (1 if $\geq$ 2mg/dl, 0 otherwise), cytoplasmic-immunoglobin heavy chain IgA and IgG (1 if present, 0 absent), kappa light chain (1 if present, 0 absent), percent plasma cells (1 if $\geq$ 0.3%, and hemoglobin (1 if $\geq$ 10g/dl, 0 otherwise); characteristics of circulating myeloma cells: plasma cell labelling index (a measure of plasma cell proliferation, 1 if $\geq$ 1, 0 otherwise), IL-6 receptor status (1 if $\geq$ 270ng/ml, 0 otherwise), and level of C-reative protein (1 if $\geq$ 2ng/ml, 0 otherwise).

All of the above variables, which were originally continuous, were dichotomized using previously published threshold values. Here we include a randomly selected group of 295 patients, on whom a particular chromosomal abnormality, the possible deletion of the short arm of chromosome 13 (denoted by 13q-), was measured by flouresecent in-

Table 13.6: R^2 analysis of the Myeloma data. Upper part of the table shows results for univariate analyzes. Lower part shows the nested multivariate coefficients.

Single covariate	$\hat{\beta}$	$se(\hat{\beta})$	R^2
Creatinine	0.66	0.16	0.05
Plasma	0.43	0.12	0.04
IL-6	0.35	0.14	0.02
C-reactive	0.53	0.17	0.02
a13q	0.22	0.12	0.01
Hemoglobin	-0.30	0.13	0.03
Albumin	-0.39	0.14	0.03
IgG	-0.15	0.12	0.01
IgA	0.16	0.14	0.01
Kappa	-0.26	0.12	0.01
Stage	-0.18	0.12	0.004
β_2 microglobin	0.48	0.13	0.03
PCL index	0.59	0.13	0.08

covariates in multivariate model	R^2
PLC	0.08
PLC, creatinine	0.11
PLC, creatinine, plasma	0.13
PLC, creatinine, plasma, a13q	0.16
PLC, creatinine, plasma, a13q, β_2 mcrglb	0.17
PLC, creatinine, plasma, a13q, β_2 mcrglb, IL-6	0.18
All 13 variables	0.20

situ hybridization (FISH) in the laboratory of R. Fonseca at the Mayo Clinic; the corresponding variable a13q was coded 1 if present, 0 absent. We also include the traditional Durie-Salmon stage (1 if I or II, 0 if III) which was routinely used to predict prognosis in multiple myeloma before the availability of assays to measure genetic and other molecular abnormalities of the myeloma cells.

Univariate Cox regression analysis indicates that all of the above 13 covariates are more or less associated with patients' survival times and most of them are highly significant. Table 13.6 shows the estimated regression effects and the standard errors, together with the univariate R^2 coefficients. Effects are not very strong. Even when all 13 variables are included in the analysis we still have an estimated eighty percent of the variance remaining unexplained. And the true figure would most certainly be higher since no accommodation has been made for the fitting biases.

The same data set was also analyzed by Huang and Harrington (2002), who proposed a penalized partial likelihood approach to the handling of high-dimensional covariates in the proportional hazards regression. The authors pointed out that because there were 270 deaths among the 295 subjects in this data set, the standard partial likelihood estimate in a Cox model with all 13 covariates should be reasonably stable. Even so, in their Table 3 one sees obvious reduction in both the magnitude of the regression effects and the standard errors of the penalized partial likelihood estimate, as compared to the unpenalized estimate. The penalty parameter in their procedure was chosen to minimize a bootstrap estimated mean squared prediction error of the prognostic index. Although the R^2 measure has been so far defined in terms of the usual estimates, it would be straightforward to extend the definition to the penalized partial likelihood estimate. In this case, it turns out that $R^2(\hat{\beta}_\lambda) = 0.198$, almost the same as the $R^2 = 0.202$ with the standard partial likelihood estimate.

Value of R^2 in applied studies

In two of the above examples effects were quite strong and in the other two, although clearly present, effects were relatively weak. This was picked up by the R^2 coefficients and the partial coefficients in particular enable us to decide how practically useful to any prognostic assessment is the inclusion of additional information. Although we have pointed out that R^2 is concerned with prediction and not fit (as often thought) the issues of fit and prediction are not orthogonal to one another. They impact one another in important but different ways. Improving a poor fit will very likely lead to increases in predictive ability as reflected in R^2. A perfect fit (in the sense that the observations are exactly generated by the supposed model) can correspond to an R^2 taking any value between zero and one. A very high value of R^2 can also correspond to a very poor fit. All of that said, in the endeavor to improve our predictive capability, we need consider, alongside one another, both fit and R^2. The fit can be improved by a relaxation of model assumptions, such as the use of a stratified model, or by the introduction of time dependent effects such as the use of changepoint models. Either way it can be worth looking at the plot indicating the quality of the fit and making sure that we are broadly satisfied with this before presenting our summary R^2 indices of prediction.

13.13 Exercises and class projects

1. A number of suggested coefficients of explained variation, adapted from linear regression, depend on the censoring even when independent of the failure mechanism. Why is this a handicap? Mostly the dependence is such that the higher the censoring the closer to zero is the adapted coefficient. It might be argued that, as the censoring increases, our ability to predict declines and, in consequence so ought a suitable coefficient. Comment on this reasoning.

2. Suppose you are the statistician analyzing the gastric cancer data. The investigating clinician, who has some rudimentary knowledge of statistics, wishes to understand just what you mean by saying that *the value of R^2 for a model with ACT and log(CEA) is 0.37 and this increases to 0.54 when stage is also included. On the other hand the corresponding partial R^2 is equal to 0.27.* How do you answer this question.

3. Using the δ-method and the expression, $1 - R^2(\beta) = \sum r_i^2(\beta) / \sum r_i^2 (0)$, derive an approximate confidence interval for $R^2(\hat{\beta})$. For the Freireich data, compare this interval with that obtained in the example on the basis of bootstrap sampling. Comment.

4. Repeat the exercise of the previous question, only applying this time the delta method to $\log[R^2(\hat{\beta})/\{1 - R^2(\hat{\beta})\}]$. What advantages, if any, are there to working with this transformation rather than working with $R^2(\hat{\beta})$ directly?

5. In the broadened definition of $R^2(\beta)$ we have

$$\mathcal{I}(b) = \sum_{i=1}^{n} \int_0^\infty \{Z_i(t) - \mathcal{E}_b(Z|t)\}^2 d\hat{F}(t),$$

where $\hat{F}(t)$ is the Kaplan-Meier estimator. Suppose that $F(t : \theta)$ is a parametric model of the marginal survival curve where θ is a parameter, possibly vector-valued. Investigate a definition of $\mathcal{I}(b)$ in which, instead of $\hat{F}(t)$, we work with $F(t : \hat{\theta})$. What might be the advantages and drawbacks of such an approach? From the results we already have is it possible to deduce properties such as consistency? If so, under what conditions?

6. For the standard linear model, the coefficient $R(\beta)$, viewed as a function of β is maximized when $\beta = \hat{\beta}$ and where $\hat{\beta}$ is the usual least squares estimate. Thus, for the linear case, a consistent estimator of β obtains by maximizing $R^2(\beta)$. Is this true for the $R^2(\beta)$ defined here for the proportional hazards model? Investigate $\sup_\beta R^2(\beta)$, given data, as an estimate for β. How does it compare with more commonly used estimates?

7. For the normal linear model the transformation $Z = \tanh^{-1}(R)$ has two advantages: the first is that $E(Z)$ provides a very close approximation to $\tanh^{-1}(\Omega)$, the second is that Var(Z) does not depend upon Z, and therefore upon Ω, to a high level of approximation. Furthermore Var$(Z) \approx 1/(n-3)$ where n is the sample size. Discuss this transformation in the context of the $R^2(\beta)$ presented in this chapter. Investigate this more deeply using simulated data.

8. In the previous question, on the basis of simulations, we anticipate that Var$(Z) \approx 1/(n-3)$ where $Z(\beta) = \tanh^{-1}\{\sqrt{R^2(\beta)}\}$. We might conjecture, in the presence of independent censoring, and where k represents the total number of failures, that Var$(Z) \approx 1/(k-3)$. Use simulated data to investigate this assertion. Present an informal argument as to why such a result might hold.

9. We know that, as $|\beta| \to \infty$ then $R^2_{\mathcal{E}}(\beta) \to 1$ and that, as $|\beta| \to \infty$ then $R^2(\beta) \to 0$. This might at first glance appear puzzling. Explain just what is taking place. Explain also why, if the model is correct, we anticipate that $R^2(\hat{\beta})$ and $R^2_{\mathcal{E}}(\hat{\beta})$ will closely agree.

10. Consider an arbitrary proportional hazards model, with unknown cumulative hazard rate $\Lambda_0(t)$, and known regression coefficient vector β_0. For two randomly chosen individuals, the first with covariate vector given by Z_j, the second with covariate vector given by Z_i, show that the probability that the second individual outlasts (survives longer) than the first is given by

$$\Pr(T_i > T_j) = \frac{\exp(\beta_0 Z_j)}{\exp(\beta_0 Z_i) + \exp(\beta_0 Z_j)}.$$

Note that this expression does not involve $\Lambda_0(t)$. How does this result help in the interpretation of R^2.

11. Refer back to Section 3.9 in order to construct a coefficient of partially explained variation from first principles. For a given data set compare this coefficient with that suggested in this chapter based on use of the multivariate coefficient defined in terms of the prognostic index. Show how we could derive an alternative definition of multivariate explained variation based on lower order partially explained variation. Comment on the advantages and disadvantages of this.

12. Suppose for some given data we are considering using an additive risk model or a multiplicative risk model, both of which employ only constant regression coefficients. Consider how we might use an R^2 measure to discriminate between these two models.

13. Using a large data set with a large number of potential risk factors, construct, on the basis of $R^2(Z)$ for some vector Z, as predictive a model as possible, noting down every step made in the construction of the model. Now, on the basis of the isolation method, calculate $R^2(Z|F)$. Compare the sizes of $R^2(Z)$ and $R^2(Z|F)$ and comment on your findings.

14. Generate or use a data set in which only very few observations are censored and in which the covariate is continuously measured. Carry out the usual analysis and calculate R^2. Next, throw away the small percentage of censored observations, replace survival time by $\log T$ and calculate the usual squared product moment correlation coefficient between the covariate and $\log T$. Compare and discuss the results.

Chapter 14

Explained randomness

14.1 Summary

We recall the main ideas concerning information gain and how it can be used to obtain a measure of dependence. This measure can be interpreted as a measure of explained randomness. We recall the work of Kent and O'Quigley (1988) in adapting such measures to proportional hazards regression. An alternative to the measure of Kent and O'Quigley, in which the order of conditioning is reversed, is presented. This alternative measure (Xu and O'Quigley 2000) fits in more naturally with the inference structure for the proportional hazards model and, in particular, makes a direct appeal to the main theorem of proportional hazards regression (Section 7.4). Among the several advantages of the Xu and O'Quigley measure is that it readily accommodates time-dependent covariates. Extensions to multiple covariates are immediate. We also indicate the straightforward extension to partial coefficients. In practice we expect a measure of explained randomness and one of explained variation to agree. We take the position that any purpose to which we may wish to put a coefficient of explained variation will be, for the most part, equally well addressed by a coefficient of explained randomness.

14.2 Motivation

As described in Chapter 2 an alternative, and arguably more general, approach to explained variation is that of explained randomness. The details of application for the proportional hazards model are now well worked out and described here. In practical examples we would anticipate a high level of agreement between explained variation and explained randomness and so the reader motivated principally by an applied problem can then be satisfied with either one of these measures. As for the motivation behind a measure of explained variation, we would like, for a measure of explained randomness; invariance to arbitrary right censoring of the response variable, provided only that such censoring be conditionally independent of the response variable, an interpretation as the proportion of explained randomness where absence of association translates to a zero value for the measure and, as strength of association increases, the measure should tend to one.

Central to the development of Kent and O'Quigley (1988) was the family of conditional distributions of the response variable, survival time T, given the covariate vector Z. However, we appeal instead to the main theorem of proportional hazards regression, working instead with the conditional distribution of Z given T. Note that: (1) for a normal model the resulting measure of explained randomness is unaltered by the way in which we do the conditioning; (2) for other models the results will often be close (Kent 1983); (3) being able to predict which subject is to fail at any given failure time is equivalent to being able to predict failure rankings of all the failed subjects; and (4) studying the conditional distribution of Z given T, does in fact correspond to the way in which inference is carried out in proportional hazards regression.

In the context of proportional hazards regression and partial likelihood, this approach is the more natural, leading to a second measure of explained randomness based on information gain. It extends readily to time-dependent covariates. In practice, for time-invariant covariates, the two measures will tend to be similar. Calculation of the second measure turns out to be easily carried out and only requires combining those quantities routinely evaluated in fitting the Cox model. Inferential procedures for the second measure are more straightforward than the earlier measure and, a key property in the context of semi-parametric survival regression, the measure remains unaltered by monotonically increasing transformations on the time axis. Furthermore, it will be interesting to see that the measure of explained randomness can be applied to other forms of relative risk.

14.3 Information gain and explained randomness

Our regression model for T given the observed value z of Z is expressed in terms of the conditional density $f(t|z)$. It is, as we know, more common to express this dependence through the hazard function but, for our purposes here, it will be helpful to re-express the proportional hazards model as

$$f(t|z; \beta) = \lambda_0(t) \exp \left\{ \beta z - e^{\beta z} \int_0^t \lambda_0(u) du \right\}. \qquad (14.1)$$

As we have seen, the baseline hazard function $\lambda_0(t)$ can be specified to be of a power form or a constant, in which cases the Weibull and exponential models are recovered (Kalbfleisch and Prentice 1980, Cox and Oakes 1984). These were the first cases studied by Kent and O'Quigley (1988). Under the model, when $\beta = 0$ there is no association between T and Z. A population measure of the strength of association, or the distance between the two models indexed by $\beta = 0$ and $\beta = \beta_0$, can be provided by twice the Kullback-Leibler information gain given via

Definition 14.1 *The information gain* Γ_1 *is* $\Gamma_1(\beta) = 2\{I_1(\beta) - I_1(0)\}$ *where*

$$I_1(\theta) = \int_{\mathcal{Z}} \int_{\mathcal{T}} \log\{f(t|z; \theta)\} f(t|z; \beta) dt \, dG(z).$$

In the above expression the domains of definition of T and Z are denoted by $\mathcal{T}$ and $\mathcal{Z}$ respectively, and $G(z)$ is the marginal distribution function of Z. Although it is common to refer to the Kullback-Leibler information as a "distance," we also can note its lack of symmetry. Thus $I_1(\theta)$ is not a distance measure in the usual geometric sense. It would be relatively straightforward to make the measure symmetric by inverting, for example, the roles of β and θ and then adding the measures together, but this is most likely not a particularly fruitful path to pursue. The lack of symmetry is quite natural, one reason being that both $I(\beta)$ and $I(0)$ can be viewed as expectations under the model with "true" regression parameter β. Another reason is that, considered as a function of β, the asymmetric $I_1(\theta)$ is maximized at $\theta = \beta$, a quantity that directly reflects the strength of association. Compared though to the regression coefficient β, an information gain measure has the advantage of not depending on scale and the simple transformation $\rho_1^2(\beta) = 1 - \exp\{-\Gamma_1(\beta)\}$ produces a coefficient with useful interpretability properties (Kent 1983, see also Section 2.8). We then simply define ρ_1^2 as the explained randomness via:

Definition 14.2 *The explained randomness of T given Z is given by;*

$$\rho_1^2(\beta) = 1 - \exp\{-\Gamma_1(\beta)\}. \tag{14.2}$$

The interpretation of $\rho_1^2(\beta)$ as the proportion of randomness explained by the regression was given by Kent (1983) and outlined in Chapter 2. For normal models and maximum likelihood or least squares estimation, $\rho_1^2(\hat{\beta})$ is the usual coefficient of correlation squared when, instead of working with $f(t|z;\beta)dt\,dG(z)$ we use the observed empirical distribution of (T, Z). Assuming no censoring, a standard estimate of information gain will be provided by n^{-1} times the usual likelihood ratio statistic. An alternative estimate, having similar statistical properties, is provided by the fitted information (Kent and O'Quigley 1988) in which $I_1(0)$ and $I_1(\beta)$ are estimated by

$$\hat{I}_1(\theta) = n^{-1} \sum_{i=1}^{n} \int_{\mathcal{T}} \log\{f(t|Z_i;\theta)\}f(t|Z_i;\hat{\beta})dt \tag{14.3}$$

with $\theta = 0$ and $\theta = \hat{\beta}$, a consistent estimate of β, respectively. In the above the marginal distribution of Z has been replaced by its empirical estimate. Kent and O'Quigley (1988) pointed out that, for the expression based on fitted information, the variable t enters into the expression as a dummy variable so that the actual values, some of which we might have not been able to observe in the presence of censoring, do not affect the calculation beyond their effect on the estimation of β. We can thus readily apply the concept of information gain to right censored data.

This observation motivated the development of the measure of dependence for censored survival data (Kent and O'Quigley 1988) in which the conditional distribution of T is taken to be of the Weibull form. The resulting integrals can be worked out explicitly (Kent and O'Quigley 1988, page 528). Thus the required expressions can be evaluated leading to an estimate of $\rho_1^2(\beta)$. Nonetheless, the procedure is not straightforward and inference for the resulting estimate is even less so (Sections 4 and 5 of Kent and O'Quigley). The approach of the next section, arising naturally in the context of the main theorem of proportional hazards regression (Section 7.4), is more straightforward. Inference is also more straightforward; in particular, confidence intervals for $\rho_1^2(\beta)$ can be obtained by inverting those obtained for β. In addition the new approach extends immediately to time-dependent covariates and general forms of relative risk.

14.4 Explained randomness in Z given T

Just as for explained variation, it appears that our main concern is to explain variation, or randomness, in T, having conditioned by Z. The above approach follows this idea. However, once again in the same way as for explained variation, in the context of proportional hazards regression, and the requirement to obtain procedures which remain rank invariant to monotonic increasing transformations on time, it is more appropriate to consider the distribution of the variable Z given T, which brings us immediately under the umbrella of the main theorem of this book (Section 7.4). We then introduce the following definition;

Definition 14.3 *Let the information for Z given T be given by*

$$I_2(\theta) = \int_T \int_Z \log\{g(z|t;\theta)\}g(z|t;\beta)dzdF(t), \qquad (14.4)$$

where $F(t)$ is the marginal distribution function of T, and $g(z|t)$ is the conditional density or conditional probability function of Z given T.

This conditional information measure for Z given T turns out to have many advantages over the more obvious definition. We assume that the censoring is independent of the failure time and, for the most part, that there is enough information in the tail of F. It would be straightforward (although rather more messy) to make a weaker assumption of conditional independence, given the covariate, between the censoring and the failure times. As before we have:

Definition 14.4 *The information gain Γ_2 is $\Gamma_2(\beta) = 2\{I_2(\beta) - I_2(0)\}$.*

Definition 14.5 *The explained randomness of Z given T is given by*

$$\rho_2^2(\beta) = 1 - \exp\{-\Gamma_2(\beta)\}.$$

Like the Kent and O'Quigley measure, $\rho_1^2(\beta)$, $\rho_2^2(\beta)$ can be naturally extended to multiple covariates cases, where $g(\cdot|\cdot)$ would be the joint conditional density or joint conditional probability function of Z given T. For the time being we stay with a single covariate in order to keep notation simple. We can proceed in a semiparametric way by working with a consistent estimate of F in the presence of censoring, such as the Kaplan-Meier (1958) estimator. Then an estimate of the information gain can be obtained through the main theorem for proportional hazards regression (Section 7.4). We can estimate the conditional distribution of Z given T by $\{\pi_j(\hat{\beta}, t)\}$, and the marginal distribution of T by the Kaplan-Meier estimate. Recall also that $W(X_i)$ is the jump of the Kaplan-Meier curve at time $t = X_i$. We then have;

Definition 14.6 *The information gain for Z given T is*

$$\Gamma_2(\beta) = 2 \int_T \int_Z \log\left\{\frac{g(z|t;\beta)}{g(z|t;0)}\right\} g(z|t;\beta) dz dF(t). \qquad (14.5)$$

From the above we see that conditioning Z on T, the computation for estimating the information gain only involves those quantities routinely calculated in a proportional hazards analysis. This leads to great simplification when compared with the approach of Kent and O'Quigley (1988). The same is true for the multivariate case where $\hat{\beta}Z_l$ would be replaced by the inner product of the vectors.

Lemma 14.1 *The information gain, $\Gamma_2(\beta)$ can be consistently estimated by*

$$\hat{\Gamma}_2(\hat{\beta}) = 2 \sum_{i=1}^k W(X_i) \sum_{j=1}^n \pi_j(\hat{\beta}, X_i) \log\left\{\frac{\pi_j(\hat{\beta}, X_i)}{\pi_j(0, X_i)}\right\}.$$

Results indicate that ρ_1^2 and ρ_2^2 may be anticipated to be close to one another as well as to some other alternatives used in this context, although there is no reason one should expect all of them to be exactly the same since we are not dealing with a normal model. There is also an interesting connection between the approach here and some other papers in the literature such as McKean and Sievers (1987) and Rousseeuw and Hubert (1997), in that all of them use the objective functions that specific procedures minimize (or maximize) to define a R^2-type measure for regression models, the same being true for the classical least squares regression.

McKean and Sievers (1987) proposed coefficients of determination for the least-absolute deviation (LAD) analysis of linear models, using the sum of absolute deviations, which the LAD minimizes, as function of the regression parameters under the model and without the model. Rousseeuw and Hubert (1997) similarly defined coefficients of determination for the least quantile of squares and least trimmed squares estimates of linear regression. In our case under the proportional hazards model the estimate of β is obtained via maximizing the partial likelihood, and the value of the partial likelihood under the model and without the model is essentially what is used above. Note that in Lemma 14.1 the expression can be shown by straightforward algebra to be equal to

$$\hat{\Gamma}_2(\hat{\beta}) = 2 \sum_{i=1}^{k} W(X_i) \left\{ \hat{\beta} \frac{\sum_1^n Y_l(X_i) Z_l e^{\hat{\beta} Z_l}}{\sum_1^n Y_l(X_i) e^{\hat{\beta} Z_l}} - \log \frac{\sum_1^n Y_l(X_i) e^{\hat{\beta} Z_l}}{\sum_1^n Y_l(X_i)} \right\}.$$

(14.6)

Whenever there is a finite upper limit τ of our observation time, intuitively no more information is to be gained beyond τ. Therefore, we can look at the information gain conditioning on $T \leq \tau$, i.e.,

$$\Gamma_2(\beta) = 2E \left\{ \log \frac{g(Z|T; \beta)}{g(Z|T; 0)} \middle| T \leq \tau \right\}.$$

The above quantity can be consistently estimated by the proposed estimator divided by $\sum_1^k W(X_i)$. It inevitably depends on the upper limit τ. This is also true of the measures of Korn and Simon (1990), and is implicitly true of the explained variation measure of O'Quigley and Flandre (1994) since the Schoenfeld residuals correspond to the observed failure times only. Some of the other proposed measures, including ρ_1^2 of Kent and O'Quigley (1988), another example being that of Schemper and Kaider (1997), do not depend on τ because they extrapolate beyond the range of the observed data; in particular, they rely on the assumption of proportional hazards being true for the whole range of $\mathcal{T}$. The dependence of ρ_2^2 on τ, on the other hand, can be viewed of as appropriate in that we have then conditioned on the time span actually studied rather than a time span of potential interest.

Properties

As for the coefficient R^2 for explained variation, we have a number of important properties that enable us to give a solid interpretation to the measures ρ_1^2 and ρ_2^2. The coefficient $\rho_2^2(\beta)$, with information I_2 defined above maintains the basic properties that we require of such a measure of dependence. For properties which hold for both ρ_1^2 and ρ_2^2 simultaneously, we write simply ρ^2. Specifically, we have:

Lemma 14.2 $\rho^2(0) = 0$, and $0 < \rho^2(\beta) < 1$ if $\beta \neq 0$.

Lemma 14.3 ρ^2 is invariant under linear transformations of Z and is also invariant under monotonically increasing transformations of T.

Definition 14.7 *Following Kent and O'Quigley (1988) we can define the randomness of Z, as a monotonic transformation of the entropy of Z, whereby, for j assuming the values 1 and 2, $D(Z) = \exp\{-2I_j(0)\}$, and the residual randomness of Z given T by $D(Z|T) = \exp\{-2I_j(\beta)\}$. Then*

$$\rho^2(\beta) = \{D(Z) - D(Z|T)\}/D(Z)$$

giving ρ_1^2 and ρ_2^2 an interpretation as the "proportion of the explained randomness."

Lemma 14.4 *For β a scalar, $\rho^2(\beta)$, as a function of β, increases with $|\beta|$.*

An immediate consequence of the lemma is that we can directly infer confidence intervals for $\rho^2(\beta)$ from those for β. Suppose that a 95% confidence interval for β is $0 < \beta_L < \beta < \beta_U$, then the interval for $\rho^2(\beta)$ should be $(\rho^2(\beta_L), \rho^2(\beta_U))$. In practice, of course, we do not know $\rho^2(\cdot)$, but, following the consistency results, we can "plug" β_L and β_U into $\hat{\rho}^2(\cdot) = 1 - \exp\{-\hat{\Gamma}_2(\cdot)\}$ to obtain an approximation of $\rho^2(\beta_L)$ and $\rho^2(\beta_U)$. The coverage properties will not be exactly the same but will be close. Further study can throw more light on this. We can similarly deal with the case where $\beta_L < \beta_U < 0$. If instead $\beta_L < 0 < \beta_U$, because $\rho^2(\beta)$ is symmetric about $\beta = 0$ we have the invariance to linear transformation of Z and monotonic increasing transformation of T, then ρ^2 is invariant under linear transformations $-Z$ of Z), 0 so that the larger in absolute value of β_L and β_U should be plugged in. This approach is illustrated in an example below. Alternatively we can always obtain confidence intervals based on asymptotic normality given by:

Lemma 14.5 *Assuming the Andersen-Gill type conditions, then $\hat{\Gamma}_2(\hat{\beta})$ consistently estimates $\Gamma_2(\beta)$ and is asymptotically normal.*

The proof of the asymptotic normality makes use of the central limit theorem of Stute (1995) for Kaplan-Meier integrals, of which the definition of information gain here is a special case. Following Stute (1996) the variance estimation of the Kaplan-Meier integrals is more complicated in general, and so it is recommended to use resampling techniques to obtain an estimate of the variance.

Boundedness

It is possible for ρ^2 to be bounded by a number strictly less than 1 as β approaches infinity. This can occur when the covariate Z is discrete and takes on very few levels. Even so, for the most extreme case of a single binary covariate, the bound is close enough to one for the phenomenon to be practically ignored. This behavior was also referred to by Kent (1983). We can illustrate such behavior in the case of an exponential model where we suppose that $\lambda_0 = 1$. We drop the index β in the notation of the hazard, density, and so on. Then $\lambda(t|z) = e^{\beta z}$ and $S(t|z) = \exp(-e^{\beta z}t)$. In particular, $S_0(t) = f_0(t) = e^{-t}$. We have

$$g(z|t) = \frac{e^{\beta z}\{S_0(t)\}^{\exp(\beta z)}g(z)}{\sum_z e^{\beta z}\{S_0(t)\}^{\exp(\beta z)}g(z)}$$

and the joint probability of T and Z is given by $f(t,z) = e^{\beta z}\{S_0(t)\}^{\exp(\beta z)}g(z)$. We can write; $\Gamma_2(\beta) = 2\sum_z \int_0^\infty \log\{g(z|t)/g(z)\}f(t,z)dt$. and, after some elementary algebra and a change of variable in the integration we have:

Lemma 14.6

$\Gamma_2(\beta)$

$$= 2\sum_i g(z_i) \int_0^\infty e^{-x} \log\left\{e^{-x} \Big/ \sum_j g(z_j)e^{\beta(z_j - z_i)} \exp(-xe^{\beta(z_j - z_i)})\right\} dx.$$

Here we use the subscripts i and j to distinguish between the two nested sums over all the atoms of the distribution of Z. Notice that for $j \neq i$, $z_j - z_i \neq 0$ and the summand indexed by j tends to 0 as $\beta \to \infty$. But this is not true with $j = i$. In particular:

Lemma 14.7 *As $\beta \to \infty$ we have;*

$$\Gamma_2(\beta) \longrightarrow 2 \sum_i g(z_i) \log \frac{1}{g(z_i)} = \Gamma_{\max}.$$

Note that $\Gamma_{\max}$ of the lemma is finite. This means that the limit of $\rho^2(\beta)$ would be strictly less than one. For discrete uniform of Kent and O'Quigley (1988) on three points, Xu and O'Quigley (1999) calculated that $\rho^2_{\max} = 0.720$ and that $\rho^2_{\max} = 0.973$ for six points. Finally note that if the distribution of Z is continuous, the sums over z in the above calculation become integrals. Then as $\beta \to \infty$, the denominator inside the log becomes zero, since the integrand is everywhere zero except for one point. Therefore, $\Gamma(\beta) \to \infty$ and $\rho^2(\beta) \to 1$. Secondly, for discrete covariates as Kent (1983) pointed out, except for very high values of correlation, ρ^2 can usually be interpreted without paying special attention to the discreteness of the distribution. The numerical study below also shows that there is not much discrepancy between the ρ^2 discussed here and the original proposal of Kent and O'Quigley's (which has the property of tending to 1 as $\beta \to \infty$). It performs well in a practical example with a binary covariate. This is detailed in the examples.

14.5   Approximation of ρ_1^2 by ρ_2^2

Under certain conditions $\rho_1^2 = \rho_2^2$. These conditions are satisfied by a bivariate normal model (itself not in the proportional hazards class) and we only expect them to be approximately met in practice. However, on the basis of this, we anticipate ρ_1^2 and ρ_2^2, as well as their sample based estimates to be generally close. If we first define $W = \beta'Z$ to be the usual prognostic index, then we have;

Theorem 14.1 *If, for each t, $E\{Z(\partial f(t|Z)/\partial W) = 0\}$ then, $\rho_1^2 = \rho_2^2$.*

The above result would follow if $I_1(\theta) = I_2(\theta) + K$ where K is a constant. Suppose then we let $dH(z,t;\beta) = g(z|t;\beta)dz\,dF(t) = f(t|z;\beta)$ $dt\,dG(z)$ then, we can write, $I_2(\theta) = \int_T \int_Z \log\{g(z|t;\theta)\}\,dH(z,t;\beta)$ and this integral in turn can be simply expressed as, $\int_T \int_Z \{\log f(t|z;\theta) + \log g(z) - \log f(t;\theta)\}\,dH(z,t;\beta)$ which simplifies to $I(\theta) + K_1 - B(\theta)$ where $B(\theta) = \int_T \int_Z \log f(t;\theta)\,dH(z,t;\beta)$ and $K_1 = \int_T \int_Z \log g(z)\,dH$ $(z,t;\beta)$. Even though H depends upon β, K_1 is a constant, depending only upon the marginal distribution of Z. Now $B(\theta) = E\log f(T;\theta)$

which depends upon the marginal distribution of T. The marginal distribution of T depends on θ via an expression for total probability and the conditional distributions of T given Z. However, writing $f(t) = \int f(t|z)g(z)dz$, then $\partial f(t)/\partial\theta = \int \partial f(t|z)/\partial w(dw/d\theta)g(z)dz$, assuming we can interchange the limiting processes. Thus, $\partial f(t)/\partial\theta = E\{Z(\partial f(t|Z)/\partial W) = 0$. Under the condition of the theorem, $B(\theta) = E\log f(T; \theta) = E\log f(T)$ does not depend upon θ and so can be written as a constant K_2. We then have $K = K_1 - K_2$.

The condition indicates, that for each fixed t, the rate of change of the conditional density with respect to βZ is uncorrelated with the covariate Z itself. For normal regression this can be verified. More generally the condition would only be approximately met. It is nonetheless a reasonable condition and, in practice, it is unlikely that we would be very far removed from the condition. Formal procedures for investigating the validity of the condition could also be developed but, at this point, no such procedures exist.

In view of the above theorem we can view $\hat{\rho}_2^2$ either as an estimator of the explained randomness in Z given T or as an approximation to ρ_1^2, the explained randomness in the ranks of T given Z. This second interpretation is the one that corresponds most closely to that required in the majority of applications. A second theorem leads to further simplification including an approximation that is very easily obtained. First we define

$$\bar{\Gamma}_2(\hat{\beta}) = 2 \sum_{i=1}^{n} W(X_i) \log \left\{ \frac{\pi_i(\hat{\beta}, X_i)}{\pi_i(0, X_i)} \right\}. \tag{14.7}$$

Again, the right-hand side of the above should be divided by $\sum_1^n W(X_i)$ if it is less than one in the same way as for $\Gamma_2(\hat{\beta})$.

14.6 Simple working approximation of ρ_1^2 and ρ_2^2

From the above we can view ρ_2^2 as a coefficient in its own right quantifying the amount by which the explanatory variables explain the rankings in survival or, we can view ρ_2^2 as an approximation to ρ_1^2. In this section we see that there is a very simple working approximation for ρ_2^2, and thereby ρ_1^2, and this approximation is widely available through the software package SAS. To see this consider:

Theorem 14.2 *Assuming the data are generated under the proportional hazards model and that the support for C and T coincides, then $|\bar{\Gamma}_2(\hat{\beta}) - \hat{\Gamma}_2(\hat{\beta})|$ converges in probability to zero.*

This is quite straightforward since:

$$\log \pi_i(\hat{\beta}, X_i) - \sum_{j=1}^{n} \pi_j(\hat{\beta}, X_i) \log \pi_j(\hat{\beta}, X_i)$$

$$= \log \frac{Y_i(X_i) \exp(\hat{\beta} Z_i(X_i))}{\sum_{l=1}^{n} Y_l(X_i) \exp(\hat{\beta} Z_l(X_i))}$$

$$- \sum_{j=1}^{n} \pi_j(\hat{\beta}, X_i) \log \frac{Y_j(X_i) \exp(\hat{\beta} Z_j(X_i))}{\sum_{l=1}^{n} Y_l(X_i) \exp(\hat{\beta} Z_l(X_i))}$$

$$= Y_i(X_i) \cdot \hat{\beta} Z_i(X_i) - \sum_{j=1}^{n} \pi_j(\hat{\beta}, X_i) \cdot Y_j(X_i) \cdot \hat{\beta} Z_j(X_i)$$

$$= \hat{\beta}\{Z_i(X_i) - \mathcal{E}_{\hat{\beta}}(Z|X_i)\} \quad \text{where} \quad \mathcal{E}_{\beta}(Z|t) = \sum_{j=1}^{n} Y_j(t) Z_j(t) \pi_j(\beta, t).$$

So the arithmetic difference between $\bar{\Gamma}_2(\hat{\beta})$ and $\hat{\Gamma}_2(\hat{\beta})$ can then be expressed as, $2\hat{\beta} \sum_{i=1}^{n} W(X_i)\{Z_i(X_i) - \mathcal{E}_{\hat{\beta}}(Z|X_i)\}$, which tends to zero (Xu and O'Quigley 2000a). Next, if we first define,

$$\bar{\Gamma}_A(\hat{\beta}) = \frac{2}{k} \sum_{i=1}^{n} \delta_i \log \left\{ \frac{\pi_i(\hat{\beta}, X_i)}{\pi_i(0, X_i)} \right\}. \tag{14.8}$$

where k is the total number of failures, then we have the following corollary.

Corollary 14.1 *In the absence of censoring $\bar{\Gamma}_A(\hat{\beta}) = \hat{\Gamma}_2(\hat{\beta})$.*

This is immediate since, in the absence of censoring, $k = n$ and $W(X_i) = W(T_i) = 1/n$. But, more generally, in the presence of independent censoring, we can take $\bar{\Gamma}_A(\hat{\beta})$ to be a good approximation to $\bar{\Gamma}(\hat{\beta})$. Both $\bar{\Gamma}_A(\hat{\beta})$ and $\bar{\Gamma}_2(\hat{\beta})$ represent empirical expectations of the same quantity, the difference being in how we assign the total probability mass of one. For most observed censoring patterns, we would not anticipate seeing much discrepancy between $\bar{\Gamma}_A(\hat{\beta})$ and $\bar{\Gamma}_2(\hat{\beta})$. Since

$\bar{\Gamma}_A(\hat{\beta})$ is particularly straightforward to evaluate, using a simple transformation of the partial likelihood ratio statistic, we would recommend it for routine use. In the absence of censoring, note that $\hat{\rho}^2(\hat{\beta})$, based upon $\bar{\Gamma}_A(\hat{\beta})$, coincides with the correlation measure provided by a SAS survival analysis book (Allison 1995). When there is censoring however, the Allison measure uses n in place of k in (14.8), and therefore will depend upon an independent censoring mechanism regardless of population effects. In particular it approaches the value zero as the percentage of censored observations approaches one. On the other hand, for all independent censoring mechanisms, $\hat{\rho}^2(\hat{\beta})$ approaches a population equivalent, this equivalent being close to $\rho^2(\beta)$, and therefore interpretable as a percentage of explained randomness. If the coefficient R_{SAS}^2 is then available from the SAS software package, we can write;

$$\hat{\rho}_2^2 \approx 1 - \exp\left\{\frac{n}{k}\log(1 - R_{SAS}^2)\right\} \qquad (14.9)$$

and use this as a simple working approximation to ρ_2^2. In the light of the previous section this would also then be a working approximation for the coefficient ρ_1^2. We might also view the above expression as a correction for the Allison SAS coefficient which has not adequately taken account of the censoring. In practice this can make quite an important difference. As a quick illustration, suppose that we have 100 subjects on which we observe a total of 30 failures. Suppose that we evaluate R_{SAS}^2 as 0.3, a value which would be indicative of moderate effects. In fact the true force of the effects are being masked by not correctly accounting for the censoring and, making the correction outlined in the above formula we find that ρ_2^2 is approximated by 0.7, a value which indicates very strong effects.

Furthermore, if "being close" is not good enough and we insist upon a consistent estimate of $\rho^2(\beta)$ then this requires little in the way of extra work. Even so, we can conclude that the measure routinely provided by the SAS package can be used and given a meaningful interpretation as long as we make the correction given by Equation 14.9. While not eliminating all of the effects of the censoring on the population parameter, the correction goes sufficiently far to reduce these effects so that, for the purposes of everyday practical data analysis, conclusions based on precise measures as opposed to those based on approximate measures are unlikely to differ very much.

14.7   Multiple coefficient of explained randomness

Returning to the definitions given at the beginning of the chapter, for instance that for $I_1(\theta)$ where:

$$I_1(\theta) = \int_{\mathcal{Z}} \int_{T} \log\{f(t|z;\theta)\} f(t|z;\beta) dt dG(z),$$

the extension to the multivariate situation appears immediate. We keep the same definitions of $f(t|z;\theta)$, $\Gamma_1(\beta)$, and $I_1(\theta)$ and, finally, $\rho_1^2(\beta)$ and take z to be vector valued. There is no extra work, at least conceptually, in obtaining this definition which is then very logical. For the definitions leading to ρ_2^2, the extension is possibly less immediate but is equally straightforward. Keeping in mind that we can estimate the conditional distribution of Z given T by $\{\pi_j(\hat{\beta}, t)\}$, and the marginal distribution of T by the Kaplan-Meier estimate and that $W(X_i)$ is the jump of the Kaplan-Meier curve at time $t = X_i$ we can then easily work with the following definition:

Definition 14.8 *The information gain for Z given T is*

$$\Gamma_2(\beta) = 2 \int_{T} \int_{\mathcal{Z}} \log \left\{ \frac{g(z|t;\beta)}{g(z|t;0)} \right\} g(z|t;\beta) dz dF(t), \qquad (14.10)$$

which appears the same as the earlier definition, only that now both z and β are vectors. The distribution $g(z|t;\beta)$ is a multivariate distribution and it can be helpful to see this as a transformation to a one-dimensional scale. In particular, we can still consistently estimate $\Gamma_2(\beta)$ by

$$\hat{\Gamma}_2(\hat{\beta}) = 2 \sum_{i=1}^{k} W(X_i) \sum_{j=1}^{n} \pi_j(\hat{\beta}, X_i) \log \left\{ \frac{\pi_j(\hat{\beta}, X_i)}{\pi_j(0, X_i)} \right\}, \qquad (14.11)$$

the only difference from the earlier expression being that, here, both β and Z_i, contained in the expression $\pi_j(\hat{\beta}, X_i)$ are vectors. Simple and multiple coefficients, based on $\Gamma_2(\beta)$, are then quite straightforward. In practice we often would like to answer questions of the form: What is the impact of some covariate, or some set of covariates, after having taken account of the effects of some other set? Such questions can be addressed by the use of partial coefficients of explained randomness and these are outlined below. One fairly immediate observation is that we can define partial coefficients in terms of simple and multiple coefficients and so there is really no need for any further work on this. The relationship between partial and multiple coefficients is an exact one for multi-normal linear models. It is not exact here but is, for practical purposes, almost certainly good enough.

14.8 Partially explained randomness

Partial measures of explained randomness in T given Z_2, after having already accounted for Z_1 can be defined directly following the outline of Section 2.8. However, it seems more satisfactory to define the partial coefficients as ratios of multiple coefficients of different orders. Aside from the multivariate normal model, the two approaches will not lead to identical results, although we would expect them to be close and not to disagree in any substantive way when confronted with real data.

In order to better present the ideas it is useful, at least for this short section, to make a change in notation. Here then, we no longer indicate the dependence of ρ^2 on β by $\rho^2(\beta)$. That notation helps in the study of the properties of ρ^2 as a function of β. Now, we wish to consider the dependence of ρ^2 as function of certain covariates after having accounted for the effects of others. Thus, we write $\rho^2(Z_1, \ldots, Z_p)$ and $\rho^2(Z_1, \ldots, Z_q)$ $(q < p)$, the multiple coefficients for models including covariates Z_1 to Z_p and covariates Z_1 to Z_q, respectively. We are then able to express, and define, the partial coefficient $\rho^2(Z_{q+1}, \ldots, Z_p | Z_1, \ldots, Z_q)$, the explained randomness provided by Z_{q+1} to Z_p after having accounted for the effects of Z_1 to Z_q by

$$
1 - \rho^2(Z_1, \ldots, Z_p)
$$
$$
= [1 - \rho^2(Z_1, \ldots, Z_q)][1 - \rho^2(Z_{q+1}, \ldots, Z_p | Z_1, \ldots, Z_q)].
$$

This is motivated by an analogous formula for the multivariate normal model and makes intuitive sense in that the value of the partial coefficient increases as the difference between the multiple coefficients increases, and takes the value zero should this difference be zero. Since simple and multivariate coefficients are readily evaluated then so is the resulting partial coefficient of explained randomness.

We can, however, proceed more directly, corresponding to the original suggestion in Kent and O'Quigley (1988). In order to do this we break down the covariate vector into two components, the one of main interest and the one we wish to adjust for. We assume then a model in which the second set of components is fixed at the value zero and we maximize over the parameter space for the other set an expression corresponding to the expected log-likelihood. Next we maximize this quantity over the whole covariate vector without constraint. Once we have these quantities we proceed as before by calculating twice the Kullback-Leibler distance between these two hypotheses. The details of this approach are described in Kent and O'Quigley (1988) and Xu and O'Quigley (1999). For routine work it is recommended to proceed on the basis of the expressions given in the above paragraph. These are simpler to evaluate, the risk of making a numerical mistake is smaller, and it is, in any event, helpful to have in mind the values of the multivariate coefficients even when we only plan to use these indirectly. A simplification, analogous to that for the ordinary coefficient, can also be worked out. For cases where there is a finite follow-up time τ, procedures similar to those for the ordinary coefficient can be used.

14.9   Isolation method for bias-reduction

The approach to bias reduction for R^2 in Section 13.11 can be equally well applied when we work with a coefficient of explained randomness. All of the points made in Section 13.11 apply in just the same way here. We do not repeat them and the reader may wish to go over Section 13.11 simply replacing each occurence of R^2 by ρ^2. In particular, we employ an entirely analogous model for overfitted effects. For a given covariate or covarariate vector Z, we calculate $\rho^2(Z|F)$ where F is as in Section 13.11, i.e., not something we measure, or observe, but a conceptual quantity indicating the sum total of all the actions taken during the fitting process. The observed multiple $\rho^2(Z)$ simultaneously involves, as well as the real effects, the fitting process, and, as in Section 13.11, we write $\rho^2(Z) = \rho^2(Z, F)$ and appeal to the expression;

$$1 - \rho^2(Z, F) = [1 - \rho^2(F)][1 - \rho^2(Z|F)] \qquad (14.12)$$

Only $\rho^2(Z, F)$, can be directly observed. We obtain a distribution for $\rho^2(F)$ as described in Section 13.11 and, again, we will label this by $H(\cdot)$. Finally, we obtain the quantity we want which is $\rho^2(Z|F)$, indicating that we have "accounted" for the fitting effects and this is;

$$\rho^2(Z|F) = \int \left(\frac{\rho^2(Z, F) - u}{1 - u} \right) dH(u). \qquad (14.13)$$

As before, a simple approximation obtains from working with the mean $\bar{u}$ across the replications of $\rho^2(F)$ and writing $\rho^2(Z|F) \approx [\rho^2(Z,F) - \bar{u}]/(1 - \bar{u})$. If our purpose is to look at one factor after having taken account of several others then we would fix at zero the coefficient corresponding to the factor of interest and allow those factors for which we are adjusting to be replaced by estimates in our replications. These would be constrained estimates in that the coefficient corresponding to the additional factor of interest is always fixed at zero. Equation 13.20 is then used with these values in order to generate the relevant distribution $H(u)$.

14.10 Simulations

The behavior of ρ_2^2 was studied by Xu and O'Quigley (1999) via some simulations, for different strength of dependence, different censoring mechanism, and different covariate distributions. To compare ρ_2^2 with other similar measures proposed under the Cox regression model, we also make use of the results from Table II of Schemper and Stare (1996). As in Schemper and Stare (1996), data are simulated with hazard function $\lambda(t) = \exp(-\beta Z)$, where $\beta = 0, \log 2, \log 4, \log 16, \log 64$, and Z distributed as either uniform $[0, \sqrt{3}]$ or dichotomous 0,1 with equal probabilities. Such choices of covariate distributions result in identical variance for Z and thus allows comparison of the results for dichotomous and continuous covariates.

Censoring mechanisms are assumed to be uniform $[0, \tau]$, where τ is chosen to achieve a certain percentage of censoring. We used 100 simulations for each entry of the results. In Table 14.1 covariate type "c" stands for "continuous", and "d" for "dichotomous." R^2 is the measure of predictive capability proposed by O'Quigley and Flandre (1994). The last four columns of the table are taken from Schemper and Stare (1996), where ρ_W^2 and $\rho_{W,A}^2$ are from Kent and O'Quigley (1988). The coefficient ρ_W^2 is their sample based estimate of ρ_1^2, and $\rho_{W,A}^2$ a simpler approximation to ρ_W^2. The measure r_{pr}^2 is from Schemper and Kaider (1997) and "KS" from Korn and Simon (1990) based on quadratic loss. Overall there is mostly good agreement among all the coefficients except for KS. We suspect there may be theoretical grounds for anticipating some level of agreement among ρ_2^2, R^2, ρ_W^2, and $\rho_{W,A}^2$. Roughly, the three versions of ρ^2 based on information gain have at their bases something like a likelihood ratio statistic whereas that of R^2 a score statistic. Large sample agreement for such statistics has been

Table 14.1: A simulated comparison of different measures ($n = 5000$).

$\exp(\beta)$	% censored	covariate	ρ_2^2	R^2	$\rho_W^2 = \rho_1^2$	$\rho_{W,A}^2$	r_{pr}^2	KS
1	0%	c	0.000	0.000	0.000	0.000	0.000	0.000
	50%	c	0.000	0.000	0.000	0.000	0.000	0.000
	90%	c	0.002	0.002	0.000	0.000	0.000	0.000
2	0%	c	0.102	0.098	0.096	0.119	0.092	0.101
	50%	c	0.108	0.105	0.089	0.122	0.093	0.088
	90%	c	0.105	0.104	0.103	0.099	0.074	0.015
	0%	d	0.102	0.099	0.113	0.118	0.096	0.095
	50%	d	0.110	0.108	0.114	0.121	0.096	0.089
	90%	d	0.106	0.108	0.125	0.100	0.076	0.016
4	0%	c	0.295	0.281	0.304	0.338	0.272	0.231
	50%	c	0.334	0.316	0.298	0.344	0.274	0.267
	90%	c	0.340	0.334	0.279	0.342	0.278	0.063
16	0%	c	0.598	0.586	0.623	0.664	0.584	0.354
	50%	c	0.690	0.644	0.622	0.668	0.584	0.564
	90%	c	0.723	0.708	0.605	0.670	0.585	0.188
64	0%	c	0.758	0.757	0.785	0.815	0.754	0.397
	50%	c	0.848	0.806	0.790	0.815	0.730	0.717
	90%	c	0.876	0.864	0.763	0.816	0.694	0.321
	0%	d	0.681	0.870	0.777	0.814	0.707	0.319
	50%	d	0.860	0.885	0.776	0.815	0.718	0.861
	90%	d	0.756	0.937	0.795	0.792	0.701	0.135

documented and further exploration here may shed light on both the behavior of measures of explained randomness and that of O'Quigley and Flandre (1994).

The values of r_{pr}^2 tend to be slightly lower than these four coefficients, although the strength of association reflected is similar. The measure r_{pr}^2 requires more computation than all the other ones in the table because of the multiple imputation technique employed. There is also noticeably close agreement between ρ_2^2 and R^2 for the majority of the cases. This may have its root in the fact that both measures are semi-parametric and essentially based on the conditional distribution of the covariates given the failure time T. The numerical results for dichotomous covariates with high hazard ratio 64 also confirm the theoretical findings where ρ_2^2 is shown to be bounded away from one as β increases, although when there is no censoring ρ_2^2 is quite close to r_{pr}^2 in this case.

Because of the censoring mechanism used in the simulation for the table, the support for time is taken to be on $(0, \tau)$. This is also reflected in the gradual although weak impact of τ on ρ_2^2. Notice that the changes are not very large even for 90% censoring, where about 80% of the observations are censored at τ itself. As mentioned before, the dependence on τ also exists for R^2 and KS, although the KS measure appears to change quite dramatically. The KS measure also does not remain invariant to monotone increasing transformation of time, unlike all the other measures included for the comparison as well as the partial likelihood estimator itself. This measure is most useful when the time variable provides more information than just an ordering. On the other hand, less agreement is seen among the last two columns and the rest of the table. This is because the measure V_2 is known to depend rather heavily on censoring, even when independent of the failure mechanism and even when there is no finite upper limit τ of the follow-up time.

Further simulations on these measures were carried out, for different strength of regression effects, different censoring percentages and different covariate distributions. Again the data were simulated via hazard functions $\lambda(t) = \exp(-\beta Z)$. The distributions of Z are standardized to have the same variances. The censoring mechanism is uniform $[0, \tau]$. Table 14.2 contains the results of a large sample $(n = 5000)$ comparison. The last three columns of the table are from Xu and O'Quigley (1999), where ρ_W^2 and $\rho_{W,A}^2$ are defined in Kent and O'Quigley (1988). The coefficient ρ_W^2 is their sample based estimate, and $\rho_{W,A}^2$ a simpler approximation to ρ_W^2. The measure denoted R_{SAS}^2 is from Allison (1995), where the partial likelihood ratio statistic is, as we've explained above, incorrectly divided by the sample size n instead of the number of events k (see Equation 14.9). Note that we would have as an approximation from Equation 14.9 obtained by rearranging the terms as;

$$R_{SAS}^2 \approx 1 - \exp\left\{ \frac{k}{n} \log(1 - \rho_2^2) \right\}. \qquad (14.14)$$

From the table we see that R_{SAS}^2 decreases dramatically with the percentage of censoring, which is consistent with our earlier discussion. The measures ρ^2 and ρ_2^2 have close agreement, as they are asymptotically equivalent. The measure ρ_k^2 appears to be a good approximation to ρ^2, even in the presence of heavy censoring. The measures from Kent and O'Quigley (1988) also have good agreement with these three measures.

Table 14.2: A simulated comparison of the measures ($n = 5000$).

$\exp(\beta)$	% censored	covariate	R^2_{SAS}	ρ^2_k	ρ^2	ρ^2_2	ρ^2_W	$\rho^2_{W,A}$
2	0%	c	0.103	0.103	0.103	0.102	0.096	0.119
	50%	c	0.053	0.106	0.097	0.108	0.089	0.122
	90%	c	0.013	0.117	0.107	0.105	0.103	0.099
	0%	d	0.094	0.094	0.094	0.102	0.113	0.118
	50%	d	0.060	0.113	0.086	0.110	0.114	0.121
	90%	d	0.012	0.114	0.116	0.106	0.125	0.100
4	0%	c	0.293	0.293	0.293	0.295	0.304	0.338
	50%	c	0.197	0.351	0.356	0.334	0.298	0.344
	90%	c	0.048	0.390	0.366	0.340	0.279	0.342
16	0%	c	0.603	0.603	0.603	0.598	0.623	0.664
	50%	c	0.473	0.713	0.679	0.690	0.622	0.668
	90%	c	0.116	0.725	0.765	0.723	0.605	0.670
64	0%	c	0.757	0.757	0.757	0.758	0.785	0.815
	50%	c	0.641	0.873	0.857	0.848	0.790	0.815
	90%	c	0.197	0.865	0.886	0.876	0.763	0.816
	0%	d	0.679	0.679	0.679	0.681	0.777	0.814
	50%	d	0.633	0.865	0.859	0.860	0.776	0.815
	90%	d	0.128	0.726	0.742	0.756	0.795	0.792

Next, simulations to evaluate the finite sample ($n = 100$) behaviors of ρ^2_n, ρ^2_k and ρ^2 (Table 14.3) were carried out. Since the difference among these three measures lies in their ways of handling censorship, for comparison purposes in the last column of the table we also provide the value of the measure for the same simulated data without censoring, ρ^2_{nc}. In the parentheses are the standard errors from the 200 simulations. As can be seen R^2_{SAS} again behaves quite poorly with increasing censoring. When compared to ρ^2_{nc}, ρ^2 is slightly less biased than ρ^2_k, but ρ^2_k turns out to have smaller standard errors than ρ^2. The mean squared errors (MSE) was also computed and, although not shown here, the MSE for ρ^2_k is generally slightly smaller than ρ^2.

14.11 Illustrations

Consider two examples. The first, as a comparison, is the example that appeared in Kent and O'Quigley's paper. The second is the data of Freireich (1963), so that we can compare the measures with other

Table 14.3: A simulated comparison of the measures ($n = 100$).

$\exp(\beta)$	% censored	covariate	R^2_{SAS}	ρ^2_k	ρ^2	ρ^2_{nc}
2	0%	c	0.105	0.105	0.105	0.105
	50%	c	0.059	0.114	0.106	0.097
	90%	c	0.021	0.173	0.149	0.106
	0%	d	0.103	0.103	0.103	0.103
	50%	d	0.065	0.122	0.118	0.106
	90%	d	0.022	0.180	0.178	0.107
4	0%	c	0.288	0.288	0.288	0.288
	50%	c	0.186	0.337	0.319	0.291
	90%	c	0.047	0.357	0.349	0.292
16	0%	c	0.587	0.587	0.587	0.587
	50%	c	0.443	0.686	0.662	0.586
	90%	c	0.118	0.695	0.658	0.591
64	0%	c	0.750	0.750	0.750	0.750
	50%	c	0.616	0.847	0.827	0.743
	90%	c	0.182	0.857	0.851	0.751
	0%	d	0.676	0.676	0.676	0.676
	50%	d	0.627	0.857	0.846	0.673
	90%	d	0.135	0.750	0.747	0.676

Table 14.4: A comparison of ρ^2_1 and ρ^2_2.

β	D1		D2		D3	
	ρ^2_2	ρ^2_1	ρ^2_2	ρ^2_1	ρ^2_2	ρ^2_1
1/4	0.039	0.058	0.054	0.055	0.060	0.061
1/2	0.135	0.193	0.172	0.177	0.195	0.208
1	0.367	0.474	0.442	0.430	0.445	0.505
2	0.699	0.770	0.695	0.722	0.704	0.795

similar measures in a well-known practical case of the proportional hazards model.

The measure of dependence ρ^2_1 proposed in Kent and O'Quigley (1988) was worked out for the Weibull regression model under several distributions of Z, namely, discrete uniform on $D1 = \{-1.195, -0.717, -0.239, 0.239, 0.717, 1.195\}$, $D2 = \{-0.707, -0.707, 1.414\}$, and $D3 = \{-1.414, 0.707, 0.707\}$. All three distributions of Z have expectation 0 and variance 1. Their results along with ρ^2_2 are compared in Table 14.4. From Table 14.4 we see that the ρ^2_1 and ρ^2_2 values appear to reflect similar strength of association between the survival variable and the covariate. Comparison of $D2$ and $D3$ shows that skewness from the distribution of Z seems to affect ρ_1^2 more than ρ_2^2. As mentioned earlier, there is no reason that the two measures should be exactly the same. Although the upper bound for ρ_2^2 was computed to be less than one, it is not far away from ρ_1^2 for all the cases studied here.

The Freireich data described in the original paper (Cox 1972) on the proportional hazards model gives an example in which there are two groups and the partial likelihood estimate of log relative risk is $\hat{\beta} = 1.65$. A visual inspection of the separation afforded by the two Kaplan-Meier curves, corresponding to the two prognostic groups, is, of itself, enough to suggest that there are quite important and strong predictive effects. We estimated $\rho^2(\beta)$ to be 0.34. Kent and O'Quigley's ρ_1^2 is estimated to be 0.37. The measure R^2 of O'Quigley and Flandre (1994) turns out to be 0.39 for the same data. Here we again see a good agreement reflecting the strength of dependence.

The measure of Schemper and Kaider (1990) also resulted in a plausible value, $r_{pr}^2 = 0.34$. The measure of Korn and Simon (1990), based on quadratic loss, gave the value 0.32. As mentioned earlier, this measure does not remain invariant to monotone increasing transformation of time. For these data the value 0.32 drops to 0.29 if the failure times are replaced by the square roots of the times. We also calculated the 95% confidence interval for ρ^2 by inverting the interval for β and it turned out to be (0.09, 0.57). The 95% confidence interval for R^2 is (0.16, 0.68) in O'Quigley and Flandre (1994), based on a simple percentile bootstrap calculation, and becomes (0.12, 0.64), an interval much closer to that obtained here, when using Efron's bias-corrected accelerated bootstrap.

The much simpler calculation, based on the modification of the coefficient introduced by Allison (1995), results in the value 0.42 giving, as we would expect, quite close agreement. The uncorrected coefficient of Allison results in the slightly depressed value 0.32 which is not all that far removed from the others but this is simply due to the fact that, for these data, only 12 out of the 42 observations were censored. As the censoring diminishes, the corrected and uncorrected Allison measures converge and, ultimately, when there is no censoring, they are identical.

A second example comes from a study of 2174 breast cancer patients, followed over a period of 15 years at the Institut Curie in Paris, France. A large number of potential and known prognostic factors were

recorded. The most important prognostic factor among those anticipated as having some prognostic importance was stage. Measuring,
not so indirectly, the evolution of the disease, our intuition tells us
that we would expect to record a high degree of explained randomness
from this covariate. Applying the uncorrected Allison coefficient in the
standard SAS program we find an estimated explained randomness of
12%. This appears rather low given what we know and, indeed, if we
calculate the corrected coefficient, as described above, we then find an
estimated 43% of the randomness in survival explained by stage. This
is relatively high but does correspond much more closely to a figure
we might expect.

14.12   Further extensions

Time-dependent covariates

The main definition can be generalized to accommodate time-dependent
covariates, a natural feature of the proportional hazards model and one
that the Kent and O'Quigley specification was not able to deal with.
For simplicity we still assume the covariates to be of dimension one.
The model of interest is then

$$\lambda(t|z(t); \beta) = \lambda_0(t) \exp\{\beta z(t)\}.$$

The covariate $Z(t)$ is generally assumed to be a stochastic process indexed by time, meaning that $Z(t)$ is a random variable at any fixed t
and may have different distributions at different t's. The conditional
distributions of Z given T, described by the main theorem (Section
7.4), then become those of $Z(t)$ given $T \geq t$ and those of $Z(t)$ given
$T = t$, respectively. The first of these can be estimated by the empirical
distribution of $Z(t)$ in the risk set at time t, because conditioning on
$T \geq t$ just as before, still has the interpretation as given all the subjects
that have survived at least until time t. For the second conditional distribution, we apply one of the corollaries to the main theorem (Section
7.4). Recalling this we have a similar result to that for time-invariant
covariates:

Theorem 14.3 *Under the model of Equation (14.15), the conditional
distribution function of $Z(t)$ given $T = t$ is consistently estimated by*

$$\hat{P}(Z(t) \leq z|T = t) = \sum_{\{j:Z_j(t)\leq z\}} \pi_j(t; \hat{\beta}),$$

where $\pi_j(t; \beta)$ is modified so that Z_j and Z_i are replaced by $Z_j(t)$ and $Z_i(t)$.

In the case of time-dependent covariates, take:

Definition 14.9

$$\Gamma_2(\beta) = 2 \int E\left\{ \log \frac{g(Z(t)|T = t; \beta)}{g(Z(t)|T = t; 0)} \middle| T = t; \beta \right\} dF(t).$$

Although this may not fit the exact definition of a Kullback-Leibler information gain as we are dealing with a stochastic process here, it can be verified that $\rho^2(\beta) = 1 - \exp\{-\Gamma_2(\beta)\}$ maintains the properties described above and therefore is suitable for use as a measure of dependence between the survival time and the covariates. In addition, estimation is essentially unchanged with Z_l replaced by $Z_l(t_i)$. The ρ^2 coefficient is no more involved computationally than is that for time-invariant covariates. Partial coefficients can also be similarly defined.

Prentice criteria for surrogate endpoints

An interesting example of time-dependent covariates arises in the study of surrogate endpoints. These are of interest in clinical trials as a potential way to reduce the overall experimentation time and thereby accelerate the testing of new, potentially more effective, treatments. The idea behind the use of surrogate variables lies in defining clinical endpoints that might equally well reflect therapeutic benefit, occurring, on average, earlier than classical clinical endpoints. The main requirement is that the surrogate represent, in some sense, a pathway so that, once we have taken account of the effect of treatment on the surrogate then little or nothing remains of the treatment effect on subsequent survival. In addition to some biological or clinical rationale, we also require that the observed value of the surrogate marker at any given point in time be predictive of ultimate survival time.

Both points have been discussed by Prentice (1989) who proposed two operational criteria for defining surrogate endpoints in clinical trials. These criteria have been discussed by other authors (Buyse and Molenberghs, 1998; Lagakos, 1993; Fleming and DeMets, 1996; De Gruttola et al. 1997, Freeman et al. 1992). and can be evaluated via regression models and likelihood ratio tests (Lin, Fischl and Schoenfeld, 1993). Other criteria have also been proposed (Buyse and Molenberghs 1998) and the purpose here is not to debate the advantages and disadvantages of the different approaches but to focus on a characterization of the Prentice criteria which turns out to be useful. Similar characterizations may exist for other proposals. It is helpful to quantify to what extent the risk changes once the surrogate variable has been observed and, more importantly, to quantify to what extent the effect of treatment can be explained by the action of the surrogate. The regression coefficients themselves are, of course, providing information on this. Measures of predictability and explained randomness, being standardized to lie between 0 and 1, and having precise meaning, can be useful in interpreting the answers to these kind of questions.

The first criterion proposed by Prentice (1989) requires the failure rate for T be independent of treatment, conditional on the surrogate variable. This notion is defined by the relation (Prentice, 1989, page 433)

$$\lambda\{t|Z_1(t), Z_2(t)\} = \lambda\{t|Z_1(t)\}, \tag{14.15}$$

where $Z_1(t)$ is a binary surrogate variable and $Z_2(t)$ is the treatment indicator, in general, not depending on t, although it can be allowed to and would enable us to analyze crossover designs for example. This criterion ensures that a surrogate for T should be able to capture the dependence of T on treatment.

The second criterion considers a model with only the surrogate variable and requires that the surrogate response have some prognostic implication for the true endpoint (Prentice, 1989, pp 434), that is

$$\lambda\{t|Z_1(t)\} \neq \lambda(t) \tag{14.16}$$

for all t. Conditions (14.15) and (14.16) were proposed as operational criteria for surrogate endpoints in clinical trials (Prentice, 1989). Recall that when a covariate Z is unrelated to survival, and the corresponding regression coefficient is zero, then $\rho^2(Z) = 0$. In the case of a surrogate endpoint having no effect where there is no gain in information concerning the intensity function by knowing $Z_1(t)$, i.e. $\lambda(t; Z_1(t)) = \lambda(t)$, then Prentice's criterion 2, as expressed in equation (14.16) is not satisfied and $\rho^2(Z_1) = 0$.

A non zero value for $\rho^2(Z_1)$ indicates that Prentice's second criterion is met. Denoting $\rho^2(Z_2|Z_1)$ to be the value of the index with respect to Z_2, after having accounted for the effects of Z_1, then, in the case that Z_2 is unrelated to survival, i.e., provides no further information on T, once we have adjusted for Z_1, we require $\rho^2(Z_2|Z_1) = 0$.

In particular, when the surrogate captures all of the treatment effect, as expressed by (14.15), then Prentice's first criterion is met and $\rho^2(Z_2|Z_1(t)) = 0$. An index of explained randomness ρ^2 can then enable us to re-express the Prentice criteria, equations (14.15) and (14.16). This alone is of interest, although, in addition, we are able to interpret $\rho^2(Z)$ as providing an ordering in terms of predictive power and that, for any specific value, say $\rho^2(Z) = 0.28$, the number 0.28 has a concrete interpretation, specifically, as it turns out, in terms of explained randomness, then such an index can be helpful in situations where the Prentice criteria are being applied. It is also helpful that $\rho^2(Z)$ be unaltered by (i) an independent censoring mechanism and (ii) monotonic increasing transformations on time since these are practical requirements in the setting of a clinical trial.

We can then use ρ^2, to re-express the two criteria of Prentice (1989) for surrogate variables. Moreover the greater is the difference between $\lambda(t; Z_1(t))$ and $\lambda(t)$ then the stronger is the effect of the surrogate variable. This difference is reflected in the observed value of $\hat{\rho}^2(\hat{\beta}_1)$ and provides a means for discriminating among rival surrogate variables of potentially different strengths. The most promising surrogate variable ought provide the largest value of $\hat{\rho}^2(\hat{\beta}_1)$.

As an illustration we looked at a study of 219 patients with resected lung carcinoma randomized three weeks after surgery (Decroix et al. 1984). In this study, time-dependent relapse appears as a strong surrogate for the grouping variable defined by stage. The staging variable describes a TNM classification between stage I or II and stage III. These data have been analyzed differently using a two-stage procedure in order to assess the predictive effects of the surrogate endpoint via survival estimation (Flandre and O'Quigley, 1995). A total of 157 deaths have been observed and for 126 patients a relapse has been recorded.

The likelihood ratio test for the first Prentice's criterion leads to a p-value of 0.68 which indicates that the staging effect appears to be largely captured by the surrogate endpoint relapse. The evaluation of the second Prentice's criterion leads to a p-value < 0.001 which confirms the strong pronostic effect of relapse on the risk of death. Therefore, in the light of the Prentice criteria, we may consider relapse to be a valid surrogate for the grouping variable stage. But we can say more. Under our estimates, and the alternative characterization via ρ^2, we find the partial coefficient estimate corresponding to Prentice's first criterion as $\hat{\rho}^2(\hat{\beta}_1, \hat{\beta}_2) = 0.00024$, in broad agreement with the likelihood ratio test and implying that when the surrogate variable is already in the model, inclusion of the staging variable does not increase, significantly, the predictive ability of the model. The predictive capability of the surrogate endpoint itself is given by $\hat{\rho}^2(\hat{\beta}_1) = 0.696$, a quantity which, while differing significantly from zero and thereby addressing Prentice's second criterion, is sufficiently large to suggest itself as a potentially powerful surrogate.

The illustration itself is rather artificial since, in a practical context, one would not anticipate relapse to be anything other than a strong indicator of subsequent survival. There would be no need to carry out any tests. It is not easy to come up with good examples in which the Prentice criteria might directly impact clinical decision making. Their greatest value is most likely in a retrospective analysis which may help underline important potential pathways for the disease and for this information to be used in the planning of further studies. The clarity of the Prentice criteria is nonetheless very compelling and the use of a measure of explained randomness can add to this.

The focus is very much on that of a pathway and the identification of other endpoints which may be used to advantage in the context of long term clinical trials. Since the work of Prentice and its focus on the idea of a pathway, an idea which can, at least in principle, be tested on a single study the concept of a surrogate endpoint has itself evolved. At least in the statistical setting, it is now more common to shine the torch away from the idea of a pathway to that of an alternative endpoint. The quest is to find endpoints which could have been assessed earlier and which are strongly correlated, usually in a multivariate linear environment, with the main endpoint. For this prescription to be carried through we would require several pairs of possible endpoints, hence the call to use meta-analysis as a way to decide on the validity or not of a surrogate endpoint.

All the same, the characterization of Prentice remains clear, powerful, and contains in two simple expressions the concepts which are the most important. In this section we see that the framework can be expanded a little further, at least in the direction of quantification.

General forms of relative risk

The relative risk $\exp(\beta z)$ can sometimes take on other forms, for example, $1 + \beta z$ for an additive model or $\exp\{\beta(t)z\}$ for time-varying regression effect. Denoting $r(t; z)$ for a general form of the relative risk, recall the form for this more general model as

$$\lambda(t|z) = \lambda_0(t)r(t; z).$$

The measure ρ^2 can be readily extended to this model as follows. Denote $g(z|t)$ the conditional density or probability function of Z given T under the true model, and $g_0(z|t)$ the same density or probability function under the null model where the relative risk is $r_0(t; z)$. We then have:

Definition 14.10 *Let $\rho^2 = 1 - \exp\{-\Gamma_2\}$ where*

$$\Gamma_2 = 2 \int_T \int_Z \log \left\{ \frac{g(z|t)}{g_0(z|t)} \right\} g(z|t) dF(t).$$

The definition coincides with the original one under a proportional hazards model. To obtain a sample based estimate of the measure, analogous we broaden the definition of $\pi_j(t)$ to:

Definition 14.11

$$\pi_j(t) = Y_j(t)\hat{r}(t; Z_j) / \sum_{i=1}^{n} Y_i(t)\hat{r}(t; Z_i), \qquad (14.17)$$

and

$$\pi_{0j}(t) = Y_j(t)\hat{r}_0(t; Z_j) / \sum_{i=1}^{n} Y_i(t)\hat{r}_0(t; Z_i), \qquad (14.18)$$

where $\hat{r}$ and $\hat{r}_0$ are consistent estimates of r and r_0, respectively.

It then follows that $\pi_j(t) = \pi_j(t; \hat{\beta})$ and $\pi_{0j}(t) = \pi_j(t; 0)$ when $r(t; z) = \exp(\beta z)$. If we replace $e^{\beta z}$ in the proof of the main theorem by $r(t; z)$, we see that:

Theorem 14.4 *Under the model of equation (14.17), the conditional distribution function of Z given T is consistently estimated by*

$$\hat{P}(Z \le z | T = t) = \sum_{\{j: Z_j \le z\}} \pi_j(t). \qquad (14.19)$$

We then have the following;

Lemma 14.8 *We can consistently estimate Γ_2 by*

$$\hat{\Gamma}_2 = 2 \sum_{i=1}^{k} P(t_i) \sum_{j=1}^{n} \pi_j(t_i) \log \left\{ \frac{\pi_j(t_i)}{\pi_{0j}(t_i)} \right\}.$$

The cases with finite follow-up time τ can again be treated as before. The measure ρ^2 defined here maintains the property of lying between 0 and 1, and is equal to zero when T and Z are independent (in which case $g(z|t) = g_0(z|t)$). The measure is invariant under monotonically increasing transformations of T, and is invariant under linear transformations of Z if the relative risk $r(t; z)$ is a function of βz. Randomness can be similarly defined so that ρ^2 still has the interpretation as the proportion of explained randomness. Boundedness might persist since this is observed in general for discrete dependent variables (Kent 1983). Partial coefficients can be obtained immediately and the measure can be applied to time-dependent covariates.

14.13 Exercises and class projects

1. Consider a question analogous to that of the previous chapter; i.e., a number of suggested coefficients of explained variation, adapted from linear regression, depend on the censoring even when independent of the failure mechanism. Would this still be a handicap for a coefficient of explained randomness? Recall from that question that the dependence is typically such that the higher the censoring the closer to zero is the adapted coefficient. It might be argued that, as the censoring increases, our ability to predict declines and, in consequence so ought a suitable coefficient of explained randomness. Comment on this reasoning.

2. Again analogous to that of the previous chapter, suppose you are the statistician analyzing the gastric cancer data. The investigating clinician, who has some rudimentary knowledge of statistics, wishes to understand just what you mean by saying that *the value of $\hat{\rho}^2$ for a model with ACT and log(CEA) is 0.36 and this increases to 0.51 when stage is also included. On the other hand the corresponding partial $\hat{\rho}^2$ is equal to 0.28.* How do you answer this question.

3. Using the δ-method and the expression for $\hat{\Gamma}_2(\hat{\beta})$, derive an approximate confidence interval for $\hat{\Gamma}_2(\hat{\beta})$. Use this interval to derive an interval for $\hat{\rho}_2^2$. Apply the resulting expression to the Freireich data and then compare this interval with that obtained on the basis of bootstrap sampling. Comment on your findings.

4. In the estimate $\hat{\Gamma}_2(\hat{\beta})$ of $\Gamma_2(\beta)$ we work with the weights $W(X_i)$ which play the role of a Kaplan-Meier estimate of the increments $dF(t)$

at $t = X_i$. Let $F(t : \theta)$ be a parametric model of the marginal survival curve where θ is a parameter, possibly vector valued. Investigate an estimator based on $dF(X_i : \hat{\theta})$. What might be the advantages and drawbacks of such an approach? From the results we already have, is it possible to deduce properties such as consistency? If so, under what conditions?

5. We know that for a continuous covariate or a discrete covariate with an unlimited number of levels that, as $|\beta| \to \infty$ then $\rho_2^2(\beta) \to 1$. Let $\bar{\rho}_2^2(\beta) = \bar{\Gamma}_2(\beta)$. Show that, as $|\beta| \to \infty$ then $\bar{\rho}^2(\beta) \to 0$. This is analogous to a result for $R^2(\beta)$ and $R_{\mathcal{E}}^2(\beta)$ and at first glance appears puzzling. Explain just what is taking place. Explain also why, if the proportional hazards model is correct, we anticipate that $\bar{\rho}_2^2(\hat{\beta})$ and $\rho_2^2(\hat{\beta})$ to closely agree.

6. The consistency results are obtained under the assumption of an independent censoring mechanism. It is common in survival analysis to make the weaker assumption of independence between the failure and the censoring mechanisms conditional on the covariates Z. Suppose that in a given application, Z takes some finite number of levels. Derive an estimator of $\Gamma_2(\beta)$ (and thereby $\rho_2^2(\beta)$) which remains consistent under the broader censoring mechanism (i.e., that of conditional independence). Which extra conditions would be needed? Would we anticipate observing much difference between estimators derived under the different censoring assumptions?

7. Consider some density $f(t, \theta_0)$, depending on a single parameter θ_0. Let $I(\theta) = E\{\log f(t, \theta)\}$. Show that $I(\theta)$, viewed as a function of θ, is maximized over the range of θ when $\theta = \theta_0$. Keep in mind that the expectation operator, $E()$ depends on θ_0.

8. Using data with several covariates, calculate simple, multivariate and partial measures of explained randomness, comparing, in particular, measures based on ρ_2^2 and those based on the approximation given in Equation 14.9.

Chapter 15

Survival given covariates

15.1 Summary

Breslow (1972, 1974), using an equivalence between the proportional hazards model and a piecewise exponential regression model, with as many parameters as there are failure times, derived a simple expression for conditional survival given covariate information. An expression for the variance of the Breslow estimate was derived by O'Quigley (1986). Appealing to Bayes' rule, O'Quigley and Xu (2000) obtained the elegant expression

$$S(t|Z \in H) = \int_t^\infty P(Z \in H|u)dF(u) \Big/ \int_0^\infty P(Z \in H|u)dF(u),$$

from which an estimate of conditional survival, making a direct appeal to the main theorem, follows (Xu and O'Quigley 2000). In most practical applications the two estimators behave similarly. We prefer the second in view of its closer association with basic inference. It is more readily generalized to deal with non-proportional hazards models, in particular the stratified model and models that include random effects. Another predictive quantity of interest is $\Pr(T_i > T_j|Z_i, Z_j)$ which gives the probability for an individual to outlast another given their respective covariate information. Finally, survival given a non-independent competing risk or non-independent censoring, is considered (Flandre and O'Quigley 1994).

15.2   Motivation

We would like to know how marginal survival is impacted by a knowledge of covariate information. In the simplest case, given such information, we might ask how much more likely is it that one individual outlast another. One of the main purposes of survival analysis is to obtain an estimate of the survivorship function given certain covariate patterns. Although inference for the proportional hazards model ignores specification of the baseline hazard rate, thereby leaving the baseline survivorship function as well as conditional survivorship functions undetermined, it is common to carry out further estimation on these quantities. The provision of such information may help guide decision making in an applied context.

While it is usually technically difficult to estimate densities and hazards (some kind of smoothing typically being appealed to), it is easier to estimate cumulative hazards and distribution (survivorship) functions. These have already been smoothed, in some sense, via the summing inherently taking place.

The question we would like to answer is expressed straightforwardly as the probability of survival time being greater than t given that Z belongs to some subset H, i.e., $\Pr(T > t | Z \in H)$. Via Bayes' rule, this probability can be immediately expressed in terms of the conditional distribution of Z at $T = t$, together with the marginal distribution of T. An estimate of this conditional distribution is available as a consequence of the main theorem. The marginal distribution of T can be taken to be the Kaplan-Meier estimate, or some other distribution should we wish to investigate effects in different contexts. We mostly limit attention to the case where the covariate Z is assumed to be time-invariant. The situation becomes more complicated in the presence of time-dependent covariates because of certain restrictions, for example, $Z(\cdot)$ should be an external covariate in order for $S(t|z)$ to be interpretable (Kalbfleisch and Prentice 1980, Keiding and Knuiman 1990, Lin et al. 1994, and Keiding 1995).

15.3   Probability that T_i is greater than T_j

If two individuals are independently sampled from the same distribution then, by simple symmetry arguments, it is clear that the probability of the first having a longer survival time than the second is just 0.5. If, instead of sampling from the same distribution, each individual is sampled from a distribution determined by the value of their covariate information, then, the stronger the impact of this covariate information, the further away from 0.5 will this probability be. When the covariates do not depend on time then this probability is very easily evaluated using:

Theorem 15.1 *For subjects i and j, having covariate values Z_i and Z_j then, under the proportional hazards model we can write*

$$\Pr\left(T_i > T_j | Z_i, Z_j\right) = \frac{\exp(\beta Z_j)}{\exp(\beta Z_i) + \exp(\beta Z_j)}.$$

An important observation to make is that the expression does not involve $\Lambda_0(t)$. If we define $\psi(a, b : \beta)$ to be $\exp(\beta b)/\{\exp(\beta b) + \exp(\beta a)\}$ we then have:

Corollary 15.1 *A consistent estimate of* $\Pr\left(T_i > T_j | Z_i, Z_j\right)$, *under the proportional hazards model is given by* $\psi(Z_i, Z_j : \hat{\beta})$ *and* $\mathrm{Var}\log\{\psi/(1-\psi)\} \approx (Z_j - Z_i)^2 \mathrm{Var}(\hat{\beta})$.

The approximation in the corollary arises from an immediate application of the mean value theorem (delta method). In the theorem and corollary it is assumed that Z_i and Z_j are scalars and that the model involves only a one dimensional covariate. Extension to the multivariate case is again immediate and, instead of $\hat{\beta}Z_i$ in $\psi(Z_i, Z_j : \hat{\beta})$ being a scalar it can be replaced by the usual inner product (prognostic index). Suppose that the dimension of β and Z is p and that we use the notation Z_{jr} to indicate, for subject j, the rth component of Z_j. An application of the delta method gives

$$\mathrm{Var}\log\{\psi/(1-\psi)\} \approx \sum_{r=1}^{p}\sum_{s=1}^{p}(Z_{jr} - Z_{ir})(Z_{js} - Z_{is})\mathrm{Cov}\left(\hat{\beta}_r, \hat{\beta}_s\right).$$

The author is not aware of any applied work making use of ψ and yet it would seem a particularly simple and transparent way in which to summarize the impact, or predictive strength, of regression effects. We all know that significance levels alone, directly dependent as they are on sample sizes (more precisely the amount of uncensored observations), are not indicative of predictive strength. The two previous chapters discuss this at length and provide workable indicators of predictive strength in the explained variation and explained randomness measures.

Whereas these measures are being averaged over some covariate distribution it is also helpful to have specific measures given two particular covariate configurations. If a patient is told that they have an unfavorable prognosis in the light of studies on their prognostic variables it can be helpful to add to that some idea on just how unfavorable. If $\psi(Z_i, Z_j : \hat{\beta})$ remains close to 0.5, then, regardless of significance which may be of importance in considering the public health effects for large groups, an individual might reasonably feel that he or she is not particularly disadvantaged by such an unfavorable prognosis. All of the above is very straightforward when dealing with the evaluation of covariate "points" as given by Z_i and Z_j.

More generally, as discussed in Section 15.5, we may wish to consider some range within the covariate space in which case we specify, $\Pr(T_i > T_j | Z_i \in H_i, Z_j \in H_j)$. To do this in practice we would need to integrate over the range of points of Z_i and Z_j, contained in H_i and H_j respectively, with respect to the densities of Z_i and Z_j within these sets.

The connection between the population explained variation Ω^2 and $\psi(Z_i, Z_j : \beta)$ would be worthy of further investigation. Xu (1996) showed that, for fixed values of the covariates, and, regarding both Ω^2 and $\psi(Z_i, Z_j : \beta)$ as functions of strength of effect as measured by β, then these two quantities are monotonic functions of one another. There is a one-to-one correspondence between them so that, clearly, in some sense, they are measuring the same phenomenon but on different scales.

Another potential application of $\psi(Z_i, Z_j : \beta)$ is in relative survival where we take Z_j to be a value across some reference group or population and Z_i to be a value for some group under study, for instance, a group having recently been treated for some particular chronic disease. The negative effects of belonging to that group, as opposed to the reference group, are then directly quantified by ψ. Further developments to these ideas immediately suggest themselves and, foremost, making use of time itself. Rather than limit our attention to $\Pr(T_i > T_j | Z_i, Z_j)$ we can explicitly introduce into this quantity the amount of time elapsed. We would not only condition on the values of Z_i and Z_j but also that both T_i and T_j are greater than t, bring our attention then to $\Pr(T_i > T_j | Z_i, Z_j, T_i > t, T_j > t)$. The resulting expressions would be more involved than in Theorem 15.1 but could be worked out. Applications to cure studies follow since we could conceive of situations in which this expression diminishes with t, becoming, at some point, sufficiently close to the value 0.5 to claim that the treated group no longer carries a disadvantage as compared to the reference group.

15.4   Estimating conditional survival given that $Z = z$

Under the proportional hazards model, the conditional survival probability $S(t|z) = \Pr(T > t|Z = z)$ can be estimated using the development of (Breslow 1972, 1974) whereby

$$\hat{S}(t|z) = \exp\left(-\hat{\Lambda}_0(t)e^{\hat{\beta}z}\right) ; \quad \hat{\Lambda}_0(t) = \sum_{X_i \leq t} \frac{\delta_i}{\sum_{j=1}^{n} Y_j(X_i)e^{\hat{\beta}Z_j}}. \quad (15.1)$$

An expression for the large sample variance of $Y = \log - \log S(t|z)$ was obtained by O'Quigley (1986). Symmetric intervals for Y can then be transformed into more plausible (at least having better coverage properties according to the arguments of O'Quigley 1986) by simply applying the exponential function twice. The Breslow estimate concerns a single point z. It is a natural question to ask what is the survival probability given that the covariates belong to some subset H. The set H may denote for example an age group, or a certain range of continuous measurement, or a combination of those.

In general we assume H to be a subset of the p-dimensional Euclidean space. A natural approach may be to take the above formula, which applied to a point, and average a set of curves over all points belonging to the set H of interest. For this we would need some distribution for the z across the set H. Keiding (1995) has a discussion on expected survival curves over a historical, or background, population, where the main approaches are to take an average of the individual survival curves obtained from the above equation. Following that one might use the equation to estimate $S(t|z)$ for all z in H, then average over an estimated distribution of Z. Xu and O'Quigley (2000) adopted a different starting point in trying to estimate directly the survival probabilities given that $Z \in H$. Apart from being direct, this approach is the more natural in view of the main theorem of Section 7.4. What is more, the method can also have application to situations in which the regression effect varies with time. In the following, for notational simplicity, we will assume $p = 1$. Extensions to $p > 1$ are immediate.

15.5   Estimating conditional survival given $Z \in H$

As for almost all of the quantities we have considered it turns out to be most useful to work with the conditional distribution of Z given $T = t$ rather than the other way around. Everything is fully specified by the joint distribution of (T, Z) and we keep in mind that this can be expressed either as the conditional distribution of T given Z, together with the marginal distribution of Z or as the conditional distribution of Z given T, together with the marginal distribution of T. Using Bayes' formula to rewrite the conditional distribution of T given information on Z, we have,

$$S(t|Z \in H) = \frac{\int_t^\infty P(Z \in H|u)dF(u)}{\int_0^\infty P(Z \in H|u)dF(u)}. \tag{15.2}$$

This is a very simple expression and we can see from it how conditioning on the covariates modifies the underlying survival distribution. If H were to be the whole domain of definition of Z, in which case Z is contained in H with probability one, then the left-hand side of the equation simply reduces to the marginal distribution of T. This is nice and, below, we will see that we have something entirely analogous when dealing with sample based estimates whereby, if we are to consider the whole of the covariate space, then we simply recover the usual empirical estimate. In particular this is just the Kaplan-Meier estimate when the increments of the right-hand side of the equation are those of the Kaplan-Meier function. The main theorem of Section 7.4 implies that $P(Z \in H|t)$ can be consistently estimated from:

Lemma 15.1 $\hat{P}(Z \in H|t)$ *is consistent for the probability* $P(Z \in H|t)$ *where;*

$$\hat{P}(Z \in H|t) = \sum_{\{j:Z_j \in H\}} \pi_j(\hat{\beta}, t) = \frac{\sum_H Y_j(t)\exp\{\hat{\beta}Z_j\}}{\sum Y_j(t)\exp\{\hat{\beta}Z_j\}}. \tag{15.3}$$

This striking, and simple, result is the main ingredient needed to obtain survival function estimates conditional on particular covariate configurations. The rest, essentially the step increments in the Kaplan-Meier curve are readily available. Unfortunately, a problem that is always present when dealing with censored data, remains and that is the possibility that the estimated survival function does not decrease all the way to zero. This will happen when the largest observation is not a failure. To look at this more closely, let $\hat{F}(\cdot) = 1 - \hat{S}(\cdot)$ be the left continuous Kaplan-Meier (KM) estimator of $F(\cdot)$. Let $0 = t_0 < t_1 < ... < t_k$ be the distinct failure times, and let $W(t_i) = d\hat{F}(t_i)$ be the stepsize of $\hat{F}$ at t_i. If the last observation is a failure, then

$$\hat{S}(t|Z \in H) = \frac{\int_t^\infty \hat{P}(Z \in H|u)d\hat{F}(u)}{\int_0^\infty \hat{P}(Z \in H|u)d\hat{F}(u)} = \frac{\sum_{t_i > t} \hat{P}(Z \in H|t_i)W(t_i)}{\sum_{i=1}^k \hat{P}(Z \in H|t_i)W(t_i)}.$$

(15.4)

When the last observation is not a failure and $\sum_1^k W(t_i) < 1$, an application of the law of total probability indicates that the quantity B_1 where $B_1 = \hat{P}(Z \in H|T > t_k)\hat{S}(t_k)$ should be added to both the numerator and the denominator in (15.4) This is due to the fact that the estimated survival distribution is not summing to one. Alternatively, we could simply reduce the support of the time frame to be less than or equal to the greatest observed failure. In addition, using the empirical estimate over all the subjects that are censored after the last observed failure, we have

$$\hat{P}(Z \in H|T > t_k) = \frac{\sum_H Y_j(t_k+)}{\sum Y_j(t_k+)},$$

(15.5)

where t_k+ denotes the moment right after time t_k. Therefore we can write

$$\hat{S}(t|Z \in H) = \frac{\sum_{t_i > t} \hat{P}(Z \in H|t_i)W(t_i) + \hat{P}(Z \in H|T > t_k)\{1 - \sum_1^k W(t_i)\}}{\sum_1^k \hat{P}(Z \in H|t_i)W(t_i) + \hat{P}(Z \in H|T > t_k)\{1 - \sum_1^k W(t_i)\}}.$$

The above estimate of the conditional survival function is readily calculated, since each term derives from standard procedures of survival analysis to fit the Cox model. An attractive aspect of the approach is that that when H includes all the possible values of z, the estimator simply becomes the Kaplan-Meier estimator of the marginal survival function. The estimate of the conditional survival probability $P(T > t + u | T > u, Z \in H)$ can also be nicely written in the form of a simple ratio where the numerator is given by $\sum_{t_i > t+u} C_i + B_1$ and the denominator by $\sum_{t_i > t} C_i + B_1$ and where $C_i = \hat{P}(Z \in H|t_i)W(t_i)$. For the gastric cancer data it is interesting to contrast the survival estimate based on these calculations for the quantity $\Pr\{T > t|Z_1 \in (0, 100)\}$, where Z_1 is the tumor marker CEA and the simple Kaplan-Meier estimator based on the subset of the data defined by tumor marker CEA

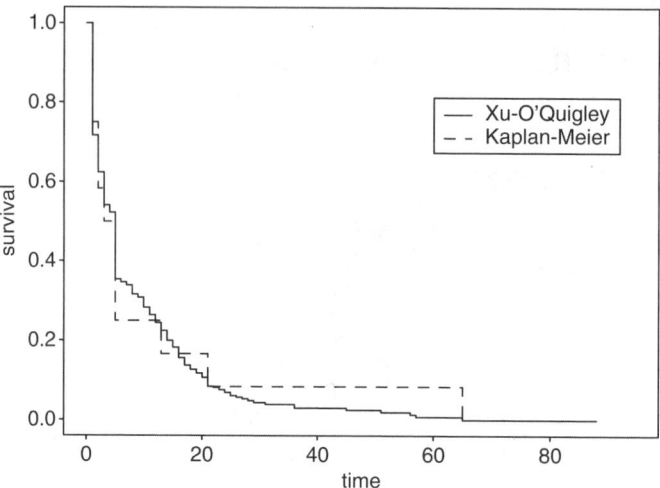

Figure 15.1: Kaplan-Meier survival plot and model based plot for subgroup based on selection of the covariate CEA to lie in interval $(0,100)$.

less than 100. This is shown in Figure 15.1 and there is good agreement between the model-based estimator of Xu-O'Quigley and the Kaplan-Meier estimate. This, although not used here as a goodness-of-fit test in its own right, can be taken to indicate that the model appears reasonable over the specified range of the covariate.

15.6   Estimating the variance of $\hat{S}(t|Z \in H)$

In this section we assume that the proportional hazards model holds with true $\beta = \beta_0$. To obtain the asymptotic variance of (15.3) at each t, we use the approach of Link (1984) and O'Quigley (1986) which is tractable and can be computed using standard packages for fitting the Cox regression model. Our experience is that this provides good estimates when compared with methods such as the bootstrap. There are two sources of variation in $\hat{S}(t|Z \in H)$, one caused by the estimate of conditional probability of survival with given β, the other by the uncertainty in $\hat{\beta}$. Using a first-order Taylor series expansion we have

$$\hat{S}(t|Z \in H) = \hat{S}(t|Z \in H)|_{\beta_0} + (\hat{\beta} - \beta_0)\frac{\partial \hat{S}(t|Z \in H)}{\partial \beta}\Big|_{\beta=\dot{\beta}}, \quad (15.6)$$

where $\dot{\beta}$ lies on the line segment between β_0 and $\hat{\beta}$. We then need to bring together some results.

Lemma 15.2 *The quantity* $\{\partial\hat{S}(t|Z \in H)/\partial\beta\}\,|_{\beta=\hat{\beta}}$ *is asymptotically constant.*

Lemma 15.3 *The quantity* $\hat{\beta} - \beta_0$ *is asymptotically uncorrelated with* $\hat{S}(t|Z \in H)|_{\beta_0}$.

Corollary 15.2 *The variance of* $\hat{S}(t|Z \in H)$ *is approximated by*

$$\text{Var}\{\hat{S}(t|Z \in H)\} \approx \text{Var}\{\hat{S}(t|Z \in H)|_{\beta_0}\}$$
$$+ \left\{ \frac{\partial\hat{S}(t|Z \in H)}{\partial\beta} \Big|_{\beta=\hat{\beta}} \right\}^2 \text{Var}(\hat{\beta}). \quad (15.7)$$

The first term in (15.7) gives the variation due to the estimation of the conditional survival, the second term the variation caused by $\hat{\beta}$. We know how to estimate the asymptotic variance of $\hat{\beta}$ under the model. So all that remains for the second term on the right-hand side of (15.7) is to calculate the partial derivative of $\hat{S}(t|Z \in H)$ with respect to β. For this we have

$$\frac{\partial}{\partial\beta}\hat{S}(t|Z \in H) = \frac{(\sum_{t_i>t} D_i)(\sum_{t_i\leq t} C_i) - (\sum_{t_i\leq t} D_i)(\sum_{t_i>t} C_i + B_1)}{(\sum_{i=1}^{k} C_i + B_1)^2},$$
$$(15.8)$$

where C_i and B_1 are the same as above and

$$D_i = \frac{\partial}{\partial\beta}\hat{P}(Z \in H|t_i)W(t_i),$$

with

$$\frac{\partial}{\partial\beta}\hat{P}(Z \in H|t) = \hat{P}(Z \in H|t)\{E_\beta(Z|t;\pi^H) - E_\beta(Z|t;\pi)\},$$

where

$$E_\beta(Z|t;\pi) = \sum Y_j(t)Z_j\exp\{\hat{\beta}Z_j\} / \sum Y_j(t)\exp\{\hat{\beta}Z_j\},$$

and

$$E_\beta(Z|t;\pi^H) = \sum_H Y_j(t)Z_j\exp\{\hat{\beta}Z_j\} / \sum_H Y_j(t)\exp\{\hat{\beta}Z_j\}.$$

The first term on the right-hand side of (15.7) can be estimated using Greenwood's formula

$$\mathrm{Var}\{\hat{S}(t|Z \in H)|_{\beta_0}\} \approx \hat{S}(t|Z \in H)|_{\beta_0}^2 \left\{ \prod_{t_i \leq t} \left(1 + \frac{\hat{q}_i}{n_i \hat{p}_i}\right) - 1 \right\}, \quad (15.9)$$

where

$$\hat{p}_i = \hat{P}(T > t_i | T > t_{i-1}, Z \in H) = \frac{\sum_{i+1}^{k} C_j + B_1}{\sum_{i}^{k} C_j + B_1},$$

$\hat{q}_i = 1 - \hat{p}_i$ and $n_i = \sum Y_j(t_i)$. Then each $\hat{p}_i$ is a binomial probability based on a sample of size n_i and $\hat{S}(t|Z \in H) = \prod_{t_i \leq t} \hat{p}_i$. The $\hat{p}_i$'s may be treated as conditionally independent given the n_i's, with β_0 fixed. Thus, Greenwood's formula applies. All the quantities involved in (15.8) and (15.9) are those routinely calculated in a Cox model analysis. In addition, we have:

Theorem 15.2 *Under the proportional hazards model $\hat{S}(t|Z \in H)$ is asymptotically normal.*

As a consequence one can use the above estimated variance to construct confidence intervals for $S(t|Z \in H)$ at each t.

Xu (1996) derived the asymptotic normality of $\hat{S}(t|Z \in H)$ under the proportional hazards model at fixed points $t = t^*$. Let's take $Q_i = 1$ if $Z_i \in H$, and 0 otherwise and follow, fairly closely, the notational set-up of Andersen and Gill (1982) from which

$$
\begin{array}{llll}
S^{(r)}(t) & = & n^{-1}\sum_{i=1}^{n} Y_i(t)e^{\beta(t)'Z_i(t)}Z_i(t)^r, & s^{(r)}(t) & = & ES^{(r)}(t), \\
S^{(r)}(\beta,t) & = & n^{-1}\sum_{i=1}^{n} Y_i(t)e^{\beta'Z_i(t)}Z_i(t)^r, & s^{(r)}(\beta,t) & = & ES^{(r)}(\beta,t), \\
S^{(H)}(\beta,t) & = & n^{-1}\sum_{1}^{n} Q_i Y_i(t)e^{\beta Z_i}, & s^{(H)}(\beta,t) & = & ES^{(H)}(\beta,t), \\
S^{(H1)}(\beta,t) & = & n^{-1}\sum_{1}^{n} Q_i Y_i(t)Z_i e^{\beta Z_i}, & s^{(H1)}(\beta,t) & = & ES^{(H1)}(\beta,t),
\end{array}
$$

for $r = 0, 1, 2$. Next rewrite

$$\hat{P}(Z \in H|t) = S^{(H)}(\hat{\beta},t)/S^{(0)}(\hat{\beta},t), \quad E_\beta(Z|t;\pi^H) = S^{(H1)}(\hat{\beta},t)/S^{(H)}(\hat{\beta},t).$$

Using the main theorem of Section 7.4 we have $s^{(H)}(\beta_0,t)/s^{(0)}(\beta_0,t) = P(Z \in H|t)$. Under the usual regularity and continuity conditions (Xu 1996) it can be shown that $\{\partial\hat{S}(t^*|Z \in H)/\partial\beta\} \mid_{\beta=\hat{\beta}}$ is asymptotically constant. Now $\hat{\beta} - \beta_0 = I^{-1}(\check{\beta})U(\beta_0)$ where $\check{\beta}$ is on the line segment between $\hat{\beta}$ and β_0, $U(\beta) = \partial \log L(\beta)/\partial \beta$ and $I(\beta) = -\partial U(\beta)/\partial \beta$. Combining these we have

$$\sqrt{n}\hat{S}(t^*|Z \in H) = \sqrt{n}\hat{S}(t^*|Z \in H)|_{\beta_0} + I^{-1}(\check{\beta})\sqrt{n}U(\beta_0)\frac{\partial \hat{S}(t^*|Z \in H)}{\partial \beta}\bigg|_{\beta=\dot{\beta}}.$$

Andersen and Gill (1982) show that $I(\check{\beta})$ converges in probability to a well-defined population parameter. In the following theorem Lin and Wei (1989) showed that $U(\beta_0)$ is asymptotically equivalent to $1/n$ times a sum of i.i.d. random variables:

Theorem 15.3 $\sqrt{n}U(\beta_0)$ *is asymptotically equivalent to* $n^{-1/2}\sum_1^n \omega_i$ (β_0), *where*

$$\omega_i(\beta) = \int_0^1 \left\{ Z_i - \frac{s^{(1)}(\beta,t)}{s^{(0)}(\beta,t)} \right\} dN_i(t) - \int_0^1 Y_i(t)e^{\beta Z_i} \left\{ Z_i - \frac{s^{(1)}(\beta,t)}{s^{(0)}(\beta,t)} \right\} \lambda_0(t)dt$$

and $N_i(t) = I\{T_i \le t, T_i \le C_i\}$.

So all that remains to show the asymptotic normality of $\hat{S}(t^*|Z \in H)$ is to show that the numerator of $\hat{S}(t^*|Z \in H)\,|_{\beta_0}$ is also asymptotically equivalent to $1/n$ times a sum of n i.i.d. random variables like the above, since the denominator of it we know is consistent for $P(Z \in H)$. To avoid becoming too cluttered we drop the subscript of β_0 in $\hat{S}(t^*|Z \in H)|_{\beta_0}$. The numerator of $\hat{S}(t^*|Z \in H)$ is $\int_{t^*}^\infty \hat{P}(Z \in H|t)d\hat{F}(t)$. Note that $\sqrt{n}\{\int_{t^*}^\infty \hat{P}(Z \in H|t)d\hat{F}(t) - P(Z \in H, T > t^*)\} = \sqrt{n}\int_{t^*}^\infty P(Z \in H|t)d\{\hat{F}(t) - F(t)\} + \sqrt{n}\int_{t^*}^\infty \{\hat{P}(Z \in H|t) - P(Z \in H|t)\}d\{\hat{F}(t) - F(t)\} + \sqrt{n}\int_{t^*}^\infty \{\hat{P}(Z \in H|t) - P(Z \in H|t)\}dF(t)$. Now $\sqrt{n}\{\hat{F}(t) - F(t)\}$ converges in distribution to a zero-mean Gaussian process. Therefore, the second term in the above expression is $o_p(1)$. The last term is $A_1 + o_p(1)$ where

$$A_1 = \sqrt{n}\int_{t^*}^1 \left\{ \frac{S^{(H)}(\beta_0,t)}{s^{(0)}(\beta_0,t)} - \frac{s^{(H)}(\beta_0,t)S^{(0)}(\beta_0,t)}{s^{(0)}(\beta_0,t)^2} \right\} dF(t)$$

$$= n^{-1/2}\sum_{i=1}^n \int_{t^*}^1 \frac{Y_i(t)e^{\beta_0 Z_i}}{s^{(0)}(\beta_0,t)} \left\{ Q_i - \frac{s^{(H)}(\beta_0,t)}{s^{(0)}(\beta_0,t)} \right\} dF(t).$$

As for the first term, we can use Theorem II.5 of Xu (1996), which is a result of Stute (1995). With $\phi(t) = 1_{[t^*,1]}(t)P(Z \in H|t)$ in the theorem, the first term is equal to $n^{-1/2}\sum_{i=1}^n \nu_i + \sqrt{n}R_n$, where $|R_n| = o_p(n^{-1/2})$ and ν's are i.i.d. with mean zero, each being a function of X_i and δ_i.

15.7   Relative merits of competing estimators

A thorough study of the relative merits of the different estimators has yet to be carried out. For any such study the first thing to consider would be the chosen yardstick with which to evaluate any estimate. For example, should the whole curve be considered or only some part of it and should a "distance" measure be an average discrepancy, the greatest discrepancy over some range or some weighted discrepancy. It is even very possible that some estimator would outperform another with respect to one distance measure and perform less well than the competitor with respect to another distance measure. In addition to this it is also quite possible for some estimator to maintain an advantage for certain population situations but to lose this advantage in other situations. In the light of these remarks it may then appear difficult to obtain a simple unequivocal finding in favor of one or another estimator. Nonetheless it would be nice to know more and further work here would be of help. In the meantime it helps to provide us with some insight by considering various situations which have arisen when looking at real data sets.

15.8   Illustrations

Rashid et al. (1982) studied a group of gastric cancer patients. The goal of the study was to determine the prognostic impact of certain acute phase reactant proteins measured pre-operatively. This biological information could then be used in conjunction with clinical information obtained at the time of surgical intervention to investigate the relative prognostic impact of the different factors. There were 104 patients and five covariates: stage (degree of invasion), ACT protein (α_1-anti chimotrypsin), CEA (carcino embryonic antigen), CRP (C-reactive protein), and AGP (alpha glyco protein).

Although it is known that stage has strong predictive capability for survival, it can only be determined after surgery. Of interest was the prediction of a patient's survival based on pre-operative measurements alone, an assessment of which might be used to help guide clinical decision making. Values of certain covariates such as CEA are very skew and have a wide range from below 1 to over 900. After a log transformation of CEA, a proportional hazards model with the four pre-operative covariates included was not rejected by the data. In fitting such a model, CRP and AGP were found to be insignificant at 0.05

level in the presence of ACT and logCEA. Therefore only the latter two were retained for subsequent analysis. The regression coefficients for ACT and logCEA were calculated to be 1.817 and 0.212, with standard errors 0.41 and 0.07, respectively. This gives a range of 0.92-4.48 for the estimated prognostic index $\beta'z$ (Andersen et al. 1983, Altman and Andersen 1986).

If we divide the patients into three groups, with low, median, and high risks, according to the prognostic index <2, 2-3 and >3, we can predict the survival probabilities in each risk group. It was then possible to estimate the survival curves for these three groups, and these are were calculated by Xu and O'Quigley (2000). The curves can be compared with the empirical Kaplan-Meier curves which make no appeal to the model. Agreement is strong. We also chose to define the set H by all those patients having values of CEA less than 50. This was not very far from the median value and provided enough observations for a good empirical Kaplan-Meier estimate. The empirical estimate and the model based estimates agree closely. For the three prognostic groups, defined on the basis of a division of the prognostic index from the multivariate model, the empirical Kaplan-Meier estimates, the Breslow estimates and the Xu and O'Quigley estimates are shown in Figures 15.2, 15.3 and 15.4. Again agreement is strong among the three estimators.

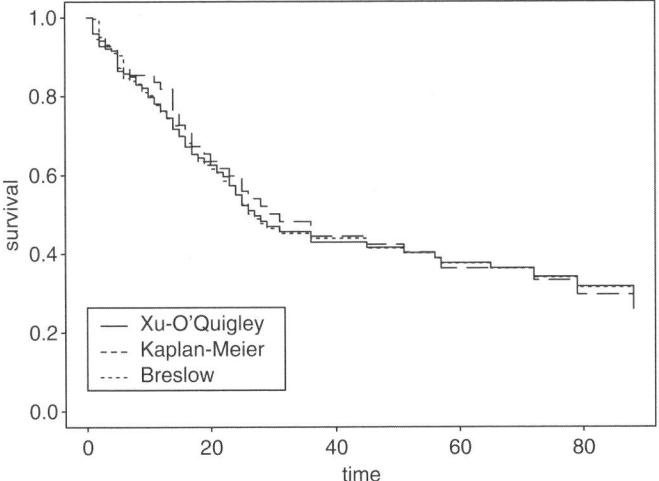

Figure 15.2: Survival probabilities based on prognostic index of lower 0.33 percentile using Kaplan-Meier, Breslow and Xu-O'Quigley estimators.

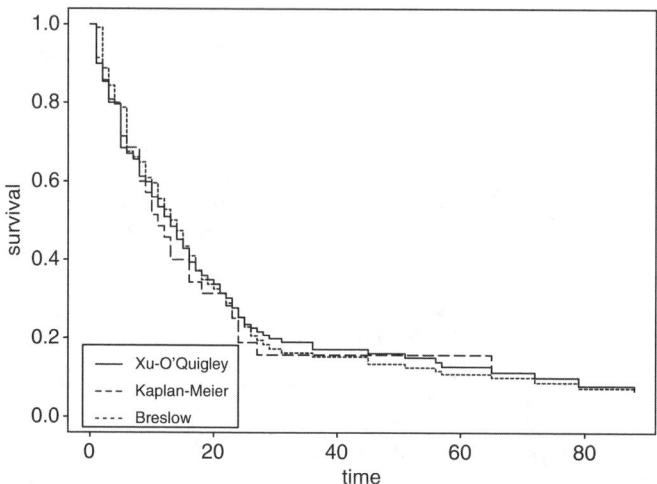

Figure 15.3: Survival probabilities based on prognostic index between 0.33 and 0.66 percentiles using Kaplan-Meier, Breslow and Xu-O'Quigley estimators.

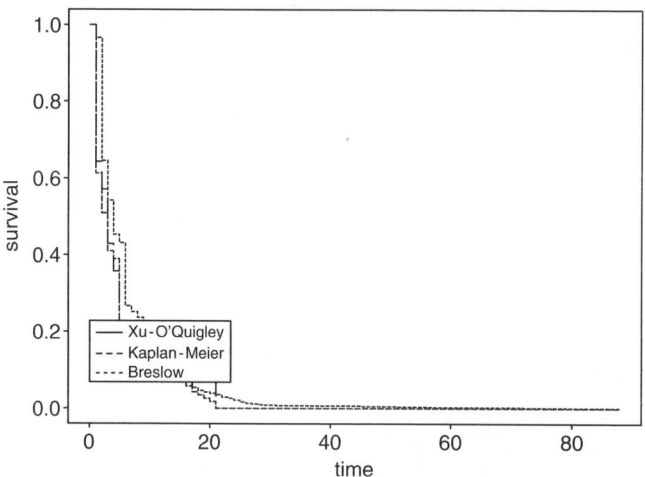

Figure 15.4: Survival probabilities based on prognostic index for values greater than the 0.33 percentile using Kaplan-Meier, Breslow and Xu-O'Quigley estimators.

15.9 Generalization under multiplicative relative risk

The estimator described above is not limited to proportional hazards models. The formula itself came from a simple application of Bayes' rule, and the marginal distribution of T can always be estimated by the Kaplan-Meier estimator or some other estimator if we wish to use other assumptions. Taking a general form of relative risk $r(t; z)$, so that $\lambda(t|z) = \lambda_0(t)r(t; z)$. Assume also that $r(t; z)$ can be estimated by $\hat{r}(t; z)$, for example, that it has a known functional form and a finite or infinite dimensional parameter that can be consistently estimated. Special cases of $r(t; z)$ are $\exp(\beta z)$, $1 + \beta z$, and $\exp\{\beta(t)z\}$. Since the main theorem of Section 7.4 extends readily to other relative risk models it is straightforward to derive analogous results to those above. We are still able estimate $S(t|Z \in H)$, with $\hat{P}(Z \in H|t) = \sum_{\{j:Z_j \in H\}} \pi_j(t)$. This is an important extension of the estimator since we may wish to directly work with some form of a non-proportional hazards model.

Stratified and random effects models

In studies where there is stratification, in particular, when there are many strata, the approach can be a relatively simple one to estimate conditional survivorship as it does not require the estimation of the baseline hazards. Suppose that V is the stratification variable, and we are interested in the survival given $Z \in H$. Note $P(Z \in H|t) = \sum_v P(Z \in H|t, V = v)P(V = v)$. We can estimate $P(Z \in H|t, V = v)$ by $\sum_{\{j:Z_j \in H\}} \pi_j^v(\hat{\beta}, t)$, where $\pi_j^v(\beta, t)$ is the conditional probability defined within strata v; and estimate $P(V = v)$ by the empirical distribution of V. Similarly to the stratified case, (15.3) can also be used to estimate survival under random effects models arising from clustered, such as genetic or familial data. The frailty models under such settings can be written

$$\lambda_{ij}(t) = \lambda_0(t)\omega_i \exp\{\beta z_{ij}\}, \tag{15.10}$$

where λ_{ij} is the hazard function of the jth individual in the ith cluster. This is the same as a stratified model, except that we do not observe the values of the "stratification variable" ω; but such values are not needed in the calculation described above for stratified models. So the procedure described above can be used to estimate $S(t|Z \in H)$. In both cases considered here, we need reasonable stratum or cluster sizes in order to get a good estimate of $P(Z \in H|t, v)$ or $P(Z \in H|t, \omega)$.

Conditional independent censorship

It is also possible to generalize (15.3) to the cases where C and T are independent given Z. First let us assume that the covariate Z is discrete with finitely many categories. Then the KM estimate can be replaced by a weighted Kaplan-Meier (WKM) estimate (Murray and Tsiatis 1996), which still consistently estimates $F(\cdot)$ under the conditional independent censorship. The WKM estimate calculates the subgroup KM estimates within each category of the covariate values, and then weights these subgroup estimates by the empirical distribution of Z. For the conditional distribution of Z given $T = t$, from the proof of the main theorem (Section 7.4);

$$f(z|T = t) = \frac{e^{\beta z} S(t|z) g(z)}{\int e^{\beta z} S(t|z) g(z) dz}. \qquad (15.11)$$

If we estimate $S(t|z)$ by the subgroup KM estimate within the category of value z, and the marginal $g(z)$ by the empirical probabilities, we are still able to consistently estimate the conditional distribution of Z given $T = t$ and thus $S(t|Z \in H)$. For other types of covariate distribution such as continuous covariates, we need to incorporate the covariates into categories, and Murray and Tsiatis (1996) suggested guidelines which could be useful in practice.

Sparse data in high dimensions

When data are sparse in high dimensions, i.e., multiple covariates, the $\pi_j(\hat{\beta}, t)$'s used to estimate the conditional distribution of Z given $T = t$ may encounter some difficulties, because they are like empirical distributions. In fact, they are obtained through the empirical distribution of Z given $T \geq t$. In this case, as seen in the gastric cancer example, we consider H as the Cartesian product of the range for that component with $(-\infty, \infty)$ for the rest components. Of course the actual range of the covariate vector is reflected in the data itself. However, sometimes we might specify H as a relatively small "block" in a p-dimensional space. For example, for the gastric cancer data, we might ask what is the survival probabilities given that CEA is greater than 10 but less than 15. In this type of situation there might be so few obseravtions in H that quite early on the risk sets may no longer contain any subjects with covariates in H, allowing for $\{\pi_j(\beta, t)\}_j$ to be a sufficiently reliable estimate of the conditional distribution of Z given $T = t$.

In many practical cases one is likely to ask for the conditional survival probabilities given a range for a single component of the covariate vector, while allowing the other components to vary freely We can proceed as follows. Denote the prognostic index $\eta = \beta' z$. Under the model two individuals should have the same survival probabilities as long as $\eta_1 = \eta_2$. That is, conditioning on $Z \in H$ is equivalent to conditioning on $\eta \in \beta' H$, Therefore, $S(t|Z \in H) = S(t|\eta \in \beta' H)$, where $\beta' H = \{\beta' z | z \in H\}$ and should contain potentially more observations than H does. This way the p-dimensional vector of covariates is reduced to the one-dimensional prognostic index, and the same number of observations in the risk set is now used to estimate the one-dimensional conditional distribution of η. Thus $\hat{S}(t|Z \in H)$ becomes;

$$\hat{S}(t|\eta \in \beta' H) = \frac{\sum_{t_i > t} \hat{P}(\eta \in \beta' H | t_i) W(t_i) + \hat{P}(\eta \in \beta' H | T > t_k) \hat{S}(t_k)}{\sum_{i=1}^{k} \hat{P}(\eta \in \beta' H | t_i) W(t_i) + \hat{P}(\eta \in \beta' H | T > t_k) \hat{S}(t_k)},$$

where $\hat{P}(\eta \in \beta' H | t) = \sum_{\{j:\eta_j \in \hat{\beta}' H\}} \pi_j(\hat{\beta}, t)$, and

$$\hat{P}(\eta \in \beta' H | T > t_k) = \sum_{\{j:\eta_j \in \hat{\beta}' H\}} Y_j(t_k+)/\sum Y_j(t_k+).$$

As before, since one can consistently estimate β, the above expression still provides a consistent estimate of $S(t|Z \in H)$. Note that when z is a single covariate, $z \in H$ is exactly the same as $\eta \in \beta H$ (unless $\beta = 0$ in which case the covariates have no predictive capability), so the above is consistent with the one-dimensional case developed earlier. While we regard the above as one possible approach under high dimensions when there are not "enough" observations falling into the ranges of covariates of interest, the variance estimation and the asymptotic properties seem to be more complicated as the estimate of β enters both η and the set $\beta' H$. When there is a need to use the estimate based on $\beta' H$, resampling methods such us bootstrap could be employed for inferential purposes. There are many potential areas for research here in order to further develop these techniques. Comparative studies would be of value. These would not be straightforward since comparing competing estimated curves requires considering the whole of the curve. For instance, at some points in time one estimator may outperform another whereas, at a later point, it may be the converse that holds. Some measure of overall distance such as the maximum or average discrepancy could be considered.

15.10   Informative censoring

Events that occur through time, alongside the main outcome of interest, may often provide prognostic information on the outcome itself. These can be viewed as time-dependent covariates and, as before, in light of the main theorem, it is still straightforward to use such information in the expression of the survivorship function. Since, in essence, we sum, or integrate, future information, it can be necessary to postulate future paths that the covariate process might take (O'Quigley and Moreau 1985).

Paths that remain constant are the easiest to interpret and, in certain cases, the simple fact of having a value tells us that the subject is still at risk for the event of interest (Kalbfleisch and Prentice 1980). The use of the main theorem in this context makes things particularly simple since the relevant probabilities themselves express themselves in terms of the conditional distribution of the covariate at given time points. We can make an immediate generalization of Equation 15.2 if we also wish to condition on the fact that $T > s$ where $s < t$. We have

$$S(t|Z \in H, T > s) = \frac{\int_t^\infty P(Z \in H|u)dF(u)}{\int_s^\infty P(Z \in H|u)dF(u)},$$

and, in exactly the same way as before, and assuming that the last observation is a failure, we replace this expression in practice by its empirical equivalent

$$\hat{S}(t|Z \in H, T > s) = \frac{\int_t^\infty \hat{P}(Z \in H|u)d\hat{F}(u)}{\int_s^\infty \hat{P}(Z \in H|u)d\hat{F}(u)} = \frac{\sum_{t_i > t} \hat{P}(Z \in H|t_i)W(t_i)}{\sum_{t_i > s} \hat{P}(Z \in H|t_i)W(t_i)}.$$

When the last observation is not a failure and $\sum_1^k W(t_i) < 1$ we can make a further adjustment for this in the same way as before.

One reason for favoring the Xu-O'Quigley estimate of survival over the Breslow estimate is the immediate extension to time-dependent covariates and to time-dependent covariate effects. Keeping in mind the discussion of Kalbfleisch and Prentice (1980), concerning internal and external time-dependent covariates, whether or not these are determined in advance or can be considered as an individual process generated through time, we can, at least formally, leaving aside interpretation questions, apply the above formulae. The interpretation questions are solved by sequentially conditioning on time as we progress along the time axis and, thereby, the further straightforward extension which seeks to quantity the probability of the event $T > t$, conditioned by $T > s$ where $s < t$, is particularly useful.

Surrogate endpoints

For certain chronic diseases, a notable example being HIV, workers have considered the need for procedures that might yield quicker results. Surrogate endpoints, viewed as time-dependent covariates, can help in addressing this issue and have received attention in the medical statistical literature (Ellenberg and Hamilton 1989, Wittes et al. 1989, Hillis and Siegel 1989). Herson (1989) wrote that "a surrogate endpoint is one that an investigator deems as correlated with an endpoint of interest but that can perhaps be measured at lower expense or at an earlier time than the endpoint of interest." Prentice (1989) defined a surrogate endpoint as "a response variable for which a test of the null hypothesis of no relationship to the treatment groups under comparison is also a valid test of the corresponding null hypothesis based on the true endpoint."

We can view a surrogate endpoint to be a time-dependent response variable of prognostic value obtained during follow-up, which indicates an objective progression of disease. In the survival context a surrogate variable is most often a discrete endpoint indicating, in some way, disease progression. Flandre and O'Quigley (1996) proposed a two-stage procedure for survival studies when a surrogate or intermediate time-dependent response variable is available for some patients. In the presence of a time-dependent intermediary event we can write

$$S(t) = \int_0^\infty S(t|c)g(c)dc = \int_0^t S(t|c < t)g(c)dc + \int_t^\infty S(t|c \geq t)g(c)dc.$$

If the effect of the occurrence of the intermediary event is to change the hazard function of death $\lambda(t)$ from $\lambda_1(t)$ to $\lambda_2(t)$; that is: $\lambda(t) = \lambda_1(t)$ if $t \leq C$ and is equal to $\lambda_2(t)$ otherwise then $\lambda_1(t) = \lambda_2(t)$ when the intermediary response variable has no influence on survival. When $\lambda_2(t) < \lambda_1(t)$ or $\lambda_2(t) > \lambda_1(t)$ then the intermediary, or surrogate, response variable carries relatively favorable or unfavorable predictions of survival. Thus the quantity $\pi(t) = \lambda_2(t)/\lambda_1(t)$ is a measure of the effect of the surrogate response on survival. When $f_1(t)$ and $f_2(t)$ are the density functions, $S_1(t)$ and $S_2(t)$ the survivorship functions corresponding to the hazard functions $\lambda_1(t)$ and $\lambda_2(t)$ respectively. then the marginal survival function is

$$S(t) = \int_0^t \exp - \left[\int_0^c \lambda_1(u)du + \int_c^t \lambda_2(u)du \right] dG(c) + \exp \left[- \int_0^t \lambda_1(u)du \right] G(t).$$

In the first stage of a two-stage design, all patients are followed to the endpoint of primary concern and for some subset of the patients there will be the surrogate information collected at an intermediate point during the follow-up. The purpose then of the first stage is to estimate the relationship between the occurrence of the surrogate response variable and the remaining survival time. This information can then be used in the second stage, at which time, for patients who reach the surrogate endpoint, follow-up is terminated. Such patients could be considered as censored under a particular dependent censorship model, the censoring being, in general "informative." The Kaplan-Meier estimator will not generally be consistent if the survival time and an informative censoring time are dependent but treated as though they were independent. Flandre and O'Quigley (1996) proposed a nonparametric estimator of the survival function for data collected in a two-stage procedure. A nonparametric permutation test for comparing the survival distributions of two treatments using the two-stage procedure is also readily derived.

The idea behind a two-stage design in the context of a time-dependent surrogate endpoint is to reduce the overall duration of the study. This potential reduction occurs at the second stage, in which follow-up is terminated on patients for whom the surrogate variable has been observed. The first stage is used to quantify the strength of the relationship between occurrence of the surrogate variable and subsequent survival. It is this information, obtained from the first stage analysis, that will enable us to make inferences on survival on the basis of, not only observed failures, but also observed occurrences of the surrogate. In the context of clinical trials, as pointed out by Prentice (1989), the surrogate variable must attempt to "capture" any relationship between the treatment and the true endpoint. We may wish to formally test the validity of a surrogate variable before proceeding to the second stage using a standard likelihood ratio test.

In the first stage N_1 patients are enrolled and followed to the endpoint of primary concern (e.g., death) or to censorship, as in a classical study, and information concerning the surrogate variable is recorded. Survival time is then either the true survival time or the informative censoring time. The information available for some patients will consist of both time until the surrogate variable and survival time, while for others (i.e. those who die or are censored without the occurrence of the surrogate variable) it consists only of survival time. In the second stage a new set of patients (N_2) is enrolled in the study. For those patients, the follow-up is completed when the surrogate variable has been observed. Thus, the information collected consists only of one time, either the time until the surrogate variable is reached or the survival time. In some cases the two stages may correspond to separate and distinct studies; the second stage being the clinical trial of immediate interest while the first stage would be an earlier trial carried out under similar conditions.

When parametric models for S and G are assumed, then the likelihood function can be obtained directly and provides the basis for inference. The main idea follows that of Lagakos (1976, 1977) who introduced a stochastic model that utilizes the information on a time-dependent event (auxiliary variable) that may be related to survival time. By taking $\lambda_0(t) = \lambda_2$, where $\lambda_0(.)$ is the hazard function for the occurrence of the surrogate response, $\lambda_1(t) = \lambda_1$ and $\lambda_2(t) = \lambda_3$, the survival function has the marginal distribution function given by Lagakos. The Lagakos model itself is a special case of the bivariate model of Freund (1961), applicable to the lifetimes of certain two-component systems where a failure of the first or second component alters the failure rate of the second or first component from β to β' or (α to α'). By taking $\alpha = \lambda_1(t)$, $\alpha' = \lambda_2(t)$ and $\beta = \beta' = \lambda_0(t)$, the Freund model can be viewed as a special case of the model described above.

Slud and Rubinstein (1983) make simple nonparametric assumptions on the joint density of (T, C) and consider the function $\rho(t)$ defined by,

$$\rho(t) = \lim_{\delta \to 0} \frac{\Pr(t < T < t + \delta | T > t, C < t)}{\Pr(t < T < t + \delta | T > t, C \geq t)}.$$

It is possible to make use of this function in order to derive a nonparametric estimation of $S(t)$ for use with the two-stage procedure with surrogate endpoints and which accounts for the dependent censoring. The authors (1983) present a class of nonparametric assumptions on the conditional distribution of T given C which leads to a consistent generalization of the Kaplan-Meier survival curve estimator. Individual values of $\rho(t) > 1$ mean that after the censoring the hazard risk of death for the patient is increased, and thus if we make an assumption of independence of T and C, the Kaplan-Meier estimator will tend to overestimate survival. On the other hand, if $\rho(t) < 1$, the independence assumption will lead to underestimation of the survival curve. Recalling the usual notation whereby $\delta_i = 1$ if $T_i \leq C_i$ and $X_i = \min(T_i, C_i)$, then X_i defines the observed portion of survival time. The Slud and Rubinstein estimator is given by

$$\hat{S}_\rho(t) = N^{-1} \left\{ n(t) + \sum_{k=0}^{d(t)-1} W_k \prod_{i=k+1}^{d(t)} \frac{n_i - 1}{n_i + \rho_i - 1} \right\}, \qquad (15.12)$$

where $n(t) = \sum I(X_i > t)$, $n_j = \sum_i I(X_i \geq X_j)$, $d(t) = \sum I(\delta_i = 1, X_i \leq t)$, $\rho_i = \rho(t_i)$ and W_j is the number of patients censored between two consecutively ordered failure times X_j and X_{j+1}. When $\rho_i = 1$ it follows that $\hat{S}_\rho(t)$ reduces to the usual Kaplan-Meier estimator. This model is a special case of the nonparametric assumption presented by Slud and Rubinstein. The focus here is not on the dependence of T and C but on the dependence of T and C_s where C_s is a dependent censoring indicator, in particular a surrogate endpoint. The function of interest is

$$\rho_s(t) = \lim_{\delta \to 0} \frac{\Pr\left(t < T < t + \delta | T > t, C_s < t\right)}{\Pr\left(t < T < t + \delta | T > t, C_s \geq t\right)}.$$

This function is equivalent to the function $\pi(t)$ and can be estimated from data from the first stage. Suppose that the conditional hazard, $\lambda(t|z)$, of death at t given $Z = z$ has the form $h_0(t)\exp(\beta z(t))$ where $z_i(t)$ takes the value 0 if $t_i \leq c_i$ and value 1 if $t_i > c_i$ then $\rho_s(t) = \rho_s = \exp(\beta)$. Thus, an estimate of ρ_s is given by $\exp(\hat{\beta})$. The estimate of β using data from the first stage, quantifies the increase in the risk of death occurring after the surrogate variable has been observed. The first stage is viewed as a training set of data to learn about the relationship between the potential surrogate endpoint and the survival time. The estimator is constructed from the entire sample N ($N = N_1 + N_2$). In the sample N, the ordered failure times X_i for which $\delta_i = 1$ are $X_1 \leq \ldots \leq X_d$, where d is the number of deaths of patients enrolled either in the first stage or in the second stage. Using the notation that $\epsilon_i = 1$ if $X_i > c_i$ and 0 otherwise, the random variable V defines either the observed survival time or the time to the surrogate variable. For patients in the second stage, let us denote the number of

X_i with $\epsilon_i = 1$ and $\delta_i = 0$ between X_j and X_{j+1} by W_j. In this way, W_j denotes the number of individuals from the second stage having a surrogate response between two consecutive failure times X_j and X_{j+1}. Let W_j' denote either the number of individuals censored in the first stage or the number of individuals censored without a surrogate response in the second stage between X_j and X_{j+1}. Clearly, W_j is the number of patients censored with "informative censoring" between two consecutively ordered failure times X_j and X_{j+1} while W_j' is the number of patients censored with "noninformative censoring" between two consecutively ordered failure times X_j and X_{j+1}. Finally n_j is the number of i with $X_i \geq X_j$. The product-limit estimator then becomes

$$\hat{S}(t; \rho_s) = N^{-1} \left\{ n(t) + \sum_{k=0}^{d(t)-1} W_k \prod_{i=k+1}^{d(t)} \frac{n_i - 1}{n_i + \rho_s - 1} + \sum_{k=0}^{d(t)-1} W_k' \prod_{i=k+1}^{d(t)} \frac{n_i - 1}{n_i} \right\}.$$

Notice that when $\rho_s = 1$ (i.e, the occurrence of the surrogate variables has no influence of survival) then $\hat{S}(t; \rho_s)$ is simply the Kaplan-Meier estimator. Considering the two-sample case, for example, a controlled clinical trial where group membership is indicated by a single binary variable Z, then the survival function, corresponding to $z = 0$, is denoted by $S_0(t)$ and the survival function corresponding to $z = 1$ is denoted by $S_1(t)$. A nonparametric permutation test (Flandre and O'Quigley 1996) for testing the null hypothesis $H_0 : S_0 = S_1$ versus $H_1 : S_0 \neq S_1$ can be easily constructed. The test accounts for the presence of dependent censoring as described and, since we would not want to assume that the effect of the surrogate is the same in both groups, then ρ_{s0} need not be equal to ρ_{s1}. A test can be based on the statistic Y defined by

$$Y(w; \psi) = \int_0^\infty w(t)\psi(\hat{S}_0(t; \hat{\rho}_{s0}), \hat{S}_1(t; \hat{\rho}_{s1}))d\hat{S}_0(t; \hat{\rho}_{s0}), \quad (15.13)$$

where $\psi(a, b)$ is some distance function (metric) between a and b at the point s and $w(s)$ is some positive weighting function, often taken to be 1 although a variance-stabilizing definition such as $w(t)^2 = \hat{S}_0(t; \hat{\rho}_{s0})(1 - \hat{S}_0(t; \hat{\rho}_{s0}))$ can be useful in certain applications. The choice $\psi(a, b) = a - b$ leads to a test with good power against alternatives of stochastic ordering, whereas the choice $\psi(a, b) = |a - b|$ or $\psi(a, b) = (a - b)^2$, for instance, would provide better power against crossing hazard alternatives. Large sample theory is simplified by assuming that the non-informative censoring distributions are identical in both groups and this may require more critical examination in practical examples. Given data we can observe some value $Y = y_0$ from which the significance level can be calculated by randomly permuting the $(0, 1)$ labels corresponding to treatment assignment. For the ith permutation $(i = 1, \ldots, n_p)$ the test statistic can be calculated, resulting in the value y_i say. Out of the n_p permutations suppose that there are n^+ values of y_i greater than or equal to y_0 and, therefore, $n_p - n^+$ values of y_i less than y_0. The significance level for a two-sided test is then given by $2 \min(n^+, n_p - n^+)/n_p$. In practice we sample from the set of all permutations so that n_p does not correspond to the total number of possible permutations but, rather, the number actually used, of which some may even be repeated. This is the same idea that is used in bootstrap resampling.

15.11   Exercises and class projects

1. For subjects i and j with covariate values Z_i and Z_j, write down the probability that the ratio of the survival time for subject i to the survival time for subject j is greater than $2/3$.

2. Calculate an estimate of the above probability for the Freireich data in which $Z_i = 1$ and $Z_j = 0$. Derive an expression for a 95% confidence interval for this quantity and use it to derive a confidence interval for use with the Freireich data.

3. Referring to the paper of Kent and O'Quigley (1988), consider the approximation suggested in that paper for the coefficient of randomness. Use this to motivate the quantity $\Pr(T_i > T_j | Z_i, Z_j)$ as a measure of dependence analogous to explained variation.

4. Consider some large data set in which there exists two prognostic groups. Divide the time scale into m non-overlapping intervals, $a_0 = 0 < a_1 < \ldots < a_m = \infty$. Calculate $\Pr(T_i > T_j | Z_i, Z_j, T_j > a_k)$ for all values of k less than m and use this information to make inferences about the impact of group effects through time.

5. Write down the likelihood for the piecewise exponential model in which, between adjacent failure times, the hazard can be any positive value (Breslow 1972). Find an expression for the cumulative hazard function and use this to obtain an expression for the survivorship

function. Although such an estimator can be shown to be consistent (Crowley and Breslow 1974) explain why the usual large sample likelihood theory would fail to apply.

6. Consider again the two group case. Suppose we are told that the survival time of a given subject is less that t_0. We also know that the groups are initially balanced. Derive an expression for the probability that the subject in question belongs to group 1. For the Freireich data, given that a subject has a survival time less than 15 weeks, estimate the probability that the subject belongs to the group receiving placebo?

7. Derive an estimator analogous to Equation 15.4 but based on the Neslon-Aalen estimator for marginal survival. By adding to the marginal cumulative hazard some arbitrary increasing unbounded function of time, obtain a simpler expression for the conditional survival estimate.

8. Use the delta method (Section 2.9) to obtain Equation 15.7.

9. Use the results of Andersen and Gill (1982) to conclude that $\hat{\beta} - \beta_0$ is asymptotically uncorrelated with $\hat{S}(t|Z \in H)|_{\beta_0}$.

10. On the basis of a large data set construct some simple prognostic indices using the most important risk factors. Divide into 3 groups the data based on the prognostic index and calculate the different survival estimates for each subgroup. Comment on the different features of these estimators as observed in this example. How would you investigate more closely the relative benefits of the different estimators?

11. Show that when T_i and T_j have the same covariate values, then $\Pr(T_i > T_j) = 0.5$.

12. For the Freireich data calculate the probability that a randomly chosen subject from the treated group lives longer than a randomly chosen subject from the control group.

13. Generate a two variate proportional hazards model in which Z_1 is binary and Z_2 is uniform on the interval $(0, 1)$. The regression coefficients β_1 and β_2 take values (i) 0.5, 0.5; (ii) 0.5, 1.0, and (iii) 1, 2, respectively. For all three situations, and taking $\lambda_0 = 1$, calculate and compare the six survival curves for $Z_1 = 0, 1$ and $Z_2 = 0.25, 0.50$ and 0.75.

Chapter 16

Proofs of theorems, lemmas and corollaries

In this chapter we provide further details, including many proofs, of the theorems, lemmas and corollaries that appear throughout the main text. For the more mathematical reader the proofs are of interest in their own right and many of these proofs either do not appear elsewhere or are not easily accessible. Many of the large sample derivations lean upon the theory of Andersen and Gill (1982) and were first worked out by Xu (1996). This chapter will be of use to the less mathematical reader who is concerned that results be well established and lean on solid reasoning. He or she, assuming some limited mathematical facility, will be in a better position to bring under scrutiny any areas of doubt. Finally, ignoring this chapter altogether will not detract from the main ideas.

Theorem 2.2, Theorem 2.7 and Corollary 2.9

The importance of the first of these two theorems is difficult to overstate. All the useful large sample results, for instance, hinge ultimately on the theorem. An elegant proof of the theorem, together with well thought out illustrations and some examples, is given in Shenk (1988). For Theorem 2.7 let T have a continuous and invertible distribution function $F(t)$ and let $U = F(T)$. The inverse function is denoted F^{-1} so that $F^{-1}\{F(t)\} = t$. Then $P(U < u) = P\{F(T) < u\} = P\{T <$

$F^{-1}(u)\} = P\{T < F^{-1}F(t)\} = P\{T < t\} = F(t) = u$. Thus U
has the distribution of a standard uniform. The proof of the corollary
leans upon some straightforward manipulation of elementary events.
An outline is provided in Section 2.5 of David (1981).

Theorem 2.13 and Corollaries 2.10 and 8.12

We have that $P\{X(t + s) > x | X(s) = x_s, X(u), 0 \le u < s\} =$
$P\{X(t + s) - X(s) > x - x_s | X(s) = x_s, X(u), 0 \le u < s\} =$
$P\{X(t + s) - X(s) > x - x_s\} = P\{X(t + s) > x | X(s) = x_s\}$.
Corollary 2.10 follows since, $f_{s|t}(x|w) = f_s(x) f_{t-s}(w - x)/f_t(w) =$
$\text{const} \exp\{-x^2/2s - (w - x)^2/2(t - s)\} = \text{const} \exp\{-t(x - ws/t)^2/$
$2s(t - s)\}$. For Corollary 8.12 note that the process $V(t)$ is a Gaussian
process in which $E\{V(t)\} = 0$ and $\text{Cov}\{V(t), V(s)\} = E\{\exp(-\alpha t/2)$
$\exp(-\alpha s/2) X(e^{\alpha t}) X(e^{\alpha s})\}$ This can be written as $\exp(-\alpha(t + s)/2) \times$
$E\{X(e^{\alpha t}) X(e^{\alpha s})\}$ This is turn is then equal to $\exp(-\alpha(t+s)/2) \exp \alpha t$
which is just $\exp(-\alpha(t - s)/2)$.

Theorem 2.14 and Corollary 2.12

For the theorem note that, $\text{Cov}(X(s), X(t) | X(1) = 0) = E(X(s), X(t) |$
$X(1) = 0)$. In turn this equals $E\{E(X(s), X(t) | X(t), X(1) = 0 | X(1) =$
$0)\}$. This is the same as $E\{X(t) E(X(s) | X(t)) | X(1) = 0\}$ which can be
written as, $E\{X(t)(s/t) X(t) | X(1) = 0\}$ which is then $(s/t) E\{X^2(t) |$
$X(1) = 0\}$ so that finally this is $(s/t) t(1 - t) = s(1 - t)$. Corollary 2.12 can be deduced from the theorem via the following simple
steps: $\text{Cov}\{W^0(s), W^0(t)\}$ $s < t$ is given by $E\{W(s) - sW(1)\}\{W(t) -$
$tW(1)\} = E\{W(s)W(t) - sW(1)W(t) + stW(1)W(1) - tW(1)W(s)\} =$
$s - st + st - st = s(1 - t)$.

Lemma 2.8

By definition the covariance of $X(s)$ and $X(t)$ is equal to $(s+1)(t+1)$
which multiplies the quantity $\text{Cov}(Z(t/(t + 1)), Z(s/(s + 1)))$. Next,
from the definition of the Brownian bridge, $Z(t) = W(t) - tW(1)$ so
that, expanding the above expression we obtain $\text{Cov}\{W(t/(t + 1)),$
$W(s/(s + 1))\} - t/(t + 1) \times \text{Cov}\{W(1), W(s/(s + 1))\} - s/(s + 1) \text{Cov}$
$\{W(1), W(t/(t + 1))\} + st/((s + 1)(t + 1)) \times \text{Cov}\{W(1), W(1)\} = s(t +$
$1) - st - st + st = s$.

Theorem 2.9 and Theorem 3.1

For Theorem 2.9 note that, $\text{Cov}\{Z(s), Z(t)\} = E\{Z(s)Z(t)\}$. We write this as $E \int_0^s X(y)dy \int_0^t X(u)du$. Bringing together the two integral operators we rewrite this as, $E \int_0^s \int_0^t X(y)X(u)dydu$ which is just $\int_0^s \int_0^t EX(y)X(u)dydu$ which we express as $\int_0^s \int_0^t \min(y, u)dydu$ and this is then $\int_0^s \int_0^u ydy + \int_u^t udydu$ which simplifies to $s^2(t/2 - s/6)$. Furthermore, for Theorem 3.1, we can calculate $\text{Cov}(S_k^*, S_m^*)$ where $k < m$ directly. We have that $\text{Cov}(S_k^*, S_m^*) = E(S_k^* S_m^*) - E(S_k^*)E(S_m^*) = (\sigma\sqrt{n})^{-2}E\{S_k S_m\} = (\sigma\sqrt{n})^{-2}E\{S_k(S_k + \sum_{k < i \le m} X_i)\}$. Developing the bracket gives, $(\sigma\sqrt{n})^{-2}\{E(S_k^2) + E(S_k)E(\sum_{k < i \le m} X_i)\}$ which is equal to $(\sigma\sqrt{n})^{-2}E(S_k^2) = (\sigma\sqrt{n})^{-2}k\sigma^2 = k/n = t$.

Theorem 3.2 and 3.6

For the first of these two theorems, note that the marginal and conditional normality of any sequence of $\sqrt{n}\{F_n(t_i) - F(t_i)\}$, $(0 < t_1 < \ldots < t_m)$, for some m, indicates the multivariate normality of $\sqrt{n}\{F_n(t_i) - F(t_i)\}$, $(0 < t_1 < \ldots < t_m)$. Thus $\sqrt{n}\{F_n(t) - F(t)\}$ is a Gaussian process. It has mean zero and variance $F(t)\{1 - F(t)\}$. To obtain the covariance, note first that the indicator variables $I(T_i \le t)$ and $I(T_j \le s)$ are independent for all t and s and $i \ne j$. It only remains to evaluate, for $s < t$ $\text{Cov}\{I(T_i \le t), I(T_i \le s)\} = E\{I(T_i \le t)I(T_i \le s)\} - E\{I(T_i \le t)\}E\{I(T_i \le s)\} = F(s) - F(t)F(s) = F(s)\{1 - F(t)\}$. For the second of these two theorems, note that, $\text{E}\{dM'(t)|\mathcal{F}_{t-}\} = \text{E}\{H(t)dM(t)|\mathcal{F}_{t-}\} = H(t)\text{E}\{dM(t)|\mathcal{F}_{t-}\} = 0$. Furthermore, $\text{Var}\{dM'(t)|\mathcal{F}_{t-}\} = \text{Var}\{H(t)dM(t)|\mathcal{F}_{t-}\} = H(t)^2\text{Var}\{dM(t)|\mathcal{F}_{t-}\} = H(t)^2 d\langle M\rangle(t)$. Similarly, $\langle \int HdM, \int KdM'\rangle(t) = \int_0^t H(s)K(s)d\langle M, M'\rangle(s)$.

Theorem 5.1 and Theorem 5.3

There are two possibilities; the observation x_i corresponds to a failure, the observation corresponds to a censoring time. For the first possibility let dx_i be an infinitesimally small interval around this point. The probability that we can associate with this event is $\text{Pr}(T \in dx_i, C > x_i) = \text{Pr}(T \in dx_i) \times \text{pr}(C > x_i)$ i.e. $f(x_i; \theta)dx_i \times G(x_i; \theta)$. For a censored observation at time x_i ($\delta_i = 0$) we have; $\text{Pr}(C \in$

dx_i , $T > x_i) = \Pr(C \in dx_i) \times \mathrm{pr}(T > x_i)$ i.e. $g(x_i; \theta)dx_i \times S(x_i; \theta)$. We can then write the likelihood as; $L(\theta) = \prod_{\delta_i=1} f(x_i; \theta)dx_i G(x_i; \theta) \times \prod_{\delta_i=0} g(x_i; \theta)dx_i S(x_i; \theta)$ In most cases the further assumption that the censoring itself does not depend on θ may be reasonable. Taking logs and ignoring constants we obtain the result. For Theorem 5.3 note that $\Pr\{X_i \geq a_\ell | X_i > a_{\ell-1}\} = \Pr\{T_i \geq a_\ell | X_i > a_{\ell-1}\} \times \Pr\{C_i \geq a_\ell | X_i > a_{\ell-1}\} = \Pr\{T_i \geq a_\ell | C_i > a_{\ell-1}, T_i > a_{\ell-1}\} = \Pr\{T_i \geq a_\ell | T_i > a_{\ell-1}\}$ by independence.

Theorem 5.4

We have $\mathrm{Var}(\log S(a_\ell) \approx \sum_{m \leq \ell} \mathrm{Var}(\log(1 - \pi_m))$. This is $\sum_{m \leq \ell}(1 - \pi_m)^{-2}\mathrm{Var}(\pi_m)$ which we write as $\sum_{m \leq \ell}(1 - \pi_m)^{-2}\pi_m(1 - \pi_m)/n_m$ where n_m corresponds to the number at risk. i.e. the denominator. We write $\sum_{m \leq \ell} d_m/r_m(r_m - d_m)$ Use log function and a further application of delta method to obtain; $\mathrm{Var}\, S(a_\ell) \approx S(a_\ell)^2 \sum_{m \leq \ell} \sum_{m \leq \ell} d_m/r_m(r_m - d_m)$ An approach not using the delta method follows Greenwood (1926). Let $t_0 < t_1 < \cdots < t_k$. An estimate of the variance of the survival probability P of the form $P = p_1 \times p_2 \times \ldots \times p_k$, where each p_i is the estimated probability of survival from time t_{i-1} to time t_i, $q_i = 1 - p_i$ and P is therefore the estimated probability of survival from t_0 to t_k. Assuming that the p_i's are independent of one another, we have $\mathrm{E}(P) = \mathrm{E}(p_1) \times \mathrm{E}(p_2) \times \ldots \times \mathrm{E}(p_k)$, as well as, $\mathrm{E}(P^2) = \mathrm{E}(p_1^2) \times \mathrm{E}(p_2^2) \times \ldots \times \mathrm{E}(p_k^2)$, and $\mathrm{E}(p_i^2) = (\mathrm{E}p_i)^2 + \sigma_i^2$, where $\sigma_i^2 = \mathrm{Var}(p_i)$. Then $\mathrm{Var}(P) = \mathrm{E}(P^2) - \{\mathrm{E}(P)\}^2 = \{\mathrm{E}(P)\}^2((1 + \sigma_1^2/(\mathrm{E}p_1)^2)(1 + \sigma_2^2/(\mathrm{E}p_2)^2) \cdots (1 + \sigma_k^2/(\mathrm{E}p_k)^2) - 1) \approx \{\mathrm{E}(P)\}^2(\sigma_1^2/(\mathrm{E}p_1)^2 + \sigma_2^2/(\mathrm{E}p_2)^2 + \ldots + \sigma_k^2/(\mathrm{E}p_k)^2)$. Now condition on $\mathrm{E}(p_i)$ and the number of observations n_i to which p_i is applied, then $\sigma_i^2 = \mathrm{E}p_i(1 - \mathrm{E}p_i)/n_i$. Substituting p_i for $\mathrm{E}(p_i)$, we obtain estimates of the variance $\hat{\mathrm{Var}}(P) = P^2\{\prod_{i=1}^k (1 + q_i/n_i p_i) - 1\} \approx P^2 \sum_{i=1}^k q_i/n_i p_i$.

Theorem 2.9 and Corollary 2.4

The theorem states that, letting $F(x) = P(X \leq x)$ and $F_r(x) = P(X_{(r)} \leq x)$ then; $F_r(x) = \sum_{i=r}^n \binom{n}{i}F^i(x)[1 - F(x)]^{n-i}$ Recall that the X_i from the parent population with distribution $F(x)$ are i.i.d. The event that $X_{(r)} \leq x$ is the event that at least r of the X_i are less than or equal to x. This is then the sum of the binomial probabilities summed over all values of i greater than or equal to r. The first part of the corollary is clear upon inspection. For the second part note that $\sum_{i=0}^{n} \binom{n}{i} F^i(x)[1 - F(x)]^{n-i} = 1$ and that $\binom{n}{0} F^0(x)[1 - F(x)]^n = [1 - F(x)]^n$.

Theorem 9.2

Suppose k individuals are observed to fail at times $t_i, ..., t_k$ and have corresponding explanatory variables z_i to z_k. Assume times are ordered. Then the probability, conditional upon the observed failure times, of obtaining a particular ordering is written $\Pr(r = (1), (2), ..., (n)) = \Pr(t_1 < t_2 < t_3) = \int_0^\infty \int_{t_1}^\infty \cdots \int_{t_{n-1}}^\infty f(t_1|z_1)...f(t_n|z_n) dt_n...dt_1 = \int_0^\infty \int_{t_1}^\infty \cdots \int_{t_{n-1}}^\infty \prod_{i=1}^{n} \{h_0(t_i)e^{\beta z_i} \exp[-H_0(t_i)e^{\beta z_i}]\} = \prod_{i=1}^{n} \{\exp(\beta Z_i)/\sum_{j=1}^{n} Y_j(X_i) \exp(\beta Z_j)\}^{\delta_i}$.

Large sample results for survivorship function

Xu (1996) derives the asymptotic normality of the survival estimate under the model for fixed $t = t^*$. Let $Q_i = 1$ if $Z_i \in H$, and 0 otherwise. Using the notation: $S^{(r)}(t) = n^{-1} \sum_{i=1}^{n} Y_i(t)e^{\beta(t)'Z_i(t)} Z_i(t)^r$, $s^{(r)}(t)$, $ES^{(r)}(t)$, $S^{(r)}(\beta, t) = n^{-1} \sum_{i=1}^{n} Y_i(t)e^{\beta' Z_i(t)} Z_i(t)^r$, $s^{(r)}(\beta, t) = ES^{(r)}(\beta, t)$, $S^{(H)}(\beta, t) = n^{-1} \sum_{i=1}^{n} Q_i Y_i(t)e^{\beta Z_i}$, $s^{(H)}(\beta, t) = ES^{(H)}(\beta, t)$, $S^{(H1)}(\beta, t) = n^{-1} \sum_1^n Q_i Y_i(t) Z_i e^{\beta Z_i}$, $s^{(H1)}(\beta, t) = ES^{(H1)}(\beta, t)$, for $r = 0, 1, 2$, we can rewrite $\hat{P}(Z \in H|t) = S^{(H)}(\hat{\beta}, t)/S^{(0)}(\hat{\beta}, t)$, and $E_\beta(Z|t; \pi^H) = S^{(H1)}(\hat{\beta}, t)/S^{(H)}(\hat{\beta}, t)$. From the main theorem we have $s^{(H)}(\beta_0, t)/s^{(0)}(\beta_0, t) = P(Z \in H|t)$. So under some elementary regularity and continuity conditions (Xu 1996) it is easy to show that $\{\partial \hat{S}(t^*|Z \in H)/\partial \beta\}|_{\beta=\hat{\beta}}$ is asymptotically constant. Now $\hat{\beta} - \beta_0 = I^{-1}(\check{\beta})U(\beta_0)$ where $\check{\beta}$ is on the line segment between $\hat{\beta}$ and β_0, $U(\beta) = \partial \log L(\beta)/\partial \beta$ and $I(\beta) = -\partial U(\beta)/\partial \beta$. Combining these results we have $\sqrt{n}\hat{S}(t^*|Z \in H) = \sqrt{n}\hat{S}(t^*|Z \in H)|_{\beta_0} + I^{-1}(\check{\beta})\sqrt{n}U(\beta_0)\partial \hat{S}(t^*|Z \in H)/\partial \beta|_{\beta=\hat{\beta}}$. It is known from for example, Andersen and Gill (1982) that $I(\check{\beta})$ converges in probability to a well-defined population parameter. Lin and Wei (1989) showed that $U(\beta_0)$ is asymptotically equivalent to $1/n$ times a sum of i.i.d. random variables: Lin and Wei (1989) show that $\sqrt{n}U(\beta_0)$ is asymptotically equivalent to $n^{-1/2} \sum_1^n \omega_i(\beta_0)$, where $\omega_i(\beta) = \int_0^1 \{Z_i - s^{(1)}(\beta, t)/s^{(0)}(\beta, t)\}dN_i(t) - \int_0^1 Y_i(t)e^{\beta Z_i}\{Z_i - s^{(1)}$

$(\beta, t)/s^{(0)}(\beta, t)\} \lambda_0(t)dt$ and $N_i(t) = I\{T_i \leq t, T_i \leq C_i\}$. For the asymptotic normality of $\hat{S}(t^*|Z \in H)$ it is only necessary to show that the numerator of $\hat{S}(t^*|Z \in H)|_{\beta_0}$ is also asymptotically equivalent to $1/n$ times a sum of n i.i.d. random variables like the above, since the denominator of it we know is consistent for $P(Z \in H)$. Now, the numerator of $\hat{S}(t^*|Z \in H)$ is $\int_{t^*}^{\infty} \hat{P}(Z \in H|t)d\hat{F}(t)$. Note that $\sqrt{n}\{\int_{t^*}^{\infty} \hat{P}(Z \in H|t)d\hat{F}(t) - P(Z \in H, T > t^*)\} = \sqrt{n}\int_{t^*}^{\infty} P(Z \in H|t)d\{\hat{F}(t) - F(t)\} + \sqrt{n}\int_{t^*}^{\infty}\{\hat{P}(Z \in H|t) - P(Z \in H|t)\}d\{\hat{F}(t) - F(t)\} + \sqrt{n}\int_{t^*}^{\infty}\{\hat{P}(Z \in H|t) - P(Z \in H|t)\}dF(t)$. Now $\sqrt{n}\{\hat{F}(t) - F(t)\}$ converges in distribution to a zero-mean Gaussian process. Therefore the second term on the right hand side of the equation is $o_p(1)$. The last term is $A_1 + o_p(1)$ (see also Xu 1996), where $A_1 = \sqrt{n}\int_{t^*}^{1}\{S^{(H)}(\beta_0, t)/s^{(0)}(\beta_0, t) - s^{(H)}(\beta_0, t)S^{(0)}(\beta_0, t)/s^{(0)}(\beta_0, t)^2\}dF(t) = n^{-1/2}\sum_{i=1}^{n}\int_{t^*}^{1} Y_i(t)e^{\beta_0 Z_i}/s^{(0)}(\beta_0, t)\{Q_i - s^{(H)}(\beta_0, t)/s^{(0)}(\beta_0, t)\}dF(t)$. As for the first term, we can use a result of Stute (1995). With $\phi(t) = 1_{[t^*, 1]}(t)P(Z \in H|t)$ in the theorem, the first term is equal to $n^{-1/2}\sum_{i=1}^{n}\nu_i + \sqrt{n}R_n$, where $|R_n| = o_p(n^{-1/2})$ and ν's are i.i.d. with mean zero, each being a function of X_i and δ_i.

Theorem 7.1

Since $\lambda(t|Z(t) = z) = \lambda_0(t)\exp\{\beta(t)z\}$, then $f(t|Z(t) = z) = \lambda_0(t)\exp\{\beta(t)z\}S(t|Z(t) = z)$, where $S(t|Z(t) = z)$ is the conditional survival function. Without loss of generality, we assume that $Z(T)$ has density function $g(z)$. The proof follows the same way if Z is discrete. By Bayes' rule, we can express the conditional density of $Z(t)$ given $T = t$ as; $f_t(z|T = t) = f(t|z)g(z)/\int f(t|z)g(z)dz = \lambda_0(t)e^{\beta(t)z}S(t|z)g(z)/\int \lambda_0(t)e^{\beta(t)z}S(t|z)g(z)dz = e^{\beta(t)z}h(z|T \geq t)/\int e^{\beta(t)z}h(z|T \geq t)$ where $h(z|T \geq t)$ is the conditional density of $Z(t)$ given $T \geq t$. From elementary probability calculus, for sets $\mathcal{A}$ and $\mathcal{B}$ we have $P(\mathcal{A}) = P(\mathcal{A}|\mathcal{B})P(\mathcal{B})/P(\mathcal{B}|\mathcal{A})$ so that we can write; $h(z|T \geq t) = f(z|T \geq t, C \geq t)P(C \geq t)/P(C \geq t|z, T \geq t)$ and, by our conditional independence assumption, the denominator simplifies so that $P(C \geq t|z, T \geq t) = P(C \geq t|z)$. Now, by the law of total probability, we have $P(C \geq t) = \int P(C \geq t|z)g(z)dz$, where the integral is over the domain of definition of Z. Next we replace $P(Z \leq z|T \geq t, C \geq t)$ by the consistent estimate $\sum_{\{Z_j(t) \leq z\}} Y_j(t)/\sum_1^n Y_j(t)$, which is simply the empirical distribution in the risk set, which leads to; $\hat{P}\{Z(t) \leq z | T = t\} = \sum_{z_i \leq z} Y_i(t) \exp\{\beta(t) z_i(t)\} \hat{\phi}(z_i, t) / \sum_{j=1}^{n} Y_j(t) \exp\{\beta(t) z_j(t)\} \hat{\phi}(z_j, t)$. An application of Slutsky's theorem enable us to claim the result continues to hold whenever $\beta(t)$ is replaced by any consistent estimator $\hat{\beta}(t)$, in particular the partial likelihood estimator when we assume the more restricted model holds.

Property (4) of R^2 (Xu 1996)

We recall the following proposition of Lin (1991): Assume the model holds. Let $\mathcal{W}(t)$ be a predictable process converging in probability to a non-negative bounded function uniformly in t. Let $\hat{\beta}_{\mathcal{W}}$ be the solution to the weighted score $U_{\mathcal{W}}(\beta) = \sum_{i=1}^{n} \delta_i \mathcal{W}(X_i) r_i(\beta) = 0$. Then $n^{1/2}(\hat{\beta}_{\mathcal{W}} - \beta)$ is asymptotically normal with mean 0. Property (4) can be seen by noting that $\partial R^2(\beta)/\partial \beta = 2/\mathcal{I}(0) \sum_{i=1}^{n} \delta_i W(X_i) r_i(\beta) V_\beta(Z|X_i)$ is a weighted score, where $V_\beta(Z|t) = \mathcal{E}_\beta(Z^2|t) - \{\mathcal{E}_\beta(Z|t)\}^2 = \sum_{j=1}^{n} Y_j(t) Z_j(t)^2 \exp\{\beta Z_j(t)\} / \sum_{j=1}^{n} Y_j(t) \exp\{\beta Z_j(t)\} - (\sum_{j=1}^{n} Y_j(t) Z_j(t) \exp\{\beta Z_j(t)\} / \sum_{j=1}^{n} Y_j(t) \exp\{\beta Z_j(t)\})^2$ is the conditional variance of $Z(t)$ with respect to the probability distribution $\{\pi_i(\beta, t)\}_i$ for fixed t. Under the model it then follows that $\partial R^2(\beta)/\partial \beta |_{\hat{\beta}} \xrightarrow{P} 0$. The second derivative is $\partial^2 R^2(\beta)/\partial \beta^2 = 2\mathcal{I}(0)\{-\sum_{i=1}^{n} \delta_i W(X_i) V_\beta(Z|X_i)^2 + \sum_{i=1}^{n} \delta_i W(X_i) r_i(\beta) \partial/\partial\beta V_\beta(Z|X_i)\}$. The second term in the braces above is again a weighted score and therefore asymptotically zero at $\beta = \hat{\beta}$. So for large enough n, $\partial^2 R^2(\beta)/\partial \beta^2$, evaluated at the point $\beta = \hat{\beta}$ is negative, which indicates that $R^2(\beta)$ reaches its maximum around $\hat{\beta}$.

Property (2) of $R^2_{\mathcal{E}}$ (Xu 1996)

Let $\psi_i(\beta) = \mathcal{E}_0(Z|X_i)^2 - 2\mathcal{E}_0(Z|X_i)\mathcal{E}_\beta(Z|X_i)$. We can then write; $R^2_{\mathcal{E}}(\beta) = \sum_{i=1}^{n} \delta_i W(X_i)\{\psi_i(\beta) + \mathcal{E}^2_\beta(Z|X_i)\} / \sum_{i=1}^{n} \delta_i W(X_i)\psi_i(\beta) + \delta_i W(X_i)\mathcal{E}_\beta(Z^2|X_i)$. With all the observed data fixed, at time X_i, as $\beta \to +\infty$ $\pi_j(\beta, X_i) \to 0$ for all j such that $z_j(X_i) < z^{(i)}_{max}$, where $z^{(i)}_{max} = \max\{z_l(X_i) : Y_l(X_i) = 1\}$. Furthermore $\mathcal{P}_\beta(Z(X_i) = z^{(i)}_{max}) = \sum_{\{j:z_j(X_i)=z^{(i)}_{max}\}} \pi_j(X_i; \beta) \to 1$ so that $\mathcal{E}^2_\beta(Z|X_i) \to [z^{(i)}_{max}]^2$ and $\mathcal{E}_\beta(Z^2|X_i) \to [z^{(i)}_{max}]^2$. Again at time X_i, as $\beta \to -\infty$ we have similarly $\mathcal{E}_\beta^2(Z|X_i) \rightarrow [z_{\min}^{(i)}]^2$ and $\mathcal{E}_\beta(Z^2|X_i) \rightarrow [z_{\min}^{(i)}]^2$, where $z_{\min}^{(i)} = \min\{z_l(X_i) : Y_l(X_i) = 1\}$. It is then clear that as $|\beta| \rightarrow \infty$, $R_\mathcal{E}^2(\beta) \rightarrow 1$.

Large sample equivalence of R_e^2 and R^2 (Xu 1996)

For showing large sample properties, it helps to make the following assumptions which are similar to those in Andersen and Gill (1982): (A) Finite interval, $\int_0^1 \lambda_0(t)dt < \infty$. (B) Asymptotic stability, there exist a neighborhood $\mathcal{B}$ of β such that 0 and β_0 belong to the interior of $\mathcal{B}$, and $\sup_{t \in [0,1]} \|nW(t) - w(t)\| \xrightarrow{P} 0$, $\sup_{t \in [0,1], \beta \in \mathcal{B}}$ $\|S^{(r)}(\beta,t) - s^{(r)}(\beta,t)\| \xrightarrow{P} 0$, for $r = 0,1,2,3,4$, where the arrows indicate convergence in probability with rate $n^{-1/2}$. (C) Asymptotic regularity conditions. All functions in B are uniformly continuous in $t \in [0,1]$; $s^{(r)}(\beta,t), r = 0,1,2,3,4$, are continuous functions of $\beta \in \mathcal{B}$, and are bounded on $\mathcal{B} \times [0,1]$; $s^{(0)}(\beta,t)$ is bounded away from zero. The difference between $\sum_{i=1}^n \delta_i W(X_i)\{r_i^2(0) - r_i^2(\beta)\}$ and the equivalent based on expectations is given by $\sum_{i=1}^n \delta_i W(X_i)\{r_i^2(0) - r_i^2(\beta) - [\mathcal{E}_\beta(Z|X_i) - \mathcal{E}_0(Z|X_i)]^2\} = \sum_{i=1}^n \delta_i W(X_i)\{2Z_i(X_i)[\mathcal{E}_\beta(Z|X_i) - \mathcal{E}_0(Z|X_i)] - 2\mathcal{E}_\beta^2(Z|X_i) + 2\mathcal{E}_\beta(Z|X_i)\mathcal{E}_0(Z|X_i)\} = 2\sum_{i=1}^n \delta_i W(X_i)r_i(\beta)\{\mathcal{E}_\beta(Z|X_i) - \mathcal{E}_0(Z|X_i)\}$. The above is a weighted score. Therefore the proposition of Lin (1991) indicates that when $\beta = \hat{\beta}$ the difference between the numerators of $R^2(\hat{\beta})$ and $R_\mathcal{E}^2(\hat{\beta})$ converges to zero in probability. Furthermore, the difference between the denominators of $R^2(\beta)$ and $R_\mathcal{E}^2(\beta)$ can be written as $\sum_{i=1}^n \delta_i W(X_i)\{r_i^2(0) - \mathcal{E}_\beta[r_i^2(0)|X_i]\} = \sum_{i=1}^n \delta_i W(X_i)\{Z_i(X_i)^2 - \mathcal{E}_\beta(Z^2|X_i)\} - 2\sum_{i=1}^n \delta_i W(X_i)\mathcal{E}_0(Z|X_i)\{Z_i(X_i) - \mathcal{E}_\beta(Z|X_i)\}$. The second term in the above is again asymptotically zero at $\beta = \hat{\beta}$ because of the proposition of Lin. The first term, is equal to $\sum_{i=1}^n \int_0^1 W(t)\{Z_i(t)^2 - S^{(2)}(\beta,t)/S^{(0)}(\beta,t)\}dN_i(t)$. This in turn can be expressed as, $\int_0^1 W(t)\sum_{i=1}^n Z_i(t)^2\alpha_i(t)dt - \int_0^1 W(t)S^{(2)}(\beta,t)/S^{(0)}(\beta,t)\bar{\alpha}(t)dt + \sum_{i=1}^n \int_0^1 W(t)(Z_i(t)^2 - S^{(2)}(\beta,t)/S^{(0)}(\beta,t))dM_i(t) = \int_0^1 nW(t)S^{(2)}(\beta_0,t)\lambda_0(t)dt - \int_0^1 nW(t)S^{(2)}(\beta,t)/S^{(0)}(\beta,t)S^{(0)}(\beta_0,t)\lambda_0(t)dt + A(1)$, where $\alpha_i(t) = Y_i(t)\lambda_i(t)dt$ is the intensity process of $N_i(t)$, $M_i(t) = N_i(t) - \int_0^t \alpha_i(s)ds$, and $\bar{\alpha}(t) = \sum_1^n \alpha_i(t)$. The total of the first two term in the last line above is zero when $\beta = \beta_0$, and $A(t) = \sum_{i=1}^n \int_0^t W(t)(Z_i(t)^2 - S^{(2)}(\beta,t)/S^{(0)}(\beta,t))dM_i(t)$. Now $\langle A(t), A(t) \rangle = \sum_{i=1}^n \int_0^t W(t)^2(Z_i(t)^2 - S^{(2)}(\beta,t)/S^{(0)}$

$(\beta,t))^2\alpha_i(t)dt = \int_0^t nW(t)^2(S^{(4)}(\beta_0,t) - 2S^{(2)}(\beta_0,t)S^{(2)}(\beta,t)/S^{(0)}(\beta,t).$
$+ S^{(2)}(\beta,t)^2 / S^{(0)}(\beta,t)^2 S^{(0)}(\beta_0,t))\lambda_0(t)dt \xrightarrow{\text{P}} 0$, under Conditions A, B and C. So from Inequality (I.2) of Andersen and Gill (1982) which is an application of Lenglart's inequality, we have that $A(1) \xrightarrow{\text{P}} 0$. Thus it is again clear that we have a convergence in probability result for the denominators. We next apply the convergence properties of transformed sequences (Serfling 1980 Ch. 1.7) to obtain convergence in probability of $R^2(\hat{\beta}) - R_{\mathcal{E}}^2(\hat{\beta})$ to 0.

Large sample results for R^2 (Xu 1996)

For $R^2(\hat{\beta})$, a first-order Taylor expansion gives $\sqrt{n}R^2(\hat{\beta}) = \sqrt{n}R^2(\beta_0) + \sqrt{n}(\hat{\beta} - \beta_0)\partial R^2(\beta)/\partial\beta$ at $\beta = \dot{\beta}$, where $\dot{\beta}$ lies on the line segment between β_0 and $\hat{\beta}$. Note that $\partial R^2(\beta)/\partial\beta$, evaluated at $\beta = \dot{\beta}$ converges to zero. We already have that $\sqrt{n}(\hat{\beta} - \beta_0)$ is asymptotically normal, so the second term on the right hand side of the equation is asymptotically zero. For the first term, $R^2(\beta_0) = 1/\mathcal{I}(0)\sum_{i=1}^n \delta_i W(X_i)\{r_i^2(0) - r_i^2(\beta_0)\}$, and we have that $\mathcal{I}(0) \xrightarrow{\text{P}} J(\beta_0, 0)$. For the rest of the proof we show that (A.3) is asymptotically equivalent to a sum of i.i.d. random variables. $\sum_{i=1}^n \delta_i W(X_i) \{r_i^2(0) - r_i^2(\beta_0)\} = \sum_{i=1}^n \delta_i W(X_i)\{\mathcal{E}_0(Z|X_i)^2 - 2Z_i(X_i)[\mathcal{E}_0(Z|X_i) - \mathcal{E}_{\beta_0}(Z|X_i)] - \mathcal{E}_{\beta_0}(Z|X_i)^2\} = 2\sum_{i=1}^n \delta_i W(X_i)Z_i(X_i)(S^{(1)}(\beta_0,X_i)/S^{(0)}(\beta_0,X_i) - S^{(1)}(0,X_i)/S^{(0)}(0,X_i)) + \int_0^1\{S^{(1)}(0,t)^2/S^{(0)}(0,t)^2 - S^{(1)}(\beta_0,t)^2/S^{(0)}(\beta_0,t)^2\}d\hat{F}(t)$. Now $\sqrt{n}\sum_{i=1}^n \delta_i W(X_i)Z_i(X_i)S^{(1)}(\beta,X_i)/S^{(0)}(\beta,X_i) = \sqrt{n}\sum_{i=1}^n \int_0^t W(t)Z_i(t)S^{(1)}(\beta,t)/S^{(0)}(\beta,t) dN_i(t)$, i.e., the sum of two quantities which we write as the sum of A_1 and A_2 where $A_1 = n^{-1/2}\sum_{i=1}^n \int_0^1 w(t)Z_i(t)s^{(1)}(\beta,t)/s^{(0)}(\beta,t)dN_i(t)$ and $A_2 = n^{-1/2}\sum_{i=1}^n \int_0^1 nW(t)S^{(1)}(\beta,t)/S^{(0)}(\beta,t) - w(t)s^{(1)}(\beta,t)/s^{(0)}(\beta,t)Z_i(t) dN_i(t)$ itself in turn the sum of two quantities which we sum A_3 and A_4. The expression for the first of these is $A_3 = \int_0^1 \sqrt{n}(nW(t)S^{(1)}(\beta,t)/S^{(0)}(\beta,t) - w(t)s^{(1)}(\beta,t)/s^{(0)}(\beta,t))S^{(1)}(\beta_0,t)\lambda_0(t)dt$. Next, this expression can be seen to be equivalent to the integral, $\int_0^1 \sqrt{n}\{nW(t) - w(t)\}S^{(1)}(\beta,t)/S^{(0)}(\beta,t)S^{(1)}(\beta_0,t)\lambda_0(t)dt + \int_0^1 \sqrt{n}(S^{(1)}(\beta,t)/S^{(0)}(\beta,t) - s^{(1)}(\beta,t)/s^{(0)}(\beta,t))w(t)S^{(1)}(\beta_0,t)\lambda_0(t)dt$ and the second term $A_4 = \int_0^1 \sqrt{n}(nW(t)S^{(1)}(\beta,t)/S^{(0)}(\beta,t) - w(t)s^{(1)}(\beta,t)/s^{(0)}(\beta,t))1/n\sum_{i=1}^n Z_i(t)dM_i(t) = o_p(1)$.

Large sample normality of R^2 and $R_{\mathcal{E}}^2$

We write, $A_5 = \int_0^1 \sqrt{n}\{nW(t) - w(t)\}s^{(1)}(\beta, t)/s^{(0)}(\beta, t)s^{(1)}(\beta_0, t)\lambda_0(t)$ $dt = \int_0^1 \sqrt{n}\,(\hat{S}(t)/s^{(0)}(0, t) - S(t)S^{(0)}(0, t)/s^{(0)}(0, t)^2)s^{(1)}(\beta, t)/s^{(0)}(\beta, t)$ $s^{(1)}(\beta_0, t)\lambda_0(t)dt + o_p(1) = n^{-1/2}\sum_{i=1}^n \int_0^1 (\xi_i(t)/s^{(0)}(0, t) - S(t)Y_i(t)/s^{(0)}$ $(0, t)^2)s^{(1)}(\beta, t)/s^{(0)}(\beta, t)s^{(1)}(\beta_0, t)$ $\quad \lambda_0(t)dt + o_p(1)$. The last two equalities arise from an application of Stute's (1995) central limit theorem for Kaplan-Meier integrals. Here the ξ's are i.i.d., each being a function of X_i and δ_i. $A_6 = \int_0^1 \sqrt{n}(S^{(1)}(\beta, t)S^{(0)}(\beta, t) - s^{(1)}(\beta, t)/s^{(0)}$ $(\beta, t))w(t)s^{(1)}(\beta_0, t)\lambda_0(t)dt = \int_0^1 \sqrt{n}(S^{(1)}(\beta, t)/s^{(0)}(\beta, t) - s^{(1)}(\beta, t)S^{(0)}$ $(\beta, t)/s^{(0)}(\beta, t)^2)w(t)s^{(1)}(\beta_0, t)\lambda_0(t)dt + o_p(1) = n^{-1/2}\sum_{i=1}^n \int_0^1 Y_i(t)\exp$ $\{\beta Z_i(t)\}/s^{(0)}(\beta, t)(Z_i(t) - s^{(1)}(\beta, t)/s^{(0)}(\beta, t))w(t)s^{(1)}(\beta_0, t)\lambda_0(t)dt +$ $o_p(1)$. The second last equality is derived in Xu (1996). Note also the connection to the techniques used in the appendix of Lin and Wei (1989). Combining A_1, A_5 and A_6, we have that $\sqrt{n}\sum_{i=1}^n \delta_i W(X_i)Z_i$ $(X_i)S^{(1)}(\beta, X_i)/S^{(0)}(\beta, X_i) = n^{-1/2}\sum_1^n \phi_i(\beta) + o_p(1), \phi_i(\beta) = \int_0^1 w(t)$ $Z_i(t)s^{(1)}(\beta, t)s^{(0)}(\beta, t)dN_i(t) + \int_0^1 ((\xi_i(t)s^{(0)}(0, t) - S(t)Y_i(t)s^{(0)}(0, t)^2)$ $s^{(1)}(\beta, t)s^{(0)}(\beta, t)). + Y_i(t)\exp\{\beta Z_i(t)\}s^{(0)}(\beta, t)(Z_i(t) - s^{(1)}(\beta, t)s^{(0)}$ $(\beta, t))w(t))s^{(1)}(\beta_0, t)\lambda_0(t)dt$. Now $\sqrt{n}\int_0^1 S^{(1)}(\beta, t)^2 S^{(0)}\ (\beta, t)^2 d\hat{F}(t) =$ $\sqrt{n}\int_0^1 s^{(1)}(\beta, t)^2 s^{(0)}(\beta, t)^2 d\hat{F}(t) + \sqrt{n}\int_0^1 (S^{(1)}(\beta, t)^2 S^{(0)}(\beta, t)^2 - s^{(1)}$ $(\beta, t)^2 s^{(0)}(\beta, t)^2)d\{\hat{F}(t) - F(t)\} + \sqrt{n}\int_0^1 (S^{(1)}(\beta, t)^2 S^{(0)}(\beta, t)^2 - s^{(1)}$ $(\beta, t)^2 s^{(0)}(\beta, t)^2)\ dF(t)$. The first term on the right hand side of is asymptotically equivalent to a sum of i.i.d. random variables namely, $n^{-1/2}\sum_1^n \zeta_i(\beta)$, following Stute (1995). The second term is $o_p(1)$. The third term is equal to $\sqrt{n}\int_0^1 (S^{(1)}(\beta, t)/S^{(0)}(\beta, t) -$ $s^{(1)}(\beta, t)/s^{(0)}(\beta, t))(S^{(1)}(\beta, t)/S^{(0)}(\beta, t) + s^{(1)}(\beta, t)/s^{(0)}(\beta, t))dF(t) =$ $\sqrt{n}\int_0^1 (S^{(1)}(\beta, t)/S^{(0)}(\beta, t) - s^{(1)}(\beta, t)/s^{(0)}(\beta, t))2s^{(1)}(\beta, t)/s^{(0)}(\beta, t)$ $dF(t) + o_p(1) = n^{-1/2}\sum_{i=1}^n \int_0^1 Y_i(t)\exp\{\beta Z_i(t)\}/s^{(0)}(\beta, t)(Z_i(t) -$ $s^{(1)}(\beta, t)/s^{(0)}(\beta, t))2s^{(1)}(\beta, t)/s^{(0)}(\beta, t)dF(t) + o_p(1)$. The second last equality above follows exactly the same way as for A_6. So we can see the asymptotic equivalence to $n^{-1/2}\sum_1^n \psi_i(\beta)$, where $\psi_i(\beta) = \zeta_i(\beta) +$ $\int_0^1 (Y_i(t)\exp\{\beta Z_i(t)\})/s^{(0)}(\beta, t)(Z_i(t) - s^{(1)}(\beta, t)/s^{(0)}(\beta, t))2s^{(1)}(\beta, t)/$ $s^{(0)}(\beta, t)dF(t)$. Then we have $\sum_{i=1}^n \delta_i W(X_i)\{r_i^2(0) - r_i^2(\beta_0)\} =$ $n^{-1/2}\sum_{i=1}^n \{2\phi_i(\beta_0) - 2\phi_i(0) + \psi_i(0) - \psi_i(\beta_0)\} + o_p(1)$. Therefore the asymptotic normality of $\sqrt{n}\{R^2(\beta_0) - \Omega^2(\beta_0)\}$ and thus $\sqrt{n}\{R^2(\hat{\beta}) - \Omega^2(\beta_0)\}$ follows from the central limit theorem. The asymptotic normality of $R_{\mathcal{E}}^2(\hat{\beta})$ can be shown similarly. A detailed proof can be found in Xu (1996).

Monotonicity convergence result (Xu 1996)

The main idea which is used frequently in many of the proofs leans on the following lemma (the proof of the lemma follows from manipulation of elementary probability inequalities).

Lemma 16.1 *Let β be real-valued. Statistics $T_n(\beta)$ converge to their population parameter $\theta(\beta)$ in probability as $n \to \infty$, and $T_n(\beta)$ increases with β for each fixed sample size n. Then $\theta(\beta)$ is also an increasing function of β.*

To see this we take $\beta_1 < \beta_2$. For arbitrary $\varepsilon > 0$ and $\delta > 0$, there exists n such that $P(|T_n(\beta_1) - \theta(\beta_1)| > \varepsilon) < \delta$ (1) $P(|T_n(\beta_2) - \theta(\beta_2)| > \varepsilon) < \delta$ (2). Let A, B be the sets on which (1) and (2) do not hold, respectively, i.e. $A = \{|T_n(\beta_1) - \theta(\beta_1)| \leq \varepsilon\}, B = \{|T_n(\beta_2) - \theta(\beta_2)| \leq \varepsilon\}$. Then on $A \cap B$, $\theta(\beta_2) - \theta(\beta_1) = \theta(\beta_2) - T_n(\beta_2) - \{\theta(\beta_1) - T_n(\beta_1)\} + T_n(\beta_2) - T_n(\beta_1) \geq -\varepsilon - \varepsilon + 0 = -2\varepsilon$. And $P(A \cap B) = P(A) + P(B) - P(A \cup B) \geq 1 - \delta + 1 - \delta - 1 = 1 - 2\delta > 0$ if we choose $\delta < 1/2$. But $\theta(\beta_2) - \theta(\beta_1)$ is non-random, and $\varepsilon > 0$ is arbitrary, which shows that $\theta(\beta_2) - \theta(\beta_1) \geq 0$. For $J(\beta, 0)$, $\sum_1^n \delta_i W(X_i) r_i^2(0)$, which does not involve β, is a consistent estimator. Then according to the lemma, $J(\beta, 0)$ is not affected by the change in β. For $J(\beta, 0) - J(\beta, \beta)$, the numerator of $\Omega^2(\beta)$, $\sum_1^n \delta_i W(X_i) \{\mathcal{E}_\beta(Z|X_i) - \mathcal{E}_0(Z|X_i)\}^2$ provides a consistent estimate, and $\partial/\partial\beta \sum_{i=1}^n \delta_i W(X_i) \{\mathcal{E}_\beta(Z|X_i) - \mathcal{E}_0(Z|X_i)\}^2 = 2\sum_{i=1}^n \delta_i W(X_i) V_\beta(Z|X_i) \{\mathcal{E}_\beta(Z|X_i) - E_0(Z|X_i)\}.(A.8)$. $V_\beta(Z|t)$ is non-negative and the derivative of $\mathcal{E}_\beta(Z|t) - \mathcal{E}_0(Z|t)$ with respect to β is again $V_\beta(Z|t)$, therefore $\mathcal{E}_\beta(Z|t) - \mathcal{E}_0(Z|t)$ is an increasing function of β. So the difference will be ≥ 0 when $\beta > 0$ and ≤ 0 when $\beta < 0$. This shows that $\sum_1^n \delta_i W(X_i) \{\mathcal{E}_\beta(Z|X_i) - \mathcal{E}_0(Z|X_i)\}^2$ increases with $|\beta|$, and so does $J(\beta, 0) - J(\beta, \beta)$, thus completing the proof.

Lemma 14.4

The proof of the main theorem implies that $\hat{\Gamma}_2(\beta)$ converges in probability to $\Gamma_2(\beta)$ for every β: For the purpose of showing monotonicity of $\rho^2(\beta)$, in the next theorem, without loss of generality, we may assume that there is no censoring when computing $\hat{\Gamma}(\beta)$, since the population parameter $\rho^2(\beta)$ is not affected by censorship. Next, take the derivative of $\hat{\Gamma}_2(\beta)$ with respect to β, $\partial\hat{\Gamma}_2(\beta)/\partial\beta = 2\sum_{i=1}^k P(t_i)(E_i + \beta$

$V_i - \sum_1^n Y_l(t_i) Z_l \exp(\beta Z_l) / \sum_1^n Y_l(t_i) \exp(\beta Z_l))$, where $E_i = \sum_1^n Y_l(t_i)$ $Z_l \exp(\beta Z_l) / \sum_1^n Y_l(t_i) \exp(\beta Z_l)$, $V_i = E_i^2 - (E_i)^2$, $E_i^2 = \sum_1^n Y_l(t_i) Z_l^2$ $\exp(\beta Z_l) / \sum_1^n Y_l(t_i) \exp(\beta Z_l)$. V_i is the variance of Z under the probability distribution $\{\pi_j(\beta, X_i)\}$, therefore non-negative.

Large sample results for explained randomness measure

The consistency follows from the consistency of $\hat{\beta}$, the main theorem and the consistency of the Kaplan-Meier estimate. To show the asymptotic normality, note that $\hat{\beta} - \beta_0 = I^{-1}(\check{\beta}) U(\beta_0)$ where $\check{\beta}$ is on the line segment between $\hat{\beta}$ and β_0, $U(\beta) = \partial \log L(\beta) / \partial \beta$, $L(\beta)$ is the log partial likelihood, and $I(\beta) = -\partial U(\beta) / \partial \beta$. Then $\sqrt{n} \hat{\Gamma}_2(\hat{\beta}) = \sqrt{n} \hat{\Gamma}_2(\beta_0) + I^{-1}(\check{\beta}) \sqrt{n} U(\beta_0) \partial \hat{\Gamma}_2(\beta) / \partial \beta$ evaluated at $\beta = \dot{\beta}$. It is known from for example, Andersen and Gill (1982), that $I(\check{\beta})$ converges in probability to the expected information under the model, $\partial \hat{\Gamma}_2(\beta) / \partial \beta$ then converges in probability to $2 \int \beta v(\beta, t) dF(t)$. Lin and Wei (1989) showed that $\sqrt{n} U(\beta_0)$ is asymptotically equivalent to $n^{-1/2}$ times a sum of i.i.d. random variables:

Theorem 16.1 *(Lin and Wei 1989)* $\sqrt{n} U(\beta_0)$ *is asymptotically equivalent to* $n^{-1/2} \sum_1^n \omega_i(\beta_0)$, *where* $\omega_i(\beta) = \int_0^1 (Z_i - s^{(1)}(\beta, t) / s^{(0)}(\beta, t))$ $dN_i(t) - \int_0^1 Y_i(t) e^{\beta Z_i} (Z_i - s^{(1)}(\beta, t) / s^{(0)}(\beta, t)) \lambda_0(t) dt$ *and* $N_i(t) = I\{T_i \leq t, T_i \leq C_i\}$.

So all that remains to show the asymptotic normality of $\hat{\Gamma}_2(\hat{\beta})$ is to show that $\sqrt{n} \hat{\Gamma}_2(\beta_0)$ is also asymptotically equivalent to $n^{-1/2}$ times a sum of n i.i.d. random variables like the above. Let $\Psi(t) = 2(\beta S^{(1)}(\beta, t) / S^{(0)}(\beta, t) - \log S^{(0)}(\beta, t) / S^{(0)}(0, t))$, and $\psi(t) = 2(\beta s^{(1)}(\beta, t) / s^{(0)}(\beta, t) - \log s^{(0)}(\beta, t) / s^{(0)}(0, t))$. Denote $\hat{F}$ the Kaplan-Meier estimate of F. Next we write, $\sqrt{n} \hat{\Gamma}_2(\beta) = \sqrt{n} \int \Psi(t) d\hat{F}(t) = \sqrt{n} \int \psi(t) d\hat{F}(t) + \sqrt{n} \int \{\Psi(t) - \psi(t)\} d\{\hat{F}(t) - F(t)\} + \sqrt{n} \int \{\Psi(t) - \psi(t)\} dF(t)$. Now $\sqrt{n} \{\hat{F}(t) - F(t)\}$ converges in distribution to a zero-mean Gaussian process. Therefore the second term on the last line is $o_p(1)$. We apply the central limit theorem under random censorship of Stute (1995) to the first term above. We state the version of the theorem that will apply directly to our cases here.

Theorem 16.2 *(Stute 1995) Let Q and H be the distribution function of C and $X = \min(T, C)$, respectively. Define $H^0(x) = P(X < x,$*

$\delta = 0$), $H^1(x) = P(X < x, \delta = 1)$ and $\tau_H = \inf\{x : H(x) = 1\}$ $\leq \infty$. Let $\psi : \mathcal{R} \to \mathcal{R}$ be any bounded measurable function. Assume that $1 - F \sim c(1 - Q)^\alpha$ in a neighborhood of τ_H, for some $c > 0$ and $\alpha > 1$. Then $\int \psi d\hat{F} = n^{-1} \sum_{i=1}^n \delta_i \gamma_0(X_i) \psi(X_i) + n^{-1} \sum_{i=1}^n (1 - \delta_i) \gamma_1(\psi(\cdot), X_i) - n^{-1} \sum_{i=1}^n \gamma_2(\psi(\cdot), X_i) + R_n$, where $|R_n| = o_p(n^{-1/2})$, $\gamma_0(x)$. In turn we can write this expression as, $\exp(\int_{-\infty}^{x^-} H^0(dy)/1 - H(y))$, $\gamma_1(\psi(\cdot), x)$ which can then be expressed as, $1/(1 - H(x)) \int 1_{\{x < w\}} \psi(w) \gamma_0(w) H^1(dw)$, $\gamma_2(\psi(\cdot), x)$ which is then written as, $\int \int 1_{\{v < x, v < w\}} \psi(w) \gamma_0(w)/([1 - H(v)]^2) H^0(dv) H^1(dw)$.

Note, as discussed by Stute (1995), that the boundedness of ψ and the assumption on the tails of F and Q ensures the convergence rate of $n^{-1/2}$. More general conditions are also given in Stute (1995). The last term above is the difference of the following two terms: $\sqrt{n} \int 2\beta(S^{(1)}(\beta, t)/S^{(0)}(\beta, t) - s^{(1)}(\beta, t)/s^{(0)}(\beta, t)) dF(t) = \sqrt{n} \int 2\beta(S^{(1)}(\beta, t)/s^{(0)}(\beta, t) - s^{(1)}(\beta, t) S^{(0)}(\beta, t)/s^{(0)}(\beta, t)^2) dF(t) + o_p(1) = n^{-1/2} \sum_{i=1}^n \int 2\beta Y_i(t) e^{\beta Z_i}/s^{(0)}(\beta, t)(Z_i - s^{(1)}(\beta, t)/s^{(0)}(\beta, t)) dF(t) + o_p(1)$, and $\sqrt{n} \int 2(\log S^{(0)}(\beta, t)/S^{(0)}(0, t) - \log s^{(0)}(\beta, t)/s^{(0)}(0, t)) dF(t) = \sqrt{n} \int 2s^{(0)}(0, t)/s^{(0)}(\beta, t)(S^{(0)}(\beta, t)/S^{(0)}(0, t) - s^{(0)}(\beta, t)/s^{(0)}(0, t)) dF(t) + o_p(1) = \sqrt{n} \int 2s^{(0)}(0, t)/s^{(0)}(\beta, t)(S^{(0)}(\beta, t)/s^{(0)}(0, t) - s^{(0)}(\beta, t) S^{(0)}(0, t)/s^{(0)}(0, t)^2) dF(t) + o_p(1) = n^{-1/2} \sum_{i=1}^n \int 2Y_i(t)/s^{(0)}(\beta, t)(e^{\beta Z_i} - s^{(0)}(\beta, t)/s^{(0)}(0, t)) dF(t) + o_p(1)$. Combining the above we have that $\sqrt{n} \hat{\Gamma}_2(\beta_0)$ is equal to $n^{-1/2}$ times a sum of n i.i.d. random variables plus a term of order $o_p(1)$. The asymptotic normality of $\hat{\Gamma}_2(\hat{\beta})$ thus follows from the central limit theorem.

Bibliography

[1] Aalen, O. (1978) Non-parametric inference for a family of counting processes. *Ann. Statist.* **6**: 701-26.

[2] Aalen O. (1987) Two examples of modelling heterogeneity in survival analysis. *Scand. J. Statist.*; **14**: 19-25.

[3] Aalen, O. (1989) A linear regression model for the analysis of life times. *Statist. Med.*; **8**: 907-25.

[4] Aalen, O. (1994) On the analysis of life tables for dependent observations. *Statist. Med.*; **13**: 2383-4.

[5] Agresti, A. and Min, Y. (2001) On small-sample confidence intervals for parameters in discrete distributions. *Biometrics*; **57**: 963-71.

[6] Ahn, H. and Loh, W. Y. (1994) Tree-structured proportional hazards regression modeling. *Biometrics*; **50**: 471-85.

[7] Ahnn, S. and Anderson, S. J. (1995) Sample size determination for comparing more than two survival distributions. *Statist. Med.* **14**: 2273-82.

[8] Akazawa, K., Nakamura, T. and Palesch, Y. (1997) Power of logrank test and Cox regression model in clinical trials with heterogeneous samples. *Statist. Med.*; **16**: 583-97.

[9] Akritas, M. G. and LaValley, M. P. (1996) Nonparametric inference in factorial designs with censored data. *Biometrics*; **52** (3): 913-24.

[10] Albert, J. M., Ioannidis, J. P., Reichelderfer, P., Conway, B., Coombs, R. W., Crane, L., Demasi, R., Dixon, D. O., Flandre, P., Hughes, M. D., Kalish, L. A., Larntz, K., Lin, D., Marschner, I. C., Munoz, A., Murray, J., Neaton, J., Pettinelli, C., Rida, W., Taylor, J. M. and Welles, S. L. (1998) Statistical issues for HIV surrogate endpoints: point/counterpoint. An NIAID workshop. *Statist. Med.*; **17**: 2435-62.

[11] Alioum, A. and Commenges, D. (1996) A proportional hazards model for arbitrarily censored and truncated data. *Biometrics*; **52**: 512-24.

[12] Allison, P. D. (1995) *Survival Analysis Using the SAS System*, SAS Institute Inc.

[13] Altman, D. G. and De Stavola, B. L. (1994) Practical problems in fitting a proportional hazards model to data with updated measurements of the covariates. *Statist. Med.*; **13**: 301-41.

[14] Andersen, P. K. and Gill, R. D. (1982) Cox's regression model for counting processes: a large sample study. *Ann. Statist.*; **10**: 1100-1120.

[15] Andersen, P. K. (1983) Comparing survival distributions via hazard ratio estimates. *Scand. J. Statist.*; **10**: 77-85.

[16] Andersen, P. K., Christensen, E., Fauerholdt, L. and Schlichting, P. (1983) Evaluating prognoses based on the proportional hazards model. *Scand. J. Statist.*; **10**: 141-144.

[17] Andersen, P. K. (1991) Survival analysis 1982-1991: the second decade of the proportional hazards regression model. *Stat Med*; **10**: 1931-41.

[18] Andersen, P. K., Borgan, O., Gill, R. D. and Keiding, N. (1993) Statistical models based on counting processes. *Springer-Verlag*. New York.

[19] Andersen, J., Goetghebeur, E. and Ryan, L. (1996) Missing cause of death information in the analysis of survival data. *Statist. Med.*; **15**: 2191-201.

[20] Andersen, P. K., Klein, J. P., Knudsen, K. M., Tabanera Y. and Palacios, R. (1997) Estimation of variance in Cox's regression model with shared gamma frailties. *Biometrics*; **53**: 1475-84.

[21] Andersen, P. K., Klein, J. P. and Zhang, M. J. (1999) Testing for centre effects in multi-centre survival studies: a Monte Carlo comparison of fixed and random effects tests. *Stat Med*; **18**: 1489-500.

[22] Andersen, P. K., Esbjerg, S. and Sorensen, T. I. (2000) Multi-state models for bleeding episodes and mortality in liver cirrhosis. *Statist. Med.*; **19**: 587-99.

[23] Andersen, P. K. and Liestol, K. (2003) Attenuation caused by infrequently updated covariates in survival analysis. *Biostatistics*; **4**: 633-49.

[24] Anderson, J. A. and Senthilselvan, A. (1980) Smooth estimates for the hazard function. *J. Roy. Statist. Soc. Ser. B*; **42**: 322-327.

[25] Anderson, J. A. and Senthilselvan, A. (1982). A two-step regression model for hazard functions. *Applied Statistics*; **31**: 44-51.

[26] Anderson, K. M. (1991) A nonproportional hazards Weibull accelerated failure time regression model. *Biometrics*; **47**: 281-8.

[27] Annesi, I., Moreau, T. and Lellouch, J. (1989) Efficiency of the logistic regression and Cox proportional hazards models in longitudinal studies. *Statist. Med.*; **8**: 1515-21.

[28] Aranda-Ordaz and Francisco J. (1983) An extension of the proportional-hazards model for grouped data. *Biometrics*; **39**: 109-117.

[29] Arjas, E. and Liu, L. (1996) Non-parametric Bayesian approach to hazard regression: a case study with a large number of missing covariate values. *Statist. Med.*; **15**: 1757-70.

[30] Aydemir, U., Aydemir, S. and Dirschedl, P. (1999) Analysis of time-dependent covariates in failure time data. *Statist. Med.*; **18**: 2123-34.

[31] Bacchetti, P. and Jewell, N. P. (1991) Nonparametric estimation of the incubation period of AIDS based on a prevalent cohort with unknown infection times. *Biometrics*; **47**: 947-60.

[32] Bacchetti, P. and Quale, C. (2002) Generalized additive models with interval-censored data and time-varying covariates: application to human immunodeficiency virus infection in hemophiliacs. *Biometrics*; **58**: 443-7.

[33] Bagdonavicius, Vilijandas B. and Nikulin, M. S. (1999) Generalized proportional hazards model based on modified partial likelihood. *Lifetime Data Anal.*; **5**: 329-350.

[34] Bagdonavicius, Vilijandas and Nikulin, M. S. (2001) On goodness-of-fit for accelerated life models. *C. R. Acad. Sci. Paris Ser. I Math.*; **332**: 171-176.

[35] Baltazar-Aban, I. and Pena, E. A. (1995) Properties of hazard-based residuals and implications in model diagnostics. *J. Amer. Statist. Assoc.*; **90**: 185-197.

[36] Barlow, W. E. and Sun, W. H. (1989) Bootstrapped confidence intervals for the Cox model using a linear relative risk form. *Statist. Med.*; **8**: 927-35.

[37] Barlow, W. E. (1997) Global measures of local influence for proportional hazards regression models. *Biometrics*; **53**: 1157-62.

[38] Barndorff-Nielsen, O. and Cox, D. R. (1994) *Inference and Asymptotics*; Chapman and Hall, London.

[39] Bartle, R. (1976). *Elements of Real Analysis.* New York: Wiley.

[40] Bartoszynski, R., Brown, B. W., McBride, C. M. and Thompson, J. R. (1981) Some nonparametric techniques for estimating the intensity function of a cancer related nonstationary Poisson process. *Ann. Statist.*; **9** (5): 1050-1060.

[41] Bebchuk, J. D. and Betensky, R. A. (2000) Multiple imputation for simple estimation of the hazard function based on interval censored data. *Statist. Med.*; **19**: 405-19.

[42] Begun, J. M., Hall, W. J., Huang, W. and Wellner, J. A. (1983) Information and asymptotic efficiency in parametric-nonparametric models. *Ann. Statist.*; **11**: 432-452.

[43] Begun, J. M. and Reid, N. (1983) Estimating the relative risk with censored data. *J. Amer. Statist. Assoc.*; **78**: 337-341.

[44] Beirlant, J., Carbonez, A. and van der Meulen, E. (1992) Long run proportional hazards models of random censorship. *J. Statist. Plann. Inference*; **32**: 25-44.

[45] Benichou, J. and Gail, M. H. (1990) Estimates of absolute cause-specific risk in cohort studies. *Biometrics*; **46**: 813-26.

[46] Berridge, D. M. (1996) An application of a marked point process in pre-clinical medicine. *Statist. Med.*; **15**: 2751-62.

[47] Berry, G., Kitchin, R. M. and Mock, P. A. (1991) A comparison of two simple hazard ratio estimators based on the logrank test. *Statist. Med.*; **10**: 749-55.

[48] Berzuini, C. and Clayton, D. (1994) Bayesian analysis of survival on multiple time scales. *Statist. Med.*; **13**: 823-38.

[49] Betensky, R. A. (1997) Conditional power calculations for early acceptance of H0 embedded in sequential tests. *Statist. Med.*; **16**: 465-77.

[50] Betensky, R. A. (1998) Multiple imputation for early stopping of a complex clinical trial. *Biometrics*; **54**: 229-42.

[51] Betensky, R. A., Lindsey, J. C., Ryan, L. M. and Wand, M. P. (1999) Local EM estimation of the hazard function for interval-censored data. *Biometrics*; **55**: 238-45.

[52] Betensky, R. A., Lindsey, J. C., Ryan, L. M. and Wand, M. P. (2002) A local likelihood proportional hazards model for interval censored data. *Statist. Med.*; **21**: 263-75.

[53] Billingsley, P. (1968) *Convergence of Probability Measures*; New York. Wiley.

[54] Binder, D. A. (1992) Fitting Cox's proportional hazards models from survey data. *Biometrika*; **79**: 139-147.

[55] Bonetti, M. and Gelber, R. D. (2000) A graphical method to assess treatment-covariate interactions using the Cox model on subsets of the data. *Statist. Med.*; **19**: 2595-609.

[56] Borgan, O., Goldstein, L. and Langholz, B. (1995) Methods for the analysis of sampled cohort data in the Cox proportional hazards model. *Ann. Statist.*; **23**: 1749-1778.

[57] Borgan, O. and Liestol, K. (1990) A note on confidence intervals and bands for the survival function based on transformations. *Scand. J. Statist.* **17**: 35-41.

[58] Braekers, R. and Veraverbeke, N. (2003) Testing for the partial Koziol-Green model with covariates. *J. Statist. Plann. Inference*; **115**: 181-192.

[59] Breslow, N. (1974) Covariance analysis of censored survival data. *Biometrics* **30**: 89-99.

[60] Breslow, N. and Crowley, J. (1974) A large sample study of the life table and product limit estimates under random censorship. *Ann. Statist.*; **2**: 437-53.

[61] Breslow, N., Edler, L. and Berger, J. (1984) A two-sample censored data rank test for acceleration. *Biometrics* **40**: 1049-62.

[62] Bretagnolle, J. and Huber-Carol, C. (1988). Effects of Omitting Covariates in Cox's Model for Survival Data. *Scandinavian J. of Statist.*; **15**: 125-138.

[63] Bristol, D. R. (1992) The analysis of failure time data in crossover studies. *Statist. Med.*; **11**: 975-7.

[64] Broet, P., Moreau, T., Lellouch, J. and Asselain, B. (1999) Unobserved covariates in the two-sample comparison of survival times: a maximum efficiency robust test. *Statist. Med.*; **18**: 1791-800.

[65] Broet, P., De Rycke, Y., Tubert-Bitter, P., Lellouch, J., Asselain, B. and Moreau, T. (2001) A semiparametric approach for the two-sample comparison of survival times with long-term survivors. *Biometrics*; **57**: 844-52.

[66] Brookmeyer, R. and Gail, M. (1987) Biases in prevalent cohorts. *Biometrics*; **43**: 739-749.

[67] Brown, C. C. (1975) On the use of indicator variables for studying the time-dependence of parameters in a response-time model. *Biometrics*; **31**: 863-72.

[68] Bryant, J. and Dignam, J. J. (2004) Semiparametric models for cumulative incidence functions. *Biometrics*; **60**: 182-90.

[69] Bunday, B. D. (1991) Statistical methods in reliability theory and practice. *Ellis Horwood*. New York.

[70] Burr, D. (1994) A comparison of certain bootstrap confidence intervals in the Cox model. *J. Amer. Statist. Assoc.*; **89**: 1290-1302.

[71] Burr, D. (1994) On inconsistency of Breslow's estimator as an estimator of the hazard rate in the Cox model. *Biometrics*; **50**: 1142-5.

[72] Burridge, J. (1981) Empirical Bayes analysis of survival time data. *J. Roy. Statist. Soc. Ser. B*; **43**: 65-75.

[73] Buyse, M. (1989) Analysis of clinical trial outcomes: some comments on subgroup analyses. *Control Clin Trials*; **10** (4 Suppl): 187S-194S.

[74] Buyse, M. and Molenberghs, G. (1998). Criteria for the validation of surrogate endpoints in randomized experiments. *Biometrics*; **54**: 1014-29.

[75] Bycott, P. W. and Taylor, J. M. (1998) An evaluation of a measure of the proportion of the treatment effect explained by a surrogate marker. *Control Clin Trials*; **19**: 555-68.

[76] Cai, J., Zhou, H. and Davis, C. E. (1997) Estimating the mean hazard ratio parameters for clustered survival data with random clusters. *Statist. Med.*; **16**: 2009-20.

[77] Cai, T., Wei, L. J. and Wilcox, M. (2000) Semiparametric regression analysis for clustered failure time data. *Biometrika*; **87**: 867-878.

[78] Cai, T., Cheng, S. C. and Wei, L. J. (2002) Semiparametric mixed-effects models for clustered failure time data. *J. Amer. Statist. Assoc.*; **97**: 514-522.

[79] Cai, T. and Betensky, R. A. (2003) Hazard regression for interval-censored data with penalized spline. *Biometrics*; **59**: 570-9.

[80] Campbell, M. K., Donner, A. and Webster, K. M. (1991) Are ordinal models useful for classification? *Statist. Med.*; **10**: 383-94.

[81] Carlin, B. P. and Hodges, J. S. (1999) Hierarchical proportional hazards regression models for highly stratified data. *Biometrics*; **55**: 1162-70.

[82] Carling, K. and Jacobson, T. (1995) Modeling unemployment duration in a dependent competing risks framework: identification and estimation. *Lifetime Data Anal.*; **1**: 111-122.

[83] Carroll, K. J. (2003) On the use and utility of the Weibull model in the analysis of survival data. *Control Clin Trials*; **24**: 682-701.

[84] Chang, I. S. and Hsiung, C. A. (1994) Information and asymptotic efficiency in some generalized proportional hazards models for counting processes. *Ann. Statist.*; **22**: 1275-1298.

[85] Chang, I. S. and Hsiung, C. A. (1996) An efficient estimator for proportional hazards models with frailties. *Scand. J. Statist.*; **23**: 13-26.

[86] Chang, M. N. (1996) Exact distribution of the Kaplan-Meier estimator under the proportional hazards model. *Statist. Probab. Lett.*; **28**: 153-157.

[87] Chang, I. S., Hsiung, C. A. and Chuang, Y. (1997) Applications of a frailty model to sequential survival analysis. *Statist. Sinica*; **7**: 127-138.

[88] Chang, C. C. and Weissfeld, L. A. (1999) Normal approximation diagnostics for the Cox model. *Biometrics*; **55**: 1114-9.

[89] Chang, I-Shou, Hsiung, C. A. and Wu, S. (2000) Estimation in a proportional hazard model for semi-Markov counting process. *Statist. Sinica*; **10**: 1257-1266.

[90] Chen, C. H. and George, S. L. (1985) The bootstrap and identification of prognostic factors via Cox's proportional hazards regression model. *Statist. Med.*; **4**: 39-46.

[91] Chen, C. H. and Wang, P. C. (1991) Diagnostic plots in Cox's regression model. *Biometrics*; **47**: 841-50.

[92] Chen, Y. Q. and Wang, M. (2000) Estimating a treatment effect with the accelerated hazards models. *Control Clin Trials*; **21**: 369-80.

[93] Chen, Y. Q. and Jewell, N. P. (2001) On a general class of semi-parametric hazards regression models. *Biometrika*; **88**: 687-702.

[94] Chen, P. Y. and Tsiatis, A. A. (2001) Causal inference on the difference of the restricted mean lifetime between two groups. *Biometrics*; **57**: 1030-8.

[95] Chen, Y. Q. (2001) Accelerated hazards regression model and its adequacy for censored survival data. *Biometrics*; **57**: 853-60.

[96] Chen, K., Jin, Z. and Ying, Z. (2002) Semiparametric analysis of transformation models with censored data. *Biometrika*; **89**: 659-668.

[97] Chen, Y. Q., Rohde, C. A. and Wang, M. C. (2002) Additive hazards models with latent treatment effectiveness lag time. *Biometrika*; **89**: 917-931.

[98] Chen, K. (2004) Statistical estimation in the proportional hazards model with risk set sampling. *Ann. Statist.*; **32**: 1513-1532.

[99] Cheng, P. E. and Lin, G. D. (1987) Maximum likelihood estimation of a survival function under the Koziol-Green proportional hazards model. *Statist. Probab. Lett.*; **5**: 75-80.

[100] Cheng, S. C., Wei, L. J. and Ying, Z. (1997) Predicting survival probabilities with semiparametric transformation models. *J. Amer. Statist. Assoc.*; **92**: 227-235.

[101] Cheng, S. C., Fine, J. P. and Wei, L. J. (1998) Prediction of cumulative incidence function under the proportional hazards model. *Biometrics*; **54**: 219-228.

[102] Cheung, Y. B., Yip, P. S. and Karlberg, J. P. (2001) Parametric modelling of neonatal mortality in relation to size at birth. *Statist. Med.*; **20**: 2455-66.

[103] Chevret, S., Leporrier, M. and Chastang, C. (2000) Measures of treatment effectiveness on tumour response and survival: a multi-state model approach. *Statist. Med.*; **19**: 837-48.

[104] Chiang, C. (1968) *Introduction to Stochastic Processes in Biostatistics*; New York. Wiley.

[105] Clayton, D. and Cuzick, J. (1985) Multivariate generalizations of the proportional hazards model. *J. Roy. Statist. Soc. Ser. A*; **148**: 82-117.

[106] Clayton, D. and Cuzick, J. (1985) The semiparametric Pareto model for regression analysis of survival times. *Proceedings of the 45th session of the International Statistical Institute, Vol. 4 (Amsterdam, 1985)*; **51**: 175-180.

[107] Clegg, L. X., Cai, J. and Sen, P. K. (1999) A marginal mixed baseline hazards model for multivariate failure time data. *Biometrics*; **55**: 805-12.

[108] Cnaan, A. and Ryan, L. (1989) Survival analysis in natural history studies of disease. *Statist. Med.*; **8**: 1255-68.

[109] Cochran, W. G. (1954) Some methods for strengthening the common chi-square tests. *Biometrics* **10**: 417-58.

[110] Cole, B. F., Gelber, R. D. and Goldhirsch, A. (1993) Cox regression models for quality adjusted survival analysis. *Statist. Med.*; **12**: 975-87.

[111] Colosimo, E. A. (1997) A note on the stratified proportional hazards model. *Int. J. Math. Stat. Sci.*; **6**: 201-209.

[112] Colosimo, E. A., Chalita, L. V. and Demetrio, C. G. (2000) Tests of proportional hazards and proportional odds models for grouped survival data. *Biometrics*; **56**: 1233-40.

[113] Commenges, D. Andersen, P. K. (1995) Score test of homogeneity for survival data. *Lifetime data analysis*; **1**: 145-156.

[114] Commenges, D., Letenneur, L., Joly, P., Alioum, A. and Dartigues, J. F. (1998) Modelling age-specific risk: application to dementia. *Statist. Med.*; **17**: 1973-88.

[115] Com-Nougue, C., Rodary, C. and Patte, C. (1993) How to establish equivalence when data are censored: a randomized trial of treatments for B non-Hodgkin lymphoma. *Statist. Med.*; **12**: 1353-64.

[116] Congdon, P. (1995) Modelling frailty in area mortality. *Statist. Med.*; **14**: 1859-74.

[117] Contal, C. and O'Quigley J. (1999) Evaluating the effect of age on survival in breast cancer using changepoint methods. *Comp. Statist. and Data Analysis*, **30**, 253-270.

[118] Coste, J., Walter, E., Wasserman, D. and Venot, A. (1997) Optimal discriminant analysis for ordinal responses. *Stat Med*; **16**: 561-9.

[119] Cox, D. R. (1958) Some problems connected with statistical inference. *Ann. Math. Statist.* **29**, 357-72.

[120] Cox, D. R. (1972) Regression models and life-tables. *J. Roy. Statist. Soc. Ser. B*; **34**: 187-220.

[121] Cox, D. R. (1972) Partial likelihood. *Biometrika*; **62**: 269-76.

[122] Cox, D. R. and Hinkley, D. (1974) *Theoretical Statistics*; Chapman and Hall; London.

[123] Cox, D. R. and Oakes, D. (1984) *Analysis of survival data*; Chapman and Hall; London.

[124] Cox, D. R. and Wermuth, N. (1992) A comment on the coefficient of determination for binary responses. *American Statist.* **46**; 1-4.

[125] Cox, C. (1995) Location-scale cumulative odds models for ordinal data: a generalized non-linear model approach. *Statist. Med.*; **14**: 1191-203.

[126] Cramer, H. (1937) *Random Variables and Probability Distributions*; Cambridge Texts in Mathematics. Cambridge. U.K.

[127] Crowder, M. J., Kimber, A. C., Smith, R. L. and Sweeting, T. J. (1991) *Statistical analysis of reliability data*. Chapman & Hall. London.

[128] Crowley, J., Liu, P. Y. and Voelkel, J. G. (1982) Estimation of the ratio of hazard functions. *Survival analysis (Columbus, Ohio, 1981)*; **2**: 56-73.

[129] Csorgo, S. (1988) Estimation in the proportional hazards model of random censorship. *Statistics*; **19**: 437-463.

[130] Csorgo, S. and Faraway, J. J. (1998) The paradoxical nature of the proportional hazards model of random censorship. *Statistics*; **31**: 67-78.

[131] Cui, J. (1999) Estimating AIDS incidence and jack-knife variance from a continuous delay distribution and incomplete data. *Statist. Med.*; **18**: 527-37.

[132] Cui, J. and Becker, N. G. (2000) Estimating HIV incidence using dates of both HIV and AIDS diagnoses. *Statist. Med.*; **19**: 1165-77.

[133] Cupples, L. A., Gagnon, D. R., Ramaswamy, R. and D'Agostino, R. B. (1995) Age-adjusted survival curves with application in the Framingham Study. *Statist. Med.*; **14**: 1731-44.

[134] D'Amico, F. and Rao, B. R. (1983) Exact maximum likelihood estimates of Cox's regression parameters based on categorized factorial data. *Biometrical J.*; **25**: 29-42.

[135] Dabrowska, D. M., Doksum, K. A., Feduska, N. J., Husing, R. and Neville, P. (1992) Methods for comparing cumulative hazard functions in a semi-proportional hazard model. *Statist. Med.*; **11**: 1465-76.

[136] Dabrowska, D. M. (1997) Smoothed Cox regression. *Ann. Statist.*; **25**: 1510-1540.

[137] Dafni, U. G. and Tsiatis, A. A. (1998) Evaluating surrogate markers of clinical outcome when measured with error. *Biometrics*; **54**: 1445-62.

[138] Daniels, H. E. (1954) Saddlepoint approximations in statistics. *Ann. Math. Statist.*; **25**: 631-50.

[139] Daniels, H. E. (1987) Tail probability approximations. *Int. Statist. Rev.*; **55**: 37-48.

[140] Datta, S., Satten, G. A. and Williamson, John M. (2000) Consistency and asymptotic normality of estimators in a proportional hazards model with interval censoring and left truncation. *Ann. Inst. Statist. Math.*; **52**: 160-172.

[141] Datta, S. and Satten, G. A. (2002) Estimation of integrated transition hazards and stage occupation probabilities for non-Markov systems under dependent censoring. *Biometrics*; **58**: 792-802.

[142] David, H. A. and Moeschberger, M. L. (1978) The theory of competing risks. *Macmillan Co.*. New York.

[143] Davies, R. B. (1977) Hypothesis testing when a nuisance parameter is present only under the alternative. *Biometrika* **64**; 247-254.

[144] Davies R. B. (1987) Hypothesis testing when a nuisance parameter is present only under the alternative. *Biometrika* **74**; 33-43.

[145] Dawson, R. and Lavori, P. W. (2002) Using inverse weighting and predictive inference to estimate the effects of time-varying treatments on the discrete-time hazard. *Statist. Med.*; **21**: 1641-61; discussion 1663-87.

[146] de Bruijne, M. H., le Cessie, S., Kluin-Nelemans, H. C. and van Houwelingen, H. C. (2001) On the use of Cox regression in the presence of an irregularly observed time-dependent covariate. *Statist. Med.*; **20**: 3817-29.

[147] Dear, K. B. (1994) Iterative generalized least squares for meta-analysis of survival data at multiple times. *Biometrics*; **50**: 989-1002.

[148] De Gruttola, V., Fleming, T., Lin, D. Y. and Coombs, R. (1997). Perspective: Validating surrogate markers - are we being naive? *Journal of Infectious Disease* **175**, 237-46.

[149] Deheuvels, P. (1996) Functional laws for small increments of empirical processes. *Statistica Neerlandica* **50**, 261-80.

[150] Deheuvels, P. (1997) Strong laws for local quantile processes. *Ann. Probab.* **25**, 2007-54.

[151] Deheuvels, P. and Einmahl, J. H. J. (2000) Functional limit laws for the increments of Kaplan-Meier product-limit processes and applications *Ann. Probab.* **28**, 1301-35.

[152] Dellaportas, P. and Smith, A. F. M. (1993) Bayesian inference for generalized linear and proportional hazards models via Gibbs sampling. *J. Roy. Statist. Soc. Ser. C*; **42**: 443-459.

[153] Desu, M. M. and Raghavarao, D. (2004) Nonparametric statistical methods for complete and censored data. *Chapman & Hall/CRC, Boca Raton.*

[154] Dewanji, A. and Sengupta, D. (2003) Estimation of competing risks with general missing pattern in failure types. *Biometrics*; **59**: 1063-70.

[155] DiCiccio, T. J. and Romano, J. P. (1988). A review of bootstrap confidence intervals *J. Roy. Statist. Soc. Series B*; **50**: 338-354.

[156] Dinse, G. E. (1994) A comparison of tumour incidence analyses applicable in single-sacrifice animal experiments. *Stat Med*; **13**: 689-708.

[157] DiRienzo, A. G. and Lagakos, S. W. (2001) Effects of model misspecification on tests of no randomized treatment effect arising from Cox's proportional hazards model. *J. R. Stat. Soc. Ser. B.*; **63**: 745-757.

[158] Dixon, D. O. and Simon, R. (1991) Bayesian subset analysis. *Biometrics*; **47**: 871-81.

[159] Dobson, A. J. (1988) Proportional hazards models for average data for groups. *Statist. Med.*; **7**: 613-8.

[160] Dobson, A. and Henderson, R. (2003) Diagnostics for joint longitudinal and dropout time modeling. *Biometrics*; **59**: 741-51.

[161] Doksum, K. A. and Nabeya, S. (1984) Estimation in proportional hazard and log-linear models. *J. Statist. Plann. Inference*; **9**: 297-303.

[162] Doksum, K. A. (1987) An extension of partial likelihood methods for proportional hazard models to general transformation models. *Ann. Statist.*; **15**: 325-345.

[163] Dpolhkabrowska, D. M. and Doksum, K. A. (1987) Estimates and confidence intervals for median and mean life in the proportional hazard model. *Biometrika*; **74**: 799-807.

[164] Dpolhkabrowska, D. M. and Doksum, Kjell A. (1988) Estimation and testing in a two-sample generalized odds-rate model. *J. Amer. Statist. Assoc.*; **83**: 744-749.

[165] Dpolhkabrowska, D. M., Doksum, K. A. and Song, J. (1989) Graphical comparison of cumulative hazards for two populations. *Biometrika*; **76**: 763-773.

[166] Draper, N. R. (1984) The Box-Wetz criterion versus R^2. *J. Roy. Statist. Soc. Series A*; **147**: 100-103.

[167] Draper, N. R. (1985) Corrections: The Box-Wetz criterion versus R^2. *J. Roy. Statist. Soc. Series A*; **148**: page 357.

[168] Dubin, J. A., Muller, H. G. and Wang, J. L. (2001) Event history graphs for censored survival data. *Statist. Med.*; **20**: 2951-64.

[169] Dunson, D. B. and Baird, D. D. (2002) A proportional hazards model for incidence and induced remission of disease. *Biometrics*; **58**: 71-78.

[170] Dunson, D. B., Chulada, P. and Arbes, S. J. Jr. (2003) Bayesian modeling of time-varying and waning exposure effects. *Biometrics*; **59**: 83-91.

[171] Dunson, D. B. and Herring, A. H. (2003) Bayesian inferences in the Cox model for order-restricted hypotheses. *Biometrics*; **59**: 916-23.

[172] Dunson, D. B. and Chen, Z. (2004) Selecting factors predictive of heterogeneity in multivariate event time data. *Biometrics*; **60**: 352-8.

[173] Dupuy, J., Grama, I. and Mesbah, M. (2006) Asymptotic theory for the Cox model with missing time-dependent covariate. *Ann. Statist.*; **34**: 903-24.

[174] Dupuy, J. and Mesbah, M. (2002) Joint modeling of event time and nonignorable missing longitudinal data. *Lifetime Data Analysis*; **8**: 99-115.

[175] Efron, B. (1975) Defining the curvature of a statistical problem. *Ann. Statist.*; **3**: 1189-1242.

[176] Efron, B. (1981a) Censored data and the bootstrap. *J. Amer. Statist. Assoc.* **76**: 312-319.

[177] Efron, B. (1981b) Nonparametric estimates of standard error: the jacknife, the bootstrap and other resampling methods. *Biometrika* **68**: 589-599.

[178] Efron, B. (2002) The two-way proportional hazards model. *J. R. Stat. Soc. Ser. B.*; **64**: 899-909.

[179] Efron, B. and Hinkley, D. V. (1978) Assessing the accuracy of the maximum likelihood estimator: Observed versus expected Fisher information. *Biometrika* **65**: 457-83.

[180] Eide, G. E., Omenaas, E. and Gulsvik, A. (1996) The semi-proportional hazards model revisited: practical reparametrizations. *Statist. Med.*; **15**: 1771-7.

[181] Elandt-Johnson, R. C. and Johnson, N. L. (1980) *Survival Models and Data Analyis*; New York. Wiley.

[182] Elashoff, M. and Lagakos, S. (1996) HIV treatment strategies utilizing virologic and immunologic markers as criteria for changing treatments. *Statist. Med.*; **15**: 2425-43; discussion 2455-8.

[183] Esteban, M. D. and Morales, D. (1995) Estimating a survival function with doubly censored data in a proportional hazard model. *Appl. Stochastic Models Data Anal.*; **11**: 145-157.

[184] Fan, J., Gijbels, I. and King, M. (1997) Local likelihood and local partial likelihood in hazard regression. *Ann. Statist.*; **25**: 1661-1690.

[185] Fan, J. and Li, R. (2002) Variable selection for Cox's proportional hazards model and frailty model. *Ann. Statist.*; **30**: 74-99.

[186] Faraggi, D. and Simon, R. (1996) A simulation study of cross-validation for selecting an optimal cutpoint in univariate survival analysis. *Statist. Med.*; **15**: 2203-13.

[187] Faraggi, D. and Simon, R. (1997) Large sample Bayesian inference on the parameters of the proportional hazard models. *Stat Med*; **16**: 2573-85.

[188] Farley, T. M., Ali, M. M. and Slaymaker, E. (2001) Competing approaches to analysis of failure times with competing risks. *Statist. Med.*; **20**: 3601-10.

[189] Farrington, C. P. (1996) Interval censored survival data: a generalized linear modelling approach. *Statist. Med.*; **15**: 283-92.

[190] Farrington, C. P. (2000) Residuals for proportional hazards models with interval-censored survival data. *Biometrics*; **56**: 473-82.

[191] Faucett, C. L., Schenker, N. and Taylor, J. M. (2002) Survival analysis using auxiliary variables via multiple imputation, with application to AIDS clinical trial data. *Biometrics*; **58**: 37-47.

[192] Fay, M. P. (1999) Comparing several score tests for interval censored data. *Statist. Med.*; **18**: 273-85.

[193] Fay, M. P. and Graubard, B. I. (2001) Small-sample adjustments for Wald-type tests using sandwich estimators. *Biometrics*; **57**: 1198-206.

[194] Feigl, P. and Zelen, M. (1965) Estimation of exponential survival probabilities with concommitant information. *Biometrics* **21**: 826-38.

[195] Figueiras, A., Domenech-Massons, J. M. and Cadarso, C. (1998) Regression models: calculating the confidence interval of effects in the presence of interactions. *Statist. Med.*; **17**: 2099-105.

[196] Fine, J. P. and Gray, R. J. (1999) A proportional hazards model for the subdistribution of a competing risk. *J. Amer. Statist. Assoc.*; **94**: 496-509.

[197] Fine, J. P. and Jiang, H. (2000) On association in a copula with time transformations. *Biometrika*; **87**: 559-571.

[198] Finkelstein, D. M. and Wolfe, R. A. (1985) A semiparametric model for regression analysis of interval-censored failure time data. *Biometrics*; **41**: 933-45.

[199] Finkelstein, D. M. (1986) A proportional hazards model for interval-censored failure time data. *Biometrics*; **42**: 845-854.

[200] Finkelstein, D. M., Moore, D. F. and Schoenfeld, D. A. (1993) A proportional hazards model for truncated AIDS data. *Biometrics*; **49**: 731-740.

[201] Finkelstein, D. M. and Schoenfeld, D. A. (1994) Analysing survival in the presence of an auxiliary variable. *Stat Med*; **13**: 1747-54.

[202] Flandre P. and O'Quigley J. (1995) A two stage design with surrogate endpoints for survival studies. *Biometrics*; **51**: 969-976.

[203] Fleming, T. R. and Harrington, D. P. (1991). *Counting Processes and Survival Analysis.*, New York: Wiley.

[204] Fleming, T. R. and DeMets D. L. (1996) Surrogate end points in clinical trials: Are we being misled. *Annals of Internal Medicine* **125**: 605-13.

[205] Ford, I., Norrie, J. and Ahmadi, S. (1995) Model inconsistency illustrated by the Cox proportional hazards model. *Statist. Med.*; **14**: 735-46.

[206] France, L. A., Lewis, J. A. and Kay, R. (1991) The analysis of failure time data in crossover studies. *Statist. Med.*; **10**: 1099-113.

[207] Frankel, P. and Longmate, J. (2002) Parametric models for accelerated and long-term survival: a comment on proportional hazards. *Statist. Med.*; **21**: 3279-89.

[208] Fraser, G. E. and Shavlik, D. J. (1999) The estimation of lifetime risk and average age at onset of a disease using a multivariate exponential hazard rate model. *Statist. Med.*; **18**: 397-410.

[209] Freedman L. S., Graubard B. I. and Schatzin A. (1992) Statistical validation of intermediate endpoints for chronic diseases. *Statistics in Med.*; **11**: 167-78.

[210] Freedman, L. S. and Spiegelhalter, D. J. (1992) Application of Bayesian statistics to decision making during a clinical trial. *Statist. Med.*; **11**: 23-35.

[211] Freireich, E. J., Gehan, E., Frei, E., Schroeder, L. R., Wolman, I. J., Anbari, R., Burgert, E., Mills, S. D., Pinkel, D., Selawry,

O. S., Moon, J. H., Gendel B. R., Spurr, C. L., Storrs, R., Haurani, F., Hoogstraten, B. and Lee, S. (1963) The Effect of 6-Mercaptopurine on the Duration of Steroid-induced Remissions in Acute Leukemia: A Model for Evaluation of Other Potentially Useful Therapy. *Blood*: **21**: 699-716.

[212] Friedman, Michael (1982) Piecewise exponential models for survival data with covariates. *Ann. Statist.*; **10**: 101-113.

[213] Fusaro, R. E., Nielsen, J. P. and Scheike, T. H. (1993) Marker-dependent hazard estimation: an application to AIDS. *Statist. Med.*; **12**: 843-65.

[214] Fusaro, R. E., Bacchetti, P. and Jewell, N. P. (1996) A competing risks analysis of presenting AIDS diagnoses trends. *Biometrics*; **52**: 211-25.

[215] Gail, M. H. (1981) Evaluating serial cancer marker studies in patients at risk of recurrent disease. *Biometrics*; **37**: 67-78.

[216] Gail, M. H., Wieand, S. and Piantadosi, S. (1984) Biased estimates of treatment effect in randomized experiments with nonlinear regressions and omitted covariates. *Biometrika*; **71**: 431-444.

[217] Galai, N., Simchen, E., Braun, D., Mandel, M. and Zitser-Gurevich, Y. (2002) Evaluating inter-hospital variability in mortality rates over time, allowing for time-varying effects. *Statist. Med.*; **21**: 21-33.

[218] Gehan, E. A. (1965) A generalized two-sample Wilcoxon test for doubly censored data. *Biometrika* **52**: 650-53.

[219] Gelfand, A. E. and Mallick, B. K. (1995) Bayesian analysis of proportional hazards models built from monotone functions. *Biometrics*; **51**: 843-52.

[220] Gelfand, A. E., Ghosh, S. K., Christiansen, C., Soumerai, S. B. and McLaughlin, Thomas J. (2000) Proportional hazards models: a latent competing risk approach. *J. Roy. Statist. Soc. Ser. C*; **49**: 385-397.

[221] Gentleman, R. and Crowley, J. (1991) Local full likelihood estimation for the proportional hazards model. *Biometrics*; **47**: 1283-1296.

[222] Ghosh, D. (2003) Goodness-of-fit methods for additive-risk models in tumorigenicity experiments. *Biometrics*; **59**: 721-6.

[223] Gilbert, P. B. (2000) Comparison of competing risks failure time methods and time-independent methods for assessing strain variations in vaccine protection. *Statist. Med.*; **19**: 3065-86.

[224] Gilbert, P. B., Wei, L. J., Kosorok, M. R. and Clemens, J. D. (2002) Simultaneous inferences on the contrast of two hazard functions with censored observations. *Biometrics*; **58**: 773-80.

[225] Gill, R. D. (1980) Censoring and stochastic integrals. Mathematical Centre Tracts 124, Mathematisch Centrum, Amsterdam.

[226] Gill, R. D. (1986) Understanding Cox's regression model: A martingale approach. *J. Amer. Statist. Assoc.*; **79**: 441-447.

[227] Gill, R. D. and Schumacher, M. (1987). A simple test of the proportional hazards assumption. *Biometrika*; **74**: 289-300.

[228] Giorgi, R., Abrahamowicz, M., Quantin, C., Bolard, P., Esteve, J., Gouvernet, J. and Faivre, J. (2003) A relative survival regression model using B-spline functions to model nonproportional hazards. *Statist. Med.*; **22**: 2767-84.

[229] Goetghebeur, E. and Ryan, L. (2000) Semiparametric regression analysis of interval-censored data. *Biometrics*; **56**: 1139-44.

[230] Goggins, W. B., Finkelstein, D. M., Schoenfeld, D. A. and Zaslavsky, A. M. (1998) A Markov chain Monte Carlo EM algorithm for analyzing interval-censored data under the Cox proportional hazards model. *Biometrics*; **54**: 1498-507.

[231] Goggins, W. B., Finkelstein, D. M. and Zaslavsky, A. M. (1999) Applying the Cox proportional hazards model when the change time of a binary time-varying covariate is interval censored. *Biometrics*; **55**: 445-451.

[232] Goggins, W. B. and Finkelstein, D. M. (2000) A proportional hazards model for multivariate interval-censored failure time data. *Biometrics*; **56**: 940-3.

[233] Gore, S. M., Pocock, S. J. and Kerr, G. R. (1984) Regression models and nonproportional hazards of the analysis of breast cancer survival. *Applied Statist.* **33**: 176-196.

[234] Gorfine, M., Hsu, L. and Prentice, R. L. (2004) Nonparametric correction for covariate measurement error in a stratified Cox model. *Biostatistics*; **5**: 75-87.

[235] Gorgens, T. (2003) Semiparametric estimation of censored transformation models. *J. Nonparametr. Stat.*; **15**: 377-393.

[236] Graf, E., Schmoor, C., Sauerbrei, W. and Schumacher, M. (1999) Assessment and comparison of prognostic classification schemes for survival data. *Statist. Med.*; **18**: 2529-45.

[237] Grambsch, P. M., Therneau, T. M. and Fleming, T. R. (1995) Diagnostic plots to reveal functional form for covariates in multiplicative intensity models. *Biometrics*; **51**: 1469-82.

[238] Gray, R. J. (1990) Some diagnostic methods for Cox regression models through hazard smoothing. *Biometrics*; **46**: 93-102.

[239] Gray, R. J. (1992) Flexible methods for analyzing survival data using splines with application to breast cancer prognosis. *J. American Statist. Assoc.*; **87**: 942-951.

[240] Gray, R. J. (1994) Spline-based tests in survival analysis. *Biometrics*; **50**: 640-52.

[241] Gray, R. J. (1994) A Bayesian analysis of institutional effects in a multicenter cancer clinical trial. *Biometrics*; **50**: 244-53.

[242] Gray, R. J. (2000) Estimation of regression parameters and the hazard function in transformed linear survival models. *Biometrics*; **56**: 571-6.

[243] Greene, T. (2001) A model for a proportional treatment effect on disease progression. *Biometrics*; **57**: 354-60.

[244] Greenland, S. (2003) Generalized conjugate priors for Bayesian analysis of risk and survival regressions. *Biometrics*; **59**: 92-9.

[245] Gustafson, P. (1995) A Bayesian analysis of bivariate survival data from a multicentre cancer clinical trial. *Statist. Med.*; **14**: 2523-35.

[246] Han, Aaron K. (1989) Asymptotic efficiency calculations of the partial likelihood estimator. *J. Econometrics*; **41**: 237-250.

[247] Hannan, P. J., Shu, X. O., Weisdorf, D. and Goldman, A. (1998) Analysis of failure times for multiple infections following bone marrow transplantation: an application of the multiple failure time proportional hazards model. *Statist. Med.*; **17**: 2371-80.

[248] Hansen, B. E., Thorogood, J., Hermans, J., Ploeg, R. J., Van Bockel, J. H. and Van Houwelingen, J. C. (1994) Multistate modelling of liver transplantation data. *Stat Med*; **13**: 2517-29.

[249] Harrell, F. E., Lee, K. L., Califf, R. M., Pryor, D. B. and Rosati, R. A. (1984) Regression modelling strategies for improved prognostic prediction. *Statist. Med.*; **3**: 143-52.

[250] Harrell, F. E., Margolis, P. A., Gove, S., Mason, K. E., Mulholland, E. K., Lehmann, D., Muhe, L., Gatchalian, S. and Eichenwald, H. F. (1998) Development of a clinical prediction model for an ordinal outcome: the World Health Organization Multicentre Study of Clinical Signs and Etiological agents of Pneumonia, Sepsis and Meningitis in Young Infants. WHO/ARI Young Infant Multicentre Study Group. *Statist. Med.*; **17**: 909-44.

[251] Harrington, D. P., Fleming, T. R. and Green, S. J. (1982). Procedures for serial testing in censored survival data. In *Institute of Mathematical Statistics Lecture Notes-Monograph Series: Survival Analysis*, Eds John Crowley and Richard Johnson, 269-286.

[252] Harrington, D. P. and Fleming, T. R. (1982) A class of rank test procedures for censored survival data. *Biometrika*: 553-566.

[253] Hashemi, R. and Commenges, D. (2002) Correction of the p-value after multiple tests in a Cox proportional hazard model. *Lifetime Data Anal.*; **8**: 335-348.

[254] Hastie, T. J. and Tibshirani, R. J. (1990) Generalized additive models. *Chapman and Hall Ltd.*. London.

[255] Hastie, T. J. and Tibshirani, R. J. (1993) Varying-coefficient models. *J. Roy. Statist. Soc. Ser. B*; **55**: 757-796.

[256] Hastings, N. and Peacock, J. (1975) *Statistical Distributions*; Butterworth. London.

[257] Hatteville, L., Mahe, C. and Hill, C. (2002) Prediction of the long-term survival in breast cancer patients according to the present oncological status. *Statist. Med.*; **21**: 2345-54.

[258] Hauck, W. W., McKee, L. J. and Turner, B. J. (1997) Two-part survival models applied to administrative data for determining rate of and predictors for maternal-child transmission of HIV. *Statist. Med.*; **16**: 1683-94.

[259] He, X. and Fung, W. K. (1999) Method of medians for lifetime data with Weibull models. *Statist. Med.*; **18**: 1993-2009.

[260] He, W. and Lawless, J. F. (2003) Flexible maximum likelihood methods for bivariate proportional hazards models. *Biometrics*; **59**: 837-848.

[261] Healy, M. (1984) The use of R^2 as a measure of goodness of fit. *J. Roy. Statist. Soc. Series A*; **147**: 608-09.

[262] Heinze, G. and Schemper, M. (2001) A solution to the problem of monotone likelihood in Cox regression. *Biometrics*; **57**: 114-9.

[263] Heinzl, H., Kaider, A. and Zlabinger, G. (1996) Assessing interactions of binary time-dependent covariates with time in Cox proportional hazards regression models using cubic spline functions. *Statist. Med.*; **15**: 2589-601.

[264] Heisey, D. M. and Foong, A. P. (1998) Modelling time-dependent interaction in a time-varying covariate and its application to rejection episodes and kidney transplant failure. *Biometrics*; **54**: 712-9.

[265] Heller, G. and Simonoff, J. S. (1992) Prediction in censored survival data: a comparison of the proportional hazards and linear regression models. *Biometrics*; **48**: 101-15.

[266] Heller, G. (2001) The Cox proportional hazards model with a partly linear relative risk function. *Lifetime Data Anal.*; **7**: 255-277.

[267] Heller, G. and Venkatraman, E. S. (2004) A nonparametric test to compare survival distributions with covariate adjustment. *J. R. Stat. Soc. Ser. B.*; **66**: 719-733.

[268] Hemyari, P. (2000) Robustness of the quartiles of survival time and survival probability. *J Biopharm Stat*; **10**: 299-318.

[269] Henderson, R. and Oman, P. (1993) Influence in linear hazard models. *Scand. J. Statist.*; **20**: 195-212.

[270] Henderson, R. (1995) Problems and prediction in survival-data analysis. *Statist. Med.*; **14**: 161-84.

[271] Henze, Norbert (1993) A quick omnibus test for the proportional hazards model of random censorship. *Statistics*; **24**: 253-263.

[272] Herndon, J. E. and Harrell, F. E. (1995) The restricted cubic spline as baseline hazard in the proportional hazards model with step function time-dependent covariables. *Statist. Med.*; **14**: 2119-29.

[273] Herring, A. H. and Ibrahim, Joseph G. (2001) Likelihood-based methods for missing covariates in the Cox proportional hazards model. *J. Amer. Statist. Assoc.*; **96**: 292-302.

[274] Hertz-Picciotto, I. and Rockhill, B. (1997) Validity and efficiency of approximation methods for tied survival times in Cox regression. *Biometrics*; **53**: 1151-6.

[275] Hess, K. R. (1994) Assessing time-by-covariate interactions in proportional hazards regression models using cubic spline functions. *Statist. Med.*; **13**: 1045-62.

[276] Hess, K. R. (1995) Graphical methods for assessing violations of the proportional hazards assumption in Cox regression. *Statist. Med.*; **14**: 1707-23.

[277] Hess, K. R., Serachitopol, D. M. and Brown, B. W. (1999) Hazard function estimators: a simulation study. *Stat Med*; **18**: 3075-88.

[278] Hill, C. (1981) Asymptotic relative efficiency of survival tests with covariates. *Biometrika*; **68**: 699-702.

[279] Hill C., Com-Nougé C., Kramar A., Moreau T., O'Quigley J. and Chastang C. (1990) *Analyse Statistique des Données de Survie*. Paris, Flammarion.

[280] Hilsenbeck, S. G. and Clark, G. M. (1996) Practical p-value adjustment for optimally selected cutpoints. *Statist. Med.*; **15**: 103-12.

[281] Hilton, J. F. (2000) Functions of oral candidiasis episodes that are highly prognostic for AIDS. *Statist. Med.*; **19**: 989-1004.

[282] Hjort, N.L. (1992) On inference in parametric survival data models. *Int. Statist. Rev.* **60**: 355-87.

[283] Holford, T. R. (1976) Life tables with concomitant information. *Biometrics*; **32**: 587-97.

[284] Hollander, M., Laird, G. and Song, K. (2001) Maximum likelihood estimation in the proportional hazards model of random censorship. *Statistics*; **35**: 245-258.

[285] Holly, A. (1995) A random linear functional approach to efficiency bounds. *J. Econometrics*; **65**: 235-261.

[286] Holt, J. D. and Prentice, R. L. (1974) Survival analyses in twin studies and matched pair experiments. *Biometrika*; **61**: 17-30.

[287] Honda, T. (2004) Nonparametric regression in proportional hazards models. *J. Japan Statist. Soc.*; **34**: 1-17.

[288] Hoover, D. R. and He, Y. (1994) Nonidentified responses in a proportional hazards setting. *Biometrics*; **50**: 1-10.

[289] Hoover, D. R., Munoz, A., He, Y., Taylor, J. M., Kingsley, L., Chmiel, J. S. and Saah, A. (1994) The effectiveness of interventions on incubation of AIDS as measured by secular increases within a population. *Statist. Med.*; **13**: 2127-39.

[290] Horowitz, J. L. and Neumann, G. R. (1992) A generalized moments specification test of the proportional hazards model. *J. Amer. Statist. Assoc.*; **87**: 234-240.

[291] Horowitz, J. L. (1999) Semiparametric estimation of a proportional hazard model with unobserved heterogeneity. *Econometrica*; **67**: 1001-1028.

[292] Horowitz, J. L. and Lee, S. (2004) Semiparametric estimation of a panel data proportional hazards model with fixed effects. *J. Econometrics*; **119**: 155-198.

[293] Hou, C. D., Chiang, J. and Tai, J. J. (2001) Identifying chromosomal fragile sites from a hierarchical-clustering point of view. *Biometrics*; **57**: 435-40.

[294] Hougaard, P. and Madsen, E. B. (1985) Dynamic evaluation of short-term prognosis after myocardial infarction. *Stat Med*; **4**: 29-38.

[295] Hougaard, P. (1986) A class of multivariate failure time distributions. *Biometrika*; **73**: 671-678.

[296] Hougaard, P. (1988) Correction to: "A class of multivariate failure time distributions". *Biometrika*; **75**: 395.

[297] Hougaard, P. (1991) Modelling heterogeneity in survival data. *J. Appl. Probab.*; **28**: 695-701.

[298] Hougaard, P., Myglegaard, P. and Borch-Johnsen, K. (1994) Heterogeneity models of disease susceptibility, with application to diabetic nephropathy. *Biometrics*; **50**: 1178-88.

[299] Hougaard, P. (1999) Fundamentals of survival data. *Biometrics*; **55**: 13-22.

[300] Hsieh, F. Y. (1992) Comparing sample size formulae for trials with unbalanced allocation using the logrank test. *Stat Med*; **11**: 1091-8.

[301] Hsieh, F. Y. and Lavori, P. W. (2000) Sample-size calculations for the Cox proportional hazards regression model with nonbinary covariates. *Control Clin Trials*; **21**: 552-60.

[302] Hu, P., Tsiatis, A. A. and Davidian, M. (1998) Estimating the parameters in the Cox model when covariate variables are measured with error. *Biometrics*; **54**: 1407-19.

[303] Hu, C. and Lin, D. Y. (2002) Cox regression with covariate measurement error. *Scand. J. Statist.*; **29**: 637-655.

[304] Hu, C. and Lin, D. Y. (2004) Semiparametric failure time regression with replicates of mismeasured covariates. *J. Amer. Statist. Assoc.*; **99**: 105-118.

[305] Huang, J. (1996) Efficient estimation for the proportional hazards model with interval censoring. *Ann. Statist.*; **24**: 540-568.

[306] Huang, X., Chen, S. and Soong, S. J. (1998) Piecewise exponential survival trees with time-dependent covariates. *Biometrics*; **54**: 1420-33.

[307] Huang, Y. (1999) The two-sample problem with induced dependent censorship. *Biometrics*; **55**: 1108-13.

[308] Huang, J. and Harrington, D. (2002) Penalized partial likelihood regression for right-censored data with bootstrap selection of the penalty parameter. *Biometrics*; **58**: 781-91.

[309] Hughes, M. D., Raskino, C. L., Pocock, S. J., Biagini, M. R. and Burroughs, A. K. (1992) Prediction of short-term survival with an application in primary biliary cirrhosis. *Statist. Med.*; **11**: 1731-45.

[310] Hughes, M. D. (1993) Regression dilution in the proportional hazards model. *Biometrics*; **49**: 1056-1066.

[311] Hughes, M. D. (1997) Power considerations for clinical trials using multivariate time-to-event data. *Statist. Med.*; **16**: 865-82.

[312] Hung, M. and Swallow, W. H. (1999) Robustness of group testing in the estimation of proportions. *Biometrics*; **55**: 231-7.

[313] Hutton, J. L. and Monaghan, P. F. (2002) Choice of parametric accelerated life and proportional hazards models for survival data: asymptotic results. *Lifetime Data Anal.*; **8**: 375-393.

[314] Huzurbazar, S. and Huzurbazar, A. V. (1999) Survival and hazard functions for progressive diseases using saddlepoint approximations. *Biometrics*; **55**: 198-203.

[315] Ibrahim, J. G. and Chen, M. (1998) Prior distibutions and Bayesian computation for proportional hazards models. *Sankhya Ser. B*; **60**: 48-64.

[316] Ibrahim, J. G., Chen, M. and MacEachern, S. N. (1999) Bayesian variable selection for proportional hazards models. *Canad. J. Statist.*; **27**: 701-717.

[317] Ingram, D. D. and Kleinman, J. C. (1989) Empirical comparisons of proportional hazards and logistic regression models. *Stat Med*; **8**: 525-38.

[318] Janssen, P. and Veraverbeke, N. (1992) The accuracy of normal approximations in censoring models. *J. Nonparametr. Statist.*; **1**: 205-217.

[319] Janssen, A. and Rahnenfuhrer, J. (2002) A hazard-based approach to dependence tests for bivariate censored models. *Math. Methods Statist.*; **11**: 297-322.

[320] Johnson, N. L. and Kotz, S. (1970) *Continuous Univariate Distributions*; New York. Wiley.

[321] Joly, P., Commenges, D. and Letenneur, L. (1998) A penalized likelihood approach for arbitrarily censored and truncated data: application to age-specific incidence of dementia. *Biometrics*; **54**: 185-94.

[322] Jones, D. and Whitehead, J. (1979) Sequential forms of the log rank and modified Wilcoxon tests for censored data. *Biometrika*; **66**: 105-113.

[323] Kalbfleisch, J. D. and Prentice, R. L. (1973) Marginal likelihoods based on Cox's regression and life model. *Biometrika*; **60**: 267-79.

[324] Kalbfleisch, J. D. and Prentice, R. L. (1980) *The statistical analysis of failure time data*. John Wiley and Sons, New York.

[325] Kalbfleisch, J. D. and Prentice, R. L. (1981) Estimation of the average hazard ratio. *Biometrika*; **68**: 105-112.

[326] Kamakura, T. and Yanagimoto, T. (1983) Evaluation of the regression parameter estimators in the proportional hazard model. *Biometrika*; **70**: 530-533.

[327] Kaplan, E. L. and Meier, P. (1958) Non-parametric estimation from incomplete observations. *JASA*; **53**: 457-481.

[328] Karrison, T. (1996) Confidence intervals for median survival times under a piecewise exponential model with proportional hazards covariate effects. *Statist. Med.*; **15**: 171-82.

[329] Kay, R. (1979) Some further asymptotic efficiency calculations for survival data regression models. *Biometrika*; **66**: 91-96.

[330] Keiding, N., Andersen, P. K. and Klein, J. P. (1997) The role of frailty models and accelerated failure time models in describing heterogeneity due to omitted covariates. *Stat Med*; **16**: 215-24.

[331] Kelly, P. J. and Lim, L. L. (2000) Survival analysis for recurrent event data: an application to childhood infectious diseases. *Statist. Med.*; **19**: 13-33.

[332] Kent, J. T. (1983) Information gain and a general measure of correlation. *Biometrika* **70**: 163-73.

[333] Kent, J. T. and O'Quigley, J. (1988) Measures of dependence for censored survival data. *Biometrika*; **75**: 525-534.

[334] Kim, M. Y., De Gruttola, V. G. and Lagakos, S. W. (1993) Analyzing doubly censored data with covariates, with application to AIDS. *Biometrics*; **49**: 13-22.

[335] Kim, D. K. (1997) Regression analysis of interval-censored survival data with covariates using log-linear models. *Biometrics*; **53**: 1274-83.

[336] Kim, H. T. and Truong, Y. K. (1998) Nonparametric regression estimates with censored data: local linear smoothers and their applications. *Biometrics*; **54**: 1434-44.

[337] Kim, S. and DeGruttola, V. (1999) Strategies for cohort sampling under the Cox proportional hazards model, application to an AIDS clinical trial. *Lifetime Data Anal.*; **5**: 149-172.

[338] Kim, M. Y. and Xue, X. (2002) The analysis of multivariate interval-censored survival data. *Statist. Med.*; **21**: 3715-26.

[339] Kim, J. S. (2003) Maximum likelihood estimation for the proportional hazards model with partly interval-censored data. *J. R. Stat. Soc. Ser. B.*; **65**: 489-502.

[340] Kim, K. and Tsiatis, A. (1990) Study duration and power consideration for clinical trials with survival response and early stopping rule. *Biometrics* **46**: 81-92.

[341] Kim, Y. and Lee, J. (2003) Bayesian analysis of proportional hazard models. *Ann. Statist.*; **31**: 493-511.

[342] Kinukawa, N., Nakamura, T., Akazawa, K. and Nose, Y. (2000) The impact of covariate imbalance on the size of the logrank test in randomized clinical trials. *Statist. Med.*; **19**: 1955-67.

[343] Kipnis, V. (1979) Forecasting short time series based on the chaotization principle (in Russian). *In: Pinsker ISh (ed). Models. Algorithms. Decision Making. Moscow*: 38-61.

[344] Klein, J. P., Keiding, N. and Kamby, C. (1989) Semiparametric Marshall-Olkin models applied to the occurrence of metastases at multiple sites after breast cancer. *Biometrics*; **45**: 1073-86.

[345] Klein, J. P. (1992) Semiparametric estimation of random effects using the Cox model based on the EM algorithm. *Biometrics*; **48**: 795-806.

[346] Klotz, J. (1982) Spline smooth estimates of survival. *Institute of Mathematical Statistics Lecture Notes-Monograph Series: Survival Analysis*, Eds John Crowley and Richard Johnson, 14-25.

[347] Kolmogorov, A. N. (1933) On the empirical determination of a distribution law. *Giornale dell'Istituto Italiano degli Attuari*; **4**: 83-91.

[348] Kong, F. H. and Slud, E. (1997) Robust covariate-adjusted logrank tests. *Biometrika*; **84**: 847-862.

[349] Kong, F. H., Huang, W. and Li, X. (1998) Estimating survival curves under proportional hazards model with covariate measurement errors. *Scand. J. Statist.*; **25**: 573-587.

[350] Kong, F. H. (1999) Adjusting regression attenuation in the Cox proportional hazards model. *J. Statist. Plann. Inference*; **79**: 31-44.

[351] Kooperberg, C. and Clarkson, D. B. (1997) Hazard regression with interval-censored data. *Biometrics*; **53**: 1485-94.

[352] Korhonen, P. A., Laird, N. M. and Palmgren, J. (1999) Correcting for non-compliance in randomized trials: an application to the ATBC Study. *Statist. Med.*; **18**: 2879-97.

[353] Korn, E. L. and Simon, R. (1990) Measures of explained variation for survival data. *Statistics in Medicine* **9**: 487-503.

[354] Korn, E. L. and Simon, R. (1991) Explained residual variation, explained risk, and goodness of fit. *The American Statistician* **45**: 201-206.

[355] Koziol, J. A. and Green, S. B. (1976) A Cramr-von Mises statistic for randomly censored data *Biometrika*; **63**: 465-74.

[356] Kronborg, D. and Aaby, P. (1990) Piecewise comparison of survival functions in stratified proportional hazards models. *Biometrics*; **46**: 375-380.

[357] Kuk, A. (1984) All subsets regression in a proportional hazards model. *Biometrika*; **71**: 587-592.

[358] Kuk, A. (1992) A semiparametric mixture model for the analysis of competing risks data. *Austral. J. Statist.*; **34**: 169-180.

[359] Kvalseth, T. (1985) Cautionary note about R^2. *American Statist.*; **39**: 279-85.

[360] Lagakos, S. W. and Schoenfeld, D. A. (1984) Properties of proportional-hazards score tests under misspecified regression models. *Biometrics*; **40**: 1037-1048.

[361] Lagakos, S. W. (1988) The loss in efficiency from misspecifying covariates in proportional hazards regression models. *Biometrika*; **75**: 156-160.

[362] Lagakos, S. W., Lim, L. L. and Robins, J. M. (1990) Adjusting for early treatment termination in comparative clinical trials. *Statist. Med.*; **9**: 1417-24; discussion 1433-7.

[363] Lagakos S. W. (1993) Surrogate markers in AIDS clinical trials: conceptual basis; validation, and uncertainties. *Clinical Infectious Disease* **16** (Suppl 1), S22-S25.

[364] Lam, K. F., Lee, Y. W. and Leung, T. L. (2002) Modeling multivariate survival data by a semiparametric random effects proportional odds model. *Biometrics*; **58**: 316-23.

[365] Lam, K. F. and Ip, D. (2003) REML and ML estimation for clustered grouped survival data. *Statist. Med.*; **22**: 2025-34.

[366] Lan, K. K. and Zucker, D. M. (1993) Sequential monitoring of clinical trials: the role of information and Brownian motion. *Statist. Med.*; **12**: 753-65.

[367] Langholz, B. and Borgan, O. (1997) Estimation of absolute risk from nested case-control data. *Biometrics*; **53**: 767-74.

[368] Lao, C. S. (1998) Survival and projection analyses of the effect of radiation on beagle dogs. *J Biopharm Stat*; **8**: 619-33.

[369] Larsen, K. (2004) Joint analysis of time-to-event and multiple binary indicators of latent classes. *Biometrics*; **60**: 85-92.

[370] Lausen, B. and Schumacher, M. (1996). Evaluating the Effect of Optimized Cutoff Values in the Assessment of Prognostic Factors, *Comp. Statist. and Data Anal.*; **21**: 307-326.

[371] Law, C. G. and Brookmeyer, R. (1992) Effects of mid-point imputation on the analysis of doubly censored data. *Stat Med*; **11**: 1569-78.

[372] Law, M. G. and Kaldor, J. M. (1996) Survival analyzes of randomized clinical trials adjusted for patients who switch treatments. *Statist. Med.*; **15**: 2069-76.

[373] Law, N. J., Taylor, J. M. and Sandler, H. (2002) The joint modeling of a longitudinal disease progression marker and the failure time process in the presence of cure. *Biostatistics*; **3**: 547-63.

[374] Lawless, J. F. (1982) Statistical models and methods for lifetime data. *John Wiley & Sons Inc.* New York.

[375] Lawless, J. F. and Fong, D. Y. (1999) State duration models in clinical and observational studies. *Statist. Med.*; **18**: 2365-76.

[376] Le, C. T. (1993) A test for linear trend in constant hazards and its application to a problem in occupational health. *Biometrics*; **49**: 1220-4.

[377] Le, C. T., Grambsch, P. M. and Louis, T. A. (1994) Association between survival time and ordinal covariates. *Biometrics*; **50**: 213-9.

[378] LeBlanc, M. and Crowley, J. (1999) Adaptive regression splines in the Cox model. *Biometrics*; **55**: 204-13.

[379] Lee, S. and Wolfe, R. A. (1995) A class of linear nonparametric rank tests for dependent censoring. *Comm. Statist. Theory Methods*; **24**: 687-712.

[380] Lee, J. W. (1995) Two-sample rank tests for acceleration in cure models. *Statist. Med.*; **14**: 2111-8.

[381] Lee, S. Y. (1996) Power calculation for a score test in the dependent censoring model. *Statist. Med.*; **15**: 1049-58.

[382] Lee, E. and Weissfeld, L. A. (1998) Assessment of covariate effects in Aalen's additive hazard model. *Statist. Med.*; **17**: 983-98.

[383] Lee, S. Y. and Wolfe, R. A. (1998) A simple test for independent censoring under the proportional hazards model. *Biometrics*; **54**: 1176-82.

[384] Leffondre, K., Abrahamowicz, M. and Siemiatycki, J. (2003) Evaluation of Cox's model and logistic regression for matched case-control data with time-dependent covariates: a simulation study. *Statist. Med.*; **22**: 3781-94.

[385] Lehmann, E. L. (1953) The power of rank tests. *Ann. Math. Statist.* **14**: 188-94.

[386] Leon, L. F. and Tsai, C. (2004) Functional form diagnostics for Cox's proportional hazards model. *Biometrics*; **60**: 75-84.

[387] Li, Q. II. and Lagakos, S. W. (1997) Use of the Wei-Lin-Weissfeld method for the analysis of a recurring and a terminating event. *Statist. Med.*; **16**: 925-40.

[388] Li, Z. (1999) A group sequential test for survival trials: an alternative to rank-based procedures. *Biometrics*; **55**: 277-83.

[389] Li, Y. P., Propert, K. J. and Rosenbaum, P. R. (2001) Balanced risk set matching. *J. Amer. Statist. Assoc.*; **96**: 870-882.

[390] Li, Z. (2001) Covariate adjustment for non-parametric tests for censored survival data. *Statist. Med.*; **20**: 1843-53.

[391] Li, Y., Betensky, R. A., Louis, D. N. and Cairncross, J. G. (2002) The use of frailty hazard models for unrecognized heterogeneity that interacts with treatment: considerations of efficiency and power. *Biometrics*; **58**: 232-6.

[392] Li, Y. and Ryan, L. (2002) Modeling spatial survival data using semiparametric frailty models. *Biometrics*; **58**: 287-97.

[393] Li, L. (2003) Linear regression analysis with observations subject to interval censoring. *Development of modern statistics and related topics*; **1**: 236-245.

[394] Liang, K. Y., Self, S. G. and Liu, X. H. (1990) The Cox proportional hazards model with change point: an epidemiologic application. *Biometrics*; **46**: 783-93.

[395] Liese, F. and Vajda, I. (2003) A general asymptotic theory of M-estimators. *Math. Methods Statist.*; **12**: 454-477 (2004).

[396] Lin, D. Y. and Wei, L. J. (1989) The robust inference for the Cox proportional hazards model. *J. Amer. Statist. Assoc.*; **84**: 1074-1078.

[397] Lin, D. Y. (1991) Goodness-of-fit analysis for the Cox regression model based on a class of parameter estimators. *J. Am. Statist. Assoc.* **86**: 725-728.

[398] Lin, D. Y., Fischl M. A. and Schoenfeld, D. A. (1993). Evaluating the role of CD4-lymphocyte counts as surrogate endpoints in human immunodeficiency virus clinical trials. *Statistics in Medicine* **12**: 835-42.

[399] Lin, D. Y., Fleming, T. R. and Wei, L. J. (1994) Confidence bands for survival curves under the proportional hazards model. *Biometrika*; **81**: 73-81.

[400] Lin, D. Y. (1994) Cox regression analysis of multivariate failure time data: the marginal approach. *Statist. Med.*; **13**: 2233-47.

[401] Lin, D. Y., Oakes, D. and Ying, Z. (1998) Additive hazards regression with current status data. *Biometrika*; **85**: 289-298.

[402] Lin, D. Y., Psaty, B. M. and Kronmal, R. A. (1998) Assessing the sensitivity of regression results to unmeasured confounders in observational studies. *Biometrics*; **54**: 948-63.

[403] Lin, D. Y., Yao, Q. and Ying, Z. (1999) A general theory on stochastic curtailment for censored survival data. *J. Amer. Statist. Assoc.*; **94**: 510-521.

[404] Lin, D. Y. (2000) On fitting Cox's proportional hazards models to survey data. *Biometrika*; **87**: 37-47.

[405] Lindkvist, M. (2000) Properties of added variable plots in Cox's regression model. *Lifetime Data Anal.*; **6**: 23-38.

[406] Lindsey, J. K., Jones, B. and Lewis, J. A. (1996) Analysis of cross-over trials for duration data. *Statist. Med.*; **15**: 527-35.

[407] Link, Carol L. (1984) Confidence intervals for the survival function using Cox's proportional-hazard model with covariates. *Biometrics*; **40**: 601-609.

[408] Lipsitz, S. R. and Parzen, M. (1996) A jackknife estimator of variance for Cox regression for correlated survival data. *Biometrics*; **52**: 291-8.

[409] Lipsitz, S. R. and Ibrahim, J. G. (1998) Estimating equations with incomplete categorical covariates in the Cox model. *Biometrics*; **54**: 1002-13.

[410] Louis, T. A. (1981) Nonparametric analysis of an accelerated failure time model. *Biometrika*; **68**: 381-390.

[411] Louzada-Neto, F. (1999) Polyhazard models for lifetime data. *Biometrics*; **55**: 1281-5.

[412] Lu, K. and Tsiatis, A. A. (2001) Multiple imputation methods for estimating regression coefficients in the competing risks model with missing cause of failure. *Biometrics*; **57**: 1191-7.

[413] Lu, S. E. and Wang, M. C. (2002) Cohort case-control design and analysis for clustered failure-time data. *Biometrics*; **58**: 764-72.

[414] Lubin, Jay H. (1980) Analysis under Cox's failure time model using weighted least squares. *Biometrics*; **36**: 307-312.

[415] Lui, K. J. and Rhodes, P. (1990) A note on point estimation of the hazard ratio in exponential distributions. *Statist. Med.*; **9**: 1167-73.

[416] Lunn, M. and McNeil, D. (1995) Applying Cox regression to competing risks. *Biometrics*; **51**: 524-32.

[417] Ma, R., Krewski, D. and Burnett, R. T. (2003) Random effects Cox models: a Poisson modelling approach. *Biometrika*; **90**: 157-169.

[418] Mahe, C. (1998) Analysis of multiple failure time data from an AIDS clinical trial. *Statist. Med.*; **17**: 1293-4.

[419] Mahe, C. and Chevret, S. (1999) Estimation of the treatment effect in a clinical trial when recurrent events define the endpoint. *Statist. Med.*; **18**: 1821-9.

[420] Mahe, C. and Chevret, S. (1999b) Estimating regression parameters and degree of dependence for multivariate failure time data. *Biometrics*; **55**: 1078-84.

[421] Mahe, C. and Chevret, S. (2001) Analysis of recurrent failure times data: should the baseline hazard be stratified? *Statist. Med.*; **20**: 3807-15.

[422] Malani, H. (1995) A modification of the redistribution to the right algorithm using disease markers. *Biometrika*; **82**: 515-26.

[423] Maller, R. A. (1987) Relating Cox's proportional hazard model to the exponential model for survival. *Biometrical J.*; **29**: 231-238.

[424] Manatunga, A. K. and Chen, S. (2000) Sample size estimation for survival outcomes in cluster-randomized studies with small cluster sizes. *Biometrics*; **56**: 616-21.

[425] Mantel, N. and Haenszel, W. (1959) Statistical aspects of the analysis of data from retrospective studies of disease. *J. Nat. Cancer Instit.* **22**: 719-48.

[426] Mantel, N. (1963) Chi-squared tests with one degree of freedom. Extension of the Mantel-Haenszel procedure. *J. Am. Statist. Assoc.* **58**: 690-700.

[427] Maples, J. J., Murphy, S. A. and Axinn, W. G. (2002) Two-level proportional hazards models. *Biometrics*; **58**: 754-763.

[428] Marzec, L. and Marzec, P. (1993) Goodness of fit inference based on stratification in Cox's regression model. *Scand. J. Statist.*; **20**: 227-238.

[429] Marzec, L. and Marzec, P. (1997) On fitting Cox's regression model with time-dependent coefficients. *Biometrika*; **84**: 901-908.

[430] Matsui, S. and Miyagishi, H. (1999) Design of clinical trials for recurrent events with periodic monitoring. *Statist. Med.*; **18**: 3005-20.

[431] Mauger, E. A., Wolfe, R. A. and Port, F. K. (1995) Transient effects in the Cox proportional hazards regression model. *Statist. Med.*; **14**: 1553-65.

[432] May, S. and Hosmer, D. W. (2004) An added variable goodness-of-fit test for the Cox proportional hazards model. *Biom. J.*; **46**: 343-350.

[433] May, S. and Hosmer, D. W. (2004) A cautionary note on the use of the Grønnesby and Borgan goodness-of-fit test for the Cox proportional hazards model. *Lifetime Data Anal.*; **10**: 283-291.

[434] Mazumdar, M., Smith, A. and Bacik, J. (2003) Methods for categorizing a prognostic variable in a multivariable setting. *Statist. Med.*; **22**: 559-71.

[435] McCall, B. P. (1996) The identifiability of the mixed proportional hazards model with time-varying coefficients. *Econometric Theory*; **12**: 733-738.

[436] McGilchrist, C. A. and Aisbett, C. W. (1991) Regression with frailty in survival analysis. *Biometrics*; **47**: 461-6.

[437] McKeague, I. W. and Utikal, K. J. (1990) Stochastic calculus as a tool in survival analysis: a review. *Appl. Math. Comput.*; **38**: 23-49.

[438] McKeague, I. W. and Utikal, Klaus J. (1990) Identifying non-linear covariate effects in semimartingale regression models. *Probab. Theory Related Fields*; **87**: 1-25.

[439] McKeague, I. W. and Utikal, Klaus J. (1991) Goodness-of-fit tests for additive hazards and proportional hazards models. *Scand. J. Statist.*; **18**: 177-195.

[440] McKeague, I. W. and Tighiouart, M. (2000) Bayesian estimators for conditional hazard functions. *Biometrics*; **56**: 1007-15.

[441] McKean, J. W. and Sievers, G. L. (1989) Rank scores suitable for analyses of linear models under asymmetric error distributions. *Technometrics*; **31**: 207-18.

[442] Mehrotra, D. V. and Roth, A. J. (2001) Relative risk estimation and inference using a generalized logrank statistic. *Stat Med*; **20**: 2099-113.

[443] Merrick, J., Soyer, R. and Mazzuchi, T. A. (2003) A Bayesian semiparametric analysis of the reliability and maintenance of machine tools. *Technometrics*; **45**: 58-69.

[444] Miller, R. and Sigmund, D. (1982) Maximally selected chi-square statistics. *Biometrics*; **38**: 1011-16.

[445] Minder, C. E. and Bednarski, T. (1996) A robust method for proportional hazards regression. *Statist. Med.*; **15**: 1033-47.

[446] Mock, P. (1990) Empirical comparisons of proportional hazards and logistic regression models. *Statist. Med.*; **9**: 463-4.

[447] Moeschberger, M. L. and Klein, J. P. (1985) A comparison of several methods of estimating the survival function when there is extreme right censoring. *Biometrics* **41**: 253-59.

[448] Molenberghs, G., Williams, P. L. and Lipsitz, S. R. (2002) Prediction of survival and opportunistic infections in HIV-infected patients: a comparison of imputation methods of incomplete CD4 counts. *Statist. Med.*; **21**: 1387-408.

[449] Molinari, N., Daures, J. P. and Durand, J. F. (2001) Regression splines for threshold selection in survival data analysis. *Statist. Med.*; **20**: 237-47.

[450] Moolgavkar, S. H. and Venzon, D. J. (1987) Confidence regions for parameters of the proportional hazards model: a simulation study. *Scand. J. Statist.*; **14**: 43-56.

[451] Moreau, T., O'Quigley, J. and Mesbah, M. (1985) A global goodness-of-fit statistic for the proportional hazards model. *J. Roy. Statist. Soc. Ser. C*; **34**: 212-218.

[452] Moreau T., O'Quigley J. and Lellouch J. (1986) Concerning Schoenfelds Chi-squared statistic for testing the proportional hazards assumption. *Biometrika*; **73**: 513-515.

[453] Muller, H. G. and Wang, J. L. (1994) Hazard rate estimation under random censoring with varying kernels and bandwidths. *Biometrics*; **50**: 61-76.

[454] Murphy, S. A. and Sen, P. K. (1991) Time-dependent coefficients in a Cox-type regression model. *Stoch. Proc. Applic.*; **39**: 153-80.

[455] Murphy, S. A. (1993) Testing for a time dependent coefficient in a Cox's regression model. *Scan. J. Statist.*; **20**: 35-50.

[456] Murphy, S. A. (1995) Asymptotic theory for the frailty model. *Ann. Statist.*; **23**: 182-198.

[457] Murphy, S. A. and van der Vaart, A. W. (2000) On profile likelihood. *J. Amer. Statist. Assoc.*; **95**: 449-485.

[458] Myers, M. H., Hankey, B. F. and Mantel, N. (1973) A logistic-exponential model for use with response time data involving regressor variables. *Biometrics* **29**: 257-69.

[459] Nagelkerke, N. J. and Hart, A. A. (1980) The sequential comparison of survival curves. *Biometrika*; **67**: 247-249.

[460] Nagelkerke, N. J., Oosting, J. and Hart, A. A. (1984). A simple test of goodness of fit of Cox's proportional hazards model. *Biometrics*; **40**: 483-486.

[461] Nakamura, T. (1992) Proportional hazards model with covariates subject to measurement error. *Biometrics*; **48**: 829-838.

[462] Nardi, A. and Schemper, M. (1999) New residuals for Cox regression and their application to outlier screening. *Biometrics*; **55**: 523-9.

[463] Natarajan, L. and O'Quigley, J. (2002) Predictive capability of stratified proportional hazards models. *J. Appl. Stat.*; **29**: 1153-1163.

[464] Nelson, W. (1969) Hazard plotting for incomplete failure data. *J. Qual. Technol*; **1**: 27-52.

[465] Newton, M. A. and Raftery, A. (1994) Approximate Bayesian inference with the weighted likelihood bootstrap (with discussion). *J. Roy. Statist. Soc. Series B* **56**: 3-48.

[466] Ng'andu, N. H. (1997) An empirical comparison of statistical tests for assessing the proportional hazards assumption of Cox's model. *Statist. Med.*; **16**: 611-26.

[467] Oakes, D. (1977) The asymptotic information in censored survival data. *Biometrika*; **64**: 441-448.

[468] Oakes, D. (1986) Semi-parametric inference in a model for association in bivariate survival data. *Biometrika* **73**: 353-61.

[469] O'Brien, P.C. (1978) A non-parameteric test for association with censored data. *Biometrics* **34**: 243-50.

[470] O'Quigley, J. (1984) Intervalles de confiance pour les estimations de courbes de survie à partir du modèle de Cox. *Revue de Statistique Appliquée* **32**: 39-45.

[471] O'Quigley, J. and Schwartz D. (1986) Comparaison de plusieurs pourcentages lorsque les effectifs théoriques sont faibles. *Revue d'Epidémiologie et de Santé Publique* **34**: 18-22.

[472] O'Quigley, J. (1986) Confidence intervals for survival estimates incorporating covariate information. *Biometrics* **42**: 219-220.

[473] O'Quigley, J. and Moreau T. (1986) Cox's Regression Model: Computing a goodness of fit statistic. *Computer Methods and Programs in Biomedicine* **22**: 253-256.

[474] O'Quigley, J. and Moreau T. (1987) Updating prognostic indices via regression models. *Applied Stochastic Models and Data Analysis* **3**: 227-236.

[475] O'Quigley, J. and Pessione, F. (1989) Score tests for homogeneity of regression effect in the proportional hazards model. *Biometrics*; **45**: 135-144.

[476] O'Quigley, J. and Pessione, F. (1991) The problem of a covariate-time qualitative interaction in a survival study. *Biometrics*; **47**: 101-15.

[477] O'Quigley, J. and Prentice, R. L. (1991) Nonparametric tests of association between survival time and continuously measured covariates: the logit-rank and associated procedures. *Biometrics*; **47**: 117-27.

[478] O'Quigley, J. and Flandre P. (1994) The predictive capability of proportional hazards regression. *Proc. Nat. Acad. Sci. USA*, **91**: 2310-2314.

[479] O'Quigley, J. (1994) Two sided tests for crossing hazards. *J. Roy. Statist. Soc. Series D*, **43**: 563-69

[480] O'Quigley, J. (1995) Daniels association measures under right censoring. *Appl. Stochastic Models and Data Analysis.* **11**: 109-119.

[481] O'Quigley, J. and Xu R. (1998) Goodness of fit in survival analysis. *Encyclopedia of Biostatistics*, Wiley, 1731-45.

[482] O'Quigley, J., Flandre P. and Reiner E. (1999) Large sample theory for Schemper's measures of explained variation in the Cox model. *J. Roy. Statist. Soc. Series D*, **48**: 53-62.

[483] O'Quigley, J. and Xu, R. (2000) Inference for the Cox model under proportional and non proportional hazards. *New Approaches in Applied Statistics.* **16**: (Eds: A. Ferligoj and A. Mrvar), 3-19.

[484] O'Quigley, J. and Xu, R. (2001) Explained Variation in Proportional Hazards Regression. *Handbook of Statistics in Clinical Oncology.* 397-409; (Eds: John Crowley), Marcel Dekker, New York.

[485] O'Quigley, J. and Stare, J. (2002) Proportional hazards models with frailties and random effects. *Statist. Med.*; **21**: 3219-33.

[486] O'Quigley, J. (2003) Khmaladze-type graphical evaluation of the proportional hazards assumption. *Biometrika*; **90**: 577-584.

[487] O'Quigley, J. and Natarajan, L. (2004) Erosion of regression effect in a survival study. *Biometrics*; **60**: 344-51.

[488] O'Quigley, J., Xu, R. and Stare, J. (2005) Explained randomness in proportional hazards models. *Statist. in Med.* Vol **24**: 479-89.

[489] O'Quigley, J. and Flandre, P. (2006) Quantification of the Prentice Criteria for Surrogate Endpoints. *Biometrics* **62**: 297-300.

[490] O'Sullivan, F. (1988) Nonparametric estimation of relative risk using splines and cross-validation. *SIAM J. Sci. Statist. Comput.*; **9**: 531-542.

[491] Oakes, D. (1994) Multivariate survival distributions. *J. Nonparametr. Statist.*; **3**: 343-354.

[492] Odell, P. M., Anderson, K. M. and D'Agostino, R. B. (1992) Maximum likelihood estimation for interval-censored data using a Weibull-based accelerated failure time model. *Biometrics*; **48**: 951-9.

[493] Orbe, J., Ferreira, E. and Nunez-Anton, V. (2002) Comparing proportional hazards and accelerated failure time models for survival analysis. *Statist. Med.*; **21**: 3493-510.

[494] Osnes, K. and Aalen, O. O. (1999) Spatial smoothing of cancer survival: a Bayesian approach. *Statist. Med.*; **18**: 2087-99.

[495] Paik, M. C. and Tsai, W. (1997) On using the Cox proportional hazards model with missing covariates. *Biometrika*; **84**: 579-593.

[496] Pan, W. and Chappell, R. (1998) Estimating survival curves with left-truncated and interval-censored data under monotone hazards. *Biometrics*; **54**: 1053-60.

[497] Pan, W. and Kooperberg, C. (1999) Linear regression for bivariate censored data via multiple imputation. *Statist. Med.*; **18**: 3111-21.

[498] Pan, W. (2000) Smooth estimation of the survival function for interval censored data. *Statist. Med.*; **19**: 2611-24.

[499] Parker, C. B. and Delong, E. R. (2000) A diagnostic for Cox regression with discrete failure-time models. *Biometrics*; **56**: 996-1001.

[500] Parmar, M. K., Torri, V. and Stewart, L. (1998) Extracting summary statistics to perform meta-analyses of the published literature for survival endpoints. *Statist. Med.*; **17**: 2815-34.

[501] Parner, E. T. and Keiding, N. (2001) Misspecified proportional hazard models and confirmatory analysis of survival data. *Biometrika*; **88**: 459-468.

[502] Parzen, M. and Lipsitz, S. R. (1999) A global goodness-of-fit statistic for Cox regression models. *Biometrics*; **55**: 580-4.

[503] Pauler, D. K. and Finkelstein, D. M. (2002) Predicting time to prostate cancer recurrence based on joint models for non-linear longitudinal biomarkers and event time outcomes. *Stat Med*; **21**: 3897-911.

[504] Pawitan, Y., Bjohle, J., Wedren, S., Humphreys, K., Skoog, L., Huang, F., Amler, L., Shaw, P., Hall, P. and Bergh, J. (2004) Gene expression profiling for prognosis using Cox regression. *Statist. Med.*; **23**: 1767-80.

[505] Peace, K. E. and Carter, W. H. (1993) Exposure analysis of dichotomous response measures in long-term studies. *J Biopharm Stat*; **3**: 129-40.

[506] Pena, E. A. and Rohatgi, V. K. (1989) Survival function estimation for a generalized proportional hazards model of random censorship. *J. Statist. Plann. Inference*; **22**: 371-389.

[507] Pena, E. A. (1998) Smooth goodness-of-fit tests for the baseline hazard in Cox's proportional hazards model. *J. Amer. Statist. Assoc.*; **93**: 673-692.

[508] Peng, Y. and Dear, K. B. (2000) A nonparametric mixture model for cure rate estimation. *Biometrics*; **56**: 237-43.

[509] Pepe, M. S. and Mori, M. (1993) Kaplan-Meier, marginal or conditional probability curves in summarizing competing risks failure time data. *Statist. Med.*; **12**: 737-51.

[510] Peto, R. and Peto, J. (1972) Asymptotically efficient rank invariant procedures. *J. Roy. Statist. Soc* **135**: 185-206.

[511] Petroni, G. R. and Wolfe, R. A. (1994) A two-sample test for stochastic ordering with interval-censored data. *Biometrics*; **50**: 77-87.

[512] Pettitt, A. N. and Bin Daud, I. (1990) Investigating time dependence in Cox's proportional hazards model. *J. Roy. Statist. Soc. Ser. C*; **39**: 313-329.

[513] Pierce, D. A., Stewart, W. H. and Kopecky, K. J. (1979) Distribution-free regression analysis of grouped survival data. *Biometrics*; **35**: 785-93.

[514] Pipper, C. B. and Martinussen, T. (2003) A likelihood based estimating equation for the Clayton-Oakes model with marginal proportional hazards. *Scand. J. Statist.*; **30**: 509-521.

[515] Politis, D. N. (1998) Computer intensive methods in statistical analysis. *Signal Processing Magazine IEEE*, **15**: 39-55.

[516] Prentice, R. L. (1978) Linear rank tests with right censored data. *Biometrika*; **65**: 167-79.

[517] Prentice, R. L. and Kalbfleisch, J. D. (1979) Hazard rate models with covariates. *Biometrics*; **35**: 25-39.

[518] Prentice, R. L. (1989) Surrogate endpoints in clinical trials: definition and operational criteria. *Statist. Med.* **8**: 431-40.

[519] Prentice, R. L. and Sheppard, L. (1995) Aggregate data studies of disease risk factors. *Biometrika*; **82**: 113-125.

[520] Prentice, R. L. and Kalbfleisch, J. D. (2003) Mixed discrete and continuous Cox regression model. *Lifetime Data Anal.*; **9**: 195-210.

[521] Qin, J. (1998) Inferences for case-control and semiparametric two-sample density ratio models. *Biometrika*; **85**: 619-630.

[522] Qiou, Z., Ravishanker, N. and Dey, D. K. (1999) Multivariate survival analysis with positive stable frailties. *Biometrics*; **55**: 637-44.

[523] Quantin, C., Moreau, T., Asselain, B., Maccario, J. and Lellouch, J. (1996) A regression survival model for testing the proportional hazards hypothesis. *Biometrics*; **52**: 874-85.

[524] Rashid, S. A. and O'Quigley, J. (1982) Plasma protein profiles and prognosis in gastric cancer. *Brit. J. Cancer*; **45**: 390-394.

[525] Rabinowitz, D. (2000) Computing the efficient score in semi-parametric problems. *Statist. Sinica*; **10**: 265-280.

[526] Randles, R. H. and Wolf, D. A. (1979) *Introduction to the Theory of Nonparametric Statistics.* New York: Wiley.

[527] Rao, B. R. and Talwalker, S. (1991) Random censorship, competing risks and a simple proportional hazards model. *Biometrical J.*; **33**: 461-483.

[528] Rao, C. R. (1973) *Linear Statistical Inference and its Applications*; New York: Wiley.

[529] Reid, N. and Crepeau, H. (1985) Influence functions for proportional hazards regression. *Biometrika*; **72**: 1-9.

[530] Ridder, G. and Woutersen, T. M. (2003) The singularity of the information matrix of the mixed proportional hazard model. *Econometrica*; **71**: 1579-1589.

[531] Ripley, B. D. and Solomon, P. J. (1995) Statistical models for prevalent cohort data. *Biometrics*; **51**: 373-5.

[532] Robins, J. M. (1997) Non-response models for the analysis of non-monotone non-ignorable missing data. *Statist. Med.*; **16**: 21-37.

[533] Rohatgi, V. K. (2003) *Statistical Inference*; Dover Publications, New York.

[534] Rousseeuw, P. and Hubert, M. (1997) Recent developments in progress, in L1-Statistical Procedures. *Institute of Mathematical Statistics Lecture Notes-Monograph Series* **31**: 201-14.

[535] Royston, P. and Parmar, M. K. (2002) Flexible parametric proportional-hazards and proportional-odds models for censored survival data, with application to prognostic modelling and estimation of treatment effects. *Statist. Med.*; **21**: 2175-97.

[536] Royston, P., Parmar, M. K. and Sylvester, R. (2004) Construction and validation of a prognostic model across several studies, with an application in superficial bladder cancer. *Statist. Med.*; **23**: 907-26.

[537] Russek-Cohen, E. and Simon, R. M. (1993) Qualitative interactions in multifactor studies. *Biometrics*; **49**: 467-77.

[538] Ryu, K. (1995) Analysis of a continuous-time proportional hazard model using discrete duration data. *Econometric Rev.*; **14**: 299-313.

[539] Salter, A. and Solomon, P. J. (1997) Truncated recurrent event survival models for methadone data. *Biometrics*; **53**: 1293-303.

[540] Samuelsen, S. O. (2003) Exact inference in the proportional hazard model: possibilities and limitations. *Lifetime Data Anal.*; **9**: 239-260.

[541] Sargent, D. J. (1998) A general framework for random effects survival analysis in the Cox proportional hazards setting. *Biometrics*; **54**: 1486-97.

[542] Sartori, N., Thomaseth, K. and Salvan, A. (2004) Local influence analysis when interfacing toxicokinetic and proportional hazard models. *Statist. Med.*; **23**: 2399-412.

[543] Sasieni, P. (1992) Information bounds for the conditional hazard ratio in a nested family of regression models. *J. Roy. Statist. Soc. Ser. B*; **54**: 617-635.

[544] Sasieni, P. (1992) Nonorthogonal projections and their application to calculating the information in a partly linear Cox model. *Scand. J. Statist.*; **19**: 215-233.

[545] Sasieni, P. (1992) A note on the presentation of matched case-control data. *Statist. Med.*; **11**: 617-20.

[546] Sato, T. (1992) A comparison of two simple hazard ratio estimators based on the logrank test. *Statist. Med.*; **11**: 847-8.

[547] Satten, G. A., Datta, S. and Williamson, J. M. (1998) Inference based on imputed failure times for the proportional hazards model with interval-censored data. *J. Amer. Statist. Assoc.*; **93**: 318-327.

[548] Sauerbrei, W. and Schumacher, M. (1992) A bootstrap resampling procedure for model building: application to the Cox regression model. *Statist. Med.*; **11**: 2093-109.

[549] Schaubel, D. E. and Cai, J. (2004) Non-parametric estimation of gap time survival functions for ordered multivariate failure time data. *Statist. Med.*; **23**: 1885-900.

[550] Scheike, T. H. and Zhang, M. J. (2003) Extensions and applications of the Cox-Aalen survival model. *Biometrics*; **59**: 1036-45.

[551] Scheike, T. H. and Martinussen, T. (2004) On estimation and tests of time-varying effects in the proportional hazards model. *Scand. J. Statist.*; **31**: 51-62.

[552] Scheike, T. H. and Juul, A. (2004) Maximum likelihood estimation for Cox's regression model under nested case-control sampling. *Biostatistics*; **5**: 193-206.

[553] Schemper, M. (1988) Non-parametric analysis of treatment-covariate interaction in the presence of censoring. *Statist. Med.*; **7**: 1257-66.

[554] Schemper, M. (1990) The explained variation in proportional hazards regression. *Biometrika*; **77**: 216-218.

[555] Schemper, M. and Smith, T. L. (1996) A note on quantifying follow-up in studies of failure time. *Control Clin Trials*; **17**: 343-6.

[556] Schemper, M. and Stare, J. (1996) Explained variation in survival analysis. *Statist. Med.*; **15**: 1999-2012.

[557] Schemper, M. and Kaider, A. (1997) A new approach to estimate correlation coefficients in the presence of censoring and proportional hazards. *Comp. Statist. & Data Analysis*; **23**: 467-76.

[558] Schemper, M. and Henderson, R. (2000) Predictive accuracy and explained variation in Cox regression. *Biometrics*; **56**: 249-55.

[559] Schemper, M. (2003) Predictive accuracy and explained variation. *Statist. Med.*; **22**: 2299-308.

[560] Schmoor, C., Ulm, K. and Schumacher, M. (1993) Comparison of the Cox model and the regression tree procedure in analysing a randomized clinical trial. *Statist. Med.*; **12**: 2351-66.

[561] Schmoor, C. and Schumacher, M. (1997) Effects of covariate omission and categorization when analysing randomized trials with the Cox model. *Statist. Med.*; **16**: 225-37.

[562] Scott, A. and Wild, C. (1991) Transformations and R^2. *American Statist.*; **45**: 127-29.

[563] Seaman, S. R. and Bird, S. M. (2001) Proportional hazards model for interval-censored failure times and time-dependent covariates: application to hazard of HIV infection of injecting drug users in prison. *Statist. Med.*; **20**: 1855-70.

[564] Seetharaman, P. B. and Chintagunta, P. K. (2003) The proportional hazard model for purchase timing: a comparison of alternative specifications. *J. Bus. Econom. Statist.*; **21**: 368-382.

[565] Self, S. G. and Prentice, R. L. (1982) Commentary on: "Cox's regression model for counting processes: a large sample study" [Ann. Statist. 10 (1982), no. 4, 1100-1120] by P. K. Andersen and R. D. Gill. *Ann. Statist.*; **10**: 1121-1124.

[566] Sellke, T. and Siegmund, D. (1983) Sequential analysis of the proportional hazards model. *Biometrika*; **70**: 315-326.

[567] Sen, P. K. (1994) Some change-point problems in survival analysis: relevance of nonparametrics in applications. *J. Appl. Statist. Sci.*; **1**: 425-444.

[568] Sen, P. K. (1995) Some change-point problems in survival analysis: relevance of nonparametrics in applications. *Applied change point problems in statistics (Baltimore, MD, 1993)*: 103-122.

[569] Sengupta, D., Bhattacharjee, A. and Rajeev, B. (1998) Testing for the proportionality of hazards in two samples against the increasing cumulative hazard ratio alternative. *Scand. J. Statist.*; **25**: 637-647.

[570] Serfling, R. (1980) *Approximation Theorems of Mathematical Statistics*; New York. Wiley.

[571] Shannon, C. E. and Weaver, W. (1949) *The Mathematical Theory of Information*; University of Illinois Press: Urbana.

[572] Shen, Y. and Fleming, T. R. (1997) Large sample properties of some survival estimators in heterogeneous samples. *J. Statist. Plann. Inference*; **60**: 123-138.

[573] Shen, Y. and Cheng, S. C. (1999) Confidence bands for cumulative incidence curves under the additive risk model. *Biometrics*; **55**: 1093-1100.

[574] Shen, Y. and Cai, J. (2001) Maximum of the weighted Kaplan-Meier tests with application to cancer prevention and screening trials. *Biometrics*; **57**: 837-43.

[575] Siegmund, K. D., Todorov, A. A. and Province, M. A. (1999) A frailty approach for modelling diseases with variable age of onset in families: the NHLBI Family Heart Study. *Stat Med*; **18**: 1517-28.

[576] Simon, R. (1999) Bayesian design and analysis of active control clinical trials. *Biometrics*; **55**: 484-7.

[577] Singh, B. and Wright, F. T. (1999) Approximating the powers of order-restricted log rank tests at local alternatives. *Statist. Probab. Lett.*; **44**: 7-17.

[578] Sinha, D. (1998) Posterior likelihood methods for multivariate survival data. *Biometrics*; **54**: 1463-74.

[579] Sinha, D., Chen, M. H. and Ghosh, S. K. (1999) Bayesian analysis and model selection for interval-censored survival data. *Biometrics*; **55**: 585-90.

[580] Sinha, D., Ibrahim, J. G. and Chen, M. (2003) A Bayesian justification of Cox's partial likelihood. *Biometrika*; **90**: 629-641.

[581] Skaug, H. J. and Schweder, T. (1999) Hazard models for line transect surveys with independent observers. *Biometrics*; **55**: 29-36.

[582] Slasor, P. and Laird, N. (2003) Joint models for efficient estimation in proportional hazards regression models. *Statist. Med.*; **22**: 2137-48.

[583] Slasor, P. and Laird, N. (2004) Categorical auxiliary data in the discrete time proportional hazards model. *Advances in survival analysis*; **23**: 363-382.

[584] Slud, E. V. (1984) Multivariate dependent renewal processes. *Adv. in Appl. Probab.*; **16**: 347-362.

[585] Slud, E. V. (1994) Analysis of factorial survival experiments. *Biometrics*; **50**: 25-38.

[586] Slud, E. V. and Rubinstein, L. V. (1983) Dependent competing risks and summary survival curves *Biometrika*; **70**: 643-49.

[587] Smith, P. J. (1993) Dispersion tests and adjustments for survival and case-control studies. *Statist. Med.*; **12**: 1683-91.

[588] Smith, F., Talwalker, S., Gracon, S. and Srirama, M. (1996) The use of survival analysis techniques in evaluating the effect of long-term tacrine (Cognex) treatment on nursing home placement and mortality in patients with Alzheimer's disease. *J Biopharm Stat*; **6**: 395-409.

[589] Snapinn, S. M. (1998) Survival analysis with uncertain endpoints. *Biometrics*; **54**: 209-18.

[590] Song, X., Davidian, M. and Tsiatis, A. A. (2002) A semiparametric likelihood approach to joint modeling of longitudinal and time-to-event data. *Biometrics*; **58**: 742-53.

[591] Sooriyarachchi, M. R. and Whitehead, J. (1998) A method for sequential analysis of survival data with nonproportional hazards. *Biometrics*; **54**: 1072-84.

[592] Spiekerman, C. F. and Lin, D. Y. (1996) Checking the marginal Cox model for correlated failure time data. *Biometrika*; **83**: 143-156.

[593] Sposto, R., Stablein, D. and Carter-Campbell, S. (1997) A partially grouped logrank test. *Statist. Med.*; **16**: 695-704.

[594] Sposto, R. (2002) Cure model analysis in cancer: an application to data from the Children's Cancer Group. *Statist. Med.*; **21**: 293-312.

[595] Stablein, D. M., Carter, W. H. and Wampler, G. L. (1980) Survival analysis of drug combinations using a hazards model with time-dependent covariates. *Biometrics*; **36**: 537-46.

[596] Stare, J. and O'Quigley J. (2004) Fit and frailties in proportional hazards regression. *Biometrical Journal*; **47**: 157-164.

[597] Storer, B. E. and Crowley, J. (1985) A diagnostic for Cox regression and general conditional likelihoods. *J. Amer. Statist. Assoc.*; **80**: 139-147.

[598] Struthers, C. A. and Kalbfleisch, J. D. (1986) Misspecified proportional hazard models. *Biometrika*; **73**: 363-369.

[599] Stuart, A. and Ord, J. K. (1994) *Kendall's Advanced Theory of Statistics*; Oxford University Press.

[600] Sturm, R. (1991) Parametric and parametrically smoothed distribution-free proportional hazard models with discrete data. *Biometrical J.*; **33**: 441-454.

[601] Stute, W. (1992) Strong consistency under the Koziol-Green model. *Statist. Probab. Lett.*; **14**: 313-320.

[602] Stute, W. (1995) The central limit theorem under random censorship. *The Annals of Statistics*; **23**: 422-439.

[603] Sueyoshi, G. T. (1992) Semiparametric proportional hazards estimation of competing risks models with time-varying covariates. *J. Econometrics*; **51**: 25-58.

[604] Sugimoto, T. and Goto, M. (2003) The proportional hazards model of time-varying type and a class of rank tests. *J. Japan Statist. Soc.*; **33**: 365-382.

[605] Sukhatme, S. and Beam, C. A. (1994) Stratification in nonparametric ROC studies. *Biometrics*; **50**: 149-63.

[606] Sun, J. (1996) A non-parametric test for interval-censored failure time data with application to AIDS studies. *Stat Med*; **15**: 1387-95.

[607] Sun, Y. and Sherman, M. (1996) Some permutation tests for survival data. *Biometrics*; **52**: 87-97.

[608] Sun, J., Liao, Q. and Pagano, M. (1999) Regression analysis of doubly censored failure time data with applications to AIDS studies. *Biometrics*; **55**: 909-14.

[609] Sun, J., Sun, L. and Flournoy, N. (2004) Additive hazards model for competing risks analysis of the case-cohort design. *Comm. Statist. Theory Methods*; **33**: 351-366.

[610] Sy, J. P. and Taylor, J. M. (2000) Estimation in a Cox proportional hazards cure model. *Biometrics*; **56**: 227-36.

[611] Tai, B. C., Machin, D., White, I. and Gebski, V. (2001) Competing risks analysis of patients with osteosarcoma: a comparison of four different approaches. *Statist. Med.*; **20**: 661-84.

[612] Tai, B. C. and Peregoudov, A. and Machin, D. (2001) A competing risk approach to the analysis of trials of alternative intra-uterine devices (IUDs) for fertility regulation. *Statist. Med.*; **20**: 3589-600.

[613] Tai, B. C., White, I. R., Gebski, V. and Machin, D. (2002) On the issue of 'multiple' first failures in competing risks analysis. *Statist. Med.*; **21**: 2243-55.

[614] Tamura, R. N., Tanaka, Y., Satoh, K. and Takahashi, M. (1996) Using the proportional odds model to assess the relationship

between a multi-item and a global item efficacy scale in a psychiatric clinical trial. *J Biopharm Stat*; **6**: 127-37.

[615] Tangen, C. M. and Koch, G. G. (1999) Nonparametric analysis of covariance for hypothesis testing with logrank and Wilcoxon scores and survival-rate estimation in a randomized clinical trial. *J Biopharm Stat*; **9**: 307-38.

[616] Tangen, C. M. and Koch, G. G. (2000) Non-parametric covariance methods for incidence density analyses of time-to-event data from a randomized clinical trial and their complementary roles to proportional hazards regression. *Statist. Med.*; **19**: 1039-58.

[617] Tangen, C. M. and Koch, G. G. (2001) Non-parametric analysis of covariance for confirmatory randomized clinical trials to evaluate dose-response relationships. *Statist. Med.*; **20**: 2585-607.

[618] Tavare, C. J., Sobel, E. L. and Gilles, F. H. (1995) Misclassification of a prognostic dichotomous variable: sample size and parameter estimate adjustment. *Statist. Med.*; **14**: 1307-14.

[619] Taylor, J. M. (1995) Semi-parametric estimation in failure time mixture models. *Biometrics*; **51**: 899-907.

[620] Taylor, J. M. and Wang, Y. (2002) Surrogate markers and joint models for longitudinal and survival data. *Control Clin Trials*; **23**: 626-34.

[621] ten Have, T. R. (1996) A mixed effects model for multivariate ordinal response data including correlated discrete failure times with ordinal responses. *Biometrics*; **52**: 473-91.

[622] ten Have, T. R. and Bixler, E. O. (1997) Modelling population heterogeneity in sensitivity and specificity of a multi-stage screen for obstructive sleep apnoea. *Statist. Med.*; **16**: 1995-2008.

[623] Terpstra, T. J. (1989) Asymptotically optimal nonparametric one- and two-way analysis of variance tests for the log linear model under censoring. *Metrika*; **36**: 63-90.

[624] Thall, P. F. and Lachin, J. M. (1986) Assessment of stratum-covariate interactions in Cox's proportional hazards regression model. *Statist. Med.*; **5**: 73-83.

[625] Thompson, W. A. (1977) On the treatment of grouped observations in life studies. *Biometrics*; **33**: 463-70.

[626] Thornquist, M. D. (1993) Proportional hazards model for repeated measures with monotonic ordinal response. *Biometrics*; **49**: 721-30.

[627] Tibshirani, R. J. and Ciampi, A. (1983) A family of proportional- and additive-hazards models for survival data. *Biometrics*; **39**: 141-147.

[628] Tod, S. and Sahdra, M. (2001) A case study of the effect of covariate adjustment in a sequential survival clinical trial. *J Biopharm Stat*; **11**: 297-311.

[629] Tsiatis, A. A. (1981) The asymptotic joint distribution of the efficient scores test for the proportional hazards model calculated over time. *Biometrika*; **68**: 311-315.

[630] Tsiatis, A. A. (1981) A large sample study of Cox's regression model. *Ann. Statist.*; **9**: 93-108.

[631] Tsiatis, A. A. and Davidian, M. (2001) A semiparametric estimator for the proportional hazards model with longitudinal covariates measured with error. *Biometrika*; **88**: 447-458.

[632] Tsodikov, A. (1998) Asymptotic efficiency of a proportional hazards model with cure. *Statist. Probab. Lett.*; **39**: 237-244.

[633] Tsodikov, A. (1998) A proportional hazards model taking account of long-term survivors. *Biometrics*; **54**: 1508-16.

[634] Tsodikov, A., Loeffler, M. and Yakovlev, A. (1998) A cure model with time-changing risk factor: an application to the analysis of secondary leukaemia. A report from the International Database on Hodgkin's Disease. *Statist. Med.*; **17**: 27-40.

[635] Tsukahara, H. (1992) A rank estimator in the two-sample transformation model with randomly censored data. *Ann. Inst. Statist. Math.*; **44**: 313-333.

[636] Tu, D. S. (1991) The Berry-Esseen theorem for the subject-years method in mortality analysis with censored data. *Metrika*; **38**: 269-283.

[637] Vaeth, M. and Skovlund, E. (2004) A simple approach to power and sample size calculations in logistic regression and Cox regression models. *Statist. Med.*; **23**: 1781-92.

[638] Vaida, F. and Xu, R. (2000) Proportional hazards model with random effects. *Statist. Med.*; **19**: 3309-24.

[639] Valenta, Z. and Weissfeld, L. (2002) Estimation of the survival function for Gray's piecewise-constant time-varying coefficients model. *Statist. Med.*; **21**: 717-27.

[640] Valsecchi, M. G., Silvestri, D. and Sasieni, P. (1996) Evaluation of long-term survival: use of diagnostics and robust estimators with Cox's proportional hazards model. *Stat Med*; **15**: 2763-80.

[641] van der Laan, M. J. and Hubbard, A. (1999) Locally efficient estimation of the quality-adjusted lifetime distribution with right-censored data and covariates. *Biometrics*; **55**: 530-6.

[642] van Houwelingen, H. C. and Thorogood, J. (1995) Construction, validation and updating of a prognostic model for kidney graft survival. *Statist. Med.*; **14**: 1999-2008.

[643] van Houwelingen, H. C. (2000) Validation, calibration, revision and combination of prognostic survival models. *Statist. Med.*; **19**: 3401-15.

[644] Vaupel, J.W., Manton, K. G. and Stallard, E. (1979) The impact of heterogeneity in individual frailty on the dynamics of mortality. *Demography* **16**: 439-54.

[645] Veraverbeke, N. (1994) Bootstrapping quantiles in the proportional hazards model of random censorship. *Comm. Statist. Theory Methods*; **23**: 997-1007.

[646] Verweij, P. J. and van Houwelingen, H. C. (1993) Cross-validation in survival analysis. *Statist. Med.*; **12**: 2305-14.

[647] Verweij, P. J. and van Houwelingen, H. C. (1994) Penalized likelihood in Cox regression. *Statist. Med.*; **13**: 2427-36.

[648] Verweij, P. J. and van Houwelingen, H. C. (1995) Time-dependent effects of fixed covariates in Cox regression. *Biometrics*; **51**: 1550-6.

[649] Verweij, P. J., van Houwelingen, H. C. and Stijnen, T. (1998) A goodness-of-fit test for Cox's proportional hazards model based on martingale residuals. *Biometrics*; **54**: 1517-1526.

[650] Voit, E. O. and Knapp, R. G. (1997) Derivation of the linear-logistic model and Cox's proportional hazard model from a canonical system description. *Statist. Med.*; **16**: 1705-29.

[651] Volinsky, C. T. and Raftery, A. E. (2000) Bayesian information criterion for censored survival models. *Biometrics*; **56**: 256-62.

[652] Vonta, F. (1996) Efficient estimation in a non-proportional hazards model in survival analysis. *Scand. J. Statist.*; **23**: 49-61.

[653] Vu, H. T. V. and Knuiman, M. W. (2002) Estimation in semiparametric marginal shared gamma frailty models. *Aust. N. Z. J. Stat.*; **44**: 489-501.

[654] Walker, S. (1998) A nonparametric approach to a survival study with surrogate endpoints. *Biometrics*; **54**: 662-72.

[655] Wallenstein, S. and Wittes, J. (1993) The power of the Mantel-Haenszel test for grouped failure time data. *Biometrics*; **49**: 1077-87.

[656] Wang, M., Brookmeyer, R. and Jewell, N. P. (1993) Statistical models for prevalent cohort data. *Biometrics*; **49**: 1-11.

[657] Wang, M. (1996) Hazards regression analysis for length-biased data. *Biometrika*; **83**: 343-354.

[658] Wang, C. Y. and Huang, W. T. (2000) Conditional-cumulant-of-exposure method in logistic missing covariate regression. *Biometrics*; **56**: 98-105.

[659] Wang, C. Y. and Chen, H. Y. (2001) Augmented inverse probability weighted estimator for Cox missing covariate regression. *Biometrics*; **57**: 414-9.

[660] Wassell, J. T. and Moeschberger, M. L. (1993) A bivariate survival model with modified gamma frailty for assessing the impact of interventions. *Statist. Med.*; **12**: 241-8.

[661] Wei, L. J. (1984) Testing goodness of fit for proportional hazards model with censored observations. *J. Amer. Statist. Assoc.*; **79**: 649-652.

[662] Wei, L. J., Lin, D. Y. and Weissfeld, L. (1989) Regression analysis of multivariate incomplete failure time data by modeling marginal distributions. *J. Amer. Statist. Assoc.*; **84**: 1065-1073.

[663] Wei, L. J. (1992) The accelerated failure time model: a useful alternative to the Cox regression model in survival analysis. *Statist. Med.*; **11**: 1871-9.

[664] Wei, W. H. and Su, J. S. (1999) Model choice and influential cases for survival studies. *Biometrics*; **55**: 1295-9.

[665] Wei, W. H. and Kosorok, M. R. (2000) Masking unmasked in the proportional hazards model. *Biometrics*; **56**: 991-995.

[666] Weissfeld, L. A. (1990) Influence diagnostics for the proportional hazards model. *Statist. Probab. Lett.*; **10**: 411-417.

[667] Wellek, S. (1993) A log-rank test for equivalence of two survivor functions. *Biometrics*; **49**: 877-81.

[668] Wells, M. T. (1994) Nonparametric kernel estimation in counting processes with explanatory variables. *Biometrika*; **81**: 795-801.

[669] White, I. R. and Pocock, S. J. (1996) Statistical reporting of clinical trials with individual changes from allocated treatment. *Statist. Med.*; **15**: 249-62.

[670] Whitehead, J. (1994) Sequential methods based on the boundaries approach for the clinical comparison of survival times. *Statist. Med.*; **13**: 1357-68; discussion 1369-70.

[671] Whitehead, J. and Thomas, P. (1997) A sequential trial of pain killers in arthritis: issues of multiple comparisons with control and of interval-censored survival data. *J Biopharm Stat*; **7**: 333-53.

[672] Whittemore, A. S. and Keller, J. B. (1986). Survival estimation using splines. *Biometrics*; **42**: 495-506.

[673] Wienke, A. (1998) An asymptotically optimal adaptive selection procedure in the proportional hazards model with conditionally independent censoring. *Biom. J.*; **40**: 963-978.

[674] Wild, C. J. (1983) Failure time models with matched data. *Biometrika*; **70**: 633-641.

[675] Willett, J. B. and Singer, J. D. (1988) Another cautionary note about R^2. Its use in weighted least squares regression analysis. *American Statist.*; **42**: 236-38.

[676] Williams, P. L. (1996) Sequential monitoring of clinical trials with multiple survival endpoints. *Statist. Med.*; **15**: 2341-57; discussion 2367-70.

[677] Williamson, P. R., Clough, H. E., Hutton, J. L., Marson, A. G. and Chadwick, D. W. (2002) Statistical issues in the assessment of the evidence for an interaction between factors in epilepsy trials. *Statist. Med.*; **21**: 2613-22.

[678] Winnett, A. and Sasieni, P. (2001) A note on scaled Schoenfeld residuals for the proportional hazards model. *Biometrika*; **88**: 565-571.

[679] Winnett, A. and Sasieni, P. (2003) Iterated residuals and time-varying covariate effects in Cox regression. *J. R. Stat. Soc. Ser. B Stat. Methodol.*; **65**: 473-488.

[680] Wong, W. H. (1986) Theory of partial likelihood. *Ann. Statist.*; **14**: 88-123.

[681] Wulfsohn, M. S. and Tsiatis, A. A. (1997) A joint model for survival and longitudinal data measured with error. *Biometrics*; **53**: 330-9.

[682] Xiang, A. H. and Langholz, B. (1999) Comparison of case-control to full cohort analyses under model misspecification. *Biometrika*; **86**: 221-226.

[683] Xiao, W., Barron, A. M. and Liu, J. (1997) Robustness of bioequivalence procedures under Box-Cox alternatives. *J Biopharm Stat*; **7**: 135-55.

[684] Xu, R. and O'Quigley, J. (1999) A measure of dependence for proportional hazards models. *J. Nonparametr. Statist.*; **12**: 83-107.

[685] Xu, R. and O'Quigley J. (2000) Estimating average regression effect in non proportional hazards regression. *Biostatistics*; **1**: 423-439.

[686] Xu R. and O'Quigley J. (2000) Proportional hazards estimate of the conditional survival function. *J. Roy. Statist. Soc. (B)*; **62**: 667-680.

[687] Xu, R. and Harrington, D. P. (2001) A semiparametric estimate of treatment effects with censored data. *Biometrics*; **57**: 875-85.

[688] Xu, R. and Adak, S. (2002) Survival analysis with time-varying regression effects using a tree-based approach. *Biometrics*; **58**: 305-15.

[689] Yakovlev, A., Goot, R. E. and Osipova, T. T. (1994) The choice of cancer treatment based on covariate information. *Statist. Med.*; **13**: 1575-81.

[690] Yakovlev, A. Y. (1994) Parametric versus non-parametric methods for estimating cure rates based on censored survival data. *Statist. Med.*; **13**: 983-6.

[691] Yamaguchi, T. and Ohashi, Y. (1999) Investigating centre effects in a multi-centre clinical trial of superficial bladder cancer. *Statist. Med.*; **18**: 1961-71.

[692] Yanagimoto, T. and Kamakura, T. (1984) The maximum full and partial likelihood estimators in the proportional hazard model. *Ann. Inst. Statist. Math.*; **36**: 363-373.

[693] Yang, S. (1998) Some scale estimators and lack-of-fit tests for the censored two-sample accelerated life model. *Biometrics*; **54**: 1040-52.

[694] Yang, Y. and Ying, Z. (2001) Marginal proportional hazards models for multiple event-time data. *Biometrika*; **88**: 581-586.

[695] Yateman, N. A. and Skene, A. M. (1992) Sample sizes for proportional hazards survival studies with arbitrary patient entry and loss to follow-up distributions. *Statist. Med.*; **11**: 1103-13.

[696] Yau, K. K. and McGilchrist, C. A. (1998) ML and REML estimation in survival analysis with time dependent correlated frailty. *Statist. Med.*; **17**: 1201-13.

[697] Yau, K. K. (2001) Multilevel models for survival analysis with random effects. *Biometrics*; **57**: 96-102.

[698] Yip, P. S., Zhou, Y., Lin, D. Y. and Fang, X. Z. (1999) Estimation of population size based on additive hazards models for continuous-time recapture experiments. *Biometrics*; **55**: 904-8.

[699] Zahl, P. H. and Tretli, S. (1997) Long-term survival of breast cancer in Norway by age and clinical stage. *Statist. Med.*; **16**: 1435-49.

[700] Zedeler, K., Keiding, N. and Kamby, C. (1992) Differential influence of prognostic factors on the occurrence of metastases at various anatomical sites in human breast cancer. *Stat Med*; **11**: 281-94.

[701] Zeger, S. L. and Qaqish, B. (1988) Markov regression models for time series: a quasi-likelihood approach. *Biometrics*; **44**: 1019-31.

[702] Zelterman, D. (1992) A statistical distribution with an unbounded hazard function and its application to a theory from demography. *Biometrics*; **48**: 807-18.

[703] Zhang, M. and Klein, J. P. (2001) Confidence bands for the difference of two survival curves under proportional hazards model. *Lifetime Data Anal.*; **7**: 243-254.

[704] Zhang, Z. (2001) Estimating hazard ratios in nested case-control studies by Mantel-Haenszel method. *Acta Math. Appl. Sinica (English Ser.)*; **17**: 457-468.

[705] Zhou, M. (1992) M-estimation in censored linear models. *Biometrika* **79**: 837-41.

[706] Zhuang, D., Schenker, N., Taylor, J. M., Mosseri, V. and Dubray, B. (2000) Analysing the effects of anaemia on local recurrence of head and neck cancer when covariate values are missing. *Statist. Med.*; **19**: 1237-49.

[707] Zippin, C. and Armitage, P. (1966) Use of concomitant variables and incomplete survival information in the estimation of an exponential survival parameter. *Biometrics*; **22**: 665-72.

[708] Zucker, D. M. and Karr, A. F. (1990) Nonparametric survival analysis with time-dependent covariate effects: a penalized partial likelihood approach. *Annals of Statistics*; **18**: 329-353.

[709] Zucker, D. M. (1992) The efficiency of a weighted log-rank test under a percent error misspecification model for the log hazard ratio. *Biometrics*; **48**: 893-9.

[710] Zucker, D. M. and Spiegelman, D. (2004) Inference for the proportional hazards model with misclassified discrete-valued covariates. *Biometrics*; **60**: 324-334.

Index